ASM COMPUTER SOURCE BOOK

*A collection of outstanding articles
from the technical literature*

ASM COMPUTER SOURCE BOOK

A collection of outstanding articles from the technical literature

**Compiled by
Consulting Editor**

Theodore K. Thomas

TK/Associates

**American Society for Metals
Metals Park, Ohio 44073**

Contributors to This Source Book

RICHARD AARONS
Business and Commercial Aviation

OSAMA T. ALBAHARNA
McGill University

JOHN J. ANDERSON
Creative Computing

STAVROS A. ARGYROPOULOS
McGill University

ERIC BANK
Consultant

ANTHONY J. BARRETT
ESDU International Ltd.

ELLEN BENOIT
Forbes Magazine

TONY BOLTON
Concord Data Systems

ALBERT BOULANGER
Bolt, Beranek and Newman Inc.

ROBERT BOWERMAN
Heliosoft

ROBERT BRAHAM
Mechanical Engineering

CARL BRANDON
Vermont Technical College

DALE H. BREEN
International Harvester Co.

GENE BYLINSKY
Fortune

RICHARD B. BYRNE
University of Southern California

PETER CALLAMARAS
U.S. Air Force

KAARE CHRISTIAN
Rockefeller University

P. R. CLARKE
BNF Metals

BERNARD M. CLOSSET
CHROMASCO Ltd.

SHEILA CUNNINGHAM
ICP Interface

GERT DATHE
VDEh-Institut fur angewandte
    Forschung GMbH

MICHAEL J. DELLA ROCCO
American Brass Div.
Arco Metals Co.

JOE DESPOSITO
Computers and Electronics

JOHN DICKINSON
PC Magazine

TERRY L. DOLLHOFF
Manufacturing Data Services Inc.

JEFF DUNTEMANN
PC Tech Journal

ROBERT J. EATON
General Motors Corp.

GARY ELFRING
Elfring Consulting Inc.

THOMAS C. ELLIOTT
Power

LEE FARRAR
Eagle Signal Controls

HENRY FERSKO-WEISS
Personal Computing

EDWARD FOSTER
Personal Computing

DAVID GABEL
Personal Software

MARVIN GROSSWIRTH

RICHARD I. GRUBER
Process Control Div.
Honeywell, Inc.

LYNN HABER
Mini-Micro Systems

JOHN HALL
I&CS

MARK HALL
Sytec Inc.

ROBERT E. HARVEY
Iron Age

LOWELL B. HAWKINSON
LISP Machine Inc.

JOHN D. HUBBARD
Hinderliter Heat Treating Inc.

LARRY JORDAN
NUS Corp.

CARL J. KEITH JR.
International Harvester Co.

DOUGLAS A. KERR
Creative Computing

CARL G. KNICKERBOCKER
LISP Machine Inc.

STEVEN M. LORD

JAMES M. MCQUEEN
American Brass Div.
Arco Metals Co.

NOTE: Affiliations given were applicable at date of contribution.

JOSEPH MARTIN
Garrick Lochhead, Inc.

PETER MASUCCI
Digital Equipment Corp.

JOHN W. MAUCHLY
Dynatrend Inc.

TED F. MEINHOLD
Plant Engineering

RICH MERRITT
I&CS

R. H. MICHAELSEN
University of Nebraska

DONALD MICHIE
Turing Institute

CHARLES MILLER
Personal Software

ROBERT L. MOORE
LISP Machine Inc.

HENRY M. MORRIS
Control Engineering

PAUL M. MUMFORD
United Calibration Corp.

CHARLES PETZOLD

C. N. REID
Open University

WILLIAM J. RIKER
General Motors Corp.

J. L. ROBERTSON
BNF Metals

WALTER G. RUDD
Louisiana State University

HENRIK SCHUTZ
Automation Controls Dept.
General Electric Co.

DONALD SMITH
University of Wyoming

EMILY T. SMITH
Business Week

JOHN T. SPONZILLI
International Harvester Co.

STEPHANIE STALLINGS
PC Magazine

JAMES SULLIVAN
Honeywell, Inc.

N. SWINDELLS
University of Liverpool

R. J. SWINDELLS
Matsel Systems Ltd.

JOHN TARTAGLIA
AMAX Materials Research Co.

ED TEJA

THEODORE K. THOMAS
Process Control Div.
Honeywell, Inc.

JAN VAN DER EIJK
NUS Corp.

GORDON H. WALTER
International Harvester Co.

VOLKER WEISS
Syracuse University

J. H. WESTBROOK
General Electric Research and
    Development Center

PREFACE

This Source Book was designed to provide the busy metallurgist and metals manufacturing engineer with a useful reference to the subject of computers in industry. The book is an outgrowth of an idea, proposed by an ASM editor, to utilize the familiar Source Book format in presenting a broad yet selective spectrum of information through authoritative trade journal and technical articles about computers and their applications.

Interest in computers, with specific attention to how they can help improve efficiency and productivity in manufacturing processes, has grown as fast as the changes in the technology of the subject itself. Literally hundreds of pages in trade and technical journals—to say nothing of bound books—each month are devoted to computers and microcomputers, with case studies, survey articles, and forecasts of future developments. This amounts to a glut of information that can be difficult for a single individual to scan and digest, even selectively.

The articles in this Source Book present a little bit of computer history, enough of the basics of computer control to enable the reader to readily relate the older and perhaps more familiar technology to the new, and applications of the computer in various aspects of metallurgical and manufacturing processes. The articles were obtained through the author's research, by computer searches of commercially available databases, and through other ASM editorial resources.

It should not surprise the reader to find that many of the articles deal with the personal computer. The so-called personal, or desktop, computer has become, in many of its current models, a more powerful machine than the minicomputer of even a few years ago and even some mainframes of a decade ago. This trend will continue.

The author wishes to acknowledge the invaluable help and support of Timothy Gall, ASM Manager of Publications Development, and his staff, together with numerous colleagues of the author at Honeywell, and the many editors and authors of the articles themselves.

The American Society for Metals extends most grateful acknowledgment to the many authors whose work is presented in this Source Book, and to their publishers.

THEODORE K. THOMAS
Willow Grove, PA 19090

CONTENTS

SECTION I: INTRODUCTION

SECTION II: FUNDAMENTALS

SECTION III: OVERVIEW OF COMPUTER HARDWARE

SECTION IV: SOFTWARE

SECTION V: GOING ON-LINE

SECTION VI: PROGRAMMABLE CONTROLLERS

SECTION VII: COMPUTERIZED HEAT TREATING

SECTION VIII: THE COMPUTER IN MATERIALS ENGINEERING

SECTION IX: LOCAL AREA NETWORKS

SECTION X: THE FUTURE

BIBLIOGRAPHY: BOOKS ON COMPUTER SUBJECTS

ASM
COMPUTER
SOURCE BOOK

*A collection of outstanding articles
from the technical literature*

SECTION I

Introduction

SECTION I
Introduction

The computer is here.

It is in our office buildings, our government bureaus, our banking houses, and our manufacturing plants. If not yet actually on *your* office desk, it has reached the office just down the hall from you. It has long since entered our homes and schools, and, in its role of space-age Pied Piper, it has attracted a growing following of fascinated, admiring young people. No other single technological development since the beginning of the Industrial Revolution has advanced so swiftly in its application and utility of use for information-handling tasks in our day-to-day business and personal lives.

Yet the function of the computer — except for a few exotic and advanced applications in research and development labs — is quite ordinary. In its imitative and repetitive registering and storing of coded information, the computer is a tireless logbook. Most of its calculations are simple: addition and subtraction, basically. But the appeal of the computer is its phenomenal speed in the execution of these mundane manipulations. It is so fast, in fact, that it can work, for all intents, in "real time." For example, it can, with electronic speed, handle thousands of airline and hotel reservations from locations all over the world, updating their status on a minute-by-minute basis. It can do the same thing for retail sales and weather forecasting.

Used in a furnace control system, a computer can detect a deviation from setpoint conditions, formulate a corrective response in accordance with the instructions stored within its memory, initiate the corrective action, and report the action to a supervisory monitor, all within a few microseconds of time.

As a research tool, the computer can search vast databases containing millions of reference items and extract the desired information within fractions of a second.

Historical

Although one can easily trace the computer concept back into history at least to the Chinese invention of the abacus in 600 B.C., the "computer age" for the purposes of this book began almost 40 years ago at the University of Pennsylvania in Philadelphia with the dedication of ENIAC (Electronic Numerical Integrator and Computer), an invention of John W. Mauchly and J. Prosper Eckert designed to perform ballistic computations involved with the firing of artillery shells.

The article "Computing Through the Ages" discusses these milestones in the development of computers.

ENIAC was a monument to vacuum tube technology. The assembled components of the device — its racks, panels, power supplies, and associated cabinetry — occupied several rooms of a large warehouse in the Moore School of Electrical Engineering. It contained 18,000 vacuum tubes, required 140,000 watts of power, and cost $400,000 to build. To us, with the enlightened virtue of hindsight, ENIAC appears primitive, clumsy, and costly, yet for its purposes it was phenomenally successful. It could compute the trajectories of artillery shells, taking into account numerous interdependent variables such as muzzle velocity, blast pressure, powder temperature, rotation of the earth, barometric pressure, wind direction and velocity, together with other factors having an effect on the flight of the shell toward its target, and solve the equations 1000 times faster than the best prior methods afforded. Delivered to Aberdeen Proving Grounds in 1946, ENIAC continued to do its specialized job for the next ten years before the tools of newer technologies replaced it.

Two articles, "Mauchly on the Trials of Building ENIAC," and "The Mauchly Legacy . . . ," provide background on this interesting period in computer development.

At about the same time that ENIAC was delighting the revenue department of the Philadelphia Electric Company with the power consumption of its 18,000 vacuum tubes, an equally revolutionary development — the transistor — was taking shape not far away in the laboratories of the Bell Telephone Company. When it appeared in 1947, the transistor not only improved tremendously upon the functions of the vacuum tube, and at a fraction of the tube's energy consumption, but more importantly for the future, it also miniaturized the functions. This aspect pointed development straight down the road to the eventual explosion of silicon chip technology and the introduction of the microprocessor by the Intel Company in 1971.

Thus a single silicon chip 1/4 in. thick is 30,000 times cheaper to produce than ENIAC, and, with the ability to do 1,000,000 calculations per second, is 200 times faster than ENIAC at its best. Today's ordinary hand-held calculator has the capabilities of two or three ENIAC's, at a fraction of its size and cost.

Read the article "Here Comes the Second Computer Revolution" for a historical treatment of Intel's development of the silicon chip processor.

Applications in Metalworking

The computer is finding four main areas of use in metalworking:

1. As a tool for data acquisition, monitoring, and manipulation in company management information systems. This has been a basic application for digital computers ever since the early days in which they were moved out of the accounting department and into broader areas of data processing.

2. As an automatic controller. In the form of a microprocessor, the computer can economically handle the job of analog loop control of temperature, pressure, flow, vacuum, and other process variables critical to the operation of furnaces and ovens. In the form of a programmable controller, a widely used microprocessor version, the computer handles sequence and logic control involved with automatic machine tool and other metalworking machinery.

3. A logical next step in control is to have the computer monitor and supervise groups of furnaces, ovens, and metalworking machinery.

4. The computer has served as a research tool for many years. Its capacious memory and great speed of data handling and manipulation allows researchers to set up mathematical models of heat processing operations, and thus simulate, in a relatively inexpensive pilot-plant application, the actual operating characteristics of a much larger and costlier process unit.

The Computer and You

This Source Book was developed for the metallurgical reader who is now or soon will be working with computers on either an academic or a vocational basis. The reader will soon find that most of the articles deal with the type of computer called "home," "personal," or "desktop" (the terms are becoming more easily interchangeable), rather than what is traditionally called the "mainframe." The fact is that hardware technology — ways of storing data, processing speed, memory size, etc. — has progressed to the point where the desktop-size machine is perfectly adequate to handle the applications once reserved for the large mainframe types.

But even a working familiarity with one type or model of computer, regardless of its size and speed classification, is often inadequate to help the reader through the bewildering labyrinth of types, models, features, and manufacturers now on the market. This Source Book will serve as a guide through this maze, and though not all-inclusive (the book is not an equipment directory), it should provide a sufficient platform for informed decision-making.

Section II will deal with the "basics" of computers for those readers who desire an orientation — or a refresher — before proceeding to the overview of actual hardware discussed in Section III, and the various applications that follow it.

Reference Bibliography

1. "Mauchly on the Trials of Building ENIAC," J. W. Mauchly, IEEE SPECTRUM, April 1975.
2. "Here Comes the Second Computer Revolution," Gene Bylinski, FORTUNE, November 1975.

Mauchly on the trials of building ENIAC

People could not see how something with 18 000 tubes and costing $500 000 could ever become practical!

Today the digital computer is as accepted a part of the business and industrial environment as fluorescent lights, copying machines, and all the other paraphernalia of the modern business world. Which is why when I remember back thirty-four years to the time J. Presper Eckert and I were discussing how to build the world's first completely electronic computer, I am struck by how terribly difficult it was to sell the idea.

Of course, the history of all innovation is largely one of resistance to change, and the digital computer is certainly not the first invention to meet resistance, nor will it be the last. However, Eckert and I did not just have trouble selling the *idea* of a digital computer. Even after we had actually built the ENIAC we found it extremely difficult to convince Wall Street of its commercial importance.

In the beginning . . .

Thus, I would like to evoke some of the highlights of these early years, between 1940 and 1946, to remind *Spectrum* readers of how hard it is to sell a major invention, and also to give some feeling for the kind of engineering that went into what, for its day, was an electronic systems project of unprecedented magnitude. Not only did ENIAC have to operate with a one in 10^{14} probability of malfunction to keep it going for 12 error-free hours, but man had never before undertaken a project of such complexity. Indeed, the computer is still man's most complex instrument.

For me, this undertaking began during the 1930s when, as a physics professor at Ursinus College near Philadelphia, I became interested in the problems of weather prediction. I was convinced that great improvements were possible if people could only find better ways of analyzing the statistical data that already existed. This naturally led me to do a lot of thinking about computing and how one might get it done faster and more cheaply than was possible with the day's crude mechanical calculating machines.

As a boy I had built some simple electronic circuits, and during the 1930s physicists were using vacuum tubes for counting and analyzing cosmic rays. Near me—at the Bartol Research Foundation and the Carnegie Institution where my father had worked —tubes were being used in scaling circuits that were hoped would accept counts as high as one million per second and divide them so that, for example, every time 1024 counts were received by the electronics, one pulse would register on a mechanical counter.

John W. Mauchly

With all of this going on, it seemed obvious to me that tube techniques could be applied to computation. As a result, I started assembling a little bit of equipment in the laboratory, which I had to pay for out of my own pocket because the $1500 annual budget for teaching physics was already being stretched as far as I could stretch it to get the simple apparatus required by Physics-I students. At any rate, I did get enough vacuum tubes and related circuit components together so that I could begin to see ways in which I might, if I took long enough, put together something that would be helpful.

At the same time as I began experimenting, the big companies were developing punch card equipment. IBM and Remington Rand each had "accounting devices" that could do things faster and better than I could with my modest funding. But they were pretty expensive. The $50 or $100 I had available was nothing compared to the several hundred dollars a month required for the most elementary punch card equip-

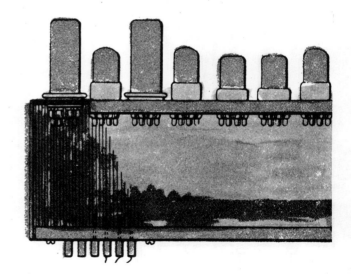

ment. Why, just to be able to punch the cards and sort them, you would be spending $100 or more a month to *rent* a card punch and a sorter. And that's not doing any multiplying or adding. If you wanted a tabulator for adding, then you were up around a thousand dollars a month in rent. And if you wanted to multiply, you had to rent a special multiplier, which took eight or ten seconds to do a multiplication by reading the factors off a punch card and punching the product back into the punch card. That kind of multiplication was only an order of magnitude faster than it would take me to put the factors into a desk calculator manually, multiply them out, and write the answer.

So for $75 I bought a second-hand desk calculator that I could operate myself, and with the aid of pencil and paper wrote down the results, fed them back in, and, by this procedure, got my answers.

To me it seemed clear that if you tried to duplicate with vacuum tubes what was done in the punch card machines with relays, you wouldn't spend any more money for the tubes than you would for the relays. They were both items that cost a dollar or two apiece. But with that same amount of equipment I could do the work a thousand times faster using tubes. So, of course, to me the basic question was what I could do to get that speed advantage of one thousand translated into a comparable economic advantage.

My first thought was to increase my salary so that I could afford to spend more money on experimental equipment. I was getting $2000 a year from college teaching, and I figured I could get something like $6000 or $7000 a year if I were an industrial scientist. But that wasn't easy. The industrial people looked at a professor as somebody who, being already geared to an easy life, would never be suitable for the competition in industry—and they told me so!

A partnership begins

The outbreak of World War II was a turning point in my fortunes. By 1940, the Government realized it needed to prepare for the crisis and, recognizing that the new art of electronics would be important, it set up programs to train people in the new field. In the summer of 1941, I was able to enroll in one of these

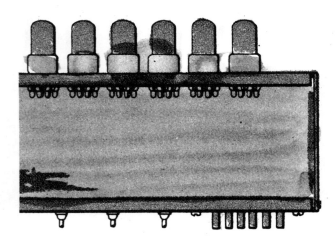

courses, which was being offered at the University of Pennsylvania. It was here that I met Eckert, a young man who had just received his bachelor's degree in electrical engineering. Eckert was acting as a lab instructor in the course. He was the man available to see that the students got whatever assistance they needed. If they had trouble connecting their appara-s, calibrating it, or understanding what they were pposed to do, it was his job to help them.

It turned out that the electronics experiments that had been arranged were of the most elementary sort, similar in many cases to those I had already arranged

for some of my advanced physics students at Ursinus. Eckert and I realized it was unnecessary for me to repeat these experiments, and so, during the times the students did not need him, the two of us had nothing to do but talk. Naturally, I talked mostly about the thing I was most interested in—the possibility of using electronic circuits to do arithmetic jobs.

I found Eckert to be very open-minded and interested in these ideas; he saw no reason at all why we couldn't devise controls on counter-circuits to do what I wanted. As it turned out, he was the one person who was encouraging in those days; most people took the point of view that my idea was just entirely too complicated and might well be too expensive. All kinds of reasons were advanced for casting the project aside as not even worth thinking about. Indeed, if it weren't for Eckert I probably wouldn't have been encouraged to proceed.

Besides my meeting with Eckert, the main result of this summer course was an invitation to join the faculty of the University of Pennsylvania's Moore School of Electrical Engineering. With professors of electrical engineering being called to active duty, the university felt it could use me since I knew something about electronics. I jumped at that chance even though it didn't pay any more than what I'd gotten as the department head of a small college. I recognized that while a professor in a liberal arts college was looked down upon as being off in a dream world, people in the engineering schools were doing "practical" things, like getting industrial contracts and, with the war imminent, taking on jobs for the Government.

Another attraction for me was a large differential analyzer that the university had built, which with wheels and gears and Rayleigh disk integrators was able to integrate ordinary differential equations. The analyzer was being used to design electrical machinery and solve network problems for General Electric.

The same type of machine had been built during the Depression for the Army Ordnance Department at Aberdeen, Md., to solve the differential equations needed to construct their artillery firing tables. Under the pressures of war, the university's differential analyzer was taken over entirely by the Army to supplement its own equipment at Aberdeen. But even it couldn't keep up with the demand, and in 1941 the Army started hiring mathematics graduates—my wife, for one—to operate desk calculators.

Well, it couldn't find enough mathematics graduates so it turned to people who had had a course in mathematics and therefore could be trained. Between 1941 and 1943, over 300 people were installed at the university and the Aberdeen Ballistics Research Laboratory to produce firing tables. But the backlog kept growing and they kept falling behind. After all, how much could a hundred women punching desk calculators add to the output—it took three weeks for one person to produce a single trajectory, and thousands of trajectories to figure all the possibilities of aiming the gun, considering the various kinds of ammunition and so forth. In contrast, the differential analyzer

With 200 other such units contained in 20 accumulators, the rather large "plug-in" decade counter illustrated on this page and throughout the article gave ENIAC a total memory capacity of 20 ten-digit numbers, making it fundamentally important for successful operation (see text).

This unique photograph of the ENIAC and its two inventors was taken at the Moore School of Electrical Engineering, Philadelphia, Pa., before the first electronic digital computer was moved to the Army Ordnance Ballistics Research Laboratory at Aberdeen, Md. Author John W. Mauchly is seen in the right foreground; coinventor J. Presper Eckert, Jr., is at the left. [UPI]

could run through the three-week job in about twenty minutes, properly adjusted and calibrated.

Getting off the ground

The first thing that Eckert (who was still working at the university for Dr. Cornelius Weygandt) did was to research the entire analyzer, including changing the torque amplifiers on the disk integrator from a purely mechanical system to a photoelectric system that used polaroid disks to sense the mismatch of angle between two positions. This increased the accuracy and decreased the wear factor so that the analyzer could be kept in more accurate calibration. But even an improved analyzer couldn't keep up with the great volume of ballistics calculations the Army required. I recognized that something entirely different was needed, and whenever someone from the Army came to visit, I tried to talk with him about my ideas for using vacuum tubes. But there was little interest.

The trouble was twofold. First of all, no one believed that a machine with hundreds, let alone thousands, of vacuum tubes would ever be reliable enough. One faculty member at the Moore School had made a study at the request of General Electric on whether some sort of digital computer would be better than the differential analyzer for doing the kinds of electrical generator and network problems in which the company was interested. His conclusion was that a digital computer made by electromechanically interconnecting desk calculators would wear out in less than one year, without even having solved a single problem of substantial size! For such a computer, this was undoubtedly correct. Unfortunately, this was translated into the idea that no digital computer would be practical.

The second obstacle to an electronics approach was

simply the belief that the War would end before a machine was built. This was the reaction of at least some people in Ordnance to the brief memo I wrote in 1942 outlining my ideas. They were pretty sure would take at least a year to build a computer and that, since the War would be over by then, they would be better off just hiring more women.

As Henry Tropp noted in *Spectrum* (Feb. 1974, pp. 70–79), this attitude had changed by 1943 and the Army decided that perhaps I had something they ought to listen to. When it turned out that no copies of my memo could be found, we reconstructed it from my secretary's shorthand notes and, from that point, things began moving very quickly.

Meeting at Aberdeen

When the Army Ordnance and university people went down to Washington and talked with the Pentagon, the response was, "Now all we need is an honest-to-God proposal for how you want to go about this, and how you will spend the money, and so on." During April 1943, we spent several long nights writing up this material for a big meeting at Aberdeen. We hadn't finished the proposal by the day of the meeting so while the discussions proceeded in the front office, Pres Eckert and I were off in a back room adding more paragraphs to explain how we could build it. At the end of the day, we got a verbal approval, which had to be followed, of course, by a signed contract with the Chief of Ordnance. It took many weeks to get the contract together, but meanwhile we were hard at work hiring people and educating them.

As liaison officer for Aberdeen, then-Lieut. Herma Goldstine's role in gaining Army support for ENIAC was of paramount importance. Not only was he instrumental in winning Army acceptance of the initial

Source: IEEE Spectrum, April 1975, 70-76

contract proposal, but he continued to act as intermediary between Army and university officials after the contract was initiated.

I should point out that one of the fortunate things about the contract was that both the Army Ordnance people and ourselves wanted to make something that was not just tied to one job and could be used in no other way. During the previous two years, when I had realized my only chance of building this machine was through the Army, I wasn't saying anything about weather prediction. I developed a whole new sales talk about how the machine could solve all the problems a differential analyzer could, only more quickly and more accurately.

The result was that we were all agreed at the outset on the need to build a general-purpose digital computer, one that would solve not only ballistics equations but any differential equations. The reason this proved so fortunate was, of course, that the War was over before the ENIAC was finished in 1945. (It was dedicated in February 1946.) The problems that were actually run on ENIAC during the roughly ten years of its life were of many different types, and they drew upon the general-purpose design of the machine (see box, pp. 74).

Worst-case design

Coming now to the actual building of the ENIAC, the biggest single task that faced us was to achieve a degree of relibility that, so far as I know, had never before been required of an electronic system of that size and complexity. After all, an erroneous digit could be produced if just one among the 18 000 tubes failed even momentarily. Eckert's approach was to try to achieve this reliability through the circuit design rather than by attempting to manufacture highly reliable components. We had estimated that the development costs would not exceed $150 000, and clearly one of our first decisions had to be to use only tubes, resistors, capacitors, and other parts that were commercially available. This meant, for example, purchasing 6SN7's at about 50 cents a piece instead of using the kinds of tubes (at about $100 each) the telephone company had developed for its underwater cable repeater systems. This meant using ±10-percent carbon resistors that were available wholesale for about 2 cents each instead of the ±1-percent wire-wound resistors some companies were then manufacturing in limited quantities for use in measuring instruments. (Even if these manufacturers could have geared up to produce the 70 000 resistors we needed, we couldn't have afforded them.)

With components of this kind it was clearly necessary to design circuits for what is now called the "worst case." As far as I know, Eckert was one of the first people to use this kind of design, at least on such a scale, and it was something he was very good at. Indeed, he deserves the major credit for the circuit design; if it hadn't been for him as chief engineer, I doubt we would have succeeded in building the ENIAC.

Without going into the technical details, I will explain that Eckert designed the circuits so they would operate within wide tolerance factors (tolerance factors of two for tubes, for example). We would breadboard a basic design for the counter circuits and find

that it would count correctly up to, say, 250 000 counts per second. This meant that we would plan to operate the counters at a maximum rate of 100 000 counts per second. Similarly, if we found that one counter would reproduce another's count accurately until the plate voltage rose from 100 volts to 400 volts, we would set the plate supply at 195 volts. In this way, we could be sure our counters would work even if one tube had twice the conductivity of the others. By using this kind of cut-and-try marginal testing to set outside limits that were then drastically backed away from, we were essentially able to make the accuracy of the computations independent of the tolerance of the components.

Another important design principle was to operate the components considerably below their normal ratings. Eckert was aiming for a tube life of at least 25 000–50 000 hours. To this end, he specified that plate voltages should be kept at not more than 50 percent of the rated maximum voltage and plate currents should not exceed 25 percent of the rated maximum. Interestingly enough, people knew what happened when you overloaded tubes—after all, radio hams had been doing this for years. So we just took flying guesses and did as much testing as we had time for. But when we started out, we had no way of knowing whether our performance would be better or worse than we hoped for. As it turned out, we got pretty much the lifetimes we were aiming for.

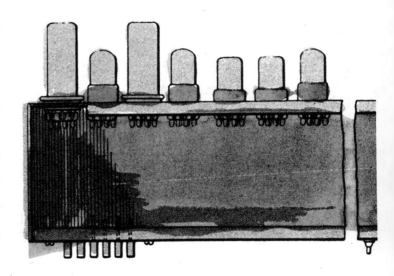

Similarly, we didn't require any of our other components to operate at their rated specifications. If we knew a resistor would have to dissipate ¼ watt, we would use a ½-watt unit—sometimes even a 1-watt resistor.

During the spring of 1943, when we were writing the actual proposal and doing the initial planning, Eckert and I naturally worked quite closely together. I tended to concentrate on defining the functions to be performed and the kinds of units we would need to perform these functions, while he concentrated on determining what it was we would actually have to build. After we got the contract in June 1943, he took the lead in designing the circuits, leaving to me the task of describing the functions to be performed and working with Eckert and the staff on the details of the logic. But when it came down to how something would actually be built, that was more or less Eckert's province. One of the first things he did was to develop a standards manual. Since there were so many repetitive circuits, we obviously couldn't have each person designing his own; Eckert laid down standard designs for the voltage supplies, counters, switches, and other circuits.

During the next 18 months, until the ENIAC was essentially finished, we had many seven-day weeks, but I can't recall any really serious difficulties. We were able to avoid devastating surprises by proceeding pretty methodically—designing, modeling, and testing as we went along.

To illustrate, one of the key circuits in the machine was an adder, which we called an "accumulator" because it could store as well as add. There were twenty of these accumulators and each one contained ten decade counters. Each counter contained ten flip-flops, hence every accumulator could store a ten-digit number. The accumulators were paired so that a 20-digit number, with sign, could be handled (a very big number for those days!). These accumulators were cumbersome pieces of equipment, consisting of several hundred tubes in a two-foot-wide panel roughly eight feet high.

Obviously we had to make sure of their design and, before we built anything else, we built two accumulators and a special power supply and signal generator so that we could test them. We called this the two-accumulator model and we looked forward to the day of its test with great anticipation. You can imagine the excitement when this crucial test went off without a hitch. Although we immediately saw ways to improve the accumulator and did a small amount of redesign, all the other units in the ENIAC ran as they were first built.

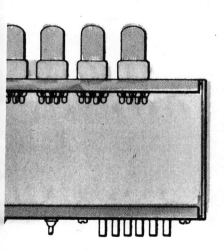

This concept of modeling was followed wherever possible. For instance, because multiplication was so important it was decided to eschew successive addition in favor of this faster operation even though it might complicate the electronics. Thus, we decided to build a single high-speed multiplier that would utilize an electrical multiplication table, But, instead of building the actual unit and all the associated circuitry, we built only the basic circuits and then loaded them with capacities representing all the other circuits that would normally be hung on. In this way, we were able to test the design more economically, finding that on the average this multiplier gave a speed advantage of at least 4½ over repeated addition.

In this manner, we proceeded relatively smoothly, step-by-step, to build the ENIAC. Of course, there were the inevitable problems of obtaining supplies.

For example, as we finished the machine and were ready to plug in 18 000 tubes, we discovered this was no small number in wartime when everything had a priority. As it developed, however, we worked through the military services to locate the tubes that were needed in a Signal Corps bank. In this way, we were able to get the proper orders through for them to release the tubes, which we couldn't possibly have gotten directly from the manufacturers.

To build better computers

The main thing that bothered us toward the end of the project, with the War drawing to a close, was how the work was going to be carried on. It should be understood that the main limitations of the ENIAC were essentially dictated by economics and the pressure to get something built as quickly as possible. At that time, it wasn't possible to store electronically all the information that might be needed in the course of a computation. Some of the data were kept on punch cards and input as needed, but the limitation that worried us the most was that, for very large and complex computations, we couldn't even store all of the instructions in the computer. And yet, to make use of the speed advantage of the vacuum tube, instructions have to be available as quickly as the execution devices can use them. A multiplication speed of a thousandth of a second is useless unless the instructions to multiply the specific factors can be supplied at the same speed.

We had to compromise here, since we knew that to provide sufficient storage capacity for all the possible instructions an operator could want would simply have required too many tubes—millions probably. Thus, some of the instructions were essentially wired up with patch cords or altered by turning switches, and we recognized this as a very important limitation on the computer—one which we wanted to remove as soon as possible. The opportunity for doing this occurred very soon after we started. In 1944. and on through 1946, whenever we were not telling somebody what they should do to get ENIAC completed, we were thinking about how to make a better computer. The uppermost priority was to find a better storage device, so that we could afford to store not only more data but a complete set of instructions as well.

The first thing we thought of to improve our storage capacity was a modification of something that Eckert had worked on for radar purposes—namely, an acoustic delay line storage tank. The idea behind this was to take the voltage signals representing actual numbers and put them through a piezoelectric quartz crystal so that they became, in effect, sound waves traveling down a tank of mercury. When these pulses impinged on another piezoelectric crystal, they would produce an electrical signal that could then be used by the electrical circuits in any way we wanted.

For storage purposes, our idea was to run the signal back through a similar delay, which sounds pretty simple except that it wouldn't work in such a basic form because the signals would die out. So an amplifier had to be provided. Having amplified the signal, it could be fed back into the same column of mercury and delayed once more. But then there would be a new enemy to fight: the fact that the path this signal follows—the column of mercury, the quartz crystals, and the amplifier—produces distortion. As a result, the original signal, which started out as a sharp pulse, gradually degenerates into a flatter and broader pulse, and sometimes even into two pulses.

This problem can be overcome by using the pulse coming out of the delay line as a trigger to enter a brand new sharp pulse into the system. Besides overcoming the problem of degeneration, this technique can also be used to overcome timing errors. A mercury tank of this sort could be economical only if many pulses could be stored in the same tank, differentiating the pulses by observing what time they were sent in and what time they came out. Obviously serious errors could occur if, say, a pulse that was supposed to represent "1" got confused with another pulse which was supposed to represent "100 million."

Once we saw how to accomplish these three things —amplification, clean pulses, and timing accuracy—the whole world opened up, so to speak, to further possibilities of making better computers.

With this kind of delay device, we were now able to reduce enormously the number of vacuum tubes required for storage. Instead of the 500 tubes per panel that had been required, we were now able to use only five tubes. This meant that with twenty such tubes we could easily store 25 ten-digit numbers. In ENIAC, 10 000–15 000 tubes were involved in the storage or retrieval of a few hundred decimal digits, equivalent to a few thousand binary digits. In Univac I, using a mercury delay line, about 1500 tubes were used to store over 100 000 bits, all of which could not only be read electronically at high speed but also be written electronically at similar speeds. Here then was an immense breakthrough from the point of view of capacity that could be handled economically—in fact, with a 500–1 improvement in cost.

We also thought that magnetic recording, probably

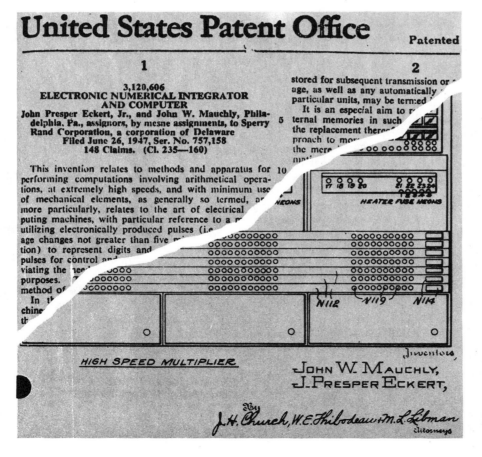

United States Patent Office

Patented

3,120,606
ELECTRONIC NUMERICAL INTEGRATOR AND COMPUTER
John Presper Eckert, Jr., and John W. Mauchly, Philadelphia, Pa., assignors, by mesne assignments, to Sperry Rand Corporation, a corporation of Delaware
Filed June 26, 1947, Ser. No. 757,158
148 Claims. (Cl. 235—160)

This invention relates to methods and apparatus for performing computations involving arithmetical operations, at extremely high speeds, and with minimum use of mechanical elements, as generally so termed, and more particularly, relates to the art of electrical puting machines, with particular reference to a utilizing electronically produced pulses (i.e. age changes not greater than five tion) to represent digits and pulses for control viating the purposes. method of In chine

HIGH SPEED MULTIPLIER

Inventors,
JOHN W. MAUCHLY,
J. PRESPER ECKERT,

By
J.H. Church, W.E. Thibodeau & M.L. Libman,
Attorneys

Filed on June 26, 1947, the original patent for the first electronic digital computer—ENIAC—did not go into effect until Feb. 4, 1964—17 years later! Based upon 148 claims, the Mauchly/Eckert invention consisted of 116 pages of descriptive text and 91 pages of illustrations.

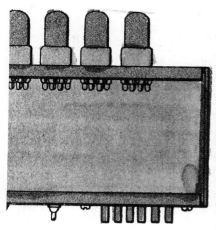

on a tape, would be a good way of managing input and output, though we didn't have any time to develop the method. This was during the infancy of magnetic recording, and there were rumors that things like this were being done in Germany toward the end of the War. As it was, all the input and output facilities of the ENIAC were handled by punch card equipment, a condition we just had to accept because we didn't have the time to go into any exhaustive research and development. But the input–output possibilities of magnetic recording appealed to us, and we planned to exploit them in a second computer.

Thus, during the years in which the ENIAC was being built and tested, we had many ideas about how we would do it better the next time. And, in fact, the potential for improvement appeared so great to me that I wondered if we shouldn't just shift from the half-built ENIAC to a computer employing these mercury delay lines and better methods.

Fortunately, we didn't abandon the original project. Although a report was written for a new machine called EDVAC, we did not advocate stopping work on ENIAC, but rather pushed it through, made it work, and got it into use. As it worked out, we must have done the right things because when the ENIAC was delivered to Aberdeen, they used it for ten years before finally deciding it was no longer economical.

Looking for support

As the War drew to a close, Eckert and I had the very strong feeling that ENIAC was the beginning of something new and important that had to be carried on. We urged that funding for continued work be arranged, but people sort of threw up their hands at the prospect—they just couldn't see how to do it.

For instance, one university official who had not had much experience with military contracts, suggested we approach the American Philosophical Society. However, its grants were usually on the order of a few hundred dollars to get secretarial help to type a manuscript or to buy an animal cage! If you counted everything, including labor and the material supplied by the Signal Corps, it was estimated that ENIAC came to a half-million dollar job. We couldn't see any way to build better machines without budgets like that and, since the military didn't appear terribly interested, we began looking elsewhere.*

During 1945 and 1946, Eckert and I spent months on Wall Street trying to interest the financial community in backing the development of a commercial machine. We didn't think it would be hard to sell them on something whose future was so obviously brilliant, but we were wrong. People could under-

---

* According to J. P. Eckert, General Electric was willing to put up $1 million to support the project, but University officials would not agree to it.

stand a radio with a few tubes, but they could not see how something with 18 000 tubes and costing $500 000 could ever become practical. After all, punched card systems existed and had proved eminently workable! The result was that we were dismissed by some of the biggest venture capitalists after a ten-minute hearing.

In hindsight, however, I must admit that I can't really blame them. When I first heard about xerography, I didn't believe it would ever succeed because I couldn't believe electric charges could be transferred to a drum to make the sharp images needed.

Thus, as with so many other inventors, Eckert and I had to finance the work essentially ourselves. When eventually I approached some friends at the Bureau of Census and they said they would like a computer for the Bureau, we had the start we needed and were able to form our own company. The rest is history!

I will conclude, however, by harking back to what started it all—my hope of finding better ways to forecast the weather. These hopes have never been fully realized. I have had no way of doing the kinds of analysis I would like to see done, and others have still not yet covered the areas I wanted to get into. Weather satellites can't tell us everything—I believe there is a great deal that can be learned from the proper kind of statistical analysis.

Perhaps with a new computer. . .   ◆

RECOMMENDED READING

Alt, F. L., "Archaeology of computers—reminiscences, 1945–1947," *Commun. ACM*, vol. 15, no. 7, pp. 693–694, 1972.

Bowden, B. V., ed., *Faster Than Thought.* New York: Pitman, 1953.

Burks, A. W., "Electronic computing circuits of the ENIAC," *Proc. IRE*, vol. 35, pp. 756–767, Aug. 1967.

Hartree, D. R., "The ENIAC, an electronic calculating engine," *Nature*, vol. 157, p. 527, Apr. 20, 1946; also "The ENIAC, an electronic computing machine," *Nature*, vol. 158, pp. 500–506, Oct. 12, 1946.

Randell, B., ed., *The Origins of Digital Computers, Selected Papers.* New York: Springer-Verlag, 1973.

Von Neumann, J., and Goldstine, H., "On the principles of large scale computing machines," in *John von Neumann: Collected Works* (vol. 5), A. H. Taub, ed. New York: Macmillan, 1963.

John W. Mauchly (F) is the coinventor of ENIAC, the world's first all-electronic digital computer. His collaboration with J. Presper Eckert, described in this article, led to the founding of the Eckert-Mauchly Computer Corp. in 1946. There they built a second digital computer, BINAC, and developed most of the UNIVAC I, the first large commercial computer. The company was purchased by Remington Rand in 1950, and Dr. Mauchly was named head of the Remington Rand Univac Applications Research Center. While there, he proved from weather statistics that the probability for large amounts of rainfall is highest right after a full or new moon. Dr. Mauchly resigned from Univac in 1959 to form Mauchly Associates, which specialized in scientific techniques for decision-making, forecasting, and management controls. In 1968, he founded Dynatrend Inc. to specialize in computer-aided stock analysis and management services. He is also a consultant for the Sperry Univac Division of Sperry Rand Corp.

Dr. Mauchly received his Ph.D. in physics from John Hopkins in 1932. The recipient of numerous medals and awards for his work with Eckert on the ENIAC and later computers, he was one of the founders of the Association for Computing Machinery and the Society for Industrial and Applied Mathematics.

THE COMPUTER'S ORIGINS: AN UNRESOLVED DRAMA

With their dials, dangling wires, and patch cords, the four dusty cabinets look like an old-fashioned telephone switchboard. It's hard to imagine that these electronic relics, shoved into a corner of a deserted foyer in the Moore School of Electrical Engineering at the University of Pennsylvania, were once part of ENIAC—the 30-ton behemoth that ushered in the computer age. Built just 40 years ago in a laboratory across the hall, ENIAC was the first electronic computer. Today, sitting in that same lab are several dozen personal computers powered by tiny silicon chips—each chip more powerful than all 18,000 of ENIAC's vacuum tubes.

Perhaps no one is more surprised at the revolution launched here than J. Presper Eckert, who in 1946 was a brash, 26-year-old engineer. He was the co-inventor—with his friend John W. Mauchly—of ENIAC, which is short for Electronic Numerical Integrator & Computer. Immaculately dressed in a stylish gray suit and looking far younger than his 65 years, he vigorously shakes his head as he muses over old memories: "I just never envisioned my life's work being put on a ¼-in. chip."

SUN SPOTS. Nor did the University of Pennsylvania. Although this school, which was founded by Benjamin Franklin in 1740, offered the first course in computer science in 1946, it was quickly eclipsed by the Massachusetts Institute of Technology. It was not until the 1970s that Eckert's picture was hung in a gallery of distinguished alumni that lines the walls outside the dean's office.

When Eckert first met Mauchly, neither man regarded himself as a revolutionary. Mauchly came to the university in 1941 with the idea for an electronic calculating machine already implanted in his mind. A physicist, he was determined to prove that solar activity—sun spots and flares—affects weather on earth. But contemporary mechanical calculating machines were not up to the complex computations he needed. Mauchly thought an electronic device would work, but he didn't understand the technology. Eckert, a precocious graduate student in electrical engineering, did.

It was not until the U. S. entered World War II that the inventors could persuade anyone to put up the money to build the device. They finally found a champion in Herman H. Goldstine, a young Army captain who was detailed to the Moore School, which served as a center for calculating artillery trajectories. It was a tedious process, with young women spending hours laboriously punching data into mechanical calcula-

ECKERT WITH PART OF ENIAC, THE FIRST COMPUTER: SOME OF THE CREDIT, LITTLE OF THE PROFIT

tors. "It was clear the machine would make human computers obsolete," says 72-year-old Goldstine, who is now executive officer of the American Philosophical Society in Philadelphia. So Goldstine, a mathematician, persuaded his superiors to finance ENIAC, despite objections from several government research and development managers. "According to the popular notion, a computer wasn't feasible with the technology of the day," Goldstine recalls.

Eckert was never troubled by such doubts. "The technology for the computer had existed for 10 years," he insists. Eckert and Mauchly completed the initial design for ENIAC in a frenzied two weeks of work. Once they received Army funding, they started building the machine without even constructing a prototype. Driven by Eckert, the ENIAC team, which eventually totaled 150 people, worked around the clock. When the electronic leviathan whirred into action on Feb. 15, 1946, it flawlessly completed

its first problem—a top-secret numerical simulation for the still untested hydrogen bomb—in 20 seconds. It would have taken existing calculating machines 40 hours to accomplish the same task.

Not surprisingly, both Eckert and Mauchly wanted to commercialize their invention. But university officials insisted that the inventors sign over their rights to the patent. They refused, and resigned just one month after ENIAC was completed.

The conflict over the patent issue was but one of many that split the ENIAC team, leaving animosities that smolder even today. In 1945, John von Neumann, a mathematician and computer theoretician then at the Institute for Advanced Study in Princeton, N. J., published a paper with Goldstine's help that laid out a fundamental concept of modern computer programming—the stored program, which allows instructions for problems to be incorporated into the machine rather than being inserted for each problem. Even today that still enrages Eckert. "Von Neumann was a damned liar," snaps Eckert, punching at the air with his hands for emphasis. "The ideas were mine and Mauchly's."

'THE UNDERDOG.' Mauchly's widow, Kay, who was a programmer on ENIAC, agrees. She says the two men first discussed the ideas in 1943 and wrote them down in a 1944 memo. Goldstine didn't bring von Neumann, a consultant to the Army, into the ENIAC project until 1944. But Goldstine recollects it differently. "Von Neumann lectured to Eckert and Mauchly about the stored program, and they saw the ideas as their own," he says. Goldstine and Arthur W. Burks, who also worked on ENIAC, supported von Neumann. Because von Neumann, who died in the 1950s, is still credited with the idea, Eckert and Mauchly "never forgave" the trio, says Kay Mauchly.

In any case, they pushed ahead to commercialize their invention. The partners spurned an offer from Thomas J. Watson, then chairman of International Business Machines Corp., to start that company's computer lab. Instead, in October, 1946, they set up their own venture, financed with a $25,000 loan from Eckert's father. The two snared government contracts to develop a large commercial computer called UNIVAC.

But they made many of the classic mistakes that often doom embryonic technology ventures. They underestimated development costs, and Eckert frequently insisted on putting technical considerations first, even at the expense of getting the product out. The final straw was the sudden death of a key financial backer. "We just kept running into a bad situation," recalls Eckert. "We were always the underdog." By 1950, on the brink of bankruptcy, they sold the Eckert-Mauchly Computer Corp. to Remington Rand Corp. for $538,000. Eckert and Mauchly together received less than $100,000 for their stock.

The modest price reflected the fact that almost no one realized the importance of the computer. IBM's Watson, who soon changed his mind, estimated on one occasion that "there is a world market for five computers." A 1950 BUSINESS WEEK story displayed the same lack of enthusiasm about Rand's acquisition: "Salesmen will find the market limited. The UNIVAC is not the kind of machine that every office could use."

UNIVAC went on to become the first commercial computer in 1951. But Rand never committed itself to the technology. By the time Rand merged with Sperry Corp. in 1955, IBM had usurped the business. Sperry's Univac division never regained lost ground, although the present Sperry Corp.'s Information Systems Group remains the descendant of Eckert and Mauchly's short-lived company.

FINAL INDIGNITY. Neither inventor's career ever really took off, either. In 1959, Mauchly left Sperry to start the first of several consulting firms. Until his death in 1985 he continued to use computers to study the relationship of solar activity and the weather, even though his scientific colleagues never accepted his work. Eckert stayed on at Sperry, but a year later he was forced out of the engineer-

ing division. He became a vice-president and technical adviser in 1963. In a final indignity, Honeywell Inc. challenged the Eckert-Mauchly patent, in an effort to end Sperry's claim to royalties.

Honeywell prevailed. In 1973 the patent courts ruled that Mauchly took his ideas from yet another computer pioneer, John Atanasoff. So the two Philadelphia scientists lost their claim to one of the world's great inventions—even though Mauchly's papers show that he tinkered with the idea of a computer as early as 1935, years before he met Atanasoff. "The trial destroyed John," Kay Mauchly remembers.

Still, in October, Eckert and other surviving members of the ENIAC team will gather at the Moore School to celebrate ENIAC's 40th anniversary. Joseph Bordogna, the school's dean, hopes they can bury the animosities and disappointments of the past in one last hurrah. Goldstine, at least, seems ready: "None of the controversy matters anymore," he says. Eckert, however, may not be so willing. "Not unless they tell the truth," he says, slapping his hand on a table.

BY EMILY T. SMITH

Staff Writer Smith covers science and technology from New York.

HERE COMES THE SECOND COMPUTER REVOLUTION

A triumph of miniaturization is represented in these pictures of old and new computers. The first electronic digital computer, called ENIAC, was completed at the Moore School of Electrical Engineering at the University of Pennsylvania in 1946. In the larger photograph, the two men in the foreground are the co-inventors of the huge machine, J. Presper Eckert Jr. and John W. Mauchly. ENIAC contained 18,000 vacuum tubes, and its power source (not shown) occupied about half as much space as the computer itself. The first microcomputer, held by its inventor, M. E. Hoff Jr. of Intel Corp., occupied a tiny chip of silicon (the small rectangle in the center). But it matched ENIAC in computational ability.

Less than thirty years ago, electrical engineer J. Presper Eckert Jr. and physicist John W. Mauchly, at times assisted by as many as fifty helpers, laboriously built the world's first electronic digital computer. Their ENIAC (Electronic Numerical Integrator and Computer) was a fickle monster that weighed thirty tons and ran on 18,000 vacuum tubes—when it ran. But it started the computer revolution.

Now under way is a new expansion of electronics into our lives, a second computer revolution that will transform ordinary products and create many new ones. The instrument of change is an electronic data-processing machine so tiny that it could easily have been lost in the socket of one of those ENIAC tubes. This remarkable device is the microcomputer, also known as the computer-on-a-chip. In its basic configuration, it consists of just that—a complex of circuits on a chip of silicon about the size of the first three letters in the word ENIAC as printed here. Yet even a medium-strength microcomputer can perform 100,000 calculations a second, twenty times as many as ENIAC could.

This smallest of all data-processing machines was invented six years ago, but its mass applications are just beginning to explode, setting off reverberations that will affect work and play, the profitability and productivity of corporations, and the nature of the computer industry itself. For the microcomputer provides an awesome amount of computer power in a package that in its simplest form costs less than $10 bought in quantity and easily fits inside a matchbox. Accessory devices bring microcomputer prices to between $50 and $250 apiece, to be sure, but that's still a lot less than the thousands of dollars a minicomputer costs.

It's cheaper to move electrons

And unlike the familiar older computers that come in their own boxes, the microcomputer is mounted on a small board that can be made to fit easily and unobtrusively into a corner of an electric typewriter, a butcher's scale, a cash register, a microwave oven, a gas pump, a traffic light, a complex scientific instrument such as a gas chromatograph, and any of a myriad other devices whose capabilities already are being enhanced by these slices of electronic brainpower. Soon microcomputers will start replacing wheels, gears, and mechanical relays in a wide variety of control applications, because it's much more efficient to move electrons around than mechanical parts.

To cite these applications and capabilities, as well as many other uses to come in the home, the factory, and the automobile, is to do only pale justice to this marvelous

invention. What sets any computer apart from every other kind of machine is its stored and alterable program, which allows one computer to perform many different tasks in response to simple program changes. Now the microcomputer can impart this power, in a compact form and at a low price, to many other machines and devices.

In the most common form of microcomputer, furthermore, a user can change the program simply by unplugging a tiny memory chip and putting a new one in its place. To show off this versatility, Pro-Log Corp. of Monterey, California, built a demonstration apparatus that in its original version is a digital clock; when a program chip that runs the clock is removed and another is put in its place, the thing suddenly starts belting out a tinny version of the theme from *The Sting*. With still another memory chip, it becomes a rudimentary piano.

An antidote for inflation

Besides providing versatility for users, the microcomputer makes possible large economies in manufacturing. Now a manufacturer can buy a standard microcomputer system for many different products and use a different program chip with each. By doing so, the manufacturer can save substantial amounts of money since a single microcomputer can replace as many as 200 individual logic chips, which cost about $3 each.

The use of microcomputers, moreover, can substantially reduce service and warranty costs because the reliability of the electronic portion of a device is increased up to tenfold. A microcomputer that replaces, say, fifty integrated circuits does away with about 1,800 interconnections—where most failures occur in electronics. The microcomputer, in other words, is one of those rare innovations that at the same time reduces the cost of manufacturing *and* enhances the capabilities and value of the product. Thus the microcomputer may be the best technological antidote for inflation to come along in quite a while.

Even the men who make and use microcomputers say that they haven't yet grasped the device's full implications, but they know the implications are large and far reaching. Fairly typical is the comment of Edward L. Gelbach, senior vice president at Intel Corp., the Santa Clara, California, semiconductor company where the tiny computer was invented. "The microcomputer," he says, "is almost too good to be true."

The microcomputer is the logical end result of the electronics industry's headlong drive to miniaturize. The industry has galloped through three generations of components in as many decades. In the late 1950's, the

> The brainpower of a digital computer is now available on a little chip of silicon. The reverberations will affect both businesses and consumers.
>
> *by Gene Bylinsky*

Research associate: Alicia Hills Moore

transistor replaced the vacuum tube. Within a few years, the transistor itself gave way to "large-scale integration," or LSI, the technique that now places thousands of microminiaturized transistors—an integrated circuit—on a sliver of silicon only a fraction of an inch on a side. LSI made possible the suitcase-sized minicomputer.

A bold technological leap

The semiconductor logic circuit, of course, contained the seed of the microcomputer, since the chip had logic elements on it—the transistors. But the individual chips were designed to perform limited tasks. Accordingly, the central processing units of large computers were made up of hundreds, or thousands, of integrated circuits.

Logic chips were also employed for control or arithmetic functions in specialized applications. In what became known as "hardwired logic" systems, chips and other individual components were soldered into a rigid pattern on a so-called printed-circuit board. The fixed interconnections served as the program. Curiously, it was even less flexible than ENIAC's primitive array of plug-in wires that could be moved around to change the program.

The electronic calculator, in all but the latest versions, uses hardwired logic. The arithmetic functions, or the operating program instructions, are embedded in the chips, while the application program is in the user's head—his instructions yield the desired calculations.

A young Intel engineer, M.E. Hoff Jr., envisioned a different way of employing the new electronic capabilities. He had received a Ph.D. in electronics from Stanford University, where he had become accustomed to solving problems with general-purpose data-processing machines. In 1969 he found himself in charge of a project that Intel took on for Busicom, a Japanese calculator company. Busicom wanted Intel to produce calculator chips of Japanese design. The logic circuits were spread around eleven chips and the complexity of the design would have taxed Intel's capabilities—it was then a small company.

Hoff saw a way to improve on the Japanese design by making a bold technological leap. Intel had pioneered in the development of semiconductor memory chips to be used in large computers. (See "How Intel Won Its Bet on Memory Chips," FORTUNE, November, 1973.) In the intricate innards of a memory chip, Hoff knew, it was possible

The intricate circuits of a microcomputer are first drawn on paper and then reduced photographically onto glass "masks" for photoengraving on a chip of silicon. Here a Motorola engineer holds plastic sheets—with circuits imprinted in various colors—that are used to check for accuracy during the reduction process.

to store a program to run a minuscule computing circuit.

In his preliminary design, Hoff condensed the layout onto three chips. He put the computer's "brain," its central processing unit, on a single chip of silicon. That was possible because the semiconductor industry had developed a means of inscribing very complex circuits on tiny surfaces. A master drawing, usually 500 times as large as the actual chip, is reduced photographically to microminiature size. The photo images are then transferred to the chip by a technique similar to photoengraving.

Hoff's CPU on a chip became known as the microprocessor. To the microprocessor, he attached two memory chips, one to move data in and out of the CPU and one to provide the program to drive the CPU. Hoff now had in hand a rudimentary general-purpose computer that not only could run a complex calculator but also could control an elevator or a set of traffic lights, and perform many other tasks, depending on its program. The microcomputer was slower than minicomputers, but it could be mass-produced as a component, on the same high-volume lines where Intel made memory chips—a surprising development that would suddenly put the semiconductor company into the computer business.

Hoff had strong backers in Intel's top executives: President Gordon E. Moore and Chairman Robert N. Noyce, the co-inventor of the integrated circuit. Unlike many other specialists, Noyce and Moore had sensed the potential of the microcomputer early on, and they lent enthusiastic support to Hoff's project. Most others had visualized a computer-on-a-chip as being something extremely expensive and far in the future. When in the late 1960's Noyce suggested at a conference that the next decade would see the development of a computer-on-a-chip, one of his fellow panelists typically remarked in all seriousness: "Gee, I certainly wouldn't want to lose my whole computer through a crack in the floor." Noyce told the man: "You have it all wrong, because you'll have 100 more sitting on your desk, so it won't matter if you lose one."

Using words to replace hardware

After other Intel engineers who took over the detailed design work got through with it, Hoff's invention contained 2,250 microminiaturized transistors on a chip slightly less than one-sixth of an inch long and one-eighth of an inch wide, and each of those microscopic transistors was roughly equal to an ENIAC vacuum tube. Intel labeled the microprocessor chip 4004, and the whole microcomputer MCS-4 (microcomputer system 4). "The 4004 will probably be as famous as the ENIAC," says an admiring Motorola executive. Despite its small size, the 4004 just about matched ENIAC's computational power. It also matched the capability of an I.B.M. machine of the early 1960's that sold for $30,000 and whose central processing unit took up the space of an office desk. If anyone had suggested in the days of ENIAC that this kind of advance would take place so soon, says Presper Eckert, now a vice president at Sperry Univac, the idea would have struck him as "outlandish."

For logic and systems designers the appearance of the microcomputer brought with it a dramatic change in the way they employed electronics. They could now replace all those rigid hard-wired logic systems with microcomputers, because they could store program sequences in the labyrinthine circuits of the memory chip instead of using individual logic chips and discrete components to implement the program. Engineers thus could substitute program code words for hardware parts.

For the semiconductor industry the arrival of the microprocessor on a chip signaled the end of a costly search for ways to reduce the complicated technology to more generalized applications. "The problem," says Moore of Intel, "was that as the technology got more complex you couldn't find any generality to the circuit functions. What customers wanted was one of this circuit, one of that circuit to build a system." Such demands threw monkey wrenches into the industry's efforts to hold down costs through mass production.

The industry kept flailing and groping for ways to master the problem. Texas Instruments, for instance, had a big program aimed at using computer-guided design to make production of integrated logic components more flexible. Fairchild Semiconductor talked about turning out as many as 500 different logic components a week to suit the requirements of different customers. In these attempts, engineers were trying to force the technology to become more flexible. Ted Hoff's solution, to make the internal design itself more flexible, was far more elegant and more powerful. Says Moore: "Now we can make a single microprocessor chip and sell it for several thousand different applications."

A rush to get on board

At first the semiconductor industry showed surprisingly little interest in this great leap in its technology. Robert Noyce recalls that when Intel introduced the microcomputer late in 1971, the industry's reaction was "ho hum." Semiconductor manufacturers had made so many extravagant promises in the past that the industry seemed to have become immune to claims of real advances. Besides, the big semiconductor companies—Texas Instruments, Motorola, and Fairchild—were pre-

occupied with their large current business, integrated circuits and calculator chips. "Looking back," says J. Fred Bucy, T.I.'s executive vice president and chief operating officer, "we probably should have started on microcomputers earlier."

Only Rockwell International and National Semiconductor got into the field early on, about a year after Intel. Fairchild came out with a microprocessor chip that it sold primarily to calculator manufacturers. It took another six months or so before the new economics of the microcomputer stung the other giants into action. By that time, hardly anyone could have missed the message: a microprocessor and its memory could replace a lot of individual logic chips—anywhere from ten to 200. To speed the adoption of microcomputers, Intel undertook to recast the thinking of industrial-design engineers—the company taught 5,000 engineers the use of the microcomputer in the early Seventies and another 5,000 or so later on. Once these engineers started ordering the tiny computers in some quantity, the big companies, as Noyce puts it, said, "We've got to get on board here."

They rushed to get on board by "second-sourcing"—i.e., copying—Intel's microcomputers. Second-sourcing is a common practice in the semiconductor industry. More often than not, it is done without the original manufacturer's permission or cooperation, but the practice is nonetheless widely accepted by the companies involved. It works to the benefit of the user in establishing a competitive source for the component as well as a backup for the original manufacturer. In fact, users normally demand second-sourcing.

Late but nimble

Second-sourcing microcomputers proved to be a complex task, however. What's more, Intel kept moving. It followed up the 4004 with a more capacious 8008 model in 1972, and toward the end of 1973 brought out its second-generation microcomputer, the 8080. This was twenty times faster than the 4004. Even then most competitors had no microcomputers of their own to offer. The first real competition to the 8080 was Motorola's 6800, which came a year afterward. The late starters finally began to catch up this year when Texas Instruments, General Instrument, and others announced microcomputer models of their own. T.I. also introduced its copy of the 8080 only this year.

To paper over the gap, some nimble competitors upgraded calculator chips and started calling them computers-on-a-chip. With memory on the same square of silicon, these basic units can perform simple and even medium-complexity control functions—running washing machines or microwave ovens, for instance. T.I., Rockwell, and others now offer such chips. The T.I. product, TMS 1000, sells for as little as $4 in large quantities.

All these companies, and many others, are battling for a market that so far is fairly small—this year it will

Source: Fortune, November 1975, 134-139, 182, 184

amount to only about $50 million. But it is expected to expand to $150 million next year, and to reach $450 million by 1980. In these estimates, the microprocessor chips account for only 15 to 20 percent of the dollar total, with memories and other components making up the bulk of the new business.

Applications of microcomputers today are tilted heavily toward data-processing equipment of various kinds, including computer terminals and other accessories. The other major market is retailing equipment—electronic cash registers and point-of-sale terminals. But the picture is expected to change drastically in a few years as microcomputers invade consumer products in force. T.I. estimates that consumer-product uses will account for about one-third of the predicted $450-million-a-year market for microprocessors in 1980.

Opportunities for upstarts

In their capabilities, microcomputers cover quite a range of applications. A simple microcomputer can act as a miniature controller, replacing an electromechanical relay or hard-wired logic systems. A more powerful model, such as the 8080, can control a computer printer, or a whole series of them. Still more powerful models begin to match—and some already exceed—minicomputers in their computational speeds.

The tiny computer is beginning to generate not only new products but new companies as well. Says Gordon Hoffman, an executive at Mostek, a Dallas semiconductor house: "A lot of big companies are going to be improperly prepared to take advantage of the microcomputer. If they don't take advantage of it, they may find themselves out in the cold when a little upstart comes along and says, 'I can do it better with a microcomputer.'"

That kind of competition has already begun, with many fast-moving small companies taking advantage of the microcomputer's mighty power. A few examples:
- Chemetrics Corp. of Burlingame, California, only two years old, has brought out an advanced blood-chemistry analyzer.
- Electro Units Corp. of San Jose has developed an electronic control system for bars; it doles out precisely measured drinks and serves as an attentive inventory controller too.
- Telesensory System Inc. of Palo Alto is introducing this fall a "talking" calculator for the blind, with a recorded vocabulary of twenty-four words for spoken verification of calculation steps and results.

Large companies, of course, are also using the capabilities of the computer-on-a-chip to turn out new products. Among them:
- General Electric, which is looking into many possible applications, recently introduced a robot industrial tool run by a tiny computer.
- AMF, with the aid of Motorola, developed an automatic scorer now being demonstrated in bowling alleys.

- Tappan Co. is designing a microwave oven with "touch-and-cook" controls; it uses the single-chip microcomputer made by Texas Instruments.

For companies large and small, instrumentation is proving to be one of the most rewarding areas of microprocessor applications. Because of its powerful data-processing capacity, a computer-on-a-chip can not only impart brand-new capabilities to an instrument but also make it much easier to operate. With the microcomputer helping out, an unskilled person can operate a complex instrument, because, as one Perkin-Elmer engineer puts it, "the skill now resides in the microcomputer." Perkin-Elmer has already introduced two different spectrophotometers incorporating the microcomputer and is working on other uses in scientific instruments.

Microcomputers will also make a lot of laboratory-type analytical equipment more readily applicable to process control. Leeds & Northrup has already produced one such instrument, a particle analyzer that uses a laser beam to measure particles and a microcomputer to figure out their size distribution. The device is being tested in a taconite (iron ore) plant, but it can be adapted to other customers' needs through a change in its program.

Semiconductor manufacturers are also looking for applications of microcomputers to appliances such as washing machines and refrigerators. The current recession has delayed new-product introduction in this field, but microcomputers are being designed into models that are expected to start showing up in about two years.

The automobile may prove to be a big user of electronics in years to come. Some electronic components are already being employed in cars to supervise ignition, measure voltages, and so on. Microcomputers are expected to start appearing in automobiles toward the end of this decade. Ford Motor Co. has found that microcomputer-run controls can cut fuel consumption by as much as 20 percent under test conditions. The company plans to introduce the tiny computers in a 1979 car. Other automakers have similar plans.

Like going from wood to nuclear fuel

In many other areas, microcomputers promise spectacular advances. In the home, microcomputer controls could result in savings on electric and heating bills. For the military, the tiny computers promise the evolution of more versatile weapons. In medical electronics, they open up possibilities for compact and less costly diagnostic instruments. There are indications that in conjunction with complex optical and mechanical devices, microcomputers could he'p restore vision for some of the blind. In one project, a microprocessor chip will be embedded in an eyeglass frame to decode visual information from artificial "eyes" and send it to the brain.

As is true with any other computer, the largest costs—and most problems—arise in writing application programs for microcomputers. Basically, a digital computer

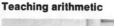
Measuring out drinks

Teaching arithmetic

Testing eyes

Weighing and displaying

Controlling traffic

THE MICROCOMPUTER'S LENGTHENING REACH

The computer-on-a-chip is beginning to enhance the capabilities of familiar products and create new ones too. A few examples are pictured here. Lower left: a traffic-flow-control system under development by TRW Inc. for the city of Baltimore; tiny data processors tied to a central computer will direct lights at 900 intersections. The items in the three upper pictures are all made by California companies. Upper left: the Dioptron, which measures the objective refraction of the eye for eyeglass prescriptions (Coherent Radiation, Palo Alto). Middle: a computerized mixed-drink dispenser for bars; it also keeps track of inventory (Electro Units Corp., San Jose). Upper right: the Digitor, a teaching aid to help youngsters learn arithmetic (Centurion Industries, Inc., Redwood City). In the foreground of the picture just above is a computerized scale that not only does the weighing, like the older scale next to it, but also converts the weights to prices and operates a display; it can also print a label with the weight and price if desired (Toledo Scale Co.).

Source: Fortune, November 1975, 134-139, 182, 184

runs in response to instructions written in the binary code of ones and zeros. That's how the first computers were programmed—with the complex instructions written out painstakingly by hand. To ease the programmers' task, the industry has over the years developed high-level computer languages in which abbreviations or even words substitute for whole series of numbers. Along with the languages came such programming aids as assemblers and compilers.

The semiconductor industry makes such aids available to microcomputer users. The machines are, in effect, small computers that utilize microprocessors. They sell for $2,500 to $10,000. Motorola calls its device the Exorciser; Intel's is called the Intellec.

Problems arise when design engineers who have previously dealt with electromechanical relays, or even hardwired logic, and are untutored in computer programming, suddenly face the complex accoutrements of data processing. For some, says one specialist, the experience is like "going from wood burning to nuclear fuel." As a result, something of an occupational obsolescence has temporarily developed in the design field because the engineers who are most skilled in product design usually have little or no experience with microprocessors and their applications.

Trying to fill the educational gap, M.I.T. and some other universities have begun intensive courses for both students and industry representatives. Reports M.I.T. Professor H.M.D. Toong: "Students go right from here out into industry and get jobs first thing heading microcomputer development and applications departments." Some specialists think that the applications of microcomputers will start expanding manyfold when the new graduates begin to enter the work force in large numbers.

The 99.9 percent price decline

For semiconductor companies the microcomputer opens another broad avenue for growth. With phenomenal price declines a way of life, the industry is a voracious consumer of new markets. Industry executives like to note that the price of an electronic function such as a transistor dropped 99.9 percent from 1960 to 1970 and is still declining. As one man puts it, "It's like putting an $8 price tag on an $8,000 Cadillac."

At the same time, each new advance in technology has brought with it a widening use of electronics. Texas Instruments calculates that during the vacuum-tube era, digital-electronic sales rose on a slope of about 10 percent a year. In the days of the transistor, the slope steepened to an 18 percent annual increase. Integrated circuits increased the sales growth rate to 38 percent. Now T.I. expects another upward tilt in the curve in the late 1970's, thanks chiefly to the microcomputer. The company anticipates that for the foreseeable future sales of electronic components will climb at a dizzying rate of 50 to 60 percent a year.

There seems to be little disagreement that the microcomputer is close to being an ultimate semiconductor circuit and that it now sets the direction for semiconductor technology. On the face of it, the principal beneficiary of this trend would appear to be Intel. The company now dominates the microcomputer market. What's more, it mainly makes semiconductor memories of the kind that go into microcomputers and does not make the integrated circuits that microcomputers replace. The principal losers would seem to be Texas Instruments, Fairchild, Motorola, and National Semiconductor, which are big in what is called transistor-transistor logic (TTL), the mainstay of the integrated-circuit business today—precisely the circuits the microcomputer replaces.

The Texans change horses

But that's not how top executives of some of those companies see the future. T.I.'s Fred Bucy envisions his company emerging as a major force in microcomputers. So does Charles E. Sporck, president of National Semiconductor. And both are probably right. Bucy stresses, and others agree, that the microcomputer's biggest use will be in applications where electronic devices have never been employed before. New applications thus will be far more important than replacement of TTL logic. Bucy also notes that T.I. is the only semiconductor company "that has lived through all the generations of electronic components. We've successfully moved from one horse to the next." Few executives in the industry would dispute T.I.'s obvious strengths as a $1.5-billion company even if it has been late in microcomputers. National Semiconductor, too, is an exceedingly clever marketer.

Everyone agrees, furthermore, that there will be a whole spectrum of microcomputers aimed at different applications, with many companies sharing the anticipated big market. And it is generally agreed that the most successful makers of microcomputers will be those that supply the best operating programs. The need to generate software to go with the tiny computers is a new activity for semiconductor companies, with the exception of T.I., which for years now has been making both minicomputers and very large machines.

Bucy and other T.I. executives feel that's another plus for their company. To keep its computers tied together, and to ease the task of users who want to employ microcomputers in conjunction with bigger machines, T.I. a few weeks ago introduced a powerful microcomputer whose software is compatible with that of the company's minis. T.I. sees a big competitive advantage in this approach, since the software of most other microcomputers does not directly match that of bigger computers.

The ability of users to operate a whole hierarchy of computers, from a big host machine to the microcomputer far down in the organization, will speed the trend toward "distributed" computer power. Bucy sees as a result a computer world polarized into giant machines and huge numbers of microcomputers, with medium-sized computers diminishing in importance.

Other specialists see computers of the future evolving into modular processor systems based on microcomputers, with many of their programs embedded in microcomputer memories, replacing expensive software. Frederick G. Withington of Arthur D. Little, Inc., predicts that ten years from now, as a result of the semiconductor industry's nonstop price erosion, the cost of even the largest CPU may come down to about $30,000.

Manufacturers of bigger mainframes are indeed beginning to incorporate microcomputers not only into terminals and minicomputers but also into their large machines, to control such functions as input and output of data. A vice president of NCR says that his company is "going to concentrate on the use of microprocessors in microcomputers, minis, and on up the line." NCR buys microcomputers from semiconductor manufacturers but it also plans to make its own. Burroughs already manufactures its own microcomputers and uses them in a variety of devices, including a small business computer. Control Data buys from Intel. I.B.M. and Honeywell do not yet make a microprocessor on a chip.

Computers by the millions

For manufacturers of big mainframes, then, the micro-computer has so far been a new component rather than a competitor. But for manufacturers of minicomputers, the arrival of the microcomputer has created a competitive danger—the micros are encroaching on the minis. To counter the threat, Digital Equipment Corp., No. 1 in minis, has made arrangements with a semiconductor company, Western Digital Corp., under which Western makes microprocessors and associated components. Digital Equipment then puts the devices on circuit boards and sells the microcomputers in direct competition with the semiconductor houses.

Some semiconductor companies, in turn, have come out with microcomputers that run on programs written by Digital Equipment and Data General Corp. for their minicomputers. These microcomputers do essentially the same job but sell for a lot less than the original minis. This blurring of dividing lines between computer and semiconductor manufacturers is expected to continue. Only half in jest, Noyce already calls Intel "the world's largest computer manufacturer."

In its impact, the microcomputer promises to rival its illustrious predecessors, the vacuum tube, the transistor, and the integrated-circuit logic chip. So far, probably no more than 10 percent of the tiny computer's potential applications have reached production stage. Today, nearly thirty years after the debut of the ENIAC, there are about 200,000 digital computers in the world. Ten years from now, thanks to the microcomputer, there may be 20 million.   END

Source: Fortune, November 1975, 134-139, 182, 184

SECTION II
Fundamentals

SECTION II
Fundamentals

What a Computer Is

A computer is an information-handling machine. What makes it so unique and useful is that it can execute instructions in a predetermined and self-directed manner. It does this by reading instructions in the form of a program that it remembers by storage in an internal memory. All subsequent data or instructions are then manipulated in accordance with this executive program. That means that the computer need not be reinstructed as the task progresses and that it can, therefore, perform tasks of considerable complexity from start to finish without any operator intervention.

General-purpose digital computers can be used for a large variety of applications, because the program stored within the computer's memory can be changed. Rapid access to the stored program, as well as the ability to follow alternative program routes according to the computer-perceived results of a series of program steps, is what makes modern computers so versatile.

Computer Language

Regardless of the technology involved, digital computers depend on the ability to represent and manipulate symbols, commonly referred to as characters. In fact, all of the information that passes through a computer, including every instruction the computer executes, is in the form of numbers. A fundamental requirement for computers, therefore, is the ability to physically represent numbers and to perform operations on the numbers thus represented.

How can the computer best deal with and represent numbers? Computer systems, which are basically composed of electronic and electromechanical devices, can readily sense or establish one of two states: positive or negative current, magnetized or not magnetized, on or off, the presence or absence of a hole. The key to the computer's ability to deal with numbers, then, is a numbering system that can be represented as a bistate condition. And that's why the simplest number representation for computers is the binary system, in which only two symbols — 1 and 0 — are required to represent any quantity.

The binary numbering system follows the same set of rules as the numbering system with which we are all familiar: the decimal numbering system. The primary difference is in the number of distinct symbols, or digits, that exist within each — ten (0 through 9) for the decimal system, and two (0 and 1) for the binary.

To understand the binary system, remember first that all modern numbering systems are based on the principle of count and carry. You count in units until you run out of symbols, then start all over again by "carrying" a one to the next place. For example, in the decimal system, each ascending digit is the result of increasing the value of the previous digit by one. In effect, the positional values of the digits in the decimal system increase to the left of the decimal point in a series represented by the powers of ten: $10 \ldots 10^3$, $10^2, 10^1, 10^0$, or N $\ldots$ 1000, 100, 10, 1. The total value of the decimal number is obtained by summing the products of each position value multiplied by the symbol values (0 through 9) in that position. For example:

$$
\begin{array}{rcl}
652 = 6 \times 10 & = & 600 \\
5 \times 10 & = & 50 \\
2 \times 10 & = & 2 \\
\hline
& & 652
\end{array}
$$

In the binary numbering system, the positional values of the digits increase to the left of the decimal point in a series represented by the powers of two; $2 \ldots 2^4, 2^3, 2^2, 2^1, 2^0$ or N $\ldots$ 16, 8, 4, 2, 1. And again, the total value of the binary number is obtained by summing the products of each position value multiplied by the symbol values (0 or 1) in that position. For example, in the binary system:

$$
\begin{array}{rcl}
1011 = 1 \times 2^3 & = & 8 \\
0 \times 2^2 & = & 0 \\
1 \times 2^1 & = & 2 \\
1 \times 2^0 & = & 1 \\
\hline
& & 11
\end{array}
$$

Remember, a number is merely a convenient symbol used to represent a quantity, and a numbering system is just a means of representing quantities using a set of numbers. All modern numbering systems use the zero to indicate no units and other symbols to indicate quantities. The number of symbols used is called the base, or radix, of a numbering system. The positional values of the numbering system increase to the left of the decimal point in a series represented by the powers of the base.

In the binary system, the system that computers can best deal with internally, a group of four binary digits is required to express the decimal digits 0 through 9 (see

Table 1. Binary Representation of Decimal Values

Decimal value	Place values	Binary notation			
		8	4	2	1
0	=	0	0	0	0
1	=	0	0	0	1
2	=	0	0	1	0
3	=	0	0	1	1
4	=	0	1	0	0
5	=	0	1	0	1
6	=	0	1	1	0
7	=	0	1	1	1
8	=	1	0	0	0
9	=	1	0	0	1
10	=	1	0	1	0
11	=	1	0	1	1
12	=	1	1	0	0
13	=	1	1	0	1

Table 1). (Incidentally, a binary digit is called a "bit," a term much bandied about in computerese.) In binary-coded decimal representation, each decimal position (and its value) is represented by a separate group of four bits. The number 30 would be written as 0011 (representing 3) followed by 0000 (representing 0), or 00110000. And the binary coded decimal representation of the decimal number 265,498 would be 001001100101010010011000 (see Fig. 1).[1]

Different Types of Computers

Computers can be generally classified as microcomputers, minicomputers, and mainframes. These classifications have recently been broadened by adding the prefix "super-" to each of the three classifications. Found within the class of microcomputers are home computers, personal computers, and programmable controllers, which are a special type of industrial microcomputer.

All microcomputers, and many minicomputers as well, can be said to be microprocessor-based in their design and construction. That is, their electronics are based on a processor that uses a specific kind of chip. For example, many early home and personal computers were based on the Intel Z-80 processor, the first real

"microprocessor chip." Recent models are based on newer chips with more functions, such as the Motorola 68000, as well as others.

The reader will note that the three basic characteristics of computer class, size, and cost are word length, memory size, and memory access speed. Microprocessors, for example, have word lengths of 4 and 8 bits. Minicomputers have word lengths of 16 and 32 bits, while the mainframes have 32, 64, and larger word lengths. Memory size, access speeds, and costs increase on a similar scale.

These classifications are further illustrated and described in Section III, on computer hardware.

Since the microprocessor is the most important element of most computers that the reader might be expected to encounter, the articles in this section deal with various aspects of microprocessor-based equipment in measurement and control applications.

Microcomputers

A microcomputer contains a microprocessor, which is a central processing unit (CPU) built on a small silicon chip. By itself, the microprocessor is of little use. But when integrated with a memory chip, a clock chip, and provisions for interfacing with a power supply, and with input/output (I/O) devices such as keyboards and visual display or readout units, it becomes a microcomputer. As we've said, most microcomputers have word lengths of 4 to 8 bits, and their memories contain from 256 to 65,535 words, depending on application requirements. At present, some microcomputers can access more than 8 million words.

At the sensor level, a microprocessor or microcomputer can increase system accuracy and the ease with which signals are transmitted. A simple microprocessor or an even simpler LSI (large scale integration) circuit in a digital converter can gather data with minimum system loss in signal strength and quality, and also can preprocess some of the data before sending it on in binary form. A microprocessor or microcomputer is "dedicated" to (designed for) a specific task or group of tasks within an application. A digital watch, for example, contains a microcomputer dedicated to computation and display of clock time; it cannot be used to figure your income tax, although it has the computing power to do so. Similarly, a digital temperature controller contains a microcomputer dedicated to measurement, display, and control of temperature over a given range, using a specific type of thermocouple or other temperature-sensing device. In an industrial control system such as a fuel-fired furnace, a microcomputer can sample and evaluate temperature

Decimal notation	2				6				5				4				9				8			
Binary notation	0010				0110				0101				0100				1001				1000			
Place value	8	4	2	1	8	4	2	1	8	4	2	1	8	4	2	1	8	4	2	1	8	4	2	1

Fig. 1. An example showing the binary-coded decimal notation for the number 265,498.

at several different points in the furnace, compare the results with acceptable limits, and make necessary adjustments. It can also continuously measure and regulate the flows of fuel and/or combustion air to maintain a preset ratio, which yields more energy-efficient control. Conventional analog instrumentation can do all of this too, but it is less accurate and involves equipment that is physically much larger, requires more power, and wears out much faster.

A microcomputer (or a minicomputer) can also integrate several instruments by acting as a data logger, a preprocessor, or a "smart front end" to a larger computer. Whereas large mainframes may have been used a few years ago for more complex control and measurement applications, microcomputers (and minicomputers) have recently found successful application as such components in larger systems approaches.

Most current applications of computer technology to metalworking involve microcomputers, although minicomputers are used for supervisory control of several furnaces where each furnace is separately con-trolled by a microcomputer system. Minicomputers are widely used in materials testing and research applications.

Minicomputers

Minicomputers are distinctly different from microcomputers in two ways: they are physically much larger, being mounted in metal racks and cabinets; and they can handle more data transactions. For example, a minicomputer can measure and control many separate variables of temperature, flow, or vacuum via high-speed multiplexing of inputs and outputs, and can also perform the supervisory function of adjusting individual control setpoints in accordance with a mathematical "optimizing" model contained in its memory. To accomplish these functions, it requires more data-storage capacity, or memory, than a microcomputer, and a more flexible process interface capability, hence its greater size. Figure 2 is a schematic diagram of a modern minicomputer.

Minicomputers have word lengths of 12 to 16 bits

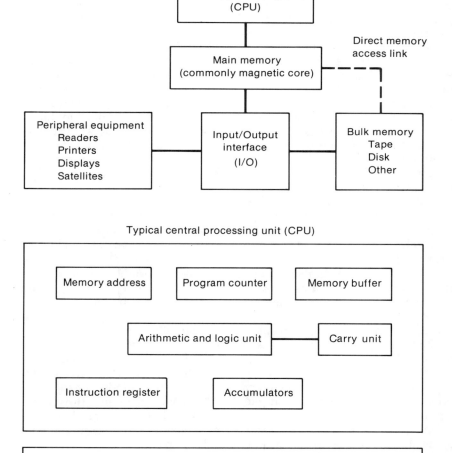

Fig. 2. Schematic diagram showing the basic structure common to all computers, with an expanded section showing the components in the central processing unit (CPU).

and minimum memory sizes of 4096 (4K) words. Mini-computers are trending toward word lengths of 32 bits, for production throughputs greater than those obtainable with microcomputers.

Mainframes

Mainframes, or large-scale computers, usually have word lengths of 32 bits or more and memory capacities of 16,384 words or more. A mainframe is usually used for after-the-fact data processing and analysis. It can also supervise or monitor the operation of a mini-computer, which is in turn connected to either a furnace process or its instrumentation. Within this system or network structure, the large computer can either monitor or augment the smaller device's capabilities.

Memory Systems

The memory system is the region of the computer that stores program instructions and data for instant use. Memories are of either the read-only (ROM) or the read-write type. If the data or instructions in a computer's memory do not need to be altered, read-only memories are used, in which case the computer cannot alter the contents. An example of this type of content is temperature-emf (electromotive force) tables for thermocouples, which give the millivolt output generated at a particular temperature. Read-write memories, on the other hand, can be altered and accessed by the computer. Read-write memories are also known as random-access memories (RAM) because any data location within the memory can be accessed as easily as any other. Data such as values of furnace temperature sensed by a thermocouple are temporarily stored in this type of memory.

Memories in modern mini- and microcomputers are of the semiconductor type, a memory matrix composed of tiny semiconductor chip-circuits. A semiconductor ROM will retain data in the event power fails, but a RAM will not, unless it is protected by a "battery backup" system. The type of memory used depends in part on the application. For example, a microcomputer-based controller with a limited number of dedicated functions would probably incorporate a ROM as the main portion of memory to hold the basic instructions and a RAM for temporary storage of data.

In control applications, computers must connect either to a binary-type signal (either an open or closed relay), to an analog signal, or to a digital signal. The connection between the digital computer and the binary circuit may involve only the matching of voltage levels. If an instrument or valve has a digital output, connecting it to the computer again involves only voltage or current matching. If the control has an analog output, the analog signal from the sensor to the computer must be converted to digital form. The interface, in this case, is an analog-to-digital (A/D) converter placed between the computer and the analog input device. If the computer must send an analog signal to a final control device, such as a valve, a digital-to-analog (D/A) converter acts as the interface. A/D and D/A converters are integral components of even small computer-based process systems because they are usually the only possible communication interfaces between the process and its control system.

Software

Software provides the computer with flexibility but also accounts for many of its problems. The problems are not necessarily due to the software but to human error. These errors result because software usually is not written in English, but in a special language that the computer can understand and translate into the binary language it uses to execute the program commands. Software language can be "machine-like," or "English-like." Machine-like languages are called "assembly," or "low-level" languages, and English-like languages are called "high-level" or "compiled" languages.

Assembly languages are collections of mnemonics that refer to the exact functions a computer must perform. In one particular assembly language, for example, CLA stands for "clear the accumulator," which is far easier to work with than the binary equivalent 1111100000000. Assembly languages sometimes are used with microcomputers because they conserve the microprocessor memory. Memory can be the largest hardware expense of a microcomputer system.

Whereas assembly languages are organized to match the way computers operate, high-level languages are organized more like spoken languages. High-level programs do not need to specify the individual steps of operations the computer performs; instead, they must indicate only what function to perform (for example, "add two fields").

While assembly languages require a special program to convert user code into binary code, high-level languages require a program to convert the high-level program into binary code. The special program, called either a compiler or an interpreter, occupies a significant portion of the memory system. This increases the size of the memory required. Greater memory capacity is also required because high-level languages are not optimized for one specific machine. Consequently, they are not as efficient in their translation as are assembly languages.

Although high-level languages are not designed for specific machines, they are designed for specific functions. Of the popular languages, BASIC (Beginner's All-Purpose Symbolic Instruction Code) and FOR-TRAN (Formula Translator) were developed for mathematic and scientific problem solving. FOR-TRAN, particularly FORTRAN IV, tends to have a greater number of features than BASIC and thus is generally more difficult to learn. High-level languages are now in very common use with computers of all sizes; however, programs in these languages require greater memory capacity than programs in assembly languages. Most computers also have standard packaged operating systems written in assembly language or high-level language. With operating systems, users need to be concerned only with the application; the operating system handles file management, program location, and programmers' "housekeeping chores."

Computers have almost unlimited use. For a computer to provide the advantages of accuracy, flexibility, and reliability, however, a potential user must also investigate available software, memory requirements, storage, peripherals (printers, plotters, cathode ray tubes, etc.), and, of course, interfaces (A/D, D/A converters, etc.) to ensure purchase of a system that will fulfill a broad range of needs, as well as to determine which system will afford the greatest possibility of flexibility and expansion. With the diversity of processor types, peripherals, and software, it is possible to apply computer operations effectively to almost any control or information management application.[2]

Reference Bibliography

1. *Computers in Manufacturing*, by the editors of *American Machinist*, McGraw-Hill Book Company, 1983, Chapter 1.
2. "Computerized Systems for Heat Treating," *Metals Handbook*, Ninth Edition, Volume 4, Heat Treating, American Society for Metals, Metals Park OH, 1981, p. 367.

Understanding microprocessors

The so-called 'computer-on-a-chip' is finding increasing application
in the power field as an adjunct to instrument and control systems. The
devices are not 'black boxes,' however, and deserve your understanding

By Thomas C Elliott, Associate Editor

POWER first talked about computers in a big way in 1965 (32-page Special Report, December, 1965). In 1971, we reported on the soaring popularity of the minicomputer (January 1971, p 34). Three years later, we talked about the coming trend toward programmable controllers (May 1974, p 72). Over the years, we've also informed readers about the solid-state revolution and the sure drift toward integrated circuits. It was only seven years ago that we extolled the virtues of printed-circuit (pc) boards, which could be ganged side by side like the pleats of an accordion.

All of these reports offered state-of-the-art coverage as we saw it at the time. During the past dozen or so years, engineers have had to cope with the continually growing influence that computers have had on the power field—and, indeed, on society as a whole. It has been especially trying for engineers out of school for 15 years or more, because they were not versed in computer technology at school and have had to upgrade themselves, sometimes willy-nilly, in the years since.

Now we are in the midst of another revolution—the plunge toward microprocessors. Just when we had grown comfortable with tape drives and blinking lights, with neatly configured com-

A microprocessor, by itself, is not much help. The microcomputer system has the greatest interest for readers

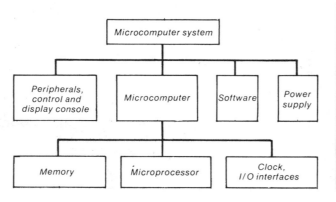

A semiconductor chip has typical dimensions shown, a data-acquisition time of 1 μsec, can operate at −55C to 125C

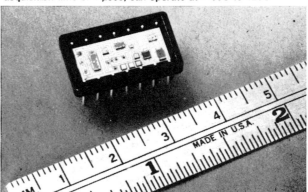

Leading IC technologies stress speed, power, density

Key advances that have made IC technology viable are (1) planar silicon crystal wafers, and (2) thin-film processing. The wafers, 2-3 in. in diameter and 16-20 mils thick, are formed by slicing them from ingots of crystalline silicon. Circuit components on the wafer are then fabricated via photolithographic means.

This thin-film processing begins by identifying individual circuits on a composite drawing, which defines the regions and layers of the circuit in terms of the various processing steps—epitaxial diffusion, isolation diffusion, metallization, contact holes, etc. To physically represent circuit regions and layers on the silicon wafer, photographic masks are applied, each of which corresponds to a single processing step of the wafer.

Circuit definition is achieved by: (1) applying a photosensitive layer to the wafer; (2) placing a mask over the layer and exposing it to light, thus transferring the mask pattern to the wafer; and (3) applying an etching solution to the wafer surface to embed the pattern within the wafer. The solution penetrates those areas exposed to light; masked areas remain. The process is repeated with different masks (and patterns) until the final circuit is embedded in the wafer.

Two basic types of electronic circuits are fabricated with silicon IC technology—*bipolar* and unipolar, or *MOS* (chart, right), from which a number of important subtechnologies have been developed. Remember, electronic circuits operate by the movement of a charge from one circuit element to another. By incorporating impurity atoms into the wafer's crystal structure, it's possible to alter the level of conductivity. Transistor diagrams (facing page) show key elements in the so-called *npn* bipolar device and the MOS field-effect transistor (MOSFET). Glossary on p 28 defines chart and diagram terms.

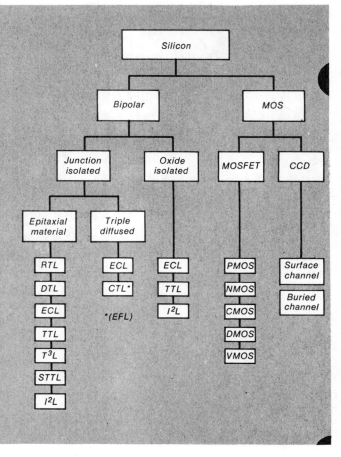

puter rooms and freestanding cabinets of pc cards, here comes another upheaval in the technological terrain, which threatens to supplant all that we've struggled to accept and learned to work with (if not love). A ray of sunshine, however: For engineers who are computer-wise, this latest incursion will be practically painless. Let's find out why.

A revolution that won't quit

Call it the IC (integrated-circuit) revolution. First it was MSI (medium-scale integration) about seven years ago, which stressed putting more digital logic gates or analog impedance elements on a chip. Because of the difficulty in interconnecting the MSI chips, however, it was almost inevitable that IC manufacturers would integrate these MSI chips into one big LSI (large-scale integration) chip (photo, previous page). They took this step about six years ago.

For our purposes, let's consider a *microprocessor* the basic CPU (central processing unit). When memory, clock, and I/O capability are added, we are talking about a *microcomputer* (chart, previous page). When software, a power supply, peripherals, control, and display capability are added to this configuration, we have a *microcomputer system,* which operates like a true computer.

Microprocessors are fabricated at present using the semiconductor technologies shown above (also see glossary). The table, opposite page, compares these technologies with respect to packing density, speed, and per-gate power dissipation. Higher packing density allows more hardware to be configured into a single chip, while faster speed means more computations can be made in a given length of time. Think of today's microcomputers as performing about 10^4 additions/sec and handling 10-30 I/O (input/output) functions.

Low power consumption—typically, 250-500 mw/chip—not only saves energy but also alleviates the heat-dissipation problem caused by high packing density or high-speed operation. These criteria control the design of microprocessors and are reflected in device characteristics to be discussed later.

As the table reveals, two emerging technologies, SOS and I²L, excel in most categories. But they are not as advanced commercially today as some other technologies. Currently, NMOS technology has gained the most acceptance in the marketplace.

Total cost vs chip cost. Today, individual chips often sell for less than $20 apiece. A basic microcomputer with one CPU chip, one ROM chip, several I/O chips, and a clock/control chip costs only about $60 or $70. By the time these component parts have been mounted on a board with a power supply, a chassis, lights, switches, and connector wiring, however, the cost has edged up to maybe $500 or so. Remember, hardware *development* costs must also be included, adding possibly another $4000.

Even for a limited-function microcomputer, the cost of developing a program with only several lines of machine-language code is typically about $5000. Testing, debugging, and verifying software and hardware will probably add another $5000. Thus, initial cost approaches $15,000 for a simple microcomputer system—say a data-acquisition device or a process monitor. For more-complex systems—microprocessor-based controllers or intelligent terminals, for example—development costs can easily zoom to the six-figure mark.

Obviously, if microcomputers are going into products, development costs can be spread over the number of units produced. For instrumentation in the power field, however, economic attraction must be based more on improved reliability, increased flexibility of operation, better control, and capability for self-diagnosis. Frequently, as we shall see, they are attractive either as compo-

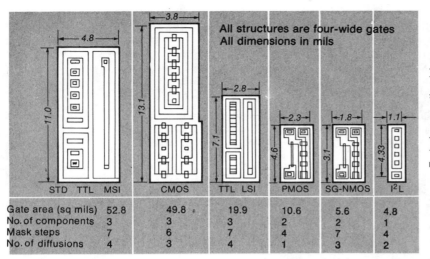

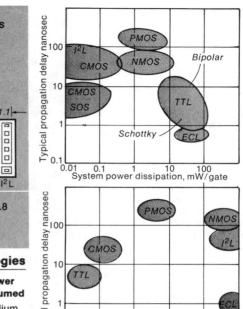

All structures are four-wide gates
All dimensions in mils

	STD TTL MSI	CMOS	TTL LSI	PMOS	SG-NMOS	I²L
Gate area (sq mils)	52.8	49.8	19.9	10.6	5.6	4.8
No. of components	3	3	3	2	2	1
Mask steps	7	6	7	4	7	4
No. of diffusions	4	3	4	1	3	2

Basic chip types (below) are bipolar (A) and unipolar, or MOS (B). Comparison of leading chip types (above) shows packing density increasing from left to right

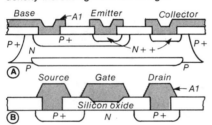

Leading solid-state technologies

Technology	Packing density	Speed	Power consumed
PMOS	G	P	Medium
NMOS	E	F	Medium
CMOS	G-E	F-G	Low
SOS	E	G	Low
Bipolar	P	G-E	High
Schottky bipolar	P	E	High
I²L	E	F-G	Low

G = good; E = excellent; G-E = good-excellent;
P = poor; F = fair; F-G = fair-good.

Main characteristics in semiconductor technology are propagation delay (speed), power dissipation, and relative packing density. Graphs relate these criteria for the leading technologies

nents of a large I&C system or in reduced-function applications that don't warrant the expense of minicomputers.

Clearly, the average power-plant engineer cannot expect to sit down and develop his own software/hardware packages. The expense would be exorbitant, even for a reduced-function device. Better that he go to a manufacturer who specializes in creative microprocessor packaging. Besides a smaller, more efficient design effective packaging can shorten the payoff period so the user isn't saddled with an obsolete system.

Main elements. For our purposes, a minicomputer consists of four main elements, as shown in the top drawing, next page. These elements are:

■ *A CPU* is the heart of the microcomputer, and is where the actual computations are executed. The CPU controls all operations of the microcomputer.

■ *The memory*—a storehouse of "1" and "0" bits—holds (1) the program, which instructs the CPU and other components of the microcomputer, and (2) the filed information, on which the program acts. The memory component makes a computer uniquely different from any other machine invented by man.

■ *I/O devices* serve as the physical interface between man and microcom-

puter. They include keyboards, teletypewriters, CRTs, printers, etc.

■ *I/O interfaces* link the CPU and I/O device. They provide hard-wired control of the I/O devices according to commands issued by the CPU.

Holding these hardware elements together, of course, like the glue in plywood, is a software item—the *program*. The program coordinates the activities of our microcomputer to achieve a useful, desired output. These hardware/software essentials are amazingly similar in function to those for mainframe computers. Let's examine them.

CPU provides the brains

The CPU is a combination of several electronic components connected in such a way as to permit logic operations to be performed. Computers of just a few years ago required good-size enclosures to accommodate the components of their CPUs. Free-standing cabinets holding row on row of pc cards, interconnected by busy bundles of cables, were (and are) a common sight. The microcomputer, on the other hand, uses a CPU that is contained in a single LSI chip; thus, cabling is reduced to zero.

Remember that microprocessors are digital devices that act on binary logic—

i.e., electronic circuits are either *on* or *off*—and that this condition (on *or* off) is designated as a *bit* of information. To be able to represent more than two conditions with binary logic, several bits can be linked in such a way as to provide a more usable logic unit, called a *word*. Although a word may contain any number of bits, microprocessors usually stick with 4-bit, 8-bit, 12-bit, and 16-bit word lengths; machines using 8-bit lengths predominate today, although 16-bit machines should soon be commercially attractive.

Microprocessors are usually fixed-word-length devices. The manufacturer builds in this capability at the factory, which means that the CPU designed to process 8-bit words cannot accommodate words of any other length. The longer the word length, the more powerful the microcomputer. With 4-bit words, only 16 distinct combinations can be represented; with 8-bit words, 256 combinations; with 16-bit words, 65,536 combinations.

Instruction sets. By using words that contain a bit pattern meaningful to the microprocessor, the unit can perform useful work. One bit pattern may be formed to tell the device to add or subtract numbers; another may tell it to print out a character or call up a CRT

Key terms in the language of microprocessors

Access time is the interval between the instant at which information is called for from storage and the instant at which delivery is completed.

ALU, arithmetic/logic unit. That part of the CPU which executes, adds, subtracts, shifts, ANDs, ORs, etc.

ASCII stands for American Standard Code for Information Interchange (pronounced as '-key). A simplified code for interfacing with the microprocessor.

Assembly language is an English-like programming language, in which bit patterns are used for each instruction.

Bit. An abbreviation of *binary digit.*

Bus. A group of lines or channels that permit memory, CPU, and I/O devices to exchange words.

CCD, charge-coupled device, is a special MOS fabrication process.

CMOS, complementary MOS. A circuit with both p- and n-channel FETs on the same MOS wafer, leading to extremely low power dissipation.

Compiler language represents numbers and letters that man works with daily.

CPU, central processing unit. See text.

CTL, complementary-transistor logic, is a semiconductor technology in which n-p-n and p-n-p transistors combine to give high-performance structures.

Diffusion. A thermal process by which minute amounts of impurities are deliberately impregnated and distributed into semiconductor material.

Doping. The addition of impurities to a semiconductor to achieve a desired characteristic; for example, to produce an n-type or p-type material.

DTL, diode-transistor logic. Logic achieved by diodes, while transistors are used as inverting amplifiers.

ECL, emitter-coupled logic. An unsaturated logic performed by EC transistors.

EFL, emitter-follower logic. Similar to CTL.

Epitaxy. The controlled growth of a layer of semiconductor material on a substrate or base layer.

EPROM, erasable PROM. Gives the user reprogramming capability.

FET, field-effect transistor, is a unipolar semiconductor device (MOS).

Floppy disk resembles a 45-rpm phonograph record. Disk has a magnetic recording surface and is packaged in a flexible plastic envelope.

Gate. A logic circuit in which one signal, generally a square wave, serves to switch another signal on or off.

HTL, high-threshold logic, is the economical answer to the need for shielding circuitry from electrical noise.

I²L, integrated-injection logic. A latecomer technology that appears to have circumvented the limitations of the low packing density and high power dissipation per gate that is common to bipolar chips.

LSI, large-scale integration. A scale of integration which permits 100 gates or more per chip.

Logic. A means of solving complex problems through the use of symbols to define basic concepts.

Machine language is the numeric form of specifying instructions, ready for loading into memory and unit execution.

Mask. A thin sheet of metal containing an open pattern, which shields selected portions of a chip base during the deposition process.

MOS, metal-oxide semiconductor. One of two key solid-state technologies for the fabrication of large, low-cost memories with very high input impedance. MOS variations: P (p-channel), N (n-channel), D (double diffused), V (V-groove), etc.

MSI, medium-scale integration. A scale of integration that permits up to 100 gates to be accommodated on one chip.

Multiplexing. The simultaneous transmission of two or more signals within a single channel.

PROM, programmable ROM. A ROM that can be field-programmed by the user.

RAM, random-access memory. Memory that has read and write capabilities.

Register. A fast-access circuit used to store bits or words in the CPU.

ROM, read-only memory. Any type of memory that cannot be rewritten.

RTL, resistor-transistor logic. Logic achieved by resistors, while transistors amplify and invert output.

Transistor. A semiconductor device with three electrodes, commonly used to amplify or switch electric current.

TTL, transistor-transistor logic. Both logic and amplification are achieved by transistors. A bipolar technology.

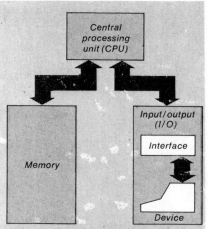

Key elements of a microcomputer parallel primary computer components. Both devices require programs to operate

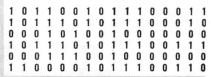

| COMPILER LANGUAGE |

$$X = Y + Z - 100$$

| ASSEMBLY LANGUAGE |

LDA	Y	Load A Register with data "Y"
LDB	Z	Load B Register with data "Z"
ADD		Add contents of B to A
LDB	HUN	Load B Register with data "HUN" (constant value of 100)
SUB		Subtract contents of B from A
STA	X	Store contents of A register in "X"

| MACHINE LANGUAGE |

```
1 0 1 1 0 0 1 0 1 1 1 0 0 0 1 1
1 0 1 1 1 0 1 0 1 1 1 0 0 0 1 0
0 0 0 1 0 1 0 0 1 0 0 0 0 0 0 0
1 0 1 1 1 0 1 0 1 1 1 0 0 1 1 1
0 0 0 1 1 1 0 0 1 0 0 0 0 0 0 0
1 1 0 0 0 0 1 0 1 1 1 0 0 1 1 0
```

Program languages that are available to communicate with the microcomputer. Each has advantages and limitations

---

symbol. Bit patterns like these are built into the CPU at the factory.

A group of interrelated patterns is called an *instruction set.* Most sets include addition, subtraction, and such commonly used logic as AND, OR, NAND, and NOR. Instruction sets will also contain special operations based on the intended use of the microprocessor. Some will concentrate on detailed I/O instructions; others on complex CPU logic.

The essential parts of the CPU are these (also see top left drawing, facing page):

■ *The instruction register.* Only one instruction can be executed by the microprocessor at one time. The sequence of instructions is usually stored in memory. As each instruction is needed, it is fetched from memory and put into the instruction register, which is a circuit that can electronically hold one word.

■ *The arithmetic/logic unit (ALU)* performs the arithmetic and logic operations defined by the instruction set. It is the most critical element in the CPU.

■ *The accumulator* is the primary register used during many CPU operations. For example, it is used by the ALU to hold one element of data during arithmetic processing.

■ *The program counter* is used by the CPU to keep track of where in the program the current operation is located.

■ *Index registers* store information that will be tapped by the program many times. They also hold addresses of reference data, such as tables of numbers.

■ *The extension register* is used in conjunction with the accumulator for performing special arithmetic modes.

■ *The status register* is generally

Source: Power, May 1977, 25-32

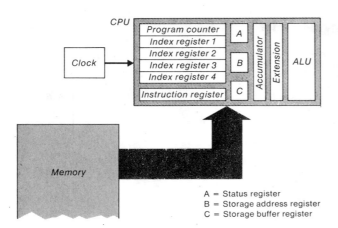

CPU contains hardware registers, besides an ALU to perform arithmetic and logic operations defined by instruction set

A = Status register
B = Storage address register
C = Storage buffer register

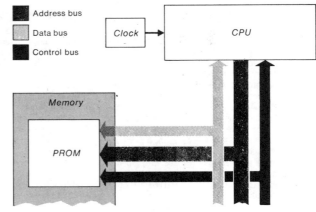

Three types of buses lead into and out of the CPU, permitting communication with I/O devices—and especially memory

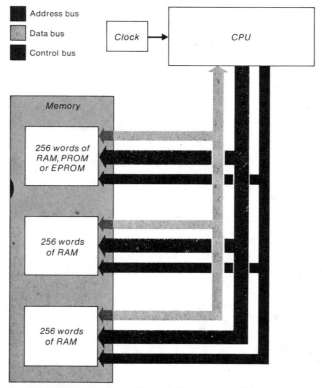

A combination of memory types is important (with proper inter-communication) to assure effective microprocessor operation

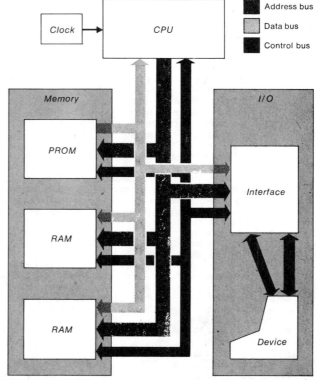

I/O interfaces and devices complete the picture. Interfaces permit hard-wired control of the devices at command of CPU

assigned to monitor the status of conditions within the microprocessor.

■ *A storage-address register* is used to hold the memory address of data that are either being loaded (read) from or stored (written) into memory.

■ *A storage buffer register* temporarily stores information, acting as a buffer against surging of data flow into and out of the microprocessor.

Communication channels

Although the CPU can be a powerful manipulator of data, it must be given all the information relating to the desired operation. It must be provided with instructions and also be told where to obtain the data to be operated upon and where to put the result after the operation has been executed.

For this purpose, communication channels capable of transmitting a binary format into and out of the CPU are necessary. If the word length is eight bits, eight channels must be connected to the CPU to transmit information in and out. Collectively, these channels are termed a *bus*. There are three types of buses (also see top right drawing, above):

■ *The data bus* is used for the transmission of pertinent data into or out of the CPU, especially to and from memory. In most cases, the same bus is used for both reading data from memory into the CPU, and for writing data from the CPU into memory.

■ *The address bus* is used to identify the individual location to or from which the transfer of data is to be made, especially in conjunction with memory. The bus carries the address of the location in

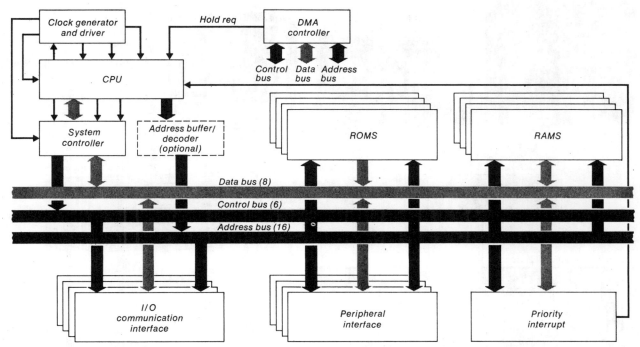

Microcomputer (Intel 8080) includes memory, a clock, and various interfaces, along with the CPU microprocessor

memory that is being accessed by the CPU. It is also possible to address I/O devices as if they were memory locations.

■ *The control bus* is a group of dedicated channels, each of which can be used for a special control purpose, such as clearing CPU registers to reset the microprocessor, or stopping the device after instructions have terminated.

Synchronizing operations. Note from the drawings on p 29 that a "clock" is always associated with the CPU—a quartz device, or perhaps an oscillator circuit to establish a universal time pulse. A clock is needed to synchronize the various operations of the microprocessor. For example, it cannot be fetching an instruction from memory at the same time it is executing an instruction already in the instruction register. Remember, all of the various operations performed do not necessarily take the same amount of time. In fact, some operations may take as much as a million times as long as others.

Memory—electronic storage

The primary function of memory is to store program instructions. The memory may also file data as future reference for some program during CPU execution. Memories may be either *volatile*, which means that, if power is removed from the chip, all memory content is lost; or *nonvolatile*, which means that the memory will retain information stored in it indefinitely—it need not have power applied or be refreshed. Leading types of

memories for microcomputers are these (see also bottom left drawing, preceding page):

■ *RAM, random-access memory.* Any word in this "scratch-pad" type of memory can be accessed without having to run through the entire gamut of stored words. Although RAM is volatile, the power these monolithic chips draw is so small (typically less than ½ watt) that it is feasible to leave them actuated all the time. To circumvent power failures, batteries can be installed across the chips.

■ *ROM, read-only memory* is similar to RAM except that it's not possible for the user to write into ROM. Instead, he must specify to the chip maker when he purchases the chip precisely what he wants to be in memory. The manufacturer uses special equipment to prepare the ROM. Thus, this type of memory is best applied for storing tables, subroutines, and frequently used programs. As long as the information is never changed, it may be read as often as desired. ROM is nonvolatile. A typical chip will store 1024 8-bit words.

■ *PROM, programmable ROM* is like a simple ROM, except that it can be user-programmed in the field. PROM chips can be purchased "blank," and then be programmed using a special machine. Once programmed, however, PROM behaves precisely the same as ROM. PROM is nonvolatile.

■ *EPROM, erasable PROM*, gives the user a reprogramming capability. He can program his ROM in the field, erase it

under UV light, and reprogram with different information. At this point, however, an EPROM behaves exactly like a ROM. Obviously, an EPROM is considerably more expensive than a PROM or ROM.

The human factor: I/O devices

Machines that can be connected to a computer and are operable by humans are called I/O devices. The machines translate human manipulations into signals that can be interpreted by a microprocessor, and/or reframe microprocessor signals into human language. The devices range from simple to complex.

Simple devices include switches and lights, with which the operator communicates with the microprocessor in binary language. A step up is a simple "hex" keyboard, which converts information back and forth between binary and another form of notation called hexadecimal—an easier language for the operator to work with.

Probably the three most-common microprocessor I/O devices are the ASCII keyboard, the teletypewriter, and the paper-tape device. The ASCII keyboard is an input device only, but is quite flexible in its operation—although to gain this ease of operation, the circuitry must be complex. The teletypewriter is a true I/O device, with a typewriter keyboard and a printer for permanent output. Pressing a key on the keyboard leads, via internal mechanics, to a binary representation of the character selected at the output. The printer reverses the

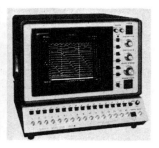

Microprocessor-based systems are installed for supervisory/control (far left) and for scanner/logger (middle). At left, logic analyzer for debugging

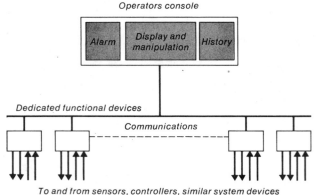

A distributed system is controlled by dedicated devices

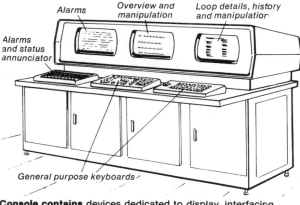

Console contains devices dedicated to display, interfacing

process, taking binary data and converting it to a keyboard character. Paper tape, used in conjunction with a tape punch and tape reader, permits the convenient storage and transmittal of coded information between operator and microprocessor.

The more complex devices speed up the rate of information exchange between the microprocessor and the operator. Modems (*mo*dulators/*dem*odulators) are such devices, used principally over telephone lines. Magnetic tape is similar to paper tape, but is much more durable. Since the tape is wound onto reels, however, information retrieval is sequential only.

Magnetic disks resemble phonograph records (without the grooves) and have the advantage of immediate accessing from or to storage via magnetic heads. These devices are much faster and more flexible than the magnetic tape devices, especially the *floppy disk* design for mass storage. Video display modules use a TV-like screen to display alphanumeric characters. Although operation is simple, the electronics are quite complex—and relatively expensive.

Machine/machine interfacing

Just as I/O devices provide a man/machine interface, I/O interfaces provide a machine/machine adaptability—namely, between the I/O device and the microprocessor (lower right drawing, p 29). Control and data buses (or lines) provide the means of getting information into and out of the micro-

processor, and of controlling the flow of that information. This interfacing is not to be confused with that between I/O devices and the outside world. Here, line surges, noise, and other electrical interferences must be screened out before reaching computer internals—an essentially serene world.

Synchronizing and timing are important, especially when *serial-mode* I/O devices are used. These machines send or

Universal digital/analog controllers

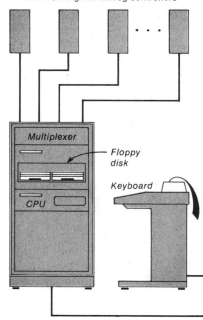

receive data as a string of bits, which are transmitted one after the other at a precise rate. The devices are usually remote from the microprocessor and rely on telephone lines or microwave. For close-in transmission, *parallel-mode* devices can be selected, in which all the bits in a word of data are sent at the same time over several lines. An ASCII keyboard is such an I/O device.

Most I/O devices require routing

Minicomputer (below) uses microprocessors within its circuitry. UDAC devices (left) use a shared-bus concept

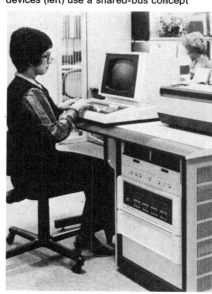

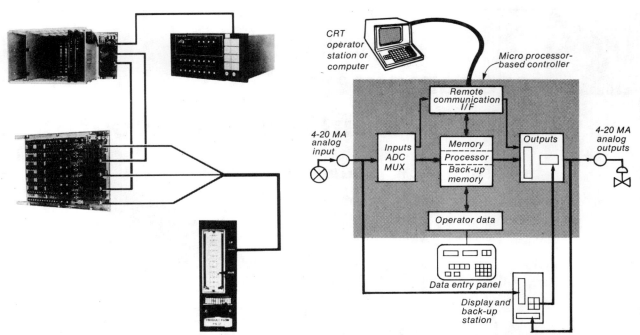

Controller file (top left), with terminal panel below, data entry panel at right, and analog display at bottom right

Microprocessor-based controller on automatic control (heavy black), on complete bypass (color), on partial bypass (dashed)

through the CPU to communicate with memory. A DMA (direct-memory-access) device, however, can communicate with memory without having to go through the CPU. Thus, the I/O interface for a DMA device must be capable of generating memory addresses as well as providing data transfer.

Putting all these elements together gives us the microcomputer system shown on p 30, which happens to be an Intel 8080. The system features a modular organization based on a bus standardized by the 8080 CPU group. There are signs that chip manufacturers are going to offer more such complete packages.

Applications across the board

Now that we have some idea of how microprocessors work, how are they applied? A review of applications in the power field reveals these categories:

■ Reduced function, where the cost of a minicomputer isn't justified.

■ Intelligent terminals, for upgrading CRTs, printers, teletypes, etc.

■ As part of a hierarchical or distributed control system.

■ As part of a communications network.

■ With equipment such as controllers, or even minicomputers.

Some experts think that microprocessors will make solid inroads into the displacement of analog instruments for reduced-function (but precise) operations, such as boiler control. Minicomputers would be too expensive, but some degree of computer control is advisable.

A microprocessor-based controller file is shown in the photo above, together with a terminal panel, a data-entry panel (for operator interfacing), and an analog display. The file contains the processing cards, interfacing cards, and independent output cards. Schematic above shows the internal architecture of a microprocessor-based controller. The internal tasks have been partitioned to allow maximum use of functions without losing manual control in the event of device failures.

"Smart" terminals. Microprocessors are also playing a part at either end of the controller loop—at the data-acquisition device and at the operator interface. "Smart" field-mounted microprocessors gather data from transmitters, convert the signals to an engineering format, and send them to the control room. The remote units handle more inputs in less time and can perform alarm functions. Top photos, previous page, show two examples of these intelligent terminal devices in use today.

Most intelligent terminals have analog-to-digital conversion capability, relaying digitized data back to the control room via a single wire, and thus avoiding the costly cabling linkup between field instruments and control room. At the operator interface in the control room, microprocessors enable the use of centralized operator consoles, complete with CRTs that call up and display the status of one or more loops at a time, as well as provide trend information and alarms.

Top drawings, previous page, show how control is carried out by dedicated devices, and how even the operator interface is achieved by equipment dedicated to display and interface tasks. A console

such as the one shown might contain these primary functions: controlled variable display and manipulation; alarm and status indication; historical and trend information. Dedicated devices offer better, more reliable control—and are feasible economically when paired with microprocessors.

At the bottom of the previous page, the drawing and photo show an on-line configuration in which a microprocessor-based controller is tied in with a mini-computer. Typically, the mini operates as a central control station as well as a "program development" station. One minicomputer can serve a large number of controllers. Such a system is quite effective when the distributed controllers are regulating similar processes, which can then be coordinated by one operator at the central control panel.

Wrapping it up. Possibilities for microprocessors in the power field, as we have seen, are excellent. Indeed, their intrusion is unstoppable, and the up-to-date engineer would be wise to accept them and learn to work with the technology.

Is this the final step in the self-imposed education of so many "precomputer" engineers, or is a whole new technology lurking on the horizon? Nothing radically new seems to be out there—except possibly bubble memories. In any event, a bubble device, which can store an unbelievable 92,000 bits of data in a 1-in.-sq package, does not appear to be aimed at harassing unreconstructed engineers. They should be able to apply the concept to current microprocessor technology without a major revision in their existing thought patterns. **TCE**

Source: Power, May 1977, 25-32

Microcomputers in instrumentation and control

First some new products— then some new systems ideas

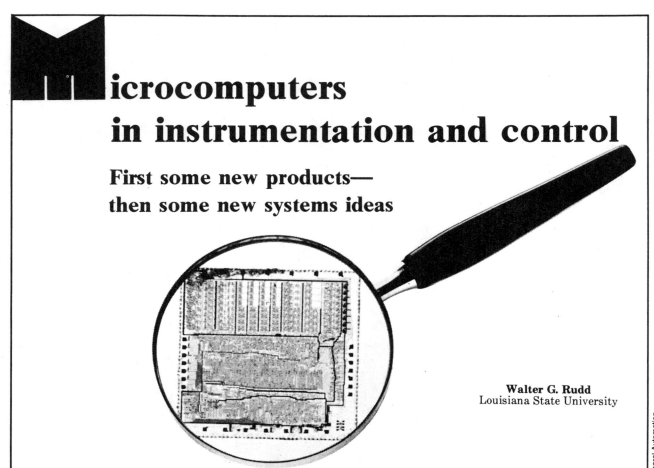

Walter G. Rudd
Louisiana State University

General Automation

MICROCOMPUTERS ARE the latest step in downward spiral in computer sizes, cost per computer function, and memory size and expense.

In the last five years, minicomputers have been developed with central processors on single standard circuit boards. Memory remained as the most expensive component of the main frame computer. To remedy this problem, the industry has used advanced semiconductor' technology. So in the last year, memory sizes and prices have taken another quantum jump downward.

Borrowing upon the new memory technology, the microcomputer manufacturers, have succeeded in developing central processors on single chips about three millimeters square. Packaged in a flat container about half an inch wide and 1½ inches long, these processors sell for between $30 and $120 each, in comparison with the single-board minicomputer processors, which cost about $2500 each.

The modularity of the structure of a microcomputer (Fig. 1) permits the individual customer to design and purchase systems with the minimum amount of hardware that will support his special task. To date, manufacturers have concentrated almost entirely upon the OEM market. Therefore, to provide for indestructible programs stored in the least expensive memory, programs are usually stored in read-only memory (ROM), mass-produced by a metal masking process. Preprogrammed ROM chips containing 2048 bits cost about $15 each in large quantities.

Random-access memory (RAM) chips for use for transient data storage cost about twice as much. To date, processor input/output facilities are rather crude, and require tailor-made interfaces for specific input/output devices.

A TYPICAL MICROCOMPUTER is compared with an average minicomputer in Table I. At a glance it can be seen that micro-

computers are not number crunchers. Long instruction times, small instruction sets, and short operand lengths combine to make microprocessors about 50 times less powerful than mini's.

But in many applications, particularly in communications and process control, minicomputers are too powerful and expensive to make their use attractive. Here is where the low cost and modularity of microcomputers will serve, not to replace minicomputers, but to add a new dimension in the use of intelligent machines in process control.

MICROCOMPUTER PROGRAMMING is presently accomplished using ROMs for specific applications and devices. Program development and testing has been done by the microcomputer manufacturers to OEM customer specifications. The vendors, having demonstrated the demand for their products, are now actively developing equipment that will permit a user to assume the system de-

Table I

	Minicomputer	Microcomputer
Processor Size	Circuit board	Single package
Instruction execution time	.5-5 microseconds	10-30 microsecond
Word length	16 bits	8 bits
Number of instructions	50-150	50
Interrupts	Multi-level automatic service	Single line
Processor cost U Unit	$1500	$120
Oem quantities	$1000	$60
Memory Cycle time	.5-1.5 microsecond	1 microsecond
Increments	4K bytes	256 bytes
Cost/byte	$.10-.50	$.06 (ROM in Oem quantities)

Table I. Comparison of minicomputer and microcomputer characteristics.

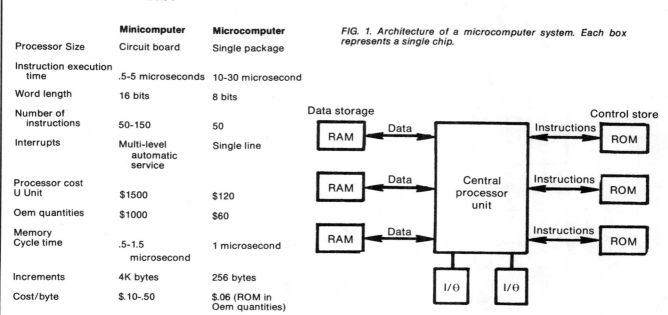

FIG. 1. Architecture of a microcomputer system. Each box represents a single chip.

sign and program development and testing responsibilities, so that the microcomputer vendors can concentrate upon product development and mass-production techniques. Getting the OEMs into system design and programming should result in an exponential increase in the number and variety of devices which use microcomputer intelligence.

Process control will be one of the areas in which we see such new products emerging. While control engineers will not go heavily into microcomputer programming, it is interesting to see what the industry is doing in order to provide OEMs with microprogramming capability.

One way to develop microcomputer programs is through the use of cross-assembler and simulator programs available from the manufacturer. These are usually written in Fortran for execution on a large-scale computer. The input to the cross-assembler is a program in the microcomputer assembly language. The microcomputer machine language produced by the cross-assembler can then be input into the simulator program. Once the user is satisfied with the program, tested through the use of the simulator, he has the microcomputer machine language code punched on paper tape, and sends this to the manufacturer for making ROMs.

For the user who wishes to build prototype computers and test them and their firmware in more realistic environments, the industry offers a wide array of prototyping and testing equipment, including machines that look and act like full-fleged minicomputers. The customer can build and field-test his own microcomputer systems using a prototyping kit that costs about $5000. The industry has even gone so far as to develop programmable read-only memory (PROM) so that a program bug does not mean a $40 chip must be discarded.

Programming a PROM or EPROM (erasable PROM) is accomplished in about 2 minutes (for 2048 bits). For this reason, PROMs and EPROMs can also be called mostly-read memories. Erasing an EPROM requires exposure of the chip to a high-intensity ultraviolet light for ten or twenty minutes.

THE IMPACT OF MICROCOMPUTERS upon the control industry as yet is neglegible. And new production plants will soon, if ever, jump into the problems involved in onsite development of microcomputer systems and programs. However, we can confidently predict that microcomputers will have a profound long-range effect upon techniques of digital control.

For example over the last decade or so, there has been a gradual migration of the intelligence centers of control systems from central positions in the plant or complex outward to positions closer to the actual processes being controlled.

IN A CENTRALIZED CONTROL system (Fig. 2) a medium- or large-scale computer is responsible for direct digital control (DDC) of all the controlled loops in a plant. The problems involved are well-documented and include the following:
- *too many eggs in one basket.* If the central computer goes down,

Source: Instruments & Control Systems, April, 1974, 63-66

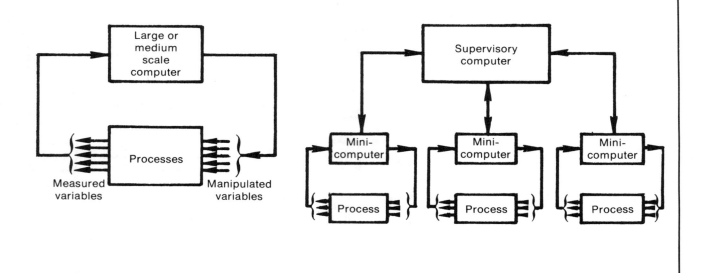

FIG. 2. A centralized control system.

FIG. 3. Minicomputer permit distribution of control intelligence among several smaller units.

digital control is lost, so many applications require complete digital or analog backup systems.

● *communications problems.* The plant might cover several square miles. Communications gear is expensive and data transmission errors must be dealt with.

● *interaction between programs.* No matter how hard we try, we never seem to be able to handle the potential catastrophes that arise in large real-time systems.

● *lack of flexibility.* Adding a new loop or modifying control algorithms requires a lot of programming and perhaps down time.

THE ANSWER to many of these problems is to distribute the control among smaller computers, which are themselves perhaps monitored by a supervisory computer. Here is where the minicomputers, each of which controls twenty or thirty related loops in a process or section of a process. This makes possible a step toward a truly hierarchical structure (Fig. 3). Since the DDC function is removed from the central computer, loss of the supervisor does

not mean complete loss of control. Loss of a DDC minicomputer does mean that local control of that process is lost, but if adequate backup is present, the entire plant is not down.

Use of minicomputers as DDCs for many loops does little to reduce total communications costs. Sampled data is normally in analog form and must often be transmitted over long distances with wiring that can cost as much as $10 per foot. Analog control signals require duplication of the wiring costs. Ideally, we should move the minicomputers closer to the loops they control and proportion the DDC among a larger number of minicomputers. But economies of scale have prevented the minicomputer from proceeding much further out into the plant floor.

Today's hardware is too expensive and too powerful to be economically attractive as controllers for small numbers of loops. For example a typical minicomputer in control of four loops is idle about 99.5 per cent of the time and costs on the order of $2500, not including analog input/

output (I/O) equipment.

THE MICROCOMPUTER is going to play a significant role at the lower levels of a heirarchical control system (Fig. 4). The slow speed and memory limitations in microcomputers are not a significant handicap in controlling a single loop or a small number of loops in most control situations. A standard PID algorithm program fits comfortably into a single 256-byte ROM chip and requires about five milliseconds for a single input-computer control-output sequence on machines currently available. We can expect this time to decrease a little in the future. And the cost for such a computer, including analog I/O converters can be as low as $500.

Moving DDC into individual loops means that sensory and process security monitoring functions should be similarly relocated. Microwatchdogs, which provide independent monitoring of process outputs and the communications and DDC computers.

Each local computer including the communications multiplexers must "check-in" periodically with

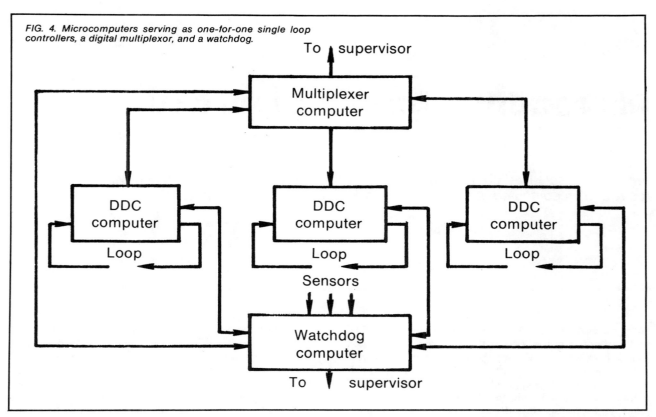

FIG. 4. Microcomputers serving as one-for-one single loop controllers, a digital multiplexor, and a watchdog.

the watchdog. The watchdogs include some diagnostic capabilities and should communicate with higher-level computers and the control panel over their own private data paths.

PRODUCTS LIKE the expensive hand-held and desk-top calculators on the market today presage a new class of new intelligent devices. Many of these are in fact microcomputers. They include a central processor, stored programs, alterable memory, and input/output facilities. Pressing a function key causes execution of firmware subroutine stored in a read-only memory (ROM). Programmability is achieved by storing programs under user control in alterable random access memory (RAM).

In the near future, we can expect to see such microcomputers appear in automobiles, communications equipment, machine tools, and process control equipment. For example, in the latter we can expect to see devices such as:
● *intelligent valves*, with built-in direct digital controllers, including A-D converters and a variety of interfaces for input of process parameters and setpoints.

● *intelligent digital multiplexers*, capable of communications monitoring and error detection and correction.
● *"smarter" control panels*, which anticipate what the operator needs to know and display it for him with a resulting reduction in display hardware.
● *intelligent programmable-gain A-D and D-A converters*, which can monitor and standardize signals and set alarm conditions.
● *intelligent self-calibrating sensors*, such as flow meters, which integrate flow in real-time and watch for sudden drops in flow.

THE POTENTIAL for new advanced instrumentation and the distribution of control intelligence are becoming economically feasible. We have indicated that the DDC microcomputers are replacements for conventional controllers. Whether or not digital controllers for single loops become economically competitive with conventional devices remains to be seen, but we are liable to spend a few extra dollars for the increased reliability and flexibility possible with programmable digital devices.

Regardless of whether the indi-

vidual controllers are digital or conventional, we expect to see microcomputers used to lower communications costs in process control. Microcomputers are already used as communications controllers in point-of-sale terminals and medium-speed communications systems. Per-line costs are substantially lower for digital communications than for analog data transmission.

For many in-plant communications requirements, direct digital data transmission can eliminate the need for signal modulation. Digital communications is not only attractive economically, but also provides for increased reliability and error-checking and correction capabilities. And the availability of inexpensive intelligent digital multiplexers will enable the use of self-checking time-shared communications lines to remote plant sites.

Much of the groundwork for providing redundancy in data paths has already been done in the development of large-scale computer communications networks. The process control industry can borrow these techniques in establishing digital micronetworks.

Source: Instruments & Control Systems, April, 1974, 63-66

Microcomputers in control

Armed with increased computing power, advanced software, and plug-in hardware, today's microcomputers are well equipped for industrial control tasks.

Rich Merritt
Senior Technical Editor

Until recently, a typical microcomputer user had to be part systems engineer and part real-time programmer. This is because few microcomputer suppliers could provide all the hardware and software needed for all industrial and process control applications. Thus, many users had to improvise solutions, interface hardware from other suppliers, and write special software drivers and application programs to make the systems work.

This situation is changing rapidly, as many microcomputer suppliers are offering a wide selection of I/O, providing high-level programming languages, and are beginning to conform to bus-board standards.

In this article, we'll look at recent developments and trends in microcomputers. (Note: For the purposes of this article, a microcomputer is a standalone, desktop-size computer based on a microprocessor CPU. It has a keyboard, CRT monitor, disk or tape storage, and is entirely self-sufficient.)

Powerful processors

The computing capabilities of microcomputers have increased by several orders of magnitude in the past few years. In fact, a microcomputer with the power of a minicomputer can often be purchased for $10,000 or less (Fig. 1). Such a system might have a 16- or 32-bit CPU, 10M byte Winchester-type "hard" disk, 1M byte of RAM (random access memory), direct memory access (DMA) ports, serial and parallel I/O interfaces, high-level programming languages, and a real-time operating system. (See the table beginning on page 21 for a complete listing of microcomputers for control and data acquisition applications.)

Because of this increased computing power, number-crunching is a rapidly developing area for microcomputers. As was pointed out in our June 1983 article on personal computers, microcomputers are beginning to replace minicomputers in laboratory, test, and analysis applications. In these applications, a sophisticated instrument—such as a chromatograph, spectrometer, or vibration analyzer—is coupled with a dedicated microcomputer. Because it's dedicated to the instrument, the microcomputer has enough power to perform the complex mathematics needed for analysis.

Another area where micros are rapidly replacing minis is data acquisition (Fig. 2). One method used here is to couple a personal or desktop microcomputer with a "smart" front end that pre-processes the data. Smart front ends (see table) have on-board microcomputers that perform I/O, signal processing, data conversion, alarm checking, and other functions in real-time. This allows the desktop computer to be used for data storage, computing, and CRT display functions, rather than I/O tasks.

Desktop mainframes

In the past few weeks, three computer manufacturers have announced microcomputers that are compatible with mainframe processors. IBM's XT/370 can run any software that operates in VM/CMS mode on an IBM 370 or 4300 mainframe. DEC's MicroVAX has about 35% of the power of a VAX-11/780 and can run all VAX software. Data General's Desktop Generation systems are compatible with Eclipse

Fig. 1: The IBM CS9000 is a good example of the powerful microcomputers available for industrial and process control use. Based on the 68000 CPU, the system is tailored for real-time applications. Here, it's being used as an environmental control system in an electronics assembly plant. The IBM personal computer (left) serves as an operator station. Photo courtesy Control Systems Specialists (Dublin, OH).

Reprinted with permission from I&CS, December 1983, 19-24, © 1983 Chilton Co.

software, and DG soon will offer a chip set for the MV series of mainframes.

Although these desktop mainframes don't have all the computing power of their big brothers, they work quite well in some applications. As Dave Rome of DG points out, a big mainframe often supports many users in a time-sharing environment. Since the micro-mainframe only has to deal with a single user, its response time and performance may be quite adequate. Rome also notes that the micro-mainframes can support some of the big, fast peripherals that mainframes use. For example, the DG Desktop Generation computer can drive two 15M byte disks, and soon will handle a 50M byte unit.

The impact these micro-mainframes will have on control applications is unknown. However, IBM 370, DEC VAX, and DG Eclipse computers are widely used in manufacturing and process control, often as host computers for a network of controllers.

Microcomputer operator stations

Another major trend is the increasing use of microcomputers as operator stations in distributed process control systems. Typically, the micro connects via an RS-232C interface, and is used to log data, prepare reports, display real-time information, and send controller setpoints to local process controllers.

For example, Bailey Controls provides a software package that allows programs in a personal computer to reach Network 90 data via subroutine calls. Jim Hoffmaster, VP of Engineering at Bailey, says personal computers will never replace operator stations, but they do work well for "casual monitoring"—by a plant manager, for example.

Similar interfaces are provided by Beckman, Inconix, KineticSystems, Moore Products, Powell Process Systems, Taylor Instruments, and Toshiba for their process control systems. Bristol-Babcock and Honeywell also have used microcomputers in this way.

It's also possible to connect a high-resolution color monitor to a microcomputer. According to Dave Winward of Aydin, several companies are combining the IBM PC-XT (which has a hard disk) with the Aydin Model 8830 Patriot color monitor. Such a combination is almost equivalent to the hardware found in many $80,000 operator stations.

Improving software

The increased computing power of today's microcomputers allows them to run advanced programming languages, such as Pascal, Ada, C, or Fortran. Previously, most micros didn't have the memory, disk storage, or op-

erating system needed to compile and run these languages.

Using advanced languages provides two advantages: one, it allows advanced control system programmers to use microcomputers where they previously had to use minicomputers; two, it allows "portable" software packages to be written for use on any of several micros.

Several portable software packages for process and industrial control are now available for a wide range of microcomputers. Typically, these are fill-in-the-blanks languages that work in batch, continuous, or sequential control applications. Examples include:
- *AIM*—For PDP-11/23 or 11/73 microcomputers. From Biles & Associates (Houston, TX).
- *Mentor*—For 8086-based micros, this package was originally developed by Exxon Chemical Co. From Canberra (Meriden, CT).
- *ABC*—Written in Pascal, ABC runs on almost all 16- and 32-bit microcomputers. From Control Systems Specialists (Dublin, OH).
- *FORTH*—For almost all 8- and 16-bit micros. From Forth, Inc. (Hermosa Beach, CA).
- *Onspec*—For the IBM PC or PC XT. From Heuristics, Inc. (Sacramento, CA).
- *WECON*—For any PDP-11 microcomputer that supports RSX-11M. From Wade Engineering Co. (Houston, TX).

For years, BASIC has been the most widely used programming language for microcomputers. Early on, this was because BASIC interpreters could be stored in read-only memory, and didn't need disks or massive memories. Also, vendors found that users quickly learned BASIC programming. Thus, many vendors of industrial computers wrote their proprietary languages in a BASIC-like format.

Such vendors improved BASIC's real-time processing capabilities immensely. Modern BASIC-like control languages now sport commands such as TURNON, TURNOFF, AIN and AOUT, which allow easier access to I/O functions than the old PEEK and POKE commands. Also, they feature PID algorithms, dimension statements, timing functions, interrupt handling, motor control commands, etc. And now that microcomputers are more powerful, compiled BASIC is available. A compiled BASIC program executes as fast as some other high-level languages.

One-stop vs bus boards

Industrial microcomputers (see Cover and Fig. 3) differ from general-purpose systems in that they are configured in hardware and software specifically for manufacturing or process control applications. Typically, they offer a real-time operating system, high-level programming lan-

Fig. 2 (left): Hewlett-Packard's Model 6901S measurement system is based on an HP 9836S industrial desktop computer and a smart front end. The computer has HPL, BASIC, and Pascal programming languages.

Fig. 3 (right): Microcomputers for harsh plant floor environments are available from many manufacturers. The RAC PAC systems from Xycom shown here have industrial enclosures and a BASIC-like programming language.

Source: I&CS, December 1983, 19-24

I&CS Guide to microcomputers for control

Manufacturer Model No.	CPU	CPU size (bits)	Max Memory (bytes)	Operating System	Programming Languages	Interfaces Available	Typical Applications	Reader Service No.
ADAC Corporation MICROBASYS	DEC FALCON	16	64K	I/O BASIC	I/O BASIC	A&D I/O	Remote DAQ and control; Front end for μCs	210
DISKBASYS, PROMBASYS	DEC LSI-11/23	16	4M	RT-11	I/O BASIC	A&D I/O	DAQ system, host for MICROBASYS	
Action Instruments A-PAC BC-3	Z-80A	8	256K	ABLE & CP/M	ABLE (BASIC)	IEEE-488, RS-232C, RS-422, STD	DAQ, PID, robotics, automatic test, annunciators, distributed controls	211
Acurex TEN/50	8088	8/16	1M	BASIC PAC	BASIC	RS-232C	DAQ, energy management, QC and QA	212
Advanced Digital Corp. SuperStar	Z-80	8	10M	CP/M, Turbo DOS	BASIC, Fortran, Pascal	S-100	Large data storage applications	213
SuperSystem II	80186	16	65M	CP/M, Turbo DOS	same as above	S-100	Multi-user tasks, up to 10 terminals	
Altos Computer Systems 586	8086	16	1M	Xenix, MP/M-86	BASIC, Pascal, Fortran	RS-232C	Business & industrial applications, DAQ	214
Ameacon Process Mate 90	Z-80	8	62K	LDOS 6.1	BASIC, Assembler	RS-232C, A&D I/O, CENT	Batch, temperature, and belt scale control; automatic weighing, DAQ	215
Analog Devices, Inc. Macsym 150	8086 & 8087	16	384	MP/M-86	MACBASIC	RS-232C, RS-422, 20mA, IEEE-488	DAQ and control in process control, product test, or laboratory experiments	216
Analogic Corp. ANDS4400	Z8671	8	64K	ANABUG	Z8 BASIC, ACL	RS-232C	DAQ, control, sequence of events, data logging	217
ANATEC PDP Micro/11	PDP 11/23+	16	4M	RSX-11M, 11M+	CRISP, Fortran, BASIC	RS-232C, A&D I/O	Unit control, graphics support for programmable controller	218
Andromeda Systems Inc. 11/M Series	LSI-11/23	16	4M	RT-11	BASIC, Fortran	RS-232C, IEEE-488, CENT	DAQ	219
Apple Computer 2E	6502A	8	128K	CP/M, Apple DOS, CP/M-86, MS-DOS	BASIC, Fortran, LOGO, Pascal	RS-232C, IEEE-488, CENT		220
August Systems, Inc. Series 300, TRI-DAC, TRI-GARD	8086 (triple redundant)	16	1M	RTTS (Real Time Task Scheduler)	Ladder, Fortran, PL/M, Pascal, BASIC	RS-232C,422,423; Modbus, Multibus	Process control and monitoring, primarily in critical or remote applications.	221
AuTech DACMASTER 6000	68000	16	1M	Multitasking	BASIC	Process I/O	DAQ, control, sequential control	222
Autocon Industries Micro CAT	CDC 114	16	1M	MP/M 86	Pascal, ladder diagrams, Fortran	RS-232C, Multibus, CENT	Process control, telemetry, data logging, DAQ	223
Bailey Controls IBM PC			256K	PC DOS	BASIC, Fortran	RS-232C	Network 90 operator station, data logging, control system support, training	224
Barber-Colman Co. EDAC	8080A, Z80	8	64K		FITB, relay ladder	RS-232C	Temperature and sequence logic control	225
Beckman Instruments Management Report Processor	Z8000	16	512K		MRP	Multibus, RS-232C	Host computer for MV8000 distributed control system	226
Bristol Babcock Inc. UCS-3000, OC 3000	8080	8	188K		ACCOL	RS-232C, RS-423	Process control	227
Canberra Industries, Inc. IMACS	8086	16	1M	OASIS-16, MENTOR	BASIC, MENTOR	RS-232C & 422, HDLC, TI, A&D I/O	Continuous and batch process control, DAQ and analysis, data base management	228
Citadel Computer Corp. CMX 250, 260	NSC 800	8	64K		XYBASIC	CIMBUS, RS-232C, 20mA		229
CMX 800	PDP-11, T-11	16	64K	DEC	DEC-compatible	same as above		
CMX-280	8088	16	1M	RMX-88, Vertis	BASIC	same as above		
Com-Trol Inc. FMS-2000	6809	8	576K		Assembler, C	RS-232C, RS-422	DAQ, communications, compressor, temp and HVAC control	230
Comptrol Inc. Comp 7000 Intelligent Controller	8085	8	8K + 2K		Assembler	RS-232C, RS-422, 20mA	Batching, motor/drive control, fault annunciation	231
Computer Systems PC/8088	8088	16	512K	DOS, CP/M-86	BASIC, Pascal, Macro, Forth, C	IEEE-488, RS-232C, CENT	IBM PC compatible applications	232
Controlex Corporation CS-105	8085	8	64K	Native Forth	Forth	RS-232C, 8-bit I/O ports, synchro	Machine control, process control, motion control, automatic test, DAQ	233
Control Systems Specialists ABC	IBM CS9000 and IBM PC	16 or 32	4M	CS/OS, PC-DOS	ABC, BASIC, Fortran	RS-232C, IEEE-488	Process control, energy management, DAQ	234
Cybersystems, Inc. CYREY 9980	Z80A	8	64K	CP/M 2.2	CyberFORTH, BASIC, Fortran, Pascal	IEEE-488, RS-232C, RS-422	DAQ, measurement and control	235
Cyborg Corp. ISAAC	68000 + Apple or IBM PC	16	2M		ICL	IEEE-488, RS-232C, Parallel	DAQ and control with IBM and Apple personal computers	236
Data Translation DT Series, PCDAX	IBM PC, Apple, Rainbow 100	8				IBM PC, Apple, DEC, A&D I/O	DAQ and control with IBM, Apple, or DEC Rainbow personal computers	237
Datricon FAC PAC IMC	6809	8/16	64K	Sphere	Sphere	RS-232C, RS-422, A&D I/O, STD-bus	Process control, DAQ, instrumentation, robots, simulators	238

I&CS Guide to microcomputers for control

Manufacturer/Model	CPU	CPU size (bits)	Main Memory (bytes)	Operating System	Programming Languages	Interfaces Available	Typical Applications	Reader Service No.
Digital Equipment Corp. PC350	DEC F-11	16	512K	P/OS	BASIC	PC3XX-AC, IEEE-488, RS-232C	DAQ	239
Micro PDP-11	Micro-PDP-11	16	2M	RSX-11M	Fortran	IEQ11-AB, IEEE-488	DAQ, control, ATE	
Dowty Electronics Zeus 8700		16			ZICOL	UART		240
Dynage CMS-2001	Commodore PET		96K		BASIC, CM-Process Control language	IEEE-488, A&D I/O, RS-422		241
CMI System 64	Commodore 64		64K	CP/M	BASIC	RS-232C, CBM bus, IEEE-488, A&D I/O		
ETI Micro, Inc. 8630	8088	16	64K	RCP-86 Multitasking	Toolkit for IBM PC	RS-232C, RS-422, modem, SBX-port	Process control and DAQ; generic host computer interface	242
E.O.I.S. Corporation EOIS 1100	8086	16	1M	CP/M	Pascal	IEEE-488, RS-232C, CENT	Automatic Vision Inspection	243
Essex Engineering Co. SX 1107	Z80A	8	512K	CP/M, MP/M	PL-1	IEEE-488, RS-232C, CENT	DAQ, communications, order picking	244
Fail-Safe Technology Corp. Fail-Safe 86	Intel 86 or 286	16	16M	RMX	PLM-86, Pascal-86, ASM-86	IEEE-488, every industry standard	Highly reliable real-time instrumentation, controls, communication, etc.	245
Fischer & Porter Co. MICRO-DCI Supervisor	8031	8	112K			RS-422, RS-232	DAQ, communications, sequential logic	246
John Fluke Mfg. Co. 1722A	TMS 99000	16	136K		BASIC, Fortran, Assembler	IEEE-488, RS-232C, RS-422	Process control, laboratories, DAQ	247
GE Intersil Systems ISB80/85				CP/M	High level	STD, A&D I/O		248
Gould Inc. IMC-4000	68000	16/32	384K	Industrial I-O (IIOS)	BASIC, Pascal, Forth	Parallel 32 bits, IEEE-488, RS-232C	Process control data concentrator, PC supervision	249
Herco AGS-100	6809	8	256K	Herco, EX6809	Fortran, BASIC	RS-232C, 20mA	Control, data collection, monitoring tank levels	250
Hewlett-Packard HP 80, 85, 9915		16	640K	Single user	BASIC	IEEE-488, RS-232C, GP-IO		251
HP 9826, 9836, 9845		16	2M	Single user	BASIC, HPL, Pascal	same as above		
HP 9845		16	1.5M	Single user	BASIC, Assembler, Pascal, Fortran	same as above		
HP 1000, Models 5, 6		16	4M	Multiuser	BASIC, Fortran 77, assembler, Pascal	same as above		
IBM Information Systems Div. IBM PC, XT	8088	16	640K	PC DOS, CP/M-86, USCD P-System	BASIC, Pascal	RS-232C		252
XT/370	68000, 8087, 8088	16/32	655K	VM/PC, PC DOS	All 370 compatible	RS-232C, 3274	Runs IBM 370 VS/CMS programs	
IBM Instruments, Inc. CS 9000	68000	32	5M	Multitasking RTOS, Xenix	BASIC, Fortran, Pascal, C	RS-232C, IEEE-488	DAQ, analysis, plotting, storage, retrieval and reporting	253
Inconix Corp. CINCH	Z80	8	64K		CINCH BASIC	RS-232C, IEEE-488	DAQ and control	254
Intel Corp. System 86/310, 286/310	8086, 8286	16	8M	RMS-286R	Fortran, Pascal, BASIC, C, PL/M	Multibus, A&D I/O	Automated test, machine control, process control, lab automation	255
System 86/380, 286/380	088C, 8286	16	1M, 16M	RMX-86, RMX-286	Fortran, Pascal, BASIC, C, Assembler	same as above	same as above	
The IPAC Group, Inc. System 15	6502	8	4M (bubble)	IPL BASIC	Interpretive BASIC	RS-232, 20mA	DAQ, communications, control	256
Ithaco CompuDAS 3	8086, 8087, Z80	8/16	768K	TMS-RTOS	DABIL	RS-232C	DAQ and control	257
Ivy Microcomputer Ivy 3000	80186		512K	Ivy DOS, PC-DOS	IBM PC compatible	RS-232C, CENT	IBM PC,XT applications	258
Kaypro Corp. Kaypro II	Z80, 8088	8 or 16	256K	MS-DOS, CP/M-86	BASIC	RS-232C		259
Keithley DAS Series 500	IBM PC, PC-XT, Apple II and IIE	16	740K	PC DOS, Apple DOS	SOFT 500, Applesoft, BASIC	IEEE-488, RS-232C, A&D I/O	Measurement and control w/IBM or Apple personal computers	260
KineticSystems Corporation LION system	LSI-11/23	16	4M	RSX-11M	Fortran	IEEE-583, RS-232C	DAQ and process control	261
Personal Computer Systems	IBM, Apple or DEC personal computers	Apple			BASIC, Fortran	IEEE-583 (CAMAC), IEEE-488, RS-232C	DAQ and process control	
Masscomp MC-500	68000, 68020	16/32	6M	UNIX	C, Fortran-77, Pascal-2, Assembler	Multibus, STD, IEEE-488, RS-232C	High speed DAQ and computing	262
Matrox Electronic Systems Macs-86	8086	16	640K	CP/M-86	Pascal, Fortran, C	RS-232C, 422 & 423;CENT,DR-11W	OEM development system	263
MACS-10	Z-80A	8	512K	CP/M		same as above	OEM development system	
Modular Integration Link/110	IBM PC				ACCESS	A&D I/O	DAQ and control using IBM PC	264
Monolithic Systems MSC 8807	80186, Z80	16/8	192K	IRMX-86, CP/M-86, CP/M 80	C, Fortran, Pascal, BASIC	Multibus, RS-232C, CENT	Communications, DAQ	265

I&CS Guide to microcomputers for control

Manufacturer Model No.	CPU	CPU size (bits)	Max. Memory (bytes)	Operating System	Programming Languages	Interfaces Available	Typical Applications	Reader Service No.
Moore Industries Model 1002	8086	16	64K		Applications	RS-232C	Distillation columns, batch reactors, etc.	266
Moore Products IBM PC	8086	16		PC DOS	PC DOS software	RS-232C to MYCRO	DAQ, optimization, analysis	267
Motorola Semiconductor VME/10	MC68010	16/32	384K	VERSADOS	Assembler, Pascal	VMEbus	Automation, process control, development	268
Multitronics SBR-700	8085	16	256K		Com-Pac CN, CYNET	RS-232C	Remote DAQ and control with host personal computer	269
Mundix Control Systems Programmable Monitoring System	8086	16	1M	RMX-88	PLM-86	RS-232C	Central monitoring of distributed industrial processes such as underground mines	270
Niagara Scientific Inc. DATKON 700 Series	M6802, M6809	8	64K	RTOS		RS-232C	DAQ and control	271
Numeridex, Inc. NICAM-III	PDP 11/23	16	4M	RSX-11M, RT-11, VMS, CP/M	BASIC, Fortran, NC languages	RS-232C	NC/CNC parts programming	272
Omnibyte Corporation OB68K/SYS	68000	16	128K	IDRIS, FORTH MSP, Regulus, MTOS	C, FORTRAN, PASCAL, FORTH, BASIC	RS-232C, CENT, parallel	Software development, process control, DAQ	273
Pixel Computer Pixel 80	68000, 68010	16/32	6M	UNIX	C, Pascal, Fortran, BASIC, APL, etc.	RS-232C, Ethernet, Multibus	Scientific, graphics	274
PolyMorphic Systems Poly 88	Z80/80186	8/16	1.25M	Exec, CP/M, CP/M-86, UNIX	BASIC, Pascal MT, C	RS-232C, IEEE-488, CENT	DAQ, communications, unit control	275
Pracsys, Inc. Micro-DACS	IBM-PC	16	512K	PC-DOS	Any IBM	RS-232C, A&D I/O	DAQ, laboratory controls	276
Pro-Log ABL-1	8085 or Z80A	8	64K	CP/M		RS-232C, parallel	SCADA, process control, communications	277
Quadrex FLIC FLICSTAR	DEC Micro-11	16	4M	RSX 11M	Fortran	RS-232C, etc.	Refinery, petrochemical, utility, food, petroleum	278
Quantum Systems Microvisory Model 16	6809	8	60K		DATADECK	RS-232C	DAQ and supervisory control	279
RI/Control Systems Micrihost Model 84000	Z-80A	8	64K	CP/M	BASIC, assembler, C	RS-232C, RS-422	Distributed control, displays, reports	280
Robertshaw Controls Co. DSM1500	6802	8	56K		Fortran	RS-232X	DAQ, communications, loading PCs	281
SBE, Inc. 200/10/F	68000	16/32	256K	Regulus, CPM-86, polyFORTH	C	Multibus, serial, parallel	Process control, communications, ATE	282
Scientific Micro Systems MDX-11	LSI-11/23	16	1M	RT-11, RSX-11M, TSX-Plus	Fortran, BASIC, all DEC software	Q-bus		283
Sharp Electronics Corp. PC-5000	8088	16	256K	MS-DOS	BASIC	RS-232C		284
Sibthorp Systems, Inc. Ultima 2000	Z80	8	108K	MICE	IGL, relay ladder, SI BASIC	RS-232C, RS-422, CENT	Automated assembly, process control	285
Stearns Computer Systems Stearns Desktop	8086, 8087	16	896K	MS-DOS, CP/M-86, ST-DOS	BASIC, Cobol, Fortran, Pascal	RS-232C, RS-422, 3270 Bisync or SNA	IBM PC-compatible tasks, communications, networking	286
Taurus Computer Products Taurus One	Z80A	8	60K		BASIC	RS-232C, IEEE-488, A&D I/O	DAQ and process control	287
Terra Computer Systems EBS-100		8	64K, 512K	CP/M 2.2	CP/M compatible	RS-232C, SASI/SCSI, IEEE-488	PC programming, process monitoring, field analysis, data collection	288
Texas Instruments Professional	8088	16	256K	MS DOS, CP/M-86, UCSD P-System	BASIC, Pascal, Fortran	RS-232C	DAQ	289
Toshiba America, Inc. T300	8088	16	512K	MS-DOS, C/PM-86	BASIC, CBASIC-86	RS-232C, CENT, TOSDIC	Process operator console in TOSDIC System, or as part of small distributed control system	290
USDATA REACT	Intel 8080	8	64K		Industrial Graphics BASIC, Fortran	RS-232C, RS-422A	Operator interface, DAQ, management reporting with programmable controllers	291
Volumetrics A-100	8085	8	64K	CP/M	CP/M compatible	RS-232C, CENT	DAQ, communications	292
Wahl Instruments, Inc. Data Force	8031/8051	8	240K	Wahl	Assembler	RS-232C	Dedicated computers, DAQ, process controller	293
Waugh Controls Corporation Model 2200	8085A	8	32K	SBC 80/24		RS-232 Ports	Flow, ratio, and blending control applications	294
Western Telecomputing ICIS 850	8085, IBM PC	16	64K	ICCS	BASIC	RS-232C, A&D I/O	DAQ and control	295
Westinghouse Combustion Control Div. Model 1700 DCS	8085	8	64K		Blockware	RS-232C, A&D I/O	Combustion control, power houses, etc.	296
Westinghouse Numa-Logic NLPL-1580	Compaq PC	16	512K	DOS	Pascal, Fortran, BASIC	RS-232C, IEEE-488	DAQ, communications, loading PCs	297
Xcel Controls, Inc. UP/DOC	Z80A/8088	8/16	256K	CPM 2.2, MS-DOS	BASIC	RS-232C, IEEE-488	PC documentation and programming	298
Xycom, Inc. 3940-IB/3941-IB	Z80B	8	128K		BASIC	RS-232C, 422A & 423A; IEEE-488	DAQ, communications, distributed process control	299

Editor's Note: Most of the information in this table came directly from questionnaires returned from the manufacturers listed. In some cases, data was taken from company literature. To the best of our knowledge, all microcomputers listed are available as standalone, desktop-sized units that can be programmed by a user.

Legend: A&D = analog and digital; CENT = Centronics printer interface; DAQ = data acquisition; FITB = fill in the blanks; I/O = input/output.

guage (as noted above), and industrial-quality CRT monitors, keyboards, and housings for the computer and I/O.

Because of their industrial-quality components, these micros are able to survive tough conditions on the factory floor or at remote sites in a process plant. For example, one microcomputer can survive temperatures from -50 to 110 °C (-58 to 230 °F), has a rugged waterproof housing, and uses military-style connectors. Other industrial micros feature membrane keyboards, sealed NEMA cabinets, DIN-type industrial connectors, vibration mounts, RFI/EMI shielding, or other special devices not found in normal computers.

In many cases, the hardware and software in an industrial computer is proprietary; that is, the rugged I/O and computer boards, application software, and programming languages will work only on the vendor's system. The advantage of such a system is that it offers one-stop shopping. Everything fits, works with other system hardware and software, and is maintained by a single vendor.

Some industrial microcomputer vendors, however, are beginning to offer standard bus interfaces for both their computer boards and I/O. Multibus, STD bus, and VMEbus systems are becoming available from several vendors. These are listed in our table, and industrial applications using bus-board systems were covered in Part 10 of our series, "Bus boards for control" (see I&CS, November 1983).

One advantage of a bus-based system is that a single vendor doesn't have to manufacture every kind of board a system might need. For example, color CRT drivers, communication handlers, robot motion controllers, vision systems, and a host of other specialized I/O boards are available on standard boards.

The move to standardization

The wheels of standardization usually grind slowly, but sometimes can be surprisingly fast. For example, I&CS Contributing Editor Jim Heaton, in a paper presented at the Factory Electronics Conference (Detroit, November 1983) says the IBM PC architecture has already become a *de facto* standard. Many companies are producing hardware and software to comply with this unwritten standard. Heaton also says that the UNIX operating system is a *de facto* standard, because it's a portable real-time operating system that works on many different microcomputers.

As we pointed out last month, communication standards are crawling toward acceptance. At present, most industrial microcomputers have proprietary communication schemes for connecting their own systems. Almost everyone seems to be waiting to see which scheme users prefer, or which scheme the leaders follow.

Another inexorable trend is "upward integration." That is, all control systems, no matter how small, someday will be accountable to another system that's higher in the hierarchy. Eventually, all systems will be accountable to the plant's mainframe computer. This trend has already started—many new products are beginning to appear that interface host computers to microcomputers.

In the end, the control engineer will be the one to benefit the most. As the move toward standardization grows, it'll be easier and easier to solve application problems with a microcomputer. ■

The Microcomputer as a Decision-Making Aid

A computer can help you make decisions at work, but only if you know what to expect

Peter Callamaras
U.S. Air Force

The phrase "decisions, decisions, decisions," used either in jest or in response to real frustrations, bears special significance. It has come to be a common, sometimes satiric, way of identifying one source of our problems. In this case, the source of irritation is the fact that the decisions we have to make have a way of piling up, demanding our time and attention, and seemingly never becoming manageable. The average person takes for granted the ability to make complex decisions. From the time we get up (what to wear today?), through the morning (which route to work?), afternoon (what's for lunch?), and evening (more dessert?), we are presented with choices, and we constantly make decisions. Throughout the day we make simple decisions about our personal well-being, even while we are engaged in professional activities. At some point this constant decision making can cause "decision overload." We get tired, our concentration suffers, and we start making mistakes.

The Problems

The combination of being overburdened with decisions and making a series of bad choices will frustrate most of us. More important, when we're at work this problem can have consequences far beyond its effect on our emotional state. A classic example of decision overload and its possible dangers is a doctor's misdiagnosis.

A second problem related to decision making is the amount of time it takes. Assimilating all the information relevant to a decision can slow the whole process. Add to that the fact that the volume of information keeps growing. In fact, things change too quickly for many of us to keep up, especially in a busy work environment.

With the need to make more and better decisions, and because of the time involved in wading through so much data, we obviously need some way to reduce the burdens of professional decision making. Microcomputers provide a means of satisfying these needs.

Microcomputers can simplify decision making, speed up the process of choosing between alternatives, and help ensure the accuracy of each decision.

Levels of Decision Making

There are three levels of decision making: operational, managerial, and strategic. Microcomputers can be of great value at all three levels.

Most operational-level decisions involve the specific needs of the decision maker. These decisions make up the majority of our routine choices. They usually are standardized in our daily activities. Operational-level decisions require detailed information, but the data is readily available and its conversion into decision-making information is often subconscious. We have a set of "canned" responses for these decisions and we often can delegate their execution. The typical advice to a cold sufferer, "take aspirin and drink plenty of liquids," is a delegated canned decision.

Managerial-level decisions require a broader base of information. The decision maker must rely on his prior experience, training, and instincts. Managerial-level decisions cannot be delegated, but they can be substantially speeded up. For instance, a lawyer about to accept a new case may have a general idea of what it concerns. However, the lawyer can't give his client any legal advice until all the data concerning the case is in. The client can help the lawyer by giving him detailed and specific information.

Strategic-level decisions require a wide range of information. These decisions usually are made after long periods of thought and planning and they often require the generation of completely new data. For example, the chief space-shuttle program engineer probably had to "imagineer" some of its aspects from technology that was either immature or still speculative when the program began. Thus, the majority of strategic decisions are heuristic (trial and error) and cannot be standardized, canned, or delegated. (For a more detailed discussion of the three levels of decision making, see *Information Processing Systems for Management*, Chapter 20, by D. Hussain and K. Hussain, Homewood, IL: Richard D. Irwin Inc., 1981.)

Time and the Microcomputer

Work time is one of our most precious resources and also one of the most difficult to conserve. Microcomputers can help us reduce the time we have to spend on the decision-making process by gathering data and converting it into usable information. Once we have all the information we need, we can concentrate on our most prudent course of action.

Computers can bound a problem and ensure that we have the information we need to make a decision at our fingertips. If the decision is routine, computers can provide a canned response and you can get on with more important matters. However, it will take time to integrate a computer into your professional life.

First, you have to decide whether you really want to add a computer to your set of professional tools. Then you have to decide what to buy.

Next, you have to learn how to operate the hardware and interact with the software. Current literature, particularly advertisements, can lead you to believe that you can become proficient at operating a microcomputer in a few hours. This is not so. While you can learn to manipulate the keyboard and turn out some useful products in short order, you will not get the full benefit of the microcomputer until operating it becomes second nature to you. Compare this to oil painting. Until you master the basic techniques of applying the paint to the canvas, shading, mixing, etc., you will not have complete freedom of creativity. The same holds true for a computer system.

It also takes time to enter necessary background information into your computer. Many ads for financial-management programs, for example, only describe the output you can generate and don't dwell on the time it takes to input the information you need to get those impressive printouts. If it took you an entire year to spend the money you are now trying to account for, you can expect that it will take a great deal of time and effort to put your spending history into the computer.

Some Helpful Solutions

In the past, two data-processing disciplines aimed at satisfying the needs of decision makers have been

---

Any professional who has to work with finances and is not using a computer system is wasting valuable time.

---

management information systems (MIS) and decision support systems (DSS). An MIS is a large data-gathering system. You define your data needs and set up a method of gathering it. The DSS is a refinement of the MIS applied to an individual's decision-making needs. An analogy can be found to a microcomputer if you imagine starting with a database management system (MIS) and then designing a set of tailored reports (DSS) based on the MIS data.

For another example, look at your annual tax return and your checkbook as an MIS and a DSS, respectively. Your tax return should contain all your financial data in one handy form. You can use it to gauge your financial health and make plans for the coming year. You use this MIS to create budget categories for the following year (checkbook/DSS). You then use these budget categories for specific financial decisions. Balancing your checkbook at the end of the month tells you how much you spent in relation to how much you had. You can then break out the totals for each budget category and take a detailed

look at your spending for the month. Then, if necessary, you can make adjustments for the next month.

In the past, both MIS and DSS systems had to be implemented on large computers. Today, most professionals can obtain the benefits of an MIS/DSS with a good microcomputer system.

How the Microcomputer Helps

Any professional who has to work on finances and is not using a computer system is wasting valuable time. This applies particularly to professionals in business for themselves. When reviewing financial activities, the computer makes it easy to compare the money that is coming in with the money that is owed. If there are discrepancies, a computerized financial-management system allows you to go back and locate the source of the problems. If things are going well, you can use a spreadsheet program to speculate on possible future directions. A microcomputer also can make tax planning an easy, ongoing exercise that maximizes income and minimizes payments.

A computerized inventory system can also be of help to the professional. For example, you can establish a set of routine procedures for ordering supplies. If it takes a week to receive a high-consumption item, you can use the computer to determine when to place the order. One way this can be done is through the application of the economic order quantity (EOQ) method. With EOQ you create a model of your consumption patterns and compare them with your ordering/receiving patterns. The result is an indication of the best time to place orders. Accurate and timely order placement ensures that a minimum amount of inventory will be on hand to satisfy operating needs and that you will never run out of something. The stock stays fresh and storage costs decline. This also turns inventory control into a set of operational-level decisions that then can be delegated to a subordinate.

Microcomputers can perform complex statistical analyses. Engineers routinely perform statistical analyses of the failure rates of materials or

Source: BYTE Magazine, May 1984, 122-124

components they want to use. The results allow them to accept or reject the materials. Once the acceptance/rejection criteria are determined, materials selection can be reduced to a canned routine.

For professionals who travel a great deal, trip planning can be made easier with a transportation model (TM). The TM can also determine the most economical route for product deliveries. For those whose business it is to move people or things around the country, the decision again can be converted into a set of canned control types and delegated.

Another type of software can aid planning and scheduling by providing a pictorial representation of the task at hand. Once the necessary events are determined, the computer generates a graph of the events along a time line. As time passes, the completion of a specific task can be tracked against the graph and corrective action can be taken as needed. One of the better known of these time-line graph programs is the critical-path method (CPM).

Teleprocessing through the phone system opens a whole new world to the microcomputerized decision maker. For those who need a great deal of information, the growing number of on-line data services can be a godsend. An on-line data service can be viewed as a specialized library in a computer. There are several medical libraries, for example, that allow doctors to make more accurate diagnoses or prescribe more effective medication.

On-line data services for lawyers, such as Westlaw, contain a vast body of judicial decisions. These services can reduce the drudgery of wading through all the material that is potentially applicable to a legal question. The search capability allows a lawyer to put in a set of key words and anything relating to those words is returned.

Conclusion

The more decisions you reduce, standardize, and delegate, the less time decision making will take. The more data you gather, the more in-formation you will have available to make the best possible decisions. The microcomputer's ability to play "what if" gives you trial-and-error results without forcing you to live with the consequences of poor decisions.

The cost of microcomputers now is generally low enough to be affordable to most professionals. Learning to use one properly takes time, but it is worth the investment. The documentation that accompanies most hardware and software is getting better. With clearer instructions, it takes less time to get "up to speed." There is plenty of software available to support decision making, and there are more decision-making packages coming out all the time. The key question is, Can you afford *not* to start using a computer to aid in your decision making?■

Peter Callamaras, an Air Force officer, can be reached at AFCC/EPPB Scott AFB, IL 62225. He recently received his master's degree in systems management. He has been interested in computers since 1966 and was the service-department manager of a computer store.

SECTION III
Overview of Computer Hardware

SECTION III
Overview of Computer Hardware

Anyone taking even a casual look at the computer world of today is struck by the proliferation of types, sizes, prices, models, and features that make up the marketplace. A few moments' attention to the classifications of computer hardware is in order, and that is the purpose of this section.

Generations of Computers

Computers can be said to be in the fourth generation, if one considers ENIAC and its contemporaries to have been the first generation. The second generation was typified by UNIVAC I, the first large transistorized computer, built in 1954 by the Sperry-Rand Corporation. Integrated circuit technology (silicon chip technology) is the technology of the third generation. Most of the industrial, business, and personal computers now in use share this technology. The fourth generation is taking shape with the development of Very Large Integrated Circuits. A shadowy fifth generation is evolving in the research laboratories, where the technology of artificial intelligence is being applied. We will dwell more on this area in Section VIII of this book.

Computer Classifications

Computers are divided into three main groups: mainframes, minicomputers, and microcomputers.

The characteristics in Table 1 are nonspecific in order to give the reader a quick framework of rough classification. The articles "Filling the Gap" and "Micro, Mini, and Mainframe Basics" will expand the subject. In "Filling the Gap," we find a further classification into "super-" subdivisions as computer makers develop more powerful models within the older classifications. The article "Superminis Shape up as Departmental CPUs" describes, for example, how a more powerful minicomputer is being developed to undertake "distributed data processing" applications in the modern business office.

Personal Computers

All "personal" and "home" computers are microcomputers. Another characteristic of this category is that they are referred to as "single-user" machines. Operators of several personal computers in a business office, for example, can all communicate with a minicomputer over a suitable "network." This application makes the mini a "multi-user" machine. However, the individual personal computer cannot be made to serve in this way.

Following is a discussion of the classification of microcomputers in the present marketplace.

Microcomputers. Include "home" computers, such as Atari, Commodore Vic-20 and 64. These models are often sold as a keyboard processor that can be coupled to the home television for readout.

Also include "portable" computers, like Osborne, Kaypro, Compaq, Panasonic, and several others whose configuration has one or more disc drives and a monitor screen built into the device. (Panasonic adds a built-in printer.) The machines are designed to be carried, like a suitcase, and so are "portable."

A new and expanding subdivision of personal computers is the "notebook," "briefcase," or "lap" version, like the Texas Instruments "Pro-Lite" and the Tandy 200. (See the article "Lap Portables. How Small is Too Small" for a full description and update on this growing category of personal computers.)

Table 1. Classifications of Computers

Division	Characteristics
Mainframes	Physically large, with floor-based cabinetry 32-bit or greater word length* Large memories, both internal and external Often have parallel processors High-speed operation Cost of $1 million and up
Minicomputers	One or two cabinets, with desk-type operator consoles 12-16 bit word length* Medium-size memories Medium speed Costs of $25,000 to $250,000
Microcomputers	Desktop, or even smaller 8-bit word length* Small internal memories Relatively low speed Costs of $500 to $5000

*The articles "It's 16 Bits, But Is That Wide or What?" and "Hello, Mr. Chips" help the reader clarify the concept of word length.

All other microcomputers fall under the classifications "desktop," "business," or "professional," terms that are appearing with increasing frequency in the literature.

Programmable Controllers and Other Microprocessor-Based Machines. Programmable controllers (called "PC's" in industry) are special forms of microcomputers used in process control applications in industrial plants. We will treat them separately in Section VI, "Programmable Controllers."

The versatile microprocessor, which is the heart of a microcomputer, has been incorporated into stand-alone measuring and control instrumentation to make these devices more accurate, more functional, and more reliable. In addition, such "smart" instruments, as they are called, can interface with mini- and mainframe computers in distributed process control systems. The role of these systems will be further described in Section VI.

The article "Getting Started: Playing 'Name That Part' " is a lighthearted but instructional article that will be of help to those readers contemplating the purchase of their first personal computer.

Playing 'Name That Part'

To switch on a personal computer and get it to run, you really don't need to know very much more than where the power button is located. So if a computer and its related equipment and programs are available to you, give that button–and yourself–a push and then *go.*

It's likely, however, that you'll want to know a lot more about this special new machine than that. Most of us feel, in general, that we should learn about a lot of things, particularly about things in which our knowledge is lacking. We just want to *know,* and that's a worthwhile motivation to learn.

Most of us were born into a world which did not contain personal computers. The first personal computer appeared less than 10 years ago, so it is a "new thing." If your kids are less than 10 years old, then personal computers are no more strange and mysterious to them than doorknobs, radios or blenders.

But for us, they are different. We don't have anything to compare them with. Personal computers are more than smart typewriters. They are more than electronic filing cabinets. They are more than game machines. And we're supposed to learn how to master them? Gimme a break!

I don't want you trying to learn too much too fast. If you know too much, you have more to be confused about. How many jalapeno peppers should you wolf down at your first real, live Texas barbeque? Not many, take my word for it. But we do need to learn a lot, and before too long.

One of the major barriers between us and competence is our fear of *how much* we will have to learn. Do we need to know all the technical vocabulary? No, not really. Columnist Jack Smith wrote about his dentist asking the technician to hand him "the little measuring thingie." "Measuring *thingie?*" But he was a terrif-

Richard B. Byrne, Ph.D., is an international lecturer, writer and consultant. He is also a professor at the Annenberg School of Communication, University of Southern California.

ic dentist. There's hope for us yet.

Do we need to understand the function and process of the guts of a computer? Again, not really. We need only know enough to get the job done. More than that is probably excess baggage.

But one thing is certain: We all want to be somewhere above the embarrassment level. If you want to buy a computer yourself, instead of relying on the experience and judgment of someone else, you need to know what to buy and when to quit, and what may go wrong.

When we buy a new car we have the feeling that we know what we are doing, although the technology has left us all in the dust. But, in a computer store, we don't know what to look at, or when we've bought a feature which is particularly wonderful.

Recently I saw a mother buying a computer for her 10-year-old, obviously computer-hip, son. When he casually tossed a box of floppy disks on the counter, she asked, "How much are those?" The answer was "$60." She said, "Why

don't we get the computer now and get those little things later?" The son rolled his eyeballs back up in his head, too mortified to comment. He just groaned out, "Mommm!," and looked at the salesperson as if his mother were beyond reach.

Start with the fundamentals. What you want or need to know will expand with time and experience. Be sure you can "re-create" what you learn here: Whip out a pencil or pen, grab a paper placemat, and draw the essential elements of a computer system, which I'm about to explain.

Over 100 years ago a wizard named Charles Babbage had an idea: a machine which would solve math problems. Just imagine, even now, that someone put the same proposition to you. "Build a box that will do math." Your response would probably be similar to that of Babbage's 19th century colleagues, which roughly paraphrased, was, "You're nuts."

But he persisted. He said that such a machine would need five elements: a mechanism to put information (input) into the machine; some kind of device (storage or memory) in which to put the information; an instrument to do calculations (now part of the Central Processing Unit, or CPU); a control unit to decide what to do, when, and how to do it (also part of the CPU); and a way to get the answer out, ideally in some printed form (output).

Guess what? Babbage was right about the route we've taken in creating and using computers. But there are two additional steps which I would like to add. I want to focus on you, and the other people who *use* computers. So the first step is the question which *you* come up with, which you want the computer to assist you with; and the last step is the *action* you take after you learn the results from the computer.

So the basic relationships which are present in any computer system are a person, a question, an input, processing, memory, control, output and action. In lots of ways this copies some of the ways the human mind works to solve problems.

When we work on a problem, questions arise. Then we gather information (input). We observe by seeing, touching, talking, listening, smelling, tasting, counting, measuring, weighing, reading.

Next we turn to the business of processing that information. We analyze, count, add, divide, evaluate, discuss, order, compare, and give and follow directions. Then we do our best to keep track of the results, to keep a record of our deliberations, or at least of the outcomes (storage or memory). Next, we spread the word. We deliver orders, or distribute those orders by memos or reports (output). Finally, we take action.

Computers are fabulously flexible machines which perform manipulations on words, numbers, images and symbols by following instructions and rules. They are accurate, precise and unbelievably fast machines.

The personal computer is probably the most versatile machine ever devised. We're all familiar with tools like a hammer, a nutcracker or a ruler. There's never any confusion about these tools, because they *do* whatever they *are*.

Ahhhhh, but not the computer. It *is* whatever it *does*. Give it new instructions, it becomes a new tool. A personal computer is at one moment a calculator, and at others an accountant, a game, an administrative assistant, a file clerk or a mail system. That's one reason for our confusion.

What can I tell you that will help clear away the fog? Let's stick with the basics for now. When you get started on anything new you need to get the big picture first. What's the essence of this new thing? Then later you master the medium and learn your tools in detail.

Let's play "Name That Part." I want you to be able to identify the key elements in a computer system, and see their relationships. In the simplest physical terms, when you go to buy a computer you need to get the computer itself, a "monitor" (or use your TV set), a "disk drive" (or a cassette tape recorder), some software programs and blank diskettes, and a printer if you want the results of your computing stored on paper.

These elements can be purchased together or separately. If you buy a portable lap model computer, you may get everything in one self-contained piece, including programs which were imbedded at the factory. A desktop computer may be purchased in many separate pieces, like a component stereo system.

What's important right now is the concept of what a computer is, not the physical reality of some specific personal computer.

The elements of computing include all of the following, and you will use them in some form. How much, how fast, and how expensive become criteria for selecting your own computer later on.

The first elements are you, a *person*, and your *question*. You should be getting or using a computer because there is some problem you want solved, or something you want to do. This is the key managerial issue, and I'll do future columns on the nature of *the question*.

Next is *input*. How do you feed your computer? Computers run on electrical current and circuits. They began with the idea that information could be represented by the presence or absence of electricity. Electricity is reduced to tiny flows of current which activate microscopic switches, or gates. The computer uses what is called binary math. All of the staggering accomplishments of computers are based on the choice of Either/Or. The current is *on* or *off*. This is sometimes referred to as "Pulse" or "No Pulse," "Yes" or "No," and most frequently as "1" or "0." This single signal, either *on* or *off*, is called a "bit" (which comes from the beginning and ending letters of "binary digit").

A single bit almost never means anything by itself. So we have to treat "bits" in series; most commonly in series of eight. Eight bits equals one "byte," or a single printed character, one alphabetic or one decimal digit. A thousand of these characters are referred to as a "kilobyte," which is where we get the infamous "K" that everyone babbles about.

The computer just adds and subtracts, multiplies and divides, compares and stores 1s and 0s. Obviously, it takes a ton of 1s and 0s to write anything. The computer handles only one signal at a time, one instruction at a time, in a precise sequence, although it works with hundreds of thousands of these signals, and processes millions of them each second. It also detects its own mistakes as it goes along. The newest personal computers are experimenting with the parallel processing of more than one instruction at a time.

This system of binary arithmetic is not natural for humans–almost nobody can think in a "0-1-0-0-1-1" fashion. So we've figured out tricks to get the job done. We have codes which convert *on-off* to become letters, numbers, symbols and signs. I'll explain these codes, and how we use them in getting work done, in later columns.

Back to input. To get those rows of 1s and 0s into the computer, we used to use punched cards, or long paper tapes with punched holes. These went through "readers" by the millions, and the computer kept track of the signals we fed it. Now we have devices which are more complex, but infinitely simpler to use.

The most common is a keyboard. Most of its keys are the same as a typewriter keyboard, But some keys are quite different, including keys labeled "Ctrl" (Control), "Alt" (Alternate), "Del" (for Delete), "Function" keys, arrows to move the "cursor" on-screen, a "Return" or "Enter" key, and often a numeric keypad like a hand calculator.

Keyboards are slow and tedious. It's much faster to put in data using a disk drive. A disk drive loads or saves programs. You insert a "floppy disk," give the computer a command, and it immediately memorizes the hundreds of thousands of instructions recorded on the floppy disk.

A series of other input devices are based on the same principle: You move them right or left, up or down, and the computer takes action of some kind. One of the first of these devices, called "game paddles," were actually two little knobs which you twisted frantically to get your "paddle" in front of the "ball" in early video games like Pong. Then, with more sophisticated outer space adventure games, the joystick appeared.

The computer uses what is called analog to digital conversion to make sense of this wiggling. It measures minute changes in voltage, and then suddenly flips to a different binary state. If you could hear the computer talking to itself while the knob is being turned, it might say, "Is that enough? No, still a zero." "Is that enough? No, still a zero." "Is that enough? *Yeah*, that's a one right there!"

A recent application of this same concept is the "mouse," a small, hand-held box which is attached to the computer. By pressing down on the mouse and moving it around you give the computer instructions.

An important part of input is the software, or programs of instructions which tell the computer what to do. There are four basic levels of software you may use. The first is an operating system. Every computer requires a disk operating system (usually called "DOS"), which issues internal commands and permits dialogue between you and the computer. This is the first program you load when you turn the computer on.

Second, you may wish to use one of many authoring languages to write your own programs. The lowest level language is called machine language, which consists of strings of 0s and 1s. The next level up is called Assembly language, which uses symbols, signs and codes. Finally, we reach the high level languages, such as BASIC, Pascal, Logo and others. In these, the program instructions begin to make sense to us. They use regular English phrases, such as "Add X to Y," or even "X + Y − Z =." Internal programs called interpreters and compilers convert these easily understood instructions into the gaggle of digits which the stupid, but lightning fast, computer can handle.

Third, you're going to want applications programs–the sets of instructions that make the computer such a productive tool. You may want to do word processing (for letters, memos and reports), spreadsheet analysis (for financial management, mathematics, business record-keeping), filing and scheduling systems (for calendars, directories, reminders, bulletin boards), graphics (for charts, illustrations, and presentations in business meetings), communications (to send messages electronically over telephone lines, and to access data bases in other locations for news, sports, airline schedules, etc.), and finally, education and entertainment.

Fourth, you will need blank diskettes

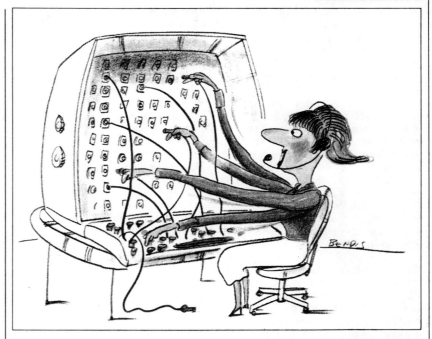

on which to write and save your own files and records.

Now let's move on to the computer's *processing* and *control* functions. At the heart of the computer is the CPU, the mysterious microchip everybody talks about and nobody can believe. It's a single "integrated circuit" on a chip of silicon, which does a startling number of things. When you peer at one through a high magnification microscope, it looks like an aerial view of a city. The most advanced microchips now have about the same complexity as a street map of an area approximately the size of Dallas and Fort Worth. These criss-crossing lines and patterns control the flow of electrical signals to and from various parts of the computer. In addition to the key microchip, there are a number of other chips–mounted on "boards" inside the computer cabinet–for storing and organizing information.

The CPU works like a switchboard. The main processor reads and decodes On/Off signals, and generates new signals to ship on to the next part of the computer. The CPU is divided into seven distinct areas and functions. Don't even imagine you're going to remember these, because you can run the biggest corporation on earth using a personal computer every day, and never encounter these phrases again. But let's hear them and let go of the nervous energy you may have about not understanding these phrases.

The CPU contains: a counter to tick off signals ("Where are we, guys?"); an instruction register ("What are we going to do next?"); a control unit ("How do we do what it says in the instruction register?"); data registers ("What words or numbers are we working on?"); a memory address register ("Where have we got that stuff stored right now?"); an arithmetic and logic unit, the ALU ("Go Do It!"); and finally, a status register ("Is everybody OK? All still together?").

These registers control how fast your computer can handle blocks of information. A computer "word" is a block of bytes which can be handled at one time. The length of the register determines how fat a block you can handle at once. The first generation of microcomputers were "8-bit micros," and handled one byte at a time. A later generation were "16-bit micros," and passed information through twice as fast. Now we're seeing the appearance of "32-bit micros." With each increase in register size and word length, the computer can run faster, and can also gain access to larger amounts of stored information.

Next, let's look at computer *memory*. This is not really memory as you know it. When we say we remember something, it means we can recall it, or re-create it, without referring to notes or files, or without having to look it up. Computers always have to look it up, and if they can't find it, they don't know that they ever knew it. Memory is really "storage," since the computer always checks and uses stored instructions or information.

The computer has both internal memory or storage, and external storage. There are several kinds of internal memory. First, the computer was taught some things in the shop. Before it ever came out of the factory, thousands of specific instructions were "burned into" chips, which are a permanent part of the computer. These instructions can't be altered by you. You can "read" them, but you can't "write on" them. For this reason, this is called "Read Only Memory," or ROM. Every computer has ROM chips which give it much of its power. The costs of ROM are plummeting, and many computers can now be purchased with extremely powerful applications programs built in.

There's also another kind of memory. You can write on it, then erase it, then write on it again. This is how you tell the computer what you want to say, how you want to organize it, where it's going to be saved, what you're going to call it. This is a dynamic memory area which lets you read and write in any sequence. You can get into it anywhere, and get out of it anywhere, so it's called "Random Access Memory," or RAM. This area only works while it is supplied with electricity. So it remembers anything you "tell" it, but it all goes bye-bye if you kick the computer's plug out of the wall, or if you lose power in your building.

External storage is usually on floppy disks. These are either 3.5" or 5.25" or 8" square paper jackets which contain a circular sheet of magnetic recording tape. If you ever rip one apart, you'll see where the "floppy" comes from, because it looks and feels like a circle of regular audiotape. Information is stored on these by recording magnetized signals on concentric tracks. Data is recorded and read back by a recording head. The floppy spins around inside its jacket, and lets you get at any piece of information in seconds. As you store information, the computer's DOS automatically catalogs the contents of the disk. It always lists the name of the file, the type of file, usually when it was created, and how long it is.

A less expensive option is to use a cassette tape recorder. But you'd better be strapped for cash to choose this option. Tape is S-L-L-O-O-W-W-W-W-W. To load a file the computer has to find it on the tape, so it says *go* and the tape starts to play. If the file is all the way at the end of the tape, good luck. With a floppy it says *go*, there's a short pause, and the computer then asks you, "What now, master?"

Hard disks are even faster and more powerful. In these devices the magnetic coating is not on a floppy piece of mylar or cellulose acetate, but is coated on an aluminum platter inside a sealed chamber. The hard disk is much faster, and stores tons more data. Many hard disks are now available which store up to 10, 20, or even 40 megabytes (that is, *40 million* typewritten characters).

Other new technologies for storage are appearing, including lasers, videodisks, and holographic systems. We'll talk more about them in later columns.

The most common form of *output* is an image on a monitor, a special cathode ray tube (CRT) which looks like an ordinary TV set. Some people actually use a TV set for home computers which are used mostly for video-type games, but monitors are always better for use at work, especially for extended periods. The On/Off signals of the computer are recoded again and generate tiny phosphorescent dots. These dots can be shaped into numbers or letters, and combined to form words. Typically, a monitor will show about 24 lines of type at one time; those lines are generally 80 characters wide. It's a full typewritten page in width, and about half a page high.

The monitor reveals 24 lines of one continuous file of text which may be many pages long. The frame of the viewing screen is like a window cut in a piece of cardboard. Slide the window in the cardboard up and down the page, and you get the illusion of the text scrolling up and down the screen.

The second most common form of output is the paper pages which come out of a printer. Having a printed copy of your work is always a good idea. Otherwise, you always have to be at the computer and have it turned on. Not very convenient for most of us.

There are many kinds of printers. One fast, relatively inexpensive, and very popular kind is the dot-matrix printer. This printer has a print head with tiny wires or pins, similar to the lines of a fork. As the head whizzes across the page, the pins spring out in the correct pattern and smack an inked ribbon against the surface of the paper. In the beginning these were rather crude, and they printed letters in which the dot pattern was all too evident. Many people refused to use them for formal business correspondence. The original dot matrix printers had only 9 pins to form the ver-tical pattern of the letters. However, in the last year printers have appeared with 21-pin print heads, and printers using many more pins will probably appear soon. These printers create a much more dense and formal-looking print style. They also produce many more different styles of type faces than were originally available.

Another kind of printer is called the letter-quality printer. It works much like a typewriter, and produces type which is indistinguishable from a typewriter, or in some cases, even from professionally set type.

Other special printers include the new ink-jet printers, which don't strike the paper at all, but spray ink out in a finely controlled stream of microscopic droplets. Laser printers and graphics plotters which print in full color are other options.

Another kind of output which is developing rapidly and creating tremendous interest is the routing of the digital signals of the computer into speakers to produce articulate simulations of human speech, and a startling explosion in synthesized music.

A final form of output which will become increasingly important to the world of business is telecommunications. Your computer can be hooked to a telephone line, and a new world of network services becomes available to you. Most of my own company business for the last three years has been conducted by electronic mail over phone lines linking the desktop computers at the office to my portable computer on the road. This area is of major relevance to business and the future, and we'll treat it in detail in future columns.

After the computer has done its job, the next *action* is up to you. The computer can record, store, analyze, plot, and communicate information for you, but the interpretation of its meaning is still up to you. You can be empowered by a personal computer, especially when you use it to do something which you already do well.

One thing is certain: The best is yet to come. The cost of computer logic is falling by 25 percent per year, and the cost of computer memory is falling by 40 percent per year. Speed has increased by a factor of 200 in 25 years, and the consumption of energy and the size of computers has gone down by a factor of 10,000 in the same time.

Storage up, size down. Speed up, cost down. Now's the time! *Get started!* ∎

It's 16 Bits, But Is That Wide or What?

Many of today's small computers are called 16-bit systems, but what does that actually mean?/**Douglas A. Kerr**

Many of today's small computer systems are described as being "16-bit" systems, and they typically use the Intel 8086 or 8088 "16-bit" microprocessors. But just what does that really mean, and how are these systems different from the "8-bit" systems we have known?

A Starting Point

Let's look at the characteristics of a typical 8-bit small computer system, one using an 8080 or Z80 microprocessor. Later, we will compare it with a typical 16-bit machine.

Memory

The memory of the 8-bit machine is organized into locations each of which can hold eight bits of data, or one byte. The byte may represent a character, such as the letter A or the numeral 3, or it may be eight bits of an instruction in machine language or of a numeric quantity represented in one of various binary formats.

So that the central processing unit (CPU) can access a particular memory location to store a byte of data there, or to read the byte already stored there, the locations have addresses in the form of 16-bit numbers. Since 2^{16} is 65,536, there can be that many distinct addresses, and thus that many memory locations. This leads to the 64 kilobyte (64K) maximum memory capacity of the typical 8-bit system.

(We often describe large quantities which are powers of two as multiples of 1024, which is 2^{10}, borrowing and stretching the scientific prefix for one thousand, kilo, abbreviated K.)

The CPU stores data in the memory, or reads it, one location at a time.

Thus, only eight bits are moved during a single memory *access cycle*. Although the CPU has instructions which move binary numbers made up of 16 bits, it does this in two steps, eight bits at a time.

Disk Storage

Data storage on either a floppy or a hard disk is also organized into 8-bit units. Transfer of data between the CPU and the disk drive is done eight bits (one byte) at a time.

Input/Output

Human input from the console keyboard and output to the console screen or to a printer is organized as sequences of characters. The characters are represented in a 7-bit coded form known as ASCII, the American National Standard Code for Information Interchange. Each 7-bit ASCII character code is carried in an 8-bit byte. Most often, the eighth bit is a binary zero and just goes along for the ride. In some systems, the eighth bit is used to expand the repertoire of characters beyond the 128 ASCII characters to provide for special graphics.

Machine Instructions

Regardless of the language in which the programmer works, the computer ultimately does its work through the CPU executing a series of *machine instructions*. Individually, these cause rather primitive actions, such as moving a byte from a memory location to a *register* in the CPU, or causing the CPU to add two 8-bit binary numbers, to be taken.

In a machine language program (one ready for the CPU to execute di-

rectly), the machine instructions are represented as sequences of bytes stored in memory. Some instructions require only one byte. Most require from two to four bytes, which are always stored in consecutive locations in memory.

Data Manipulation

The 8080 or Z80 CPU has seven principal *data registers*, which are used for data manipulation. Any of these can hold eight bits, or one byte, of data. Six of these registers are grouped into pairs. Any of these three *register pairs* can be used to hold a 16-bit number. As the program does its work, the machine instructions successively cause data to be moved between the registers and memory locations, or cause various operations to be performed on the data in the registers.

The CPU also has a 16-bit register, the Program Counter, which holds the memory address where the next machine instruction begins. Since the instructions in a program are most often executed in sequence, this register is normally set automatically to the address immediately after the current instruction. However, if the current instruction is one which calls for a *jump* to another part of the program, the Program Counter is reset to the corresponding new address.

Another 16-bit register, the Stack Pointer, keeps track of the next location scheduled for use in a memory area assigned by the programmer as the *machine stack*. This area is used to store temporarily the contents of registers while they are "borrowed" by the program for other purposes.

The CPU can perform a number of arithmetic and logical operations on bi-

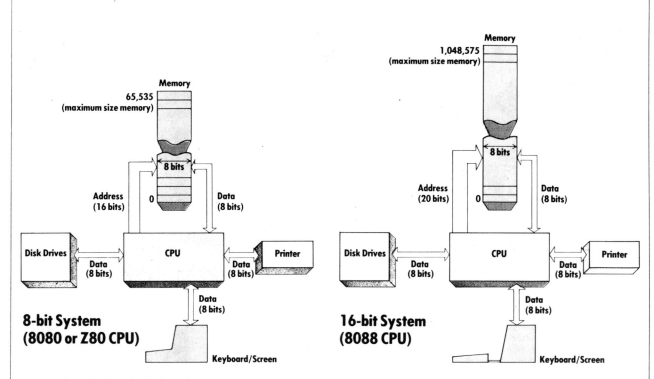

Figure 1. Organization of typical small computers.

nary numbers. It can, for example, add or subtract two 8-bit numbers. These functions take place in the one data register which is not paired with another, the A Register, or *accumulator*. The CPU can perform many of the same operations on 16-bit numbers. In this case, one of the register pairs is the host, the one known as HL (made up of the H and L registers).

The internal structure of the 8080 or Z80 microprocessor is such that operations on 16-bit numbers are done, in effect, eight bits at a time. For this reason, 16-bit operations take substantially more time than the same operations on 8-bit numbers.

A 16-Bit System

Now let's look at a typical 16-bit small computer system, such as the IBM PC, based on the Intel 8088 microprocessor.

Memory

As in the case of the 8-bit machine, the memory is organized into locations which hold eight bits or one byte of data. The CPU can still store or read only one byte at a time; when it stores or reads 16-bit numbers, it does so in two steps, eight bits at a time.

In the 16-bit machine, however, memory locations have 20-bit addresses (see sidebar). As 2^{20} is 1,048,576, the

CPU can access up to that many locations, if the memory is in fact that large. (Since that number is 1024 x 1024, and since we treat 1024 as if it were one thousand, we call that amount of memory one megabyte, borrowing the scientific prefix for one million, abbreviated M.) It is

> The 8088 CPU can perform a wider range of arithmetic and logical functions than the 8080 or Z80.

the ability to use such large amounts of memory that is a major advantage of the 8088 microprocessor.

Disk Storage

As in the case of the 8-bit machine, storage on floppy or hard disk is organized into units of eight bits, or one byte.

Input/Output

Also as in the case of the 8-bit machine, input from the keyboard and output to the screen or a printer is

conducted in the form of ASCII characters carried within 8-bit bytes.

Machine Instructions

Again as in the 8-bit machine, the machine instructions used by the 16-bit CPU are coded as sequences of 8-bit bytes, from one to six bytes being required depending upon the instruction. The bytes which represent one instruction are stored in consecutive memory locations.

The 8088 CPU, however, has a valuable feature called *instruction prefetch*. While the CPU is executing one instruction, it brings from memory the next several bytes (which we would expect to include the next instruction) and places them in a special "pipeline" buffer within the CPU. In this way, when the time comes to work on the next instruction, it is already close at hand. This significantly speeds up the operation of the system.

Of course, if the current instruction requires a jump to an instruction other than the next one in sequence, the information in the pipeline is useless: the required instruction is somewhere else in memory. In that case the pipeline is emptied, the bytes of the instruction are brought in from the new locations, and the pipeline is refilled with subsequent bytes.

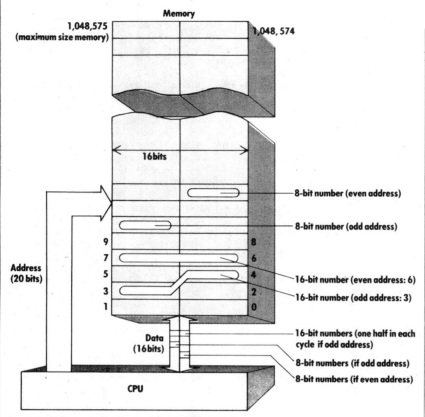

Figure 2. Memory organization with the 8086 CPU.

Data Manipulation

The 8088 CPU has eight principal data registers, each capable of holding a 16-bit number. Each is divided into two portions, either of which can be used to hold an 8-bit number (one byte).

The 8088 CPU can perform a wider range of arithmetic and logic functions than the 8080 or Z80. For example, it can directly perform multiplication and division of 8- or 16-bit binary numbers, important functions in the multiplication and division of decimal numbers. With the 8080 or Z80, these functions must be performed by long sequences of program steps which involve addition or subtraction of the binary numbers, plus shifting the bits of a number left or right to multiply or divide it by powers of two—in effect, doing "binary long multiplication" (or division).

All arithmetic and logic operations can be performed on either 8- or 16-bit numbers. Most operations can be performed in any register which is convenient, not just in certain ones, as in the 8-bit machine. In addition, arithmetic and logical operations can be performed on numbers stored in memory locations without first bringing them into a CPU register. The internal structure of the CPU performs 16-bit operations just as quickly as 8-bit ones. These features

How does a 16-bit machine come up with 20-bit addresses?

Just how does a system that operates mainly with 8- and 16-bit numbers come up with the 20-bit address required for access to a memory location?

Suppose we identify an arbitrary region of the memory consisting of 64K (2^{16}) consecutive locations, including the one in which we are interested. We will call such a region a *memory segment*. We can then completely identify a particular location by giving the address at which the segment begins (called the *segment base address*) plus a 16-bit number which tells how far the location is beyond that point. That 16-bit number is called the *location offset*.

In the 8086 or 8088 CPU, the base address of the currently defined segment is specified by a number stored in a special 16-bit segment register. The segment base address is 16 times that number (and thus must be expressed in 20 bits).

When a memory location is to be accessed, the *offset* portion of its address, typically coming from one of the CPU data registers, is added to the segment base address developed from the number in the segment register. The result, also a 20-bit number, is the address of the desired location.

As long as memory operations continue to use locations in the same segment, the base address in the segment register remains constant, and addressing instructions need specify only the 16-bit offset portion of the location address.

Note that the total memory is not divided up into fixed segments starting every 64K locations. A segment can be defined as starting at any location whose address is a multiple of 16. Therefore, the programmer can

define a segment that will contain all the locations used during a certain stage of program operations. In this way, the need to change the segment base address can be minimized.

The CPU actually has several segment registers. One holds the segment base address which pertains to the location of the next instruction. It augments the 16-bit register called *Instruction Pointer*, which is equivalent to the Program Counter in our 8-bit machine. A second segment register identifies the segment to be used in addresses for storing or reading data. A third one provides the segment base address for the machine stack. It augments the Stack Pointer register, which is like the one in the 8-bit machine.

By having these separate segment registers, it is possible to take the current instructions from one

greatly speed most data manipulation compared to the 8-bit machine.

The 8086 CPU

The Intel 8086 microprocessor is also used by certain modern 16-bit computer systems (like the ACT Apricot). It is almost identical in function to the 8088 with one exception: the memory used in 8086-based systems is organized into 16-bit *words* rather than the 8-bit *bytes* we have previously encountered. Nevertheless, for addressing purposes, each word is considered to be two consecutive 8-bit locations, each of which has its own 20-bit location address.

When the CPU wants to store or read an 8-bit number (one byte), it gives the 20-bit address of the location, along with an extra electrical signal which tells the memory to access only the eight bits from that location, not all 16 bits which are stored together.

When a 16-bit number is stored in memory by the CPU, it may end up in either of two situations, depending on the address assigned to it. It may occupy an entire 16-bit word, or it may occupy the second half of one word and the first half of the next.

In the first case, the CPU will store or read the entire 16-bit number in one memory access cycle, giving the address for the location of the first part of the word but signaling the memory to access the entire word. In the second case, the CPU must access the memory twice, in each case accessing only one of the two locations, joining the two returned bytes (in the case of a read) to form the entire 16-bit number.

For this reason, if a program needs to store or read a long sequence of 16-bit numbers, it is worthwhile for the pro-

> ## An 8-bit machine uses 16-bit addresses, while a 16-bit machine uses 20-bit addresses.

grammer to arrange the address of each to be "even" so that each number cleanly occupies one memory word. If this is done, the speed of an 8086-based system, when handling 16-bit data, can be significantly greater than for an 8088-based system.

In the case of program instructions, this consideration does not arise. When the CPU "prefetches" additional bytes, it always does so an entire memory word (containing two consecutive bytes) at a time.

What Does All This Mean To The User?

We have seen that a 16-bit machine works with both 8- and 16-bit data, just as an "8-bit" machine does. We have also seen that an 8-bit machine uses 16-bit addresses, while a 16-bit machine uses 20-bit addresses. And, we have seen that, with the 8088 CPU, the 16-bit machine uses an 8-bit memory. No wonder it has been hard to answer the question, "16-bit? What does that mean?"

Much more important is that the 16-bit systems offer substantial advantages to the programmer and user. Direct access to as much as one megabyte of memory, as a result of the 20-bit address structure, allows large application programs and large arrays of data to reside in memory. This minimizes the need to move program portions and data records from disk. Various other new features of the CPUs, not necessarily related to their 16-bit orientation, provide additional speed and programming power.

We now hear that 32-bit machines are coming into use in the personal/ professional computer world. Let's see, will they have 32-bit memories? Maybe not. But we can be certain that they will reflect the continuing progress of the computer industry. ∎

memory segment, store and read the data in another segment, and use an area in a third segment as the machine stack.

When it is necessary to address a location in a new segment for one of these purposes, the instruction must give both the offset and the new segment base address. For this reason, there are usually at least two forms of many of the machine instructions.

Fortunately, it is not always necessary for the programmer to worry about the differences, even when working in assembly language. The normal mnemonic assembly code for the instruction is the same regardless of the addressing type needed. The assembler can tell when the intended address is in a different segment from the current one and will code the proper type of instruction in the object code.

For example, if the specified operation is an unconditional jump to a memory location identified by a symbolic label, the programmer can always use the same assembly code.

The assembler, however, chooses one of three types of machine instructions to perform the jump:

● If the location represented by the label is within 127 locations of the current one and within the segment whose base address is now in the instruction segment register, the assembler codes a jump instruction which specifies the new location in terms of the distance (forward or back) from the present one, stated in eight bits. This corresponds to the jump relative instruction used with the Z80 CPU. This instruction requires two memory locations, one carrying the code for the instruction (called in this case JUMP SHORT) and one carrying the distance to be jumped.

● If the location represented by the label is farther than that from the current location, but within the current segment, the assembler puts in a jump instruction which specifies the 16-bit offset portion of the new address. This instruction (called JUMP NEAR) requires three memory loca-

tions, one to identify the instruction and two to carry the offset. The base address is taken from the instruction segment register.

● If the location represented by the label is not in the current segment, the assembler puts in a jump instruction which includes both the offset address and the new segment base address (carried in another 16-bit number). This type of instruction (called JUMP FAR), therefore, requires five memory locations.

Of course, there are many different segment base addresses which could be used to accommodate the desired location. If not instructed otherwise by the programmer, the assembler uses the segment base address which was originally used when the address represented by the label was established.

This is only one example of the way that improved program development tools (such as the 8086/8088 assembler) aid the programmer in exploiting the enhanced features of the new CPUs. —*DAK*

Hello, Mr. Chips

The word bit, or straying off the data path   **John J. Anderson**

In the beginning there was the vacuum tube, and with that innovation electricity got its first real chance to become electronics. Circuit complexity translated into bulk, however, and if you wanted that new-fangled toy called a computer, you needed a building to devote to it, and the riches of Croesus to acquire it and keep it healthy.

The vacuum tube begat the transistor, and we saw it was good. Circuits of greater complexity could be designed more reliably, cooler, and in much less space. Central processing units (CPUs), the brainstems of computer circuitry, shrank to the size of mere refrigerators. And prices came down.

The transistor begat the integrated circuit, and we saw it was very good. A single chip of silicon could contain multiple transistors. There but for the grace of the integrated circuit went the aerospace advances of the sixties—things like walking on the moon. And prices came down.

But up until the end of that turbulent decade, digital IC technology was limited to arithmetic, logic, I/O controller, and memory chips. The CPU on a chip, and its ancillary development, the microcomputer, were children of the '70s.

Ironically, the first integrated circuit to closely resemble a CPU was developed in the U.S. by Intel, while under contract to a Japanese calculator company. ETI, a Japanese manufacturer of expensive desktop calculators, specified a new type of IC to spearhead a new line of machines. Marcian "Ted" Hoff, of Intel, envisioned extending ETI's specifications to include programmable characteristics. The result was the Intel 4004, which incorporated on a single chip the equivalent of more than 4000 transistors.

This was the genesis of the microprocessor. Under one cover, in a miniscule package, the business of computing now takes place. Nowadays even mini- and mainframe computers use IC-based central processing units, called microprocessors (MPUs) in place of multicomponent CPUs. One result of the MPU was the microcomputer; another was *Creative Computing*.

Pegging Power

There are four basic criteria typically considered in judging the power of a microprocessor. They are:

Speed: The cycling rate at which instructions can be executed within the MPU.

Addressable memory: The maximum RAM size the MPU can access from a single state.

Instruction set: Includes both the number and complexity of instructions that can be invoked.

Word width: The "swatch" of bits (binary digits) upon which the MPU can act at one time.

It is impossible to put these criteria into an indisputable hierarchy, but without a doubt, word width is a very significant entity. The speed, addressable memory, and instruction set of a microprocessor are architecturally tied to its word width.

Unfortunately, the concept of word width has been popularized in a fashion that obscures rather than clarifies its importance. It is easy to state that a 16-bit MPU is twice as powerful as an 8-bit, and a 32-bit MPU twice again as powerful—easy to state, and perhaps a powerful sales tool, but somewhat incorrect. At the least, such reasoning leads to serious oversimplification.

First off, let us consider speed. A 16-bit processor running at 2 MHz is certainly not twice as powerful as an 8-bit processor that runs at 5 MHz. How much more powerful one is than the other runs us immediately into some nasty shoals. Our quantification approach becomes marooned in value judgments more likely to reflect the biases of the arguers than the merits of the arguments.

Then we may consider the natures of instruction sets. These vary among chips and especially among families of chips. You can write the same assembly language program for different kinds of microprocessors, but the code itself, and more importantly the ease of writing such code, varies greatly. Programmers tend toward vehement chauvinism when it comes to MPUs, assuredly as a direct result of the effort they have put into learning a system that works in a certain way. They may naturally resist the stress of change, even when a new slant makes things easier overall. Although we can say that the instruction set of one MPU is larger and more powerful than that of another, we cannot quantify the appeal of any one instruction set. A chip that makes one type of task easier might make another more arduous.

Passing the Word

Quantification of chip power is made even more difficult by the fact that word width can vary, even within a single chip. Typically the term word width is used to indicate the width of the *registers* within a processor. Any and all CPUs pull information out of memory, act upon it, then return it to RAM in a process known as fetch, alter, and store. Upon the execution of a fetch, the CPU loads the word fetched into a storage register. Then a specific instruction can be

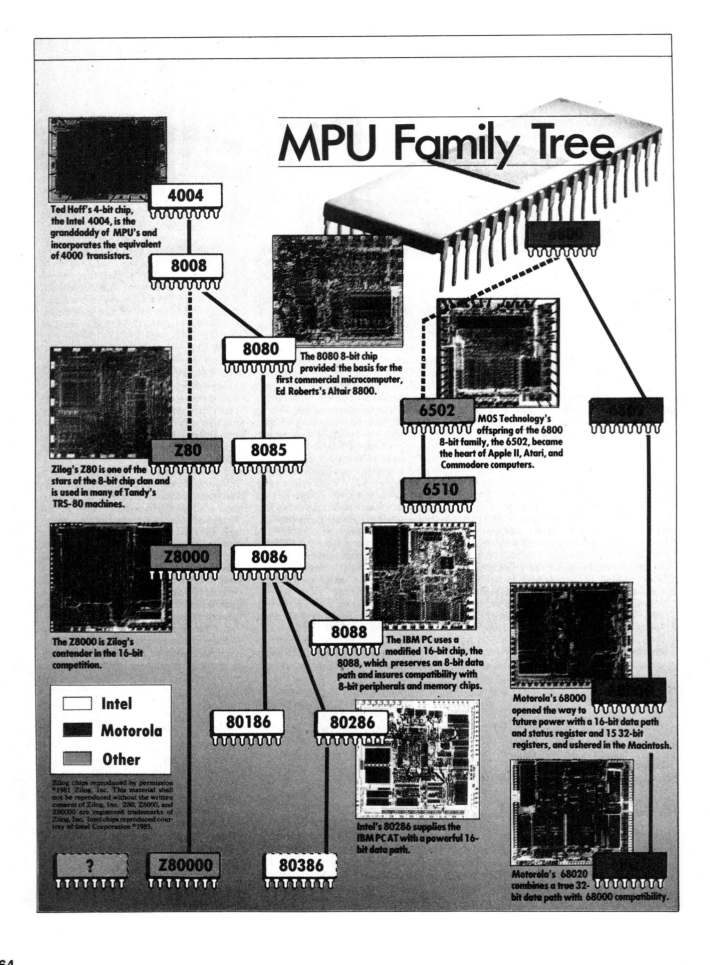

MPU Family Tree

4004

Ted Hoff's 4-bit chip, the Intel 4004, is the granddaddy of MPU's and incorporates the equivalent of 4000 transistors.

8008

8080

The 8080 8-bit chip provided the basis for the first commercial microcomputer, Ed Roberts's Altair 8800.

6502

MOS Technology's offspring of the 6800 8-bit family, the 6502, became the heart of Apple II, Atari, and Commodore computers.

Z80

Zilog's Z80 is one of the stars of the 8-bit chip clan and is used in many of Tandy's TRS-80 machines.

8085

6510

Z8000

The Z8000 is Zilog's contender in the 16-bit competition.

8086

8088

The IBM PC uses a modified 16-bit chip, the 8088, which preserves an 8-bit data path and insures compatibility with 8-bit peripherals and memory chips.

Motorola's 68000 opened the way to future power with a 16-bit data path and status register and 15 32-bit registers, and ushered in the Macintosh.

	Intel
	Motorola
	Other

Zilog chips reproduced by permission ©1981 Zilog, Inc. This material shall not be reproduced without the written consent of Zilog, Inc. Z80, Z8000, and Z80000 are registered trademarks of Zilog, Inc. Intel chips reproduced courtesy of Intel Corporation ©1985.

80186

80286

Intel's 80286 supplies the IBM PC AT with a powerful 16-bit data path.

?

Z80000

80386

Motorola's 68020 combines a true 32-bit data path with 68000 compatibility.

invoked to act upon data stored in that register.

Most mainframe computers make use of 32-bit registers, and as we shall see, micros are also moving quickly into the 32-bit realm. In college I occasionally had the dubious honor of programming a mainframe known as the CDC Cyber, which made use of whopping 60-bit words. I could never quite fathom any good reason for making an address register that wide. It made calculating offsets, the number of bits away from a given reference point, an easy way to lose track of reality.

Hoff's Intel 4004, the granddaddy of microprocessors, was a 4-bit machine. It could operate upon or transfer only four bits at a time. As a result, it could handle numbers but not alphabetic characters in a single "gulp." Although wider registers could be simulated through piggybacked instructions, this made programming the 4004 rather convoluted. It quickly became obvious that for alphanumeric processing, a more powerful chip was necessary. The natural step was to 8-bit words, which could handle alphabetic coding as well as a capable instruction set relatively straightforwardly.

The Intel 8008, originally developed for the Computer Terminal Corp. (now Datapoint) was introduced in 1973. It was simply an 8-bit version of the 4004. This processor was still rather Byzantine in its architecture—actually better suited as a machine controller than a general purpose MPU. Intel then followed up with an improved chip, the 8080. Although most of the design philosophy of the 8080 came directly from the 8008, its instruction set could act as a bonafide CPU. The 8080 soon became

the mind of Ed Roberts' Altair 8800, generally acknowledged to be the first commercial microcomputer.

The 8080 has six general purpose 8-bit registers with a stack pointer and program counter both 16 bits wide. Like many others, the 8080 is a stack-oriented MPU, which is used for temporary storage without the need for address pointers. The *data path*, which is the number of bits that a microprocessor can fetch from or store to memory in a single swatch, is eight bits wide on the Intel 8080.

Here we encounter one of the real rubs of the width myth. The width of the internal registers of an MPU often exceeds the width of its data path. By way of analogy, we might imagine loading six-packs into a carton to make a case of beer. The case holds 24 cans of beer—therefore the address register for our Molson computer is 24 bits wide. However we will put in and lift out the cans by the six-pack, so the data path of our brew is six bits wide. If Molson made computers, it might claim it had a 24-bit beer. Moosehead would be quick to point out, however, that Molson was not a "true" 24-bit beer, as it has a six-bit data path. Doubtless as well, their marketing department would quickly point out that Moosehead is available in eight-packs.

As the 8080 has 8-bit registers and an 8-bit data path, it may be termed a "true" 8-bit processor. One might suspect that implies the existence of "false" processors, but it is better not to linger over that point.

Between 1973 and 1981, quite a bit of begetting went on in the 8-bit realm until the 8-bit processor began to yield to a new crop of 16-bit chips. We can trace the genealogy (see page 49) of two prominent families: those that trace their roots from the Intel 4004 and those that trace back to the Motorola 6800, first introduced in 1974. It should be noted, however, that in both genealogies there appear important contributions from people "outside the family."

In fact, the Z80, a second generaton 8080, and the 6502, a very close cousin to the 6800, were developed outside of Intel and Motorola, respectively. These chips went on to become the most important 8-bit processors—the ones that sowed the seeds of the microcomputer revolution. The Z80 made its way into successive generations of TRS-80 machines, while the 6502 was to form the heart of the Apple II, Atari, and Commodore computers.

The Better Bitters

Although the heyday of 8-bit processors is now behind us, they will remain important for years to come. They are still very capable chips, each with an established core of loyal programmers, and most important, they are now dirt cheap.

They do, however, pose certain limitations. An instruction set with an 8-bit width is limited to 256 total instructions. Although on some chips prefix bytes and other devices are introduced to get around this stricture, they again make the task of programming more burdensome. Certainly, the next logical step was a 16-bit microprocessor on a chip. An instruction set with a 16-bit width is capable of 65,536 discrete instructions. As that is far more than generous, and a 9-bit instruction width is quite reasonable to imagine, remaining word width can be used for data.

As explained above, more instruc-

The analogy is a little far-fetched, but beer bottles, six-packs, and cases can be used to illustrate some aspects of data width. Initially, each bottle (bit) comes down the assembly line (RAM) singly.

Before storage or shipment, however, each drops into a six-pack as it continues down the belt. We might think of this as a six-bit data width. The bottles are now moved solely in sets of six.

Further down the belt, six-packs are dropped in groups of four into cases. We might stretch the analogy to the point of imagining each case as a 24-bit register, as each holds four times six bottles. When finally sold, the product will again move in sets of six. So our beer has a 24-bit register, but still a six-bit path. To have a "true" 24-bit beer, we would have to sell it in 24-packs only. If that were the case, we might or might not choose to pack it in cases holding 48 or 72 bottles (two or three 24-packs). Often MPUs use registers wider than their data paths.

tions mean a more powerful MPU. Instead of treating multiplication as recursive addition, or division as recursive subtraction, for example, a multiply or divide instruction can be added to the instruction set. (The processor may still treat the instruction recursively, but the programmer need not.) And through the addition of memory management logic, 16-bit processors can cross the address boundary of a single 64K chunk of RAM.

When, in 1981, IBM announced it would use the Intel 8088 in its first microcomputer, Intel was able to reassert itself as a major player in the microprocessor game. The 8088 is a special case in itself; it is a 16-bit processor in 8-bit clothing. Its registers are a uniform 16-bits wide, while its data path is 8-bits. The 8088 is a version of Intel's true 16-bit chip, the 8086, with special bus hardware added. This ensured that the 8088 would remain compatible with the 8-bit memory and peripheral chips that proliferated at the time it was introduced. The downside of this customization is that the 8088 is slowed down substantially by overhead transfer time. The *Creative Computing* benchmark, in fact, logged the IBM PC as significantly slower than a number of 8-bit machines with quite decent cycle rates.

The 8086 and 8088 can address up to 1Mb of RAM (in segments of 64K), and include multiply and divide instructions. They were among the first to use multiplexed address and data lines, wherein more than one signal shares a common circuit to reduce chip size and cost.

Zilog's answer to the 16-bit challenge was the Z8000. This chip might have been a much more serious contender if its introduction had not been plagued by delays, and in its early days by a lack of support. The chip is clearly superior to the 8086/8088, but in an industry in which timing is crucial, the effort misfired. Motorola's 6809 is also a powerful chip which upgraded in size and versatility the instruction set of the 6800 series in an MPU with 16-bit registers and an 8-bit data bus. It found its way into the highly underrated TRS-80 Color Computer, but not much else.

By far the most interesting Motorola entry is the MC68000, which in 1982 first levered a foot into the door of the 32-bit world. In one fell swoop, the 68000 launched Motorola right back into the fray. The data path and status register of the 68000 are 16-bits wide, but the other 15 available registers are all 32-bits. The 24-bit address bus allows fully 16 megabytes of RAM to be addressed linearly. The instruction set of the 68000 contains over 90 instructions, and the memory addressing configuration makes debugging assembly code on the 68000 much less painful than on the 8086.

Certainly the first 68000-based microcomputer to come to mind is the Apple Macintosh, and the Mac does serve as a good example of the power of the 68000—juggling programs, data, and a highly-refined user interface simultaneously. But the Mac was not the first 68000-based micro. That honor belongs to the Fortune Systems 32:16, which due to ill fortune, is no longer with us.

The 68000 does not multiplex signals, and so appears in a 64-pin package, as opposed to the 8086/8088 which is in 40-pin DIP configuration. The trade off is a bigger chip, but one that requires less external logic. Intel introduced a hybrid

8086 in 1982, called the 80186, which incorporates a substantial amount of support logic onboard—a move toward truly manufacturing a computer on a single silicon chip. The 80186 represented a significant step, offering better performance for substantially less cost than an 8086 with the requisite bevy of support chips required to drive it. Lowered chip count results not only in decreased manufacturing costs, but increased hardware reliability. The 80286 introduced by Intel last year took things a step further, and now finds itself ensconced in the muscular IBM PC AT.

There are other 16-bit processors, like the National Semiconductor 16032 and Texas Instruments TMS 9900, some with quite admirable specifications, but none of these has had much real impact on the microcomputer market. One might observe a striking parallel between the two major 8-bit contenders of the past, the Z80 and the 6502, and their combatant 16-bit progeny—Intel's 8086/8088 propped up by hordes of IBM PCs and PC clones, and the Motorola 68000 residing within the Apple Lisa and Macintosh.

The Bit Goes On

Already, however, the field is being cleared for the next big battle. This one will be fought by the true 32-bit titans, and never has the competition been so fierce. Semiconductor manufacturers are scratching, biting, and scrambling for position in a race that again will probably result in two big winners and a slew of battered also-rans. As the ante in developing a new chip is typically over $50 million, that represents a gamble indeed.

Why a 32-bit processor, you ask?

We can also stretch the analogy to suggest the logic of stack-oriented processing. Imagine cases of beer being stacked in a storeroom. The first cases stacked will be the last to move out of the stack, so in accounting parlance, this is a LIFO (last-in first-out) stack. If you tried to pull out a case that wasn't on the top of the stack, you would probably cause an avalanche. Hence sales are made from the top down. Many MPUs work in the same way, using registers to stack data temporarily for subsequent processing.

The answers are as follows: speed, multi-tasking, multiuser capability, mini- and mainframe compatibility, the ability to tackle enormous tasks, like expert systems programming, artificial intelligence, voice recognition, and perhaps most important, Unix compatibility.

Like it or not, the microcomputer industry is rushing headlong toward the Unix operating system, and it will take a 32-bit processor to implement it in all its glory [sic]. Programmers are talking more and more about this marvelous and mysterious language cryptically called C, and even the behemoth IBM has acknowledged that its 8086/8088 software base will be effectively neutralized in a few short years. And so the race is on.

No fewer than 14 American companies have announced that they are or will be entrants in the 32-bit fracas. Intel and Motorola, the Hatfields and McCoys of the microprocessor industry, are siring the next generation of combatants. Intel is readying its 80386, which will combine Unix compatibility with PC compatibility on a single chip. Motorola has already introduced its MC68020, which is entirely compatible with the instruction set of the 68000, with a true 32-bit data path.

But it is important to note that 32-bit processors are architecturally much more independent than their ancestors. The software bases built around the 16-bit processors of Intel and Motorola are, therefore, not as likely to give those companies pole position in the 32-bit race. A dark horse has a better chance now than ever before to usurp pre-eminence in the microprocessor market.

Zilog is back and in the running with the aptly named Z80000. It of course is downwardly compatible with the Z8000. NCR is in the running with the 32-000, which packs the equivalent of 40,000 transistors on a single chip. In conjunction with an Address Translation Chip, the NCR microprocessor can address up to 300 Mb.

Also announced are entries from Inmos, Fairchild Camera, National Semiconductor, Texas Instruments, and Western Electric (no longer to be confused with AT&T). And chip manufacturers are no longer alone in their pursuits. Mini- and mainframe makers like DEC, IBM, and Data General are reverse engineering MPUs compatible with their existing machines. Even Hewlett Packard and AT&T are in the on-deck circle.

Let us not ignore Japan in the coming equation. NEC has announced it is working on a 32-bit MPU with no fewer than 700,000 transistors on board. Hitachi has announced the completion of a proprietary 32-bit chip. The timeframe announced for shipment of the Hitachi chip is 1986; for completion of NEC's superchip, 1987.

The Racing Form

It is difficult to predict how the 32-bit race will take shape and impossible to predict the winners. But there are a few predictions that can possibly be made more safely.

As in any marathon, many of the entrants will not finish. The micro-

> **N**o fewer than 14 American companies have announced that they are or will be entrants in the 32-bit fracas.

processor industry makes for strange bedfellows, and second-sourcing has resulted in some unholy alliances indeed. To sell a chip in quantity, a manufacturer typically accepts more orders than it can fill. It authorizes another company to manufacture its chips and gives "masks" of the chip to that company so it can do so. If a chip is very popular, it may be second-sourced to multiple companies. The 8086 was, in its golden years, manufactured by no fewer than seven companies.

Second-sourcing agreements in the 32-bit arena could be made into a soap opera for TV. Often second-sourcing is a major means of gleaning technology, and we find second-sources suddenly announcing their own chips. Marketing divisions live on Maalox, and fickleness is rampant. Fairchild had committed to second-sourcing for National, but now is pursuing design of its own proprietary CMOS 32-bit chip. Texas Instruments initially announced its own proprietary chip, but now has committed as a second-source for National. Fujitsu and Toshiba are second-sourcing for Intel. Don't be surprised when these companies break off and announce proprietary chips of their own. The bottom line is to take any and all announcements of 32-bit plans with a grain of Alka-Seltzer.

My predictions are as follows. You might be able to narrow the field down to four: the Motorola 68020 will find a niche, owing to a growing degree of loyalty to the power and elegance of the 68000 family; National Semiconductor will find itself back in the big leagues with the 32000 because of its suitability with Unix and C and its proximity to the ultra-powerful VAX minicomputer; Intel's new chip is bound to find its way into IBM's 32-bit micro, as IBM now holds 20% ownership of Intel; and the field might be big enough, at least early on, to allow for one dark horse candidate, conceivably Japanese. In the end there will be one or two survivors. And I'm not about to guess who they might be.

The race is sure to continue even from there, but at a greatly slowed pace. Although there will undoubtedly appear 64-bit microprocessors toward the end of this decade, my experience with the Cyber leaves me with the hunch that we will hit a point of diminishing returns in that realm. My guess is that development will continue much more strongly along the lines of incorporating support chips onto the MPU, and even at some point including RAM memory. Sooner or later we will rid ourselves of circuit boards entirely. Fairchild is moving in the right direction with CMOS technology—that will find a niche in the future of microprocessors and RAM technology. A day will come when we look back at today's micros as dinosaurs of power consumption. And RISC chips (for "reduced instruction set chip"), as pursued by Inmos, DEC, and HP, pose an interesting angle. Their philosophy is that conventional microprocessors are burdened by many instructions that are rarely or never used. Chips can be made faster, cheaper, and better by tailoring them more carefully.

Finally, I expect parallel processing to come into its own by the end of the decade. In our entire discussion here we have conceived of computing in a traditionally serial manner; though it may happen at incredible speed, only one instruction is executed at a time. The next major breakthrough in computing will be the advent of machines with multiple 32-bit processors, each operating in its own bailiwick, while in full communication through some hierarchical structure with the other processors onboard. What might a machine of this kind be capable of doing? Well, among other things, it just might be able to grasp the English language. Perhaps then I'll ask it what exactly makes one MPU superior to another. I'll program it to laugh. ∎

Edited by Stephen Kindel

There are superminicomputers and there are minisupercomputers. The difference between the two is a major market. Maybe.

Filling the gap

By Ellen Benoit

FIGURING OUT WHAT the weather will be ten days from now requires almost endless, complex calculations: a half-trillion of them, more or less, according to the European Center for Medium Range Weather Forecasts in Reading, England. How does the center produce such forecasts before the weather arrives? With lots of help from its Cray Research, Inc. supercomputer.

Supercomputers are generally defined as the fastest processors available at any given time. But with speed, you pay for what you get. Su-

percomputers that can perform in the 1,000-megaflop range—that's 1 billion floating point operations per second—run as high as $14 million each. Pay less and you get a lot less: superminicomputers, often Digital Equipment Corp.'s 32-bit VAX family, or mainframes, usually from IBM. These are often equipped with a special processor for scientific calculations, run in the $500,000-to-$2 million range and are typically 1/100th as fast. Small wonder, then, that while there are more than 35,000 VAXs toiling away in the world, there are just over 100 supercomputers installed, mainly in generously funded government re-

search institutions.

Such a wide gap in price and performance *(see illustration)* means that a good part of the scientific community makes do with substantially less computing power than it wants. The result is tortured research schedules to use the large chunks of needed minicomputer time usually available only on nights and weekends.

But one man's need is another's opportunity, offering a potential market for the half-dozen or so companies now introducing what might be called "minisupercomputers." These machines are not much more costly than superminicomputers but have at least 20 times the power.

Potential, note well, is the operative word for the market. The companies developing minisupers are small and still private; and commercial and technological imponderables remain to be resolved.

Who is developing minisupers, and how? One firm, Scientific Computer Systems Inc. of Wilsonville, Ore., decided simply to shrink a Cray. Says SCS President Robert Schuhmann, "Basically, there are three approaches you could take if you don't want to use a Cray. You can take a VAX and try to make it run faster. You can take an IBM mainframe and provide it with a set of vectoring instructions, or you could start from scratch with a totally new architecture and a new instruction set." According to Schuhmann, formerly vice president of marketing and sales at Floating Point Systems Inc., a manufacturer of add-on proces-

Getting what you pay for

Comparisons among the different classes of computer are, perhaps surprisingly, rough at best. This is because the industry lacks a standard measure for speed, making do with the shorthand MIPS, or millions of instructions per second. Still, it can be seen that price and performance change dramatically as you move up the scale.

Microcomputer	**Supermicrocomputer**	**Minicomputer**	**Superminicomputer**
Apple IIe	Digital Equipment Corp.	Digital Equipment Corp.	Digital Equipment Corp.
.0005 MIPS[1]	MicroVAX I	PDP-11/44	VAX-11/780
$1,795	.36 MIPS	400 nanoseconds per	1.1 MIPS
	$11,245	instruction cycle.[2]	$125,000 to $500,000
		$29,950	

[1]MIPS equivalent. Actual rate: 500,000 8-bit operations per second. [2]Nanosecond: one-billionth of one second.

sors, each one of those tasks is equally daunting. So Schuhmann talked with Cray Research and came away with permission to build a machine that was compatible with Cray's most powerful supercomputer, the X-MP.

Schuhmann believes his minisupers will provide "satellites" for existing Cray users, to make more efficient use of available computer time. Models or experiments could be developed on a Cray, for example, then off-loaded to an SCS minisupercomputer for manipulation. The SCS-40, when it's finally available in 1986, will execute operations at about one-quarter the speed of a Cray-1, but 20 to 40 times the power of Digital's VAX series. Price? About $500,000, which is close to that of a loaded, high-end VAX.

Schuhmann is quick to point out that the SCS-40 will be able to run all the application software written for the Cray. This compatibility, he believes, will raise demand for supercomputer power to $7 billion a year by 1990. But that's a long leap. Currently, total supercomputer sales, chiefly of Cray machines and Control Data's Cyber series, run only about $500 million annually. Where will all the additional sales come from? In Cray installations alone, Schuhmann sees 30,000 potential customers, or 300 "I-wish-I-had-a-Cray" users, as he calls them, for each machine.

While SCS boasts of its close relationship with Cray, Convex Computer Corp. of Richardson, Tex. considers it a drawback. Convex' approach, ac-

The Convex C-1 and its inventor, Steven Wallach
For the scientist who has everything—but a Cray.

Ben Weaver/Camera 5

cording to marketing director Stephen Campbell, is to exploit the program development tools available with AT&T's Unix, the operating system most widely used by existing superminicomputers, such as the VAX series. The Convex C-1, with some jiggering, will also run Cray application software written in Fortran, but Campbell dismisses the Cray installed base as too small a market. Instead, he's counting on VAX users to upgrade. Like the SCS-40, the Convex C-1 runs at about one-quarter the speed of a Cray. Cost is similar, too, at $495,000.

If both SCS and Convex are right, will minisupers spell doom for super-

minis? Not necessarily. The two varieties of machines do their jobs in different ways, which accounts for the difference in performance. Supercomputers are vector processors, which means they organize similar problems in strings, or vectors, in order to perform many operations at once; for example, whenever possible, all multiplication tasks are done together and all addition tasks done together. Vector processing is not done by most minicomputers and mainframes; such machines use scalar processing, in which each operation is done in sequence. Some problems work better with vector processing, as when you need to multiply thousands of num-

Laszlo Kubinyi

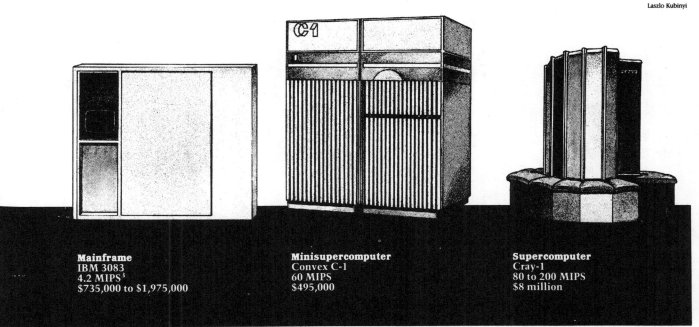

Mainframe
IBM 3083
4.2 MIPS[3]
$735,000 to $1,975,000

Minisupercomputer
Convex C-1
60 MIPS
$495,000

Supercomputer
Cray-1
80 to 200 MIPS
$8 million

[3]MIPS equivalent. Actual rate: 24 nanoseconds per instruction cycle.   Note: computers not drawn to scale.

Source: Forbes Magazine, March 11, 1985, 166-167, 170

bers by thousands of numbers. Some work better with scalar processing—high-energy physics, for example, where determining particle properties must be done in sequence, not simultaneously. Some operations are done with a mix of the two.

To select the right computer, therefore, you must determine the kinds of problems you intend to solve. That variability is what has, until now, been the major market for companies such as Floating Point Systems in Beaverton, Ore. Since 1981 Floating Point has offered processors that are not free-standing like Convex', but attach to mainframes and superminicomputers. Floating Point's 32- and 38-bit machines are known as array processors, because they can handle two- or three-dimensional arrays of numbers, as well as vectors.

Floating Point bets supermini users won't want to defect to Cray machines when they can simply attach a processor to their existing equipment. The company's most recent product, the $640,000 FPS-264, can make an IBM mainframe or a DEC minicomputer about one-third as powerful as a Cray-1. But the speed of the 32- and 38-bit machines may suffer somewhat from the travel time of data between the two processors; hence the price/performance ratio is slightly lower than that of the SCS and Convex machines.

That doesn't faze Floating Point President Lloyd Turner, who says, "People buy supercomputers for two reasons: to get the price/performance they want, or for the total dollar amount that it costs to do the computing. Even though a Cray will give them better price/performance, most people simply can't afford it. But there are lots of applications that can be done on a 264, if you have enough time."

That's the catch. If you have enough time. Minisupercomputers are still slower than supercomputers, so scientists are likely to continue to wish for Crays. But in the meantime, the gap has narrowed considerably. Says Peter Gregory, an SCS director and former Cray vice president: "To get some idea of the scale, consider that if a minisupercomputer is even 60 times faster than a supermini, you can do in one minute what used to take one hour to do. The difference is electrifying."

More important, the difference means flexibility in applications. Gregory sees a movement away from experimentation and toward computation in the laboratory. Aerodynamicists, for example, use wind tunnels to study the effects of air currents on airplanes. With a supercomputer, says Gregory, the job is done better in several ways: "In a wind tunnel, you must use a cut-down model of a plane. With a computer, researchers can simulate events and study the results in detail not available in the wind tunnel.

"Besides," he points out, "you can change the experiment more easily in a computer, and it's cheaper."

If Gregory is correct, minisupercomputers may be a step in the right direction. "Lots of people now can't afford access to a supercomputer," he says. "The transition to computational methods in research will only come about when researchers have the opportunity to do real-world problems on machines."

Micro, mini and mainframe basics

Each computer "generation" is explored in terms of application and memory, interface and software requirements.

Peter Masucci,
Digital Equipment Corp.

Traditionally, most instruments and their associated controls have been based on analog techniques. In fact, analog measuring techniques are so tried and proven that they often are taken for granted as the only method to use.

Actually, there are many cases where digital logic is superior. Digital techniques have been used for years in computers and associated equipment. However, until the early 1960s, using computers directly with instruments or for control functions was an expensive proposition. Then, with the introduction of the minicomputer, it became practical to hook computers up to instruments in the laboratory, or connect them to a number of control loops. And now, with microcomputers, a computer can be buried inside instruments or placed in control loops.

Each new generation of computer equipment has meant many new applications. And as computers were applied, many interfaces, peripherals and auxiliary equipment were developed so computers could be used more easily. So to discuss computer basics, we can start with computers (processors with memory), but we must also deal with the other aspects of computer technology needed to put a digital processor in a system.

Micro, mini or mainframe

Every computer has a processor, memory and some type of i/o. From there, computers can be classified as microcomputers, minicomputers or mainframes. Although each class overlaps the other, here are the "generalized" boundaries. A microcomputer contains a microprocessor, usually located on one LSI (large scale integrated) chip or a small LSI chip set. The processor's word length is between 4 and 8 bits.

Most microcomputers have at least 256 words of memory. A minicomputer's word length is between 12 and 16 bits. And its minimum memory size is 4096 (4K) words. A mainframe, or so called "large-scale" computer, usually has a word length of 32 bits or more and a minimum memory capacity of 16,384 words.

A microprocessor or microcomputer at the sensor level can increase system accuracy and the ease of transmitting signals. Whereas analog signals may suffer from variations in amplification or other losses when transmitted over long distances, digital data is transmitted in binary (1's and 0's), which is less subject to error. A simple microprocessor or even simpler LSI circuit in a digital converter can gather data with minimum system loss and can also preprocess some of the data before sending it on in binary form. Nevertheless, the microprocessor or microcomputer is dedicated to a specific task or group of tasks within an application.

In an industrial control system, for example, a microcomputer can sample and evaluate the temperature and pH balance of a solution, compare the results with acceptable limits, and make necessary adjustments, such as increasing the temperature or adding replenisher. Because the microcomputer is in control, it's possible to relate temperature and pH together rather than separately, which yields more efficient control.

A microcomputer or a minicomputer can also tie together a number of instruments. In this case, the computer can act as a data logger, preprocessor or "a smart front end" to a larger computer.

Whereas a large mainframe may have been used a few years ago for more complex control or measurement applications, today's minicomputer usually fills the bill. Directly

Microcomputer

Minicomputer

Mainframe Computer

Today, three sizes of computers are used in control systems: microcomputers (top), minicomputers (center), and mainframe computer systems (bottom). Depending on individual control requirements, any or all of these computers may be used. The trend today is toward systems using distributed control whereby several computers are distributed throughout the system and interconnected to one another.

loading a large-scale computer down with many instruments or controllers is a waste of its resources. More commonly, a large-scale computer is used for after-the-fact analysis, or is connected to a smaller computer. This is in turn, connected either to a process or instrumentation. In this network fashion, a large computer either monitors the small computer's activities, or it can back up the smaller devices. Along the oil pipeline in Canada, for example, a large DEC system 10 keeps track of PDP-8 minicomputers that monitor and control pipeline pumping stations.

A processor is just the beginning

Using any size computer in an application involves other devices along with the central processor. A computing/control system also includes memory, interfaces to attach the computer to the application, programs or "software" to make the system run and devices to store or communicate data (peripherals). This article will cover memory, interfaces and software. A future article will deal with the subject of peripherals.

Memory stores instructions and data

In any computer, the region which stores program instructions and data for instant use is the memory. Memories can be either read-only or read/write. If the data or instructions in a computer's memory do not need to be altered, read-only memories (ROM's) are used since the computer can't alter their contents. Read/write memories, on the other hand, can be altered and accessed by the computer. Read/write memories are also known as random access memories (RAM's) since any data location within memory can be accessed as easily as any other.

Memories can also be classified according to the way they are constructed—either with ferrite core or semiconductors. Technically, core memory is a matrix whose elements are composed of tiny toroids of ferrite materials. These elements are magnetized to store information. Semiconductor memory is a memory matrix composed of tiny semiconductor circuits.

All core and semiconductor ROM memories retain data whether power is off or on. Such is not the case with semiconductor RAM's. If they lose power, you lose memory. RAM can be protected by connecting batteries to the memory package for use in the event of a power failure.

Semiconductor memory is usually faster, more compact and less expensive than core memory. So in applications where volatility is not a key factor, or where battery backup can be used, semiconductor memory is more often used. Some computers use a combination of semiconductor and core memory.

Which type of memory is used with a computer depends in part on the application. For instance, a microcomputer-based controller with a limited number of functions would probably incorporate a ROM as the main part of memory and a small RAM to temporarily store data.

Interfaces connect the computer to the outside world

Computers and their memory can not yet connect directly to analog transducers or other analog devices. An interface is needed to connect the worlds of digital and analog together. Some standard interfaces are available, although even with them, users often must add some special hardware and software.

In control applications, computers must connect either to a binary-type signal (i.e., open or closed relay), an analog signal, or in some cases a digital signal. Because binary signals are a simple form of digital signal, sometimes the connection between the digital computer and the binary circuit involves only matching voltage levels. If an instrument or valve has a digital output, connecting it to the computer again involves only voltage or current matching.

If the control has an analog output, which is usually the case, the analog signal from the sensor to the computer must be converted to digital form. The interface, in this case, is an analog-to-digital (A/D) converter placed between the computer and the analog control device.

If the computer must send an analog signal to a control loop, then a digital-to-analog (D/A) converter acts as the interface. The D/A converter constructs an analog signal from the digital information from the computer. Both A/D and D/A interfaces can be found with even very small processors, since they are usually the only communication links possible between processors and analog control devices.

Connecting lab or test instruments to a computer or terminal has been made simpler with the development of a standard interface bus—the IEEE 488. The IEEE 488 standard is a bus that connects numerous instruments. To connect to the bus, both computers and instruments must be designed with the bus connection. Digital's IBV11-A, for example, connects the LSI-11 microcomputer to the IEEE-488 bus. On the instrument side, as many as 15 multimeters, signal generators and counters can be attached.

Other interfaces connect storage devices to the computer. In most cases, these are controllers (disk, tape or printer) supplied by the peripheral or an independent vendor. Terminals usually use a current loop (or TTY) interface to tie into the computer.

Another interface links computers, terminals or both over standard telephone lines. These devices, known as modems (modulator/demodulator), convert the digital information to audio signals that can be transmitted over telephone lines. A modem on the other end converts the signals back.

Software tells the computer what to do

Without software, the computer, no matter how complex, will not work. Software gives the computer its flexibility and also many of its problems. The problems are not necessarily due to the software, but to human error. These human errors result because software is usually not written in English, but in a special language that the computer can translate into the binary language it recognizes. The software language written by the user can range from "machine-like" to "English-like." Machine-like languages are called assembly or low-level languages, and English-like languages are called high-level or compiled languages.

An assembly language is a collection of mnemonics that refers to the exact functions a computer must perform. In PDP-8 assembly language, for example, CLA stands for "clear the accumulator," which is far easier to work with than its binary equivalent of 111110000000.

Each processor has a unique assembly language and assembler that translates the language into binary code. The assembler can be stored on paper tape in low priced systems or on tape or disk in higher priced systems. After the assembler is read into the computer, the user submits his assembly language application program, which the assembler then translates into binary code.

Assembly language programming is efficient as far as using memory. But it's not particularly efficient when it comes to programming time. With assembly language, programmers must write an instruction for every single operation of the computer.

Assembly languages are used mostly with microcomputers because it conserves the micro's memory. Memory can be the largest hardware expense of a microcomputer system.

While assembly languages are organized to match the way computers operate, high-level languages are organized more along the lines of a spoken language. High-level programs do not need to specify every single operation the computer performs; instead, they must only indicate what function to perform (e.g., add two fields).

Just as assembly languages require a special program to convert user code into binary code, high-level languages require one to convert the high-level program into binary. The special program, called either a compiler or interpreter, occupies a significant portion of memory. This increases the size of memory required. More memory is also needed because high-level languages are not optimized for one specific machine. Hence, they are not as efficient in their translation as the assembler.

Although high-level languages are not designed for specific machines, they are designed for specific functions. Of the popular languages, BASIC and FORTRAN were developed for math and scientific problem solving. FORTRAN, particularly FORTRAN IV, tends to have higher arithmetic precision than an equivalent BASIC, but BASIC is easier to learn.

High-level languages are usually associated with minicomputers and larger computers because they require more memory, and also because they are only starting to be available for microcomputers. Minis and larger computers also have standard packaged "operating systems" written in assembly language or high-level language. By using an operating system, users need to be concerned only with the application.

A system solution

Computers can be used nearly everywhere. But for a computer to provide the advantages of accuracy, flexibility and reliability, users must not only choose the right processor for the job, but must also look at available software, memory requirements, storage, peripherals, and of course, interfaces.

The important thing to keep in mind is that with the diversity of processor types, peripherals and software, it is possible to apply computer operations effectively to almost any control or instrumentation application. ∎

SUPERMINIS SHAPE UP AS DEPARTMENTAL CPUs

Functional enhancements and competitive price/performance characteristics lead major minicomputer vendors into battle

Lynn Haber, Associate Editor

Distributed data processing, an idea of the 1970s, never materialized, primarily because good networking did not exist. But, today, with the de facto standardization of Ethernet and the availability of superminicomputer hardware and software at price/performance levels that are sometimes more cost-effective than microcomputer-to-mainframe links, distributed data processing is a reality.

To witness this change, note the vociferous marketing efforts and flurry of product announcements by major minicomputer vendors, such as Digital Equipment Corp., Data General Corp., Wang Laboratories Inc., Prime Computer Inc. and Hewlett-Packard Co. These companies offer integrated office solutions based on the superminicomputer as a departmental CPU. Even IBM Corp., the granddaddy of centralized data processing, is touting the System/36 as the cornerstone of its departmental strategy.

As the departmental CPU, some of the functions that the superminicomputer must perform include database management and control, networking and communications, workstation support, report generation and more advanced services such as document image capture and processing.

Combined with a proliferation of desktop machines, an increase in the amount of data produced and a growing need to share that information with co-workers, minicomputer companies are looking to satisfy a bottom-to-top drive toward office integration.

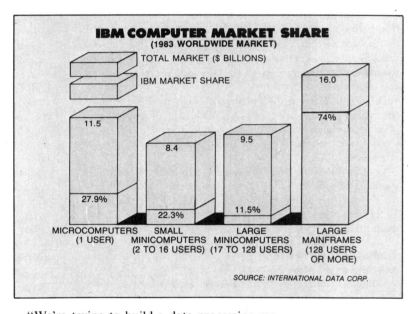

IBM COMPUTER MARKET SHARE
(1983 WORLDWIDE MARKET)

TOTAL MARKET ($ BILLIONS)
IBM MARKET SHARE

MICROCOMPUTERS (1 USER): 11.5, 27.9%
SMALL MINICOMPUTERS (2 TO 16 USERS): 8.4, 22.3%
LARGE MINICOMPUTERS (17 TO 128 USERS): 9.5, 11.5%
LARGE MAINFRAMES (128 USERS OR MORE): 16.0, 74%

SOURCE: INTERNATIONAL DATA CORP.

"We're trying to build a data-processing machine that meets the needs of users and connects to every terminal and processor on the desktop," says John Thibault, vice president of systems product management at Wang.

Wang promotes what it calls Integrated Information Processing, a multipart strategy to integrate data processing, communications and office automation via its VS line of minicomputers. They range from the low-end VS/15, with 256K to 2M bytes of main memory, to the high-end VS/300 with 4M to 16M bytes of main memory. Prices for base models of the VS line range from $13,500 to $170,000.

Reprinted with permission from Mini-Micro Systems, April 1985, 109-110, 113-114, © 1985 Cahners Publishing Co.

"The function of the superminicomputer at the departmental level is to support personal computers in one of a number of activities, providing power, flexibility and software sup-

Superminicomputers are becoming *increasingly popular as the "middlemen" in distributed data networks.*

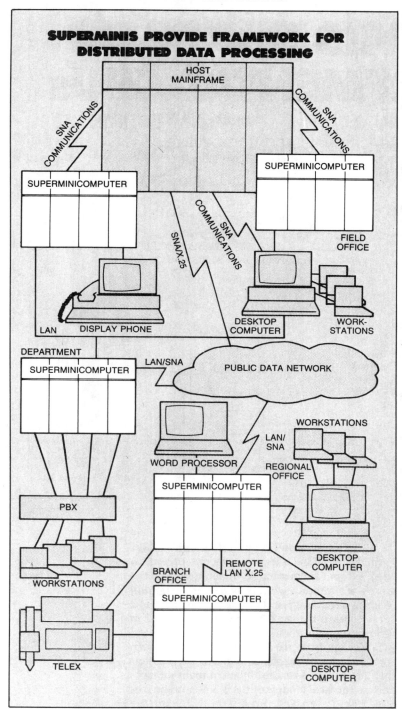

SUPERMINIS PROVIDE FRAMEWORK FOR DISTRIBUTED DATA PROCESSING

port," says William C. Rosser, vice president of the Gartner Group Inc., a research and consulting company in Stamford, Conn.

As the processing power of the superminicomputer matures, evidenced by recent product introductions such as DEC's VAX/8600—reportedly capable of processing 4.4 million instructions per second (MIPS)—an evolutionary change in commercial data-processing environments results.

A three-tier office environment

In a three-tier business structure, the low end is occupied by the personal computer, the middle tier, by the superminicomputer and the upper level, by the mainframe, or host, computer.

Compared to some of the first attempts to satisfy the user's need to process and share information, which included micro-to-mainframe links and microcomputer-based clusters, the superminicomputer as the departmental CPU can be a better dollar-per-job-provided solution. According to Rosser, the cost per MIPS for superminicomputers ranges from $70,000 to $120,000, compared with mainframe computing (based on an IBM 3080-family machine) at $210,000 per MIPS.

Major superminicomputer manufacturers have identified and are going after the middle tier. According to market research company International Data Corp., Framingham, Mass., IBM, which claims 74 percent of the mainframe market, 27.9 percent of the personal computer market, 11.5 percent of the large minicomputer market and 22.3 percent of the small minicomputer market, has largely ignored data processing at the departmental level. What many of these minicomputer companies offer that IBM currently does not, is a smooth migration path across their product lines.

"This is a big area of opportunity," says Rosser. According to the industry consultant, if the number of personal computers reaches 50 percent of the white-collar work force, as predicted, and the three-tier approach to integration proves to be a good solution for data processing, then the number of superminicomputers shipped will escalate. "These companies recognize that they've got to get out there, even if the personal computer at the bottom is IBM and the host at the top is IBM," he explains.

According to market research company Dataquest Inc., San Jose, Calif., the number of superminicomputers shipped in 1989 is expected to exceed 199,000 units, compared with 15,500 in 1983. The dollar value for superminicomputer shipments is expected to exceed $19 billion by

1989, compared with $3.65 billion in 1983.

Recognizing the inevitability of a multivendor environment, minicomputer manufacturers are committing to product compatibility for IBM 3270 emulation (between mainframe, minicomputers and desktops), X.25 (to the public-switched networks), Ethernet protocols, the Navy Document Exchange (between multivendor desktops), IBM's Systems Network Architecture (allowing access to an IBM host), the Document Interchange Architecture/Document Content Architecture (DIA/DCA) and the UNIX operating system.

Many manufacturers will purchase from independent vendors, products and services that they cannot provide themselves. "Our philosophy on joint ventures and company alliances," says a Prime spokesman, "is that where we can add true benefit and have product uniqueness, we'll develop products internally, but it makes no sense to reinvent the wheel."

As products get more complex, development cycles become longer and product life cyles become shorter, manufacturers' research and development allocations must be wisely calculated. "The nature of the computer business is such that you see decreased product life, increased capital investment and decreased profit margins," says Brad Smith, associate director of research for the computer industry service at Dataquest. "The expectations of the past are no longer applicable."

Integration is important

Currently, IBM's biggest disadvantage in approaching the middle tier is that the company offers multiple products that are incompatible (e.g., the Series 1, System/36, System/38, the 4300 and the 8100). In contrast, companies like DEC, DG and Wang offer users a single, consistent architecture that covers a wide performance range. A consistent operating system and user interface across products allows departments to swap machines or upgrade when necessary without the burden and expense of having to regenerate software.

"The user wants to have a machine that will be the appropriate investment to meet the requirements of different-sized departments,"says Gartner's Rosser. "If you run out of gas using the IBM System/36, what do you do beyond that? Multiple [System/] 36s are an awkward approach to solving the departmental problem."

IBM does have powerful systems within discrete groups and offers good upgrades for systems within a family. A sign of the company's marketing strategy became evident last October when it announced Displaywrite 1, 2 and 3 for

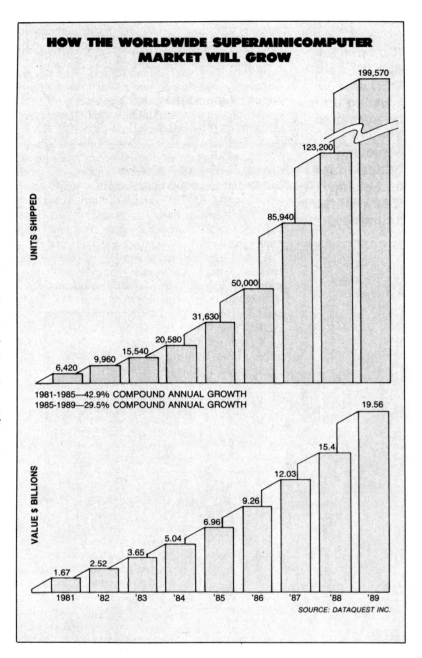

HOW THE WORLDWIDE SUPERMINICOMPUTER MARKET WILL GROW

UNITS SHIPPED

199,570
123,200
85,940
50,000
31,630
20,580
15,540
9,960
6,420

1981-1985—42.9% COMPOUND ANNUAL GROWTH
1985-1989—29.5% COMPOUND ANNUAL GROWTH

VALUE $ BILLIONS

19.56
15.4
12.03
9.26
6.96
5.04
3.65
2.52
1.67

1981 '82 '83 '84 '85 '86 '87 '88 '89

SOURCE: DATAQUEST INC.

the PC, Displaywrite/36 and intentions to offer Displaywrite/370—software that, according to IBM, offers complementary functions and common user interfaces to provide an integrated office-system network. IBM also offers Personal Service/PC, 36 and 370, which also provide office integration.

DEC's All-in-1 Office and Information System software offers the user a menu of applications that can be accessed from the company's terminals or personal computers. All-in-1 applications also run on VAXes, from the 11/725 to the

> **'Companies like DG and DEC are shaping up to be sure bets. They're achieving critical mass and giving IBM a run for its money.'**

recently introduced 8600. All-in-1 cannot run on the MicroVax, but DEC plans to announce compatibility sometime in the future (see "Integrated software spurs mini market," Page 95).

Although DEC has not publicly announced support for the IBM PC, third-party VT100 emulation software is available that allows users to batch data between an IBM PC and a VAX. Additionally, with DEC's recent announcement to halt production of the company's Rainbow personal computer, industry analysts suspect that minicomputer manufacturers will offer some kind of IBM PC compatibility. According to a DEC spokesman, the company will soon announce products that will allow IBM and DEC products to function together more effectively.

All-in-1 applications include word processing, electronic mail, desk management and integrated publishing. The system also allows users to develop their own programs or tailor third-party software to the DEC system.

DEC also offers the DECnet/SNA gateway. VAX systems use gateway-access software to communicate with the gateway and perform the functions of 3270 terminal emulation, program-to-program data exchange, remote job entry, file transfer and network management. DEC has also announced an interface to IBM's Distributed Office Systems Support and plans to introduce the product by midyear.

Last October, DEC announced a working agreement with Cullinet Software Inc., Westwood, Mass., for joint development of a product which would allow VAX computers to access Cullinet's Information Data Base (IDB), a micro-to-mainframe software product.

In February, DG introduced the AOS/VS Decision Link, reportedly the first software package that enables users to access data in IBM mainframes via DG's Comprehensive Electronic Office (CEO). The product resulted from an agreement betweeen DG and Cullinet and allows CEO users access to Cullinet's IDB software via an SNA communications link.

The key trends

In terms of hardware competition, Dataquest's Smith expects the next shoot-out to occur shortly, when superminicomputer manufacturers take on IBM and the "BUNCH" (Burroughs Corp., Univac Corp., NCR Corp., Control Data Corp. and Honeywell Information Systems Inc.). "Companies like DG and DEC are shaping up to be sure bets," says Smith. "They're achieving critical mass and giving IBM a run for its money." Smith refers specifically to DEC's

8600 with its clustering capability, which he believes will "take DEC into the 1990s."

Last year's minicomputer trend was UNIX compatibility. This year's buzzwords are emitter-coupled logic (ECL), gate-array circuit technology, clustering and fault-tolerance.

Though more expensive than Schottky transistor-to-transistor logic or reduced-instruction-set-computer architecture, ECL reportedly offers more reliability and faster processing. Such vendors as DEC, Prime, DG, Gould Inc. and Harris Corp. have implemented ECL circuitry in their products.

Clustering allows users to add computers incrementally in a transparent manner. This arrangement provides multiple resources and greater reliability and speed. For example, the DEC VAXcluster, introduced in April 1983, is a network of loosely coupled processors and storage systems that appears to the user as a single accessible computer. In the VAX line, the 11/750, 11/780, 11/785 and the 8600 have cluster capabilities.

Fault-tolerant systems

Although not synonymous, both cluster and fault-tolerant systems are high-priority concepts for many minicomputer manufacturers because of increased system reliability. Fault-tolerant architecture protects the computer system from failure either through hardware, redundant components or software that directs one processor to assume the work of a faulty processor. Fault-tolerant systems are most often used in applications that require on-line, real-time transaction processing, such as banking.

Traditional computer manufacturers have not built fault-tolerant machines because of incompatibility with their existing product line. In addition, uncertainty exists as to whether users can justify the cost of the fail-safe features of a fault-tolerant system.

Many long-standing minicomputer vendors are keeping a watchful eye on the fault-tolerant market and admit it is a technology that users will soon request. In February an agreement was announced that IBM would buy an unspecified number of fault-tolerant computers from Stratus Computer Inc., Marlborough, Mass.

Without public comment from IBM, the agreement has analysts predicting that the computer giant will build a machine based on the Stratus system and offer it to its large customers who are either considering—or demanding—the benefits of fault-tolerant technology. □

'Lap' Portables

HOW SMALL IS TOO SMALL?

They're small wonders, alright, but has technology transformed them into real personal computers?

by Edward Foster

REGARD THE INCREDIBLE SHRINKING COMPUTER: YOU CAN ALREADY HAVE the equivalent of a room-size mainframe of the Sixties on your desktop, but no one is stopping there. The push for true portability in personal computers goes on. What could be more personal, after all, than a computer that can go wherever you go and be used on the spot?

Ever since integrated circuits became available, it has made sense to expect that some day we would have a system that could do everything and do it anywhere. The idea has tended to attract pioneers on both sides of the fence–sellers and buyers. Expectations have occasionally been too high on the part of both, and as a result there have been some well-publicized financial debacles in the "ultimate portable" marketplace as well as a few dissatisfied users.

In spite of the early difficulties, a number of advances are making it possible for computers to go and work where they never have before. Thin liquid crystal displays (LCDs) can now be made a full 25 lines deep by 80 characters wide. Microprocessor and memory chips manufactured with the latest CMOS (Complementary Metal-Oxide Semiconductor) technology have minimal power requirements, making battery-operated computers a reality. Disks and disk drives continue to shrink, sometimes disappearing altogether in favor of circuitry built into the system.

So, smaller and smaller computers are showing up in the real world: A college student takes notes at a lecture with what looks to be no more than a keyboard; an insurance salesman pounds away at his prospect by showing him policy alternatives on a LCD screen that looks like an overgrown digital watch. Who hasn't noticed at least one typically accoutered businessman hurrying through an airport with a portable computer tucked under his arm?

So much for appearances—but how useful do the people who sport these small wonders find them to be, really? Are they novelties, good for only limited, specialized jobs? Or are they true personal computers?

At this stage in the development of what are commonly called "lap" or "lap-top" portable computers, any simple answer is debatable.

The most natural applications for battery-powered portables involve professionals who spend most of their time in the field. For example, the 30-man internal audit staff of Singer Company, the manufacturer of sewing machines and other consumer products, spends much of its time visiting different Singer facilities throughout the world. Audit teams had been sharing three Compaq transportable systems to take to the field with them, but three DG/Ones have now been added to the pool. "The Compaq is quite good if, say, you're going to a facility you can drive to, but it's too bulky for an airplane when you're going overseas," says Wolf Kasparek, general auditor for Singer.

Preparation for the field audit begins at headquarters when the audit team downloads financial data concerning the facility to be audited from one of the corporate mainframes into the portable. This information is taken into Lotus 1-2-3, with auditors eliminating

Tandy Model 100 Portable Computer

PRICE: $399

SCREEN SIZE: 8 lines by 40 characters

AVAILABLE USER MEMORY: 8k

WEIGHT: 3.9 lbs.

When IS a Computer a Portable?

In no other area have computers spawned more colorful, and more misleading, nomenclature than when it comes to describing the variety of systems professing to be portables. Lap-top, knee-top, thigh-high, hand-held, mobile, commuter, briefcase, notebook, transportable, toteable, luggable . . . all these terms have been advanced as generic terms for one class of machine or another.

Portability is much on the mind of current personal computer buyers, although a great deal of uncertainty exists as to which systems really have it. In a survey conducted by *Personal Computing* of our subscribers, just before Christmas 1984, 18 percent had already made the decision to buy a portable computer, or at least a computer which they understood to be portable. Naturally, many transportable systems such as the Compaq, Kaypro and IBM "Portables" were mentioned, as well as a number of desktop systems and home computers which offer no built-in monitor of any sort. The system most commonly mentioned as a possible portable computer buy was the Apple IIc—ironically, one for which a portable display is not yet available at the time of this writing. Of the current lap-top systems on the market, Tandy Model 100, the HP 110 and the DG/One, in that order, received the most attention.

The term "portable" itself has unfortunately been used most often, at least by manufacturers, to describe systems weighing as much as 30 pounds or more. These systems are not battery-powered and certainly are not meant to be used in anybody's lap.

Devices which have been called portable computers actually fall into three relatively distinct classes.

*The smallest, sometimes referred to as notebook or hand-held computers, generally weigh less than a pound, have a one-line display from 12 to 40 characters long, a miniaturized keyboard, and from 4k to 16k memory.

*The largest, called transportable, luggable or suitcase computers, weigh in from about 20 pounds on up, possess a full-fledged keyboard and a somewhat down-sized but full

24-line by 80-column CRT display, from 64k to 256k or more as well as built-in 5.25" floppy and/or hard disk drives. Neither of these two categories need really concern us here, the first because they have more in common with programmable calculators than computers, the later because they are actual desktop systems which offer only as a secondary feature the ability to be hauled from place to place, as long as the places aren't too far apart.

*The middle category has had almost more terms coined to describe it than actual sales of units: lap-top, knee-top, and briefcase are some of the more common ones. Minimum and maximum specifications for these systems overlap the other two categories, but in general they fall into a range of 4 to 12 pounds in weight, an LCD or other flat-panel display offering 40 or 80 columns and 8, 16 or 24 lines, a typewriter-like keyboard with reasonably wide keys but few dedicated to special functions, and abbreviated memory capabilities which at least emulate the capacities of full-fledged personal computers. Some but not all have built-in disk drives, with the 3.5" microfloppy on its way to becoming the standard. Others employ tape cartridges, bubble-memories or built-in ROM cartridges for data storage and/or software delivery. Most offer a built-in serial port or modem for uploading and downloading information with a host computer and driving non-portable peripherals like a full CRT display, printers or 5.25" disk drives.

What characterizes this class of computer is a common aim: to offer computing performance which is at least comparable to desktops in a package which can be used practically anywhere. To this end, most have a rechargeable battery built-in or a battery option. Thus the term lap-top or lap portable may convey the idea the best, but it seems likely that eventually this class of machine will take over the word "portable" for itself. Unless otherwise noted, "portable" used here refers to the lap-top variety of system and not the transportable or notebook machines.

—E.F.

columns and rows they will not need for their analysis. "We use it just like a paper accounting pad for totaling cash, inventory and receivables, and then doing our calculations," says Kasparek. The team then uses Multi-Mate on the DG/One to generate a first draft of their audit report while still in the field so that they can review their findings with the facility managers.

Mitchell Johnson, a producer of television films based in Ft. Worth, Texas, who goes to Europe frequently on business, has also found a portable computer to be a handy item for traveling. He uses a Radio Shack Model 100 while on the road, particularly for keeping up with electronic mail from his staff back in Texas and to write letters. "I can write a letter, get in touch with my office via the built-in modem and have the text transferred directly into the word processing program we use on our IBM XT," he says. "I'll even use the portable at the office for dashing off some correspondence and just hand it to my secretary to transfer."

Johnson remembers one incident in particular where the Model 100 proved invaluable. "I was in Europe on a sales trip and unexpectedly discovered that we might have an opportunity to sell one of our films to Austrian television," he recalls. Johnson wrote out a proposed contract for the deal while on the train from Berlin to Vienna, and was lucky enough to find a friend with a printer he could use to produce a draft of the contract. Negotiating on the basis of that draft, he was able to make the agreed-upon revision in the contract in a matter of hours and close the deal. "We probably would have gotten the deal eventually in any case," he believes, "but being able to present them with the contract so fast while they were in the mood to make the deal allowed us to come to an agreement in a matter of days instead of months."

Another TV executive, Chan Mahon, an associate producer with Doris Keating Productions in Burbank, Calif., believes the newest portables have great capability. Mahon, an experienced desktop personal computer user, recently bought a Data General/One portable for his office. "We have it set up to do our word processing, for handling the daily typing chores and putting in scripts to edit," Mahon says.

Like some other users' portables, Mahon's functions like a desktop much of the time. "It does stay in the office a lot," he admits, "but it's nice that if I have to, I can unplug the adapter, take it home or up to the mountains, and set it up to work. We'll be going to Singapore in a few weeks to shoot a TV movie, for example, and the first thing we'll have to do is preproduction tasks like scouting locations."

OK, so if you're a movie producer, the portable allows you to take notes at possible shooting locations and later to do a scene-by-scene breakdown in the data base as to the cast members, time-of-day, props, etc., needed for each shooting. Somewhat esoteric? Yes, but it's becoming common in mainstream businesses to see the portable being used in the field as a secondary system, sharing data with the desktop or even the mainframe back at the office, as the Singer audit team does.

Some individuals who take their portables into "the field" do so quite literally. Software programs written specifically for agricultural applications take on a new dimension when used on a system that can go to work for a farmer.

"You can pick up the computer, note a problem you've got with soil erosion in one part of your field, then track another region that has a pest infestation, and later take it home to do crop modeling," explains Richard Eittreim, president of Program Design Corp. (Fairfield, Calif.), a software company which recently began selling a version of its agricultural software for the Epson Geneva portable.

Data General/One

PRICE: $2895

SCREEN SIZE: 25 lines by 80 columns

AVAILABLE USER MEMORY: 128k

WEIGHT: 9.1 lbs. without battery, 10+ lbs with battery

SPECIAL FEATURES: IBM PC compatible, AC adaptor or battery, MS-DOS or CP/M-86, 1 double-density 3.5" disk drive, 2 double-density 3.5" disks of 720k each.

Grid Compass II Model 1139

PRICE: $7995

SCREEN SIZE: 25 lines by 128 characters

AVAILABLE USER MEMORY: 512k, 384k bubble, 512k user configured ROM

WEIGHT: 10 lbs.

SPECIAL FEATURES: built-in 300/1200-baud modem, electroluminescent screen with wide viewing angle, magnesium case.

It offers a farmer the ability to "map" his fields while he's in them; he can categorize problems he wants to track–such as erosion, nematodes, cutworms, soil deficiencies, and so on–and assign a number to each. As he's moving through a field he can pinpoint each problem he encounters simply by entering its number equivalent for the proper coordinates. This gives him a continuing record of problems he's had in each plot, which can help in his crop management decisions.

The day has not yet dawned where no door-to-door salesman will ring your doorbell without a computer in hand, but it might not be far off. Portable systems designed for the sales force in a particular industry are attractive to potential users. Mary Lou Williams of the San Fernando Valley Board of Realtors in Los Angeles pondered what portable computers could do for realtors in her area, and helped her colleagues develop a system. Now, agents visit a potential home buyer with an HP 110 in hand. After talking to the buyer and discovering what his or her needs are, they use the system to call up the Board of Realtors' mainframe to get the pertinent data on homes which meet the customers' requirements.

If the buyer is interested in pursuing matters on a particular home, the broker uses the Memomaker word processor which is built into the HP 110 to fill in a template for a letter printed out on the HP Thinkjet printer, which realtors can also bring with them.

"Multiple listing information is copyrighted, so the broker can't just give the customer a copy of the listing for the home," explains Williams. "The printer lets them print out a letter so that they can leave the customer a hard copy with details about the house."

It seems that a growing army of users of portables is still discovering the range and limits of this equipment. "A lot of things that wouldn't get done otherwise do get done when you can carry your computer with you," asserts Roger Parker, a Seattle, Wash., advertising consultant. Parker uses an HP 110 portable when he calls on clients to help plan advertising campaigns and write ad copy.

"Using Lotus 1-2-3, we can put together a media plan, removing or adding ads quickly to see the differences," he explains. He uses WordStar for making revisions in ad copy at the client's office, and can print out drafts on a Thinkjet printer. "Not only are the drafts more immediately available this way," notes Parker, "but the client is more involved in the process, which saves additional rewrites."

While it's obvious that the new portables can be put to a wide variety of uses, some users prefer to dedicate them to a specialized task. William Bartel, an attorney in Fair Lawn, N.J., has adopted a Radio Shack TRS-80 Model 100 for a time-and-billing application using the Time Manager software program from Traveling Software. This could be really bad news for heavy users of legal services–since it keeps the "meter" running, constantly.

Bartel keeps the Model 100 with him by the phone or during consultations to record the time he spends. "With just a few keystrokes I can make a record of who I talked to and how long we talked, and add a short comment on what the subject was," he says. "Later on I can print out a complete report to submit to the

clients as an invoice, and they can see exactly what they're getting charged for and how much time was involved with each step."

All these new and enthusiastic users make it very clear that portables are already serving some very productive ends, whether as a mobile desktop, general in-the-field application tool or specialized auxiliary machine. But before we consign all our old-fashioned, deskbound systems to the scrap heap, we should first examine just where the limits of technology stand now, and how far they might yet go.

The ideal portable in many people's eyes would be a system which duplicates the performance of, and is compatible with, currently popular personal computers. If price and performance were equal, it seems likely that everyone would choose a true portable over a desktop system. So far, at least, the technology of small portables has required some significant compromises with this goal.

Display technology has doubtless been the most significant factor in separating the portables from the desktops. Chan Mahon says that the LCD display on his DG/One is "its only drawback . . . if you don't have proper lighting, it's difficult to read." And one of the reasons that Roger Parker always lugs the printer around with his HP 110 is that he doesn't feel comfortable asking clients to try to read that computer's 16-line screen.

CRT displays have too much weight and bulk to be appropriate for true portables, so the only answer has been to find a "flat-panel technology." Most developers have chosen liquid crystal display as the practical solution. LCD displays have been growing in size at a rapid rate from a few years ago when a one-line readout only a few characters long was just about the maximum to be had. Its use has progressed from the eight-line, 40-column display common to the Radio Shack Model 100, Docutel/Olivetti M-10 and NEC PC-8201A, through the 16-line, 80-column display on the HP 110 and Morrow Pivot, to the 24- or 25-line displays introduced with the recent Data General/One and Texas Instruments' ProLite.

Since 24 lines by 80 columns is the standard display on most personal computers, it would seem at first glance that the portable display problem has been licked. A closer look–or a farther look, or even a glance from a few degrees to the side–reveals that this is not so.

LCD displays do not have anywhere near the contrast of a CRT screen; most require a strong, direct external light source to be usable. The larger the LCD display, the more likely it is to suffer from inherent lack of contrast. This has led some companies, such as Hewlett-Packard and Zenith, to go with 16-line displays as the best alternative until the technology improves. "If you can't read them, 25 lines aren't particularly worthwhile," says Corey Staton-Smith, product manager at Hewlett-Packard. One problem is that full-screen LCD displays can look excellent in a store with the lighting just right, but a fundamental requirement of portable computing is that the system be usable where the light conditions are not controllable.

Still, others feel that full 25-line displays are obligatory. "Twenty-five lines are a must; anyone who doesn't have it won't be taken seriously," says David Kay, vice-president of marketing for Kaypro, a company that does not yet offer a lap-top portable but is known to be close to introducing one after considering several designs. Kay is not convinced that the distinction in quality between different-sized displays is all that great. "I've heard people complain about HP's 16-liner, and I've been pretty impressed with the full screen on the ProLite," he says.

While 16 versus 24 lines is a very real issue to someone buying a portable right now, the pace of change in display technology is sure to present buyers with different choices in the not-too-distant future. Not only are significant improvements expected in LCD displays, but other flat-panel technologies may become viable alternatives. New technologies are likely, however, to bring their own sets of problems with them.

The one portable computer manufacturer not currently using an LCD display, Grid Systems, employs electroluminescent technology for its Compass II line. The display itself is vastly superior in readability to any LCD screen, but it helps jack the price of the system up to more than $7,000 for the top-of-the-line model. In addition, the EL display draws more current, making battery-powered operations dependent on an optional battery pack which detracts from the system's portability and still requires frequent recharging.

The importance of being able to run a portable computer on batteries is another area where manufacturers have different philosophies, but the results are a little less obvious to the user than with displays. By definition a truly portable system doesn't need a wall socket, so the main difference between portable models is whether the battery comes as an integrated part of the system or as an extra piece of equipment that must be attached to the unit or carried around separately. The Grid Compass II and the TI Prolite are examples of systems which require an optional battery pack, while the Model 100 and 200 from Tandy and the Hewlett-Packard HP 110 have integrated batteries. Rechargeable battery life must also be considered, particularly for those users who know their systems will be getting a full day's workout without an opportunity to recharge. The Grid system can run off its external battery pack for about two hours before recharging, while the Tandy and HP systems can run 12 to 18 hours without a recharge.

"In our portable terminal business we found that many people don't really require the battery option," says Faris Gaffney, manager of product marketing for Texas Instruments, which used lessons from its portable terminal business in designing the Prolite portable. Alan Lefkof, vice-president of marketing for Grid Systems, agrees with Gaffney. The ability to actually use a system while it sits on your lap, he says, "is more advertising hype than real consumer demand. The real reason to buy this kind of machine is the ability to carry it easily and unobtrusively."

Chan Mahon and lawyer William Bartel might disagree: Both have leaned heavily on battery use. Bartel occasionally uses his Model 100 in court to keep track of detailed legal points in complex financial cases. Roger Parker uses his HP 110 on battery power most of the time because, as an advertising man, he knows the value of image.

"When you walk in carrying a system like this, you're saying something positive about your credibility and professionalism," he

Displays At A Glance

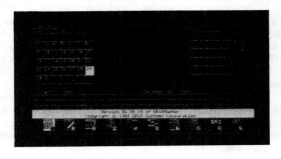

Laptop liquid crystal display formats vary: Data General (left) offers a full 24 lines. Zenith's Pal (below) is 16 lines deep. The more costly electroluminescent screen on the Grid (bottom) is the only one usable in marginal lighting.

comments. "If you have to spend the first few moments with your rear in the air looking for a place to plug your machine in, you lose something."

The area where there is the greatest divergence of approach from one portable computer vendor to the next is memory storage–still another critical factor because of its relationship to software. While microcassettes or bubble memories are available on a number of systems, the major schools of thought fall into two basic camps: those who feel that a portable must have disk drives like a personal computer, even if they are not of the 5.25″ standard, and those who see the wave of the future in RAM disks and software (firmware) delivered on ROM chips. The DG/One is the trendsetter for the disk-drive advocates with one 3.5″ drive standard and a second integrated drive available as an option, while the new Zenith Satellite emphasizes the solid-state memory approach.

Building an increasing amount of software into personal computer systems is a general

Source: Personal Computing, March 1985, 83, 85, 87-89, 91, 93, 95, 97

trend, the most obvious example being the Apple Macintosh with its entire operating system loaded on board in ROM. In portable computers, however, that idea has been taken a step further by actually delivering applications programs in ROM, sometimes in the form of interchangeable cartridges. The Radio Shack Model 100 has long had BASIC, communications, a text editor and other utilities built in. The HP 110 includes an in-ROM version of Lotus 1-2-3, and the Zenith system offers an integrated set of Microsoft application programs.

Andrew Czernek, director of marketing for Zenith Data Systems, believes that the trend is toward software built into the machine. "The market is moving toward a machine that is lighter, more compact but with professional functions, and in our view that means putting software in firmware," he says. "People don't want to have to carry removable media with them, and they want non-volatile RAM."

HP's Staton-Smith goes even further. "Having your software in ROM is at the essence of computer portability," he asserts. "The reliability issues associated with the physical integrity of drives in the conditions portables will be exposed to drove us to our decision (not to build a disk drive into the HP 110)." Staton-Smith also points to the speed advantage that ROM-based programs have over loading from a disk.

An additional advantage with some systems such as the Grid and Zenith portables is the ability to execute programs directly from ROM without loading the program into RAM, thereby freeing additional memory space for the user.

But for all of the technical advantages of ROM-based software, there are those who feel strongly that portable users will prefer to have disk drives built into the system. "We don't want to limit our customers; the broadest range of software is what they need," says Alan Oppenheimer, director of dealer marketing for Data General. "The market changes so fast and there's always new stuff coming out that people are going to want." He emphasizes the importance of compatibility with the IBM Personal Computer world, a somewhat tricky proposition in that there is no IBM standard at the moment for the 3.5″ microfloppy drives that the DG/One and other portable systems employ. (The DG/One is capable of using MS-DOS based software programs from standard 5.25″ disks in an external disk drive.)

The most common argument in favor of disk drives is that they simply make data storage easier and more independent of other systems. A RAM disk can only store data so long before it has to be saved to a more permanent media such as an external disk drive or another computer. Systems like the HP 110 that rely on RAM disks do, of course, provide such options, but saving your data to a disk that is part of your portable is clearly an easier process.

If software compatibility range is your final arbiter, diskettes still have a clear edge over ROM-based software, which is almost inevitably machine-specific. At this point, however, all lap–portable software has to be rewritten for each system. While the 3.5″ disk media supported by Data General, Texas Instruments, Hewlett-Packard and others seems to be evolving as a standard for portables, that does not yet mean that programs or files can be shared between the systems, even though all the above use an MS-DOS operating system. There is as yet no standard for software developers to write for, although porting a program for one system to the next is not an overwhelming task. When will a standard disk operating system appear for the portables? The best bet is that it will happen about the same time as when IBM introduces a lap portable with 3.5″ disk drives, an occurrence which many industry observers anticipate in the near future.

Zenith PAL (Portable Advanced Laptop)

"Prototype"

SCREEN SIZE: 16 lines by 80 characters

AVAILABLE USER MEMORY: 32k

WEIGHT: 7.7 lbs. (3.47 kg)

SPECIAL FEATURES: Built-in 300-baud modem, software in ROM, clock/calendar.

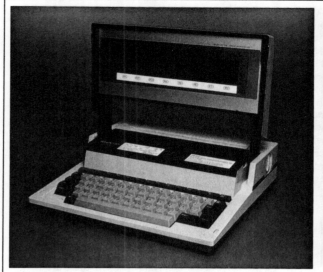

One final tradeoff manufacturer and consumer alike must ponder is the most fundamental of all: price. At the low end of the market, machines like the Model 100, currently selling for under $400, are inexpensive enough that users will overlook capabilities it lacks. "There are so many more things that can get done in an office when you can pass out a machine like this to anyone who needs it," says attorney Bartel.

"One major company has over 2000 Model 100s . . . they can almost look at them as a throwaway because of its cost," says Stewart Weinstock of Tandy Corp. "It helps get people computerized to a degree who might not otherwise be able to use one, so they can at least use it for one function like electronic mail or getting mainframe data."

At the other end of the spectrum, manufacturers frankly admit that they have to charge a premium for portability. "You can say that the systems are too expensive, but compared to what?" argues Grid's Lefkof. "It is expensive compared to an IBM Personal Computer, but when you are getting performance that is equal," Lefkof says, "to that of a Personal Computer with the electronics packed into a much smaller package, it's a lot easier to justify paying a premium for portability."

Should we be willing to pay a premium for portability? "I don't think portables have to be competitive with desktops on price," says Tom Billadeau, who tracks portable computers for the Gartner Group (Cambridge, Mass.), a market research firm. "People expect the price relationship of portables to desktops to be the same as that of the IBM Personal Computer to the Compaq, and I don't think that's a reasonable analogy."

Quite a few others are convinced that portables will not be generally accepted until they do not require a premium. "Why should I buy this little dinky thing, without real disk drives,

when I could spend the same money and get a Kaypro, or a Compaq, or whatever?" argues Kay. "As long as we're stuck with less-than-CRT display quality, the market is going to be looking for a lap-top that isn't much more expensive than a full-size computer."

Everyone agrees, however, that there are certain types of people for whom portables, even at premium prices, are a natural. Yet underneath this consensus is a deep philosophical schism between those who see the portable as still a personal computer, perhaps the personal computer of the future, and those who feel it represents a new genre of computer.

Those who see portables as the dominant type of personal computer in the future feel that they represent the natural culmination of the personal computer's evolution. "It's the direction that computers have been going in for 30 years, and there's no reason to think it will stop," says Peter Teige, a market analyst for Dataquest in San Jose, California. Teige feels that ultimately the true portable will also be more truly personal than today's desktops. "It's not a piece of office equipment: When you can carry it around with you it becomes a personal possession like your briefcase."

Not all share Teige's view of the portable as the future personal computer. Billadeau sees the portable's primary role as the field component in corporate office automation systems. HP's Corey Staton-Smith believes that the portable represents a new market category, one which requires a different philosophy than imitating the popular personal computers in a smaller box. "What consumers think they want right now is essentially a shrunk-down transportable, but as people get experience in the field they'll realize that the true portable requires a different approach," he asserts.

Do you really need a system that can be used in your lap? Can you live with a display that is not of the quality found in non-portable systems? How important will it be

Sharp PC-5000

PRICE: $1695

SCREEN SIZE: 8 lines by 80 characters

AVAILABLE USER MEMORY: 128k

WEIGHT: 9.74 lbs. with battery and bubble cartridge

SPECIAL FEATURES: built-in battery and bubble memory cartridge.

Epson Geneva/PX-8

PRICE: $995

SCREEN SIZE: 8 lines by 80 characters

AVAILABLE USER MEMORY: 64k

WEIGHT: 4 lbs.

SPECIAL FEATURES: LCD display, CP/M 2.2, software in ROM, internal power supply.

Source: Personal Computing, March 1985, 83, 85, 87-89, 91, 93, 95, 97

for you to get new programs in the future? Will a portable give you all the functionality you need or would a transportable desktop model give you all the mobility you'll really require? Can you afford to wait for the next generation of portable systems? These are some of the questions that have to be asked before making a decision about portables, and for the time being there is no answer to any of them that will be true for everyone.

Whatever the rhetoric, a true portable computer still entails compromises, in one respect or another, over what is offered in a desktop or transportable system. Perhaps they always will. When the day comes that you can hold a computer in your lap with as much power and for the same price as today's personal computers, then how far will desktop technology have advanced? Yet it is clear that portables are already being used in a variety of useful ways, some which closely resemble personal computer functions and some which are quite unique. At some point, the ability to use your computer anywhere is a feature that is worth the premium, whether in higher price or limited performance, that you pay to get it. For some, that point has already been reached; for the rest of us it may well be worth the wait. ■

Tandy 200

PRICE: $999

The Interface Question

A critical factor to consider before you buy a lap portable is how it will interface to other equipment. It is a rare user who will not have some need to exchange information between his portable and another system–either a personal computer or a larger corporate system. And since portables vary widely in the types of communication they support, you will need to find the system which offers the right level of interface capability without causing you to pay for unnecessary features.

The lowest common denominator for the system link is the RS-232 serial interface, which comes standard on all of the serious portables. Just as with a desktop system, the serial port can be used for a wide variety of connections such as hooking up an external modem, sending data to a printer or making a direct cable link with another computer. One obvious consideration is to make sure you won't need more than one serial port; you could not, for example, use your portable in teletype-fashion, printing out information as it comes in over the phone, if you have just one port for your modem and your printer.

Most portables can be equipped with an integrated modem which leaves the serial interface free for other duties. With the exception of Grid, which has a 1200-baud option, all portables use the slower 300-baud transmission rate with their modems. In many cases, the telecommunications software for the modem is limited to a machine-specific program supplied by the manufacturer.

The Tandy/ Radio Shack Model 100, the landmark product in this category, comes standard with a built-in modem and the telecommunications software required to use it. The Model 100 is often employed by large companies as a remote access device via the modem to data housed on a mainframe system. Such applications can put a tremendous amount of information at the fingertips of people in the field, but it usually requires that programming be done on the mainframe side to have the data available in a format that an 8-bit system like the Model 100 can use.

A major drawback to the interface capabilities of 8-bit systems for some users is the lack of software compatibility with the 16-bit world, especially if you don't have the services of a programmer to help with data transfer problems. While it is certainly possible to take information stored in a Model 100 and get it into a file format that IBM Personal Computers programs will accept, it might require some fancy footwork. If you're going to be sharing a great deal of information with a Personal Computer or compatible system, it makes sense to consider an IBM-compatible.

Among the portables which run MS-DOS, there are several different approaches to exchanging information with a desktop. While Hewlett-Packard does not offer a high degree of program compatibility with the MS-DOS world due to its 16-line screen, the company did take pains to support MS-DOS file compatibility through its Portable-Desktop Link product, retailing for $125, which is a combination of software and an add-on board for the Personal Computer and compatibles. Using the link, the HP 110 essentially takes over the desktop system and its peripherals, saving or reading its data from both hard and floppy disks.

The compatibility Data General emphasizes with the DG/One also facilitates file transfers with MS-DOS desktop machines. The optional 5.25″ disk drive, which retails for $600, makes it possible to use the same disk in either system's drive just as you would with two compatible desktops. Grid Systems, along with offering selected MS-DOS programs for the Compass II, offers the Grid Server, priced at $10,500, which can connect 32 or more users of the portable or IBM-compatible desktops in a network to share data, disk storage and printers.

Software is available for most of the popular portables to support terminal emulation, although frequently portable manufacturers support terminal connections only to their own large computers. Data General's CEO Connection enables the DG/One to link to Data General's office automation system for operations such as electronic mail, calendar scheduling and word processing. The DG/One can also serve as a terminal with the company's Desktop Generation multiuser microcomputers. The HP 110 has a built-in terminal emulation program in ROM for accessing a number of systems remotely via the modem. As lap-top portables grow in popularity, however, it is likely that terminal emulation for most major computer environments will be supported, either directly by the manufacturer or through third parties.

The limitations of portable displays being what they are, a very desirable capability would be to run a standard, non-portable CRT device at your office, for example, while using the LCD screen at home or on the road. The Model 100 has an optional expansion chassis for $799 which includes video output for a CRT monitor and a floppy drive. A number of other portable manufacturers have announced plans to offer a similar feature.

–E.F.

SECTION IV
Software

SECTION IV
Software

Software is the written material that, when inserted into a computer's memory system, tells the computer what to do next. Without software, the computer is helpless. With the software to provide its operating instructions and functional programs, the computer becomes a useful and powerful tool.

As we learned in Section II, software is divided into two parts: (1) assembly and machine languages, and (2) the programs written in those languages. It is not necessary for users to know the languages unless they intend to write their own programs. It is more important that users find the program best suited to the task, whether it is a program for doing personal accounting or a program for controlling a rolling mill. In the world of personal and business computers, the current practice is for the user to select the program first, then to select the computer best suited to work with the programs desired. Until that distant day when software programs and computer hardware are universally interchangeable, the "software first" approach has much to recommend it.

The articles in this section have been selected to give practical information on current software programs, as well as news-type descriptions of programs for a variety of applications.

100
Things You Need To Know Before Buying Software

— *Sheila Cunningham* —

In some ways, buying software is like buying anything else. There must be: awareness of the need or desire for a specific product, a knowledge of the economics involved, a thorough evaluation of the available products, and a willingness to do a little shopping around. It goes without saying that the entire process should be preceded by the phrase *caveat emptor*.

In other ways, however, buying a software product is totally unlike buying any other product. In part, this is because of the incredible variety of products available, thus making the alternatives somewhat overwhelming as well as confusing. An additional factor is the level of sophistication of the products available, making evaluation and selection a difficult and complicated process. Add to this the consequences of an unsuccessful software function and it is evident why the entire process must be approached in a deliberate, planned, step-by-step manner, with an even greater emphasis on buyer beware.

There are several critical steps involved in the acquisition of a software product. As has been noted, the need or desire for the product must be documented, economic considerations must be taken into account, and a little comparison shopping must be done. Patterns and guidelines for conducting a successful search for software have emerged to aid the prospective buyer. As the proprietary software industry grows more competitive and refined, it will offer increasingly more products and the conscientious buyer will have to know how to take advantage of the software alternative.

DETERMINING NEED

Ask:

Is an automated solution necessary?
Does an EDP solution exist elsewhere in-house?
Is there user need or interest in a new system?
Is there DP interest or need for a new system?

Conduct:

a discussion of the needs for the product among:
Technicians
Management
Users

Determine:

Who's in charge?

Before embarking on the sometimes rough road of software product purchasing, it is always best to know a product is needed. Usually the buyer finds that a new product is needed because there is a new function, there is a shortcoming in a current program, or there is a new function for which there are no programs available within the company.

Buying a software product should be the concern of three areas within the company. The technician, who is concerned with the technical aspects of the product, the end users, who are mainly concerned with product performance and managers, who are concerned with bottom line impact. Three levels of the corporation, therefore, have three different concerns with the product. If the buyer gathers information from each of these levels, he will avoid buying a package that might interface well (fine for the technician), and that might be well within corporate guidelines as far as expenditure (and will get management's blessing), but one that is unsatisfactory as far as, say, reporting requirements (in which case the user will be displeased). There are any number of combinations that can be derived, and none of them are good, except one, when everyone concerned is happy with the product. Everyone's happiness can be assured by evaluating the product at all three levels.

Reprinted with permission from ICP Interface, Spring 1981, 27-30, © 1981 International Computer Programs, Inc.

ALTERNATIVES

Look At The Market:

Is there a package that solves this problem?

Does it run on the existing equipment?

Does it cost less than an in-house developed system?

Conduct:

A cost/price analysis to determine if a purchased system would be a viable choice

Request:

Brochures and specifications from likely vendors.

Request user lists.

A thorough examination of the alternatives is one of the preliminary steps that must be taken before buying a software product. This is the time to become attuned to the marketplace by examining what is available in trade magazines and directories, direct mail, through advertising, and by asking others.

The prospective buyer should ask: Is there a product available that is compatible with the existing operating environment? Will that same product meet all the needs of the users? Will it generate reports in the way most desired without a lot of costly enhancements?

Additionally, a cost/price study should be conducted and a deter-

mination made on what it would cost to develop the product in-house. Ask: Can this software product be implemented and maintained within the figures laid out in the cost/price study?

This economic review and evaluation of the proposed project should be exercised for every project. It is essential that two areas be reviewed: the total cost involved in a project, and the economic impact on the organization. The total cost of the project includes not only the cost of the software to be acquired and the resultant installation expenses as well, but also the cost incurred internally to accommodate the system in question and the cost of conversion.

EVALUATION

Check Vendor Characteristics for:

Length of time in business

Size of the organization

Primary markets served

Financial statements

Number of products

Number of customer service employees

Technical writing department

Number of installations

Active user group

Check User Satisfaction With:

Overall product

Implementation

Support

Promised enhancements

System Reliability

Ease of Use

Documentation

Training

A good vendor will be able to give a detailed definition and illustration of the difference between maintenance and enhancement for the product. It doesn't really matter where he draws the line between the two, as long as the line is drawn somewhere. An inability on his part to distinguish between the two is, if not enough to disqualify him, then certainly enough to discount him.

The vendor will also have detailed and firm pricing for the product. That's not to say negotiations shouldn't be conducted, particularly in the case of a relatively new product in the marketplace, one that is untried and unproven but it's also fair to say that a vendor who knows his prices and sticks with them probably also knows his costs and economics.

A financial analysis of the product as it applies to the buyer's environment should be provided. Possibly the vendor has to gather the information from the buyer, but he should be thoroughly familiar with his side of the equation.

User references should be willingly supplied as well and that includes names, addresses, and telephone numbers of local users that the prospective buyer can call for a first-hand reference.

REQUEST FOR PROPOSAL

Should Address:

Proposed software product including specifications, language, prices, terms, and time frame

Cost and terms to upgrade the system

Terms of license or contractual constraints

Delineation of responsibility for software development and maintenance, training, conversion, and program modification

Guarantees

Users of product

Number of support people

Important features of the product

Financial statements indicating financial responsibility

Information about the company

Once the buyer has done his evaluation, it is time to choose two or three "finalists" and move on to the next stage. Now a method of comparison is needed to consider the vendor's proposals and claims. The buyer needs a formal Request For Proposal (RFP), which spells out the questions each vendor must answer in order to be considered a viable bidder. It is the only way to compare the proposals of several vendors. Each vendor should bid comparable systems, with the features and volume handling capacity to do the job, with the present and future costs clearly spelled out, and with a clear-cut commitment to all the mandatory terms and conditions under which the buyer is willing to do business.

In return, the buyer must take the lead by spelling out exactly what is wanted. The vendor needs a brief description of the hardware environment, a statement of what the software need is, an invitation to supply information and the type of information needed for a first level evaluation. Additionally, the vendor will need to know the time schedule under which the buyer is operating, and a direct statement that the buyer either does or does not want a salesman to call.

The written RFP is the first step toward forcing a vendor to be specific about features, performance claims, support, training, things that should eventually be incorporated into the contract.

PRICE

Ask:

What pricing arrangements are available?

What is included in the purchase or lease price?

Is there an additional cost for multiple installations?

What is the length and cost of package warranty?

Is there a cost for system maintenance during and/or after the warranty period?

Are there any additional costs? (Manuals, support, improvements, training)

Dollar evaluations:

Compare make or buy cost

Compare costs of several buyable systems

Evaluate the importance of the system on the total organization

The make or buy cost was evaluated earlier. In comparing the prices of several companies software products, it is necessary to know the entire price structure of the product. The variables to consider are: terms of the lease, terms upon expiration of the lease (if applicable), installation assistance provided, amount of documentation, training, maintenance cost, and conditions under which enhancements are provided. An intangible element that must be assigned some quantitative value is the number of previous sales the product has had. The more the better, because the seller will have had more experience and consequently will be better equipped to respond to the buyers' needs.

The economic impact should be estimated as early in the process as possible. The impact should be viewed in as many ways as possible, *i.e.*, unit production costs, cents per share in earnings, expended manhours, elapsed time, percentage of overhead rate.

INSTALLATION

The Buyer Should Know:

Will the vendor create all master files and interfacing requirements?

How long will the vendor provide assistance without cost?

How long will the vendor maintain the system?

Will the buyer share in all improvements or enhancements to the product?

Are any modifications to the package in progress or planned?

The vendor should be required to describe the installation process including elapsed time, buyer assistance required, computer time requirements, as well as an estimate of the man-time required to provide for implementation, continuing operations, control, and support.

The installation phase begins the important user training phase. Training will be conducted by the vendor's service personnel, who will teach the customer's personnel (both user and data processing departments) how to use the system. Educational sessions may be held to answer questions from the customer's personnel directed to their specific needs. Any problems that arise during the installation phase should be answered by the vendor's personnel.

The installation should be warranted for a specified period of time and if any "bugs" are encountered during that period, it is the responsibility of the vendor to correct them within a reasonable period of time and without cost to the user. Additionally, the vendor may offer an extended maintenance agreement which entitles the user to additional modifications and enhancements to the product.

DOCUMENTATION

Should include:

Systems flowchart and narrative

Program listings

All input-output file layouts

Tape control instructions

Distribution instructions

Control instructions

Step operating instructions

Step narrative

Users manual

JCL deck and list

Estimated step run times

Report examples

Restart procedures

Documentation should not, someone once said, be measured by the pound. Just because the documentation manual is thick, don't think it is good. Ask: Is the vendor's documentation thorough? Does it give lucid examples of problem-solving techniques? Is the documentation insufficient in any way? How?

The documentation should be read by the people who will be using the system. This is absolutely vital. Operator instructions should be read by the operators, management instructions read by management, and the user department should read their section.

Documentation should meet quality standards for completeness and ease-of-use. In order to obtain these standards, the vendor must maintain an on-going stream of communication with the user base. Additionally, any changes or enhancements to the product must follow with changes to the documentation.

BENCHMARKING

The Benchmark Should:

Have enough volume to be meaningful.

Produce elapsed time for throughput, broken down into the smallest segments possible.

Consider:

Input preparation

Setup time

Run time

Output dissemination

Be run in a pure environment, isolated from a job stream, particularly if comparative timings are to be produced.

Conducting a benchmark test of the package can be a valuable step in the evaluation process. The benchmark test can provide insight into the relative merits of the package, and when properly performed will provide data on which to base comparisons. Since this is a somewhat expensive and elaborate way to evaluate the product, it should be done only with the last few products being considered. The information provided by the benchmark will be a close approximation of the results that will be achieved once the package is installed.

The people involved should include the technical support function, which is often considered the best area to oversee the project; the operations and development functions; members of the software vendor's technical staff; and applications personnel. The benchmark should measure peformance against a predetermined set of criteria for the products that represent actual user requirements. The purpose of the benchmark is to determine the "best of the best."

CONTRACTS

Should Include:

Package requirements (Per RFP)

Anticipated performance

Specifications and desired modifications

Purchase/lease/license prices (Lease/licenses should run about 3% of purchase price per month)

Sign-off procedure

Payment terms

Conditions of acceptance

Warranties, waivers and penalties (Liability limitation, responsibility for errors, compatibility, compensation or credits for late delivery and support)

Free trial

Method of terminating contract

Availability of source code

Vendor support (emergency and ongoing)

Vendor-supplied training (initial and ongoing)

Documentation (User manuals, technical manuals, updates)

Sales literature and correspondence

Enter the contract negotiations stage with: clearly defined goals and objectives; clarification that all commitments are to be in writing; a negotiating team of specialists (DP specialist, users, financial individual, lawyer, and an experienced negotiator). One individual should be assigned responsibility for the negotiations.

Contract terms should be evaluated. Remember, once the contract is signed, the advantage immediately shifts to the vendor. Evaluating contract terms may be difficult, but it is a vital step in the final process. Don't buy a product that isn't warranted against defects for at least one year and make sure the vendor's obligation to pay is clearly stated. Be certain installation responsibilities and acceptance procedures are in writing and that maintenance terms are clearly established and include enhancements and upgrades as well as cost. Specified dates should be incorporated into the contract for guaranteed maintenance — usually from one to twelve months — in

which the vendor promises to get all of the bugs out of the system.

Don't sign a contract unless the vendor guarantees that he has clear title to the product and is willing to defend against any trade secret, patent, trademark or copyright infringement claims.

Above all, a buyer should not thwart his own best interests by merely asking, "Where do I sign?" An unsuccessful software acquisition can spell disaster for the buyer organization. In the end, the contract should be one on which both parties can build a solid and profitable business relationship.

There is no question that purchased software products — whether off the shelf, custom built, timeshared or in a turnkey box — will be one of information management's most valuable resources in the 1980s. With over 7,000 software products* available in the marketplace today it is imperative for effective managers to know how to sift through the marketplace, locate the right product, evaluate it, test it, and contract for it. The rewards will be time, money and people saved, satisfied users and managements, and a sound comfort in the knowledge of a job well done.

* According to the current editions of The ICP Directory.

CHOOSING A PROGRAMMING LANGUAGE

BY GARY ELFRING

It's a three-step process

IF YOU WERE a carpenter building a new house, the first thing you would do would be to collect your tools. The tools you'd select would vary depending on the type of job. The same thing should be true if you are a programmer. You have a wide range of tools available, and you just choose the right tools for the job. Your tools are the languages that you program in and the environments needed to support those languages.

How do you go about selecting the right tool for the job? There are more programming languages available for microprocessors than most people could learn in a lifetime. What you need is a methodology that can be used to select one language from all the rest for a given application.

This article presents a practical method for comparing programming languages. It has an inherent bias toward compiled high-level languages. Compiled languages are faster than interpreted ones, and most interpreted languages also offer a compiler version. Since program speed is often an issue, I chose compilers over interpreters.

The actual process of evaluating a group of programming languages can be broken down into three major steps. The first step is to characterize the application the language is being selected for. Then you must identify the features that a language should have in order to deal with the previously described application. Finally, you should take into account practical considerations to further narrow down the language selection.

THE APPLICATION

You can't choose a tool unless you know what you intend to do with it. You have to describe your application. Once you have this information you can then proceed to determine whether or not the existing language choices are the right tools for the job.

To describe an application, you must consider both the type and size of the application. These questions must be answered before you can proceed any further in the language evaluation:

What is the type or class of application?
What level of language is needed?

There are a number of different classes of program applications. An application can belong to a single class or several. Identifying the class of your application is relatively simple and helps narrow the list of acceptable languages. Some of the more common classes include scientific, business, and system programming; text processing; expert systems; and real-time control.

Most programming languages are better suited to solving one particular class of problem than another. COBOL is one example. While it is easy to write maintainable business programs with COBOL, no one would expect to use this language to solve real-time control problems.

Another consideration is the level of programming that the application will require. If you need low-level control of various machine-dependent features, then a very high level language

(continued)

Gary Elfring (4N899 West Mary Dr., St. Charles, IL 60174) is the president of Elfring Consulting Inc., a microprocessor consulting firm. His interests include robotics and artificial intelligence.

would be a bad choice.

Is it too big to be expressed as one module?
Is it too big to be fully understood by one programmer?

Just how big is the potential application? Large programs should not be squeezed into a single module. This implies that the language chosen must support separate program module compilation. It is always easier to compile many small programs and combine them into one large version than to compile one gigantic program.

Related side questions consider the complexity of the application itself. Can a single programmer understand it? If not, then the language chosen must support multiple program modules and some way of managing them. Also, remember that a number of programs start out small and end up growing quite large.

LANGUAGE FEATURES

After characterizing the nature of the software application, the next step is to identify what features are required or desired to implement it. This list of features can then be used to rate each available language. The result of this process will be a short list of acceptable programming languages.

What audience was the language designed for?
What class of problems was the language designed to solve?

Some languages were designed for a specific audience instead of a class of applications. This type of language was typically designed to do something like teach programming techniques. A language that was designed to solve a specific class of problems will generally do a much better job at that task than one designed to teach the techniques of programming.

Most languages were designed with some class of applications in mind. This inherent bias toward certain classes of problems will affect the way a language is used on other classes of problems. A short history of each language under consideration is necessary to aid in the evaluation of that language. For example, knowing the

history of COBOL and APL will allow you to reject them immediately as languages for programming a real-time control application.

Can the syntax be understood?
Is it terse or verbose?
Is it consistent?

The syntax of a language should be both readable and intelligible. A syntax should aid the mechanical aspects of reading it and help you understand what the program is doing.

A syntax should be concise, yet expressive. Verbose languages can turn what should be a one-page program into a multipage listing. Since, generally, the longer a program is, the harder it is to follow, such verbose syntax can actually defeat its own purpose of increasing the understandability of a program.

Inconsistency in syntax makes a language hard to learn. It also increases the possibility of error significantly. The syntax should be an aid to programming, not a hindrance.

What data types are supported?
How are data types treated?

The organization and representation of data is an important part of programming. Some basic data types on microprocessors are 8-, 16-, 32-bit integers, single- and double-precision floating point, records/structures, pointers, bit fields, and arrays of all data types. Some compilers even allow you to create user-defined data types.

Structures or records, in particular, are important data-handling tools. They let you group items of different data types together so they can be referenced as a unit. The more control a language gives over the use and construction of structures, the easier it will be to handle data.

The use of data types not only gives you great flexibility in how you manipulate data but also lets a good compiler do a considerable amount of error checking. The compiler can check for mismatches in the use of data types and flag them as errors. This will catch a large percentage of the simple errors that a programmer

makes, well before the program enters the debugging stage. For these reasons a language should offer as rich a selection of data types as is possible.

Does the language support structured programming?
Are exceptions possible?

Languages that support structured programming are inherently more safe to use than those that do not. A programming language should at least offer you the choice of using structured programming techniques. Modern structured programming generally requires, at a minimum, a grouping syntax (either functions or procedures), two types of decision statements (generally the IF and CASE statements), and two types of loops (counted and uncounted).

Sooner or later exceptions will arise that must be dealt with in some manner. Languages that allow no exceptions can make it quite difficult to write certain kinds of programs. Some languages totally rule out exceptions or can deal with them only in an uncontrolled fashion such as with a GOTO. One structured way of handling exceptions, the BREAK command, is used as an escape from counted and uncounted loops. It is preferable to the use of a GOTO and is featured in languages like Ada, C, and Modula-2.

Is portability needed?
How portable is the language?

The portability of programming languages is becoming more and more important in the world of microprocessors. Large amounts of time and money are spent developing software that becomes obsolete when the next microprocessor comes out. Some method is needed to protect this large investment in software. The answer is portability.

There are several levels of portability. At the lowest level, the language is portable from compiler to compiler on the same microprocessor. The next level of portability covers the ability to port code from one micro-

Source: BYTE Magazine, June 1985, 235, 236, 238, 240

processor to another of the same level, for example, between 16-bit microprocessors. A final level of portability is between any two microprocessors.

If an application can be ported from an 8-bit microprocessor to a 32-bit one (say from Intel's 8051 to Motorola's 68020), then the language is truly portable. Note that it is probably unreasonable to expect that any application be ported all the way down from a 32-bit microprocessor to an 8-bit one, but you should expect a language to offer upwardly compatible extensions.

How is I/O (input/output) handled?
Is access to other programming languages needed?
Is stand-alone product support required?
Is real-time control needed?

Not all programs need file or terminal I/O. However, almost anyone would agree it is a useful function. Device-independent I/O is preferable to any other kind. Languages that make no distinction between different types of I/O devices are easier to program in. They are inherently more consistent since they don't differentiate between hardware devices.

Some applications require the use of more than one language. The second language is typically assembly language. If your application needs an interface to another language, then the amount and type of support for the other language must be assessed.

Is the software application going to be for a stand-alone product (such as a microprocessor-controlled blender) or a product run on a computer with an operating system (such as a database-management system)? A language for a stand-alone application must be able to get "close" to the hardware; that is, it will need features that allow absolute control over addressing, I/O, and interrupts. It will also generally have to run from some kind of ROM (read-only memory). However, a product such as a database-management system interacts with the hardware through the oper-

ating system and runs from RAM (random-access read/write memory).

Yet another consideration is real-time control. Does the intended language support it, or will it at least allow itself to be modified or extended to do so?

PRACTICAL CONSIDERATIONS

There is more to selecting a programming language than a simple comparison of features. In the real world a number of practical considerations enter into the picture. They range from considerations about existing compilers to questions about the development tools for a particular language. This series of qualifications is used to further narrow the language choices based on real-world criteria.

How available is the language?
How popular is the language?

How many different compilers are available for this language? Are they compatible with each other? How hard is it to locate compilers for this language? Answers to these questions describe the availability of a language. Generally, the more available a language is, the safer it will be to use.

It is important to choose a language that has an established history and predictable future. Will the language you choose be around tomorrow or does it belong to the "Language of the Month" club? Brand-new languages may offer many desirable features. But the future of both the language and its new features can be uncertain. If it doesn't catch on, the language and its special features may die.

How do you learn the language?
What is the source of this information?

If the language chosen is not in your programming repertoire, you are going to have to learn it. How do you learn to program in the language? Are there good reference materials or outside help available to teach you the language? The best language in the world is no help to you if you can't figure out how to use it.

What are the characteristics of the compiler?
Is the code produced quick, compact, and predictable?

Theory and practice must merge in the creation of a compiler. The compiler should operate quickly and be reliable. It should not require a great deal of memory or disk space. Finally, the amount of support from the manufacturer is important.

Compilers translate high-level instructions into code that a particular computer can execute. The code that they produce must be efficient both in size and speed. Furthermore, the execution time and size of the code produced by the compiler should be reasonably predictable. That is, the compiler should be consistent in the quality of the code it produces.

What hardware-development tools are available?
What software-development tools are available?
What kinds of software libraries are available?

The types of tools associated with a programming language are important considerations in that language's evaluation. Tools make you significantly more productive and ease your work load.

Tools come in many different forms. Hardware-development tools, such as an in-circuit emulator, are indispensable in the development of software-controlled products. The availability of this type of tool could easily alter a choice of a programming language.

Another form of tool is software-oriented. Software-development, debugging, and management tools are quite popular but not compatible with all programming languages. A good directory program for file maintenance is also a tool. The purpose of these tools is to make the job of programming as easy as possible.

One final type of tool is the software library. It is often more feasible to purchase a library of software functions than to write your own. Examples of software libraries include graphics and real-time control packages. The availability of good software libraries at low cost is an important practical consideration in the choice of a language.

LANGUAGE SUITABILITY

Now that you know how to describe the application, programming features, and practical considerations, the selection process can begin. The end result of this evaluation process is a list of from one to three languages that are ideally suited to your particular application. If you end up with more than one language on your list, then any one of these can be selected. This is the only point in the language-selection process where you can apply your own personal bias and not affect the quality of the final language choice. ∎

Source: BYTE Magazine, June 1985, 235, 236, 238, 240

Software Sources I: Directories and Data Bases

Let your fingers do the walking. Here is a quick survey of micro, mini, and mainframe software references. Excluded are the third-party catalogs usually supplied by hardware vendors. The directories cover commercially available packages.

Generally, the listings in each book (and data base search) give details on vendor contacts, application areas, operating system, hardware/memory required, and prices. The more detailed give source code availability, and report on customer support, updates, and warranties. Some works list company financial data, and number of installed packages to date. Product descriptions range from nonexistent to short blurbs supplied by the vendors, or, in the more comprehensive (and expensive) books and supplements, detailed analyses and user ratings. Frequency of updates should also be considered.

Most of the reference works cover all software, including utility software (language development tools, systems software); the lion's share of applications software goes to management, accounting, and financial packages.

Different books have different subcategories for mechanical engineering, if they have them at all—what's called piping in one book may be listed under HVAC in another, stress analysis packages under civil engineering, materials processing under manufacturing, etc. CAD is usually CAD. "Manufacturing" covers a wide field. It may take some doing to sort out the programs for inventory control from the software directly relating to whatever aspect of technology you're looking for.

Without a good index any work over five pages long is put to best use as a prop for a table leg. Slogging through some of these volumes is time consuming; truckloads of information are useless if they lie hidden under more truckloads of information. I advise you to question each publisher about their index before investing in a book. It's probably most important to see how far you can narrow down the categories that can be searched.

Next month I will list indexes and distributors solely for science and technology software, including rich federal government sources. (The ICP directory is the only one with a separately available volume for engineering; of course, you can be as selective as you please when searching a data base.) Before you get your hopes up, I am unaware of distributors only for *mechanical* engineering—a vertical market too precipitous to make it worthwhile even for clearinghouses. The only consistent ME software reference is the Software Exchange (four-five pages) in *CIME,* ASME's superb bimonthly computer-applications magazine. It costs $30 a year; if you so much as glance at this column, it seems obvious you should buy *CIME.*

International Computer Programs (ICP) is considered the "grandaddy" of mass market directories—the first edition came out in 1967 with 110 listings. The current directory has 17,000 listings in seven volumes, for $550 with two updates a year. Volume V is devoted to engineering and manufacturing software (1600 entries, 790 mechanical engineering) for mainframes and minis. Micro software—all categories—gets its own book, Volume VII, with 290 listings for ME out of 4750 total. Individual volumes cost $95 with biennial updates, $75 a single copy. ICP also publishes a manufacturing software quarterly magazine that is sent free to "qualified users" (so far about 25,000 of them).

Auerbach puts out 14 volumes on more than you'd ever want to know. Volume GI covers main and mini applications software (800 engineering, graphics, and statistics); Volume JI, systems software. Each costs $395, the pair $670. Excerpts from the 14-volume opus are compiled in the more manageable three-volume *Data World,* at $650. Two volumes cover hardware and communications; the software volume has 1500 listings, about 400 for engineering and science. Micro hardware and software are listed in the two-volume *Micro World,* at $295. Every book by Auerbach is updated monthly.

Some time ago part of Auerbach's marketing team left and founded **Datapro** (which in turn had defectors set up Data Decisions—all three stayed in South Jersey, the Silicon Valley of data directories). Datapro's main and mini software book, with 6500 entries, costs $520; the micro book has 3500 entries, and costs $550. User ratings are included, and each book is updated monthly. Newsletters and phone consulting are available at no charge.

Data Decisions (DD) and **Data Sources** (DS) are sister companies that publish separate directories that relate to each other "as an encyclopedia does a dictionary," according to a spokesman. (Some directories need directories, it seems. Mine eyes glaze.) DD's three volumes, for main and mini, cost $485, and have user ratings. For $39 you can order "in-depth analyses" of individual products. About once a year DD publishes three software reports/surveys of interest, on manufacturing, graphics and CAD/CAM, and building and construction. They can be ordered separately for $39 each. DD just released its three-volume micro catalog, at $625, which contains ratings prepared in-house based on reliability, ease of use, documentation, etc. The company believes this to be the first consistent rating system in use—as opposed to the idiosyncratic, biased, or just plain cranky reviews that may appear in various journals. Services include monthly updates, newsletters, and phone consulting.

DS's offering, the "dictionary," has a volume apiece for main-to-micro hardware and software. About 22,000 programs are listed. The set costs $120 and is republished four times a year.

International Directory of Computer Programs is a well-respected hardback with about 5100 main/mini listings from 1000 U.S. and 700 overseas vendors. It costs $301 with three updates a year, $254 a single copy. Their micro volume ($39) only covers business and education software, but expanded editions will appear shortly.

Computerworld Buyer's Guide is offered as a bonus or freebie (depending on point of view) when you take out a subscription to the *Computerworld* weekly newspaper. The newspaper is very good—you can look at it as a weekly freebie if you buy the 800-page *Buyer's Guide.* The guide lists 10,000 micro-to-main products, referenced to vendor profiles. With the $44 annual subscription you also get directories for communications, large systems hardware, large systems software, and micro hardware.

Automated Searches. When a directory can be searched by computer—presto!—a data base. No doubt the companies mentioned above maintain their files on computer tape, but the following two companies will search their files on request, and thus save you the cost of buying information on things you don't need. It's also possible that the mix-'n-match computer search will unearth packages you might have missed just by looking. It certainly is faster.

International Software Database is a software shopper's dream come true. Automated searching can be done either on- or off-line through the "Menu" service. The ISD data base of 50,000 programs is searchable by machine, operating system, source language, application, vendor, price, and date of release, among others, and combinations therof. When the company does a search for you, reports cost $25 for the first 10 programs found, $1 per program up to 50 programs, and 25 cents for additional ones. The reports are mailed in 24 hours. If you see some likely candidates in the report, you can request ISD to search for published third-party reviews and "customer feedback" that could assist you. Finally, ISD will order any package selected, crediting the price of the search toward the order.

ISD is also the only search service that is on-line (although SofSearch plans to be), through Dialog and the Knowledge Index. Dialog search charges are $60 per connect hour and 15 cents per file printed off-line. The Knowledge Index's hourly rate is $24 (nighttime jobs), but you can't get off-line prints. A spokesman for ISD said that even if you are a subscriber to Dialog, it's more economical to have ISD do extensive searches. Also, the ISD data base that Dialog lets you into is updated once a month; the home base is updated daily.

The ISD catalog can be read by humans. Two versions are published twice a year and updated twice. The microcomputer subscription is $142.50, a single copy, $69. The mini software books are $191.50 or $95.

Less grand, but definitely worth looking into, is the **SofSearch** Software Locator Service. A quick overall search of a data base of over 36,000 programs (110,000 operating versions) turned up 131 programs in ME, 40 in industrial engineering, 67 in metals fabrication, and 14 in primary materials processing. At last count there were about 1500 programs in the "engineering/scientific" grouping. The company will perform searches using 150 functional categories and 85 application areas—12 of them devoted to subfields of manufacturing. Cost: In addition to a $40 annual service fee, $35 for a search by category and hardware or operating system, or $70 for a range search, e.g., everything they've got in a category for micros. The prices are the same if your report comes back (normally in 48 hours) with two listings or 200. A $150 "corporate account" entitles you to six searches of either type per year. The company also provides phone consultation, a quarterly newsletter, and "pass-through discounts" on selected packages that you order from the vendor.

Software Sources

Auerbach Information Co.
6560 N. Park Drive
Pennsauken, N.J. 08109
(800) 257-8162
(609) 662-2070

Computerworld
Paramus Plaza 1
140 Rte. 17 North
Paramus, N.J. 07652
(201) 967-1350

Data Decisions
20 Brace Rd.
Cherry Hill, N.J. 08034
(609) 429-7100

Data Sources
20 Brace Rd.
Cherry Hill, N.J. 08034
(609) 429-2100

Datapro
1805 Underwood Blvd.
Delran, N.J. 08075
(609) 764-0100

International Computer Programs
9000 Keystone Crossing
P.O. Box 40946

Indianapolis, Ind. 46240
(317) 844-7461

International Directory of Software
CUYB Publications Ltd.
First Federal Bldg., Suite 401
Pottstown, Pa. 19464
(215) 326-5188

International Software Database
"Menu" Search:
1520 South College Ave.
Fort Collins, Colo. 80524
(800) 525-4955
(303) 482-5000

Printed Volumes:
Elsevier Science Publishing Co.
52 Vanderbilt Ave.
New York, N.Y. 10017
(800) 223-2115
(212) 867-9040 Ext. 307

SofSearch International
P.O. Box 5276
San Antonio, Tex. 78201
(800) 531-5955
(512) 340-8735

Ada Language Moving Closer to Market

The Ada programming language is attracting attention this year as compilers for micro, mini, and mainframes make their debuts. Commissioned by the Department of Defense and designed specifically for real-time applications, Ada has many features that distinguish it from other languages. Although commissioned for the military, it is anticipated that Ada will also eventually have a widespread industrial impact. This article reviews the current status of Ada development.

CARL BRANDON, Vermont Technical College, Randolph, VT
JOSEPH MARTIN, Garrick Lochhead, Inc., South Deerfield, MA

By the early 1970s, it had beome evident to the U.S. military that computer language proliferation and unstructured code were major factors in the rapid growth of military software costs. In 1975, the Department of Defense (DoD) decided to target the large body of languages then being used in embedded computer applications for replacement with a single high level language. A committee soon determined, however, that none of the existing languages was suitable, and recommended that DoD contract for the creation of an entirely new language.

In 1978, following several rounds of proposals and revisions, a proposal from Honeywell was chosen as the basis for the new language, and in the Spring of 1979 the full language design was completed. It was given the name "Ada" for Augusta Ada, Countess of Lovelace, in recognition of her pioneering contributions to the art of computer programming in the early 1800s.

The design was subjected to slight revision over the next year or two, and was firmly established in 1981. In the time since, companies have striven to meet the rigorous requirements of the Ada design, but military and commercial applications are only just now beginning to become reality.

Ada compilers

The U. S. Government Ada Joint Program Office has registered the name Ada as its trademark. This agency is responsible for setting industry standards covering use and applications of the new language. For Ada to be successfully implemented as a programming tool, compilers are needed, and these, too, are subject to control by the Ada Joint Program Office.

For any compiler to merit rating as an Ada compiler, it must meet validation criteria consisting of nearly 2,000 tests covering all features of Ada. Only compilers passing the entire "suite" can be designated with the name Ada or any name that has Ada in it. Validation must be repeated on a yearly basis. This standardization greatly increases program portability; for example, one company reported that in moving a 12,500 line program from a Digital Equipment Vax computer to a Motorola 68000 microprocessor-based system, only six line changes were necessary.

The Ada development process itself was carried out largely in abstraction, i.e., the properties desired in the new language were specified and refined long before compilers were adduced or a computer actually operated by means of it.

As of June, 1984, four compilers have been validated and are ready for commercial use: the New York University Ada/ED version for the Digital Equipment Vax; the Rolm/Data General version for the Data General Eclipse and the Rolm MSE 800; Gensoft's Ada (a recent spinoff from Western Digital) designed for the Western Digital Microengine and Delphi 100; and Telesoft's Ada for four Motorola 68000 systems.

There are also at least five completed compilers awaiting validation: Dansk Datamatic (Denmark), Alsys, Vertix, DEC, and University of Karlsruhe (Germany). Others nearing completion include Mills International, U.S.A.F. Armament Labs, Softech, Intermetrics, Intel, and Irvine Computer Science. These and other partial compilers include implementations for Intel 8086/8088, Motorola 68000, Zilog Z80 and Z8000, Digital Vax, DEC 10, DEC 20, IBM 370, CDC Cyber, HP 1000, 3000, and also Amdahl, Burroughs, Honeywell, Univac, Siemens, and Olivetti computers.

An Ada compiler has two stages or parts: the "front-end" takes the Ada program and derives an intermediate representation; the second part translates the first, by a process known as code generation, into a symbol system the hardware can understand. Both Telesoft and Alsys have developed their compilers by writing a single front-end and several code generators. It can be expected that as this strategy becomes more widely adopted, compiler costs will be reduced and new compilers will become available for new hardware with less development time.

Features and applications

The lengthy development period and careful design of Ada have produced a language rich in constructs suitable for embedded computer systems—systems in which the computer is not used for computation alone, but as part of a larger system that includes external sensors, control devices, and I/O devices with which it must interact in real time. The software for these systems is characterized by its large size (tens of thousands of lines of code), longevity (during which there are constant revisions and improvements in the software and changes in the hardware), and the hardware-intimacy that real-time applications demand.

Embedded computers for control applications are commonly required to be capable of parallel processing or performance of several tasks at once, and real-time processing which takes place simultaneously with events happening in real time.

Since embedded computers analyze processes unfolding in real time, they must be able to handle exceptional situations without "crashing." In other words, such programs must be supplied with an "exception handler" that will enable the program to "recover" when confronted with exceptional situations or data. For example, a division by zero will make the average program stop functioning. But this will not do for a real-time system, which must continue to operate as long as does the process of which it is a part.

The Ada language is particularly suited, then, to real-time programs embedded in larger systems. Ada will tell you when you have exceeded a "normal" range and makes it easy for individual users to write exception handlers. For these reasons, Ada language programs are particularly useful for the following:
• embedded systems;
• laboratory instrument control;
• data collection;
• industrial process control.

A superficial look at Ada suggests that it is very similar to Pascal. All four of the companies which submitted preliminary designs did, in fact, use Pascal as their starting point. Like Pascal, Ada has strong data typing and software modularity, which allow powerful error prevention capability within the compiler. Ada, however, goes beyond the design goals of Pascal:

Reprinted with permission from Control Engineering, November 1984, 89-90, © 1984 Technical Publishing Co.

- Typing is even stronger in Ada; it is also sufficiently flexible to leave direct access to specialized hardware unhindered.
- Ada encourages modular programming; not only with subroutine modules, but also with data abstraction modules called packages.
- General packages and subprograms can be used for many purposes without being rewritten, since they may be parameterized with either types and subroutines or values and objects.
- Parallel process (tasks) are well supported.

In the most general sense, an Ada package is a means of dividing a very complex system into separate concepts. A package in Ada may define a data abstraction, like a stack or a linked list, by describing a data structure *and* defining the possible operations that could be performed on the data. What is particularly powerful about Ada packages is their ability to be used "generically"—i.e., without predefining the type of data from which an abstraction might be composed. For example, a single Ada-linked list package could be used to create linked lists of integers, reals, files, or other created types. Each diverse use of a package is known as an "instantiation."

When gathered together in a library, previously written and generic packages are unprecedented programming tools. According to one estimate, these packages can increase programming productivity significantly by avoiding redundant coding and by simplifying maintenance. It is possible, in fact, for a useful Ada program to consist almost entirely of a collection of modules from a library. Several companies using Ada have reported programmer productivity figures already three-to-five times the industry average.

The package concept is one example of the rich programmer-support environment that is an integral part of the Ada design. Known as the APSE (Ada Programming Support Environment), this environment includes a program library facility that catalogs user and system program modules, and standard packages of procedures.

Information hiding is another important component of modern structured software design. Information hiding refers to making one level of detail invisible at other levels (as an automatic transmission hides unwanted details about shifting gears from the driver of a car). Ada accomplishes this through packages and private types.

Unlike most high level languages, Ada is specifically designed for real-time, parallel processing, and hardware intimate applications. In such cases, Ada programs can be made up of "tasks", each capable of simultaneous execution. Such tasks can be asynchronous, or may be synchronized by means of constructs called "rendezvous". Ada can also access a real-time clock, and, like the language C, has constructs that make possible very direct and efficient use of hardware.

Ada can be used for almost any large application in general. The cost-benefits of Ada are significant—particularly in the case of larger programs, and several early implementations have indicated that the cost-benefit ratio improves over time.

From the management perspective, the major disadvantages of Ada are its large size and complexity, which make it relatively difficult to learn, and the present scarcity of compilers. Its advantages include its cost efficiency in performing large and complex tasks, and the fact that changes can be confined to local dimensions within the program. Since the availability of compilers is on the increase, and Ada is becoming more accessible, management's task is to begin evaluating the uses of Ada for future large-scale, long-lived applications.

Training and maintenance

The Ada language incorporates many of the features of Pascal, but adds new ones which increase its power and its range of possible applications. It will be easier to get the seasoned programmer started in Ada if he or she has already been trained in Pascal. While learning Ada is but a short process for those already trained in Pascal, it should be noted that there are fundamental "philosophical" differences between the two languages and that it is necessary to train the programmer beyond the point where he has mastered the Ada vocabulary. It is not sufficient to write programs in the Ada language that are based on Pascal syntax. Since Ada has special strengths, strengths lacking in Pascal, it will be necessary that the retrained programmers come to understand it thoroughly.

Program maintenance is costly and time consuming. It is usually a consequence of changes in existing programs that follow the need to correct errors in an existing program or program changes mandated by changing environmental or task conditions.

Compared to older languages, programs written in Ada are maintained and corrected at relatively low cost. Ada packages are written in discrete modules, which means that if a change is made in one area, unanticipated and undesired changes are not likely to arise and plague another area. This simplifies the process and reduces the costs of making changes.

Further, Ada packages are so constructed that they operate through a separate specification that tells the user what the package does and a "body" that actually executes the code. Since these are separate, it is possible to alter the Ada program—comprised of many such packages—by replacing the body of a given package without also having to change the underlying specification program. The program interacts with its specifications and these need not be changed if alterations are made on the level of the "body". Consequently, the frequency of recompilation diminishes as well.

Maintenance is also reduced because people who have not actually written Ada programs can understand and operate them with relative ease. In a realistic environment, the user must grasp only that part of the Ada program which actually impinges upon or is involved with his immediate task. He need not internalize a complex whole.

Future Ada trends

The Defense Department is currently expending billions of dollars on Ada programs, thereby guaranteeing it a large initial market. It is clear that this initial thrust will have widespread implications. If future developments follow past patterns, we can expect to see a proliferation of uses and users of Ada.

For example, as more companies write programs in Ada for the Defense Department, they will become increasingly familiar with its advantages, and their staffs will become increasingly comfortable with it. From that point, it should be no great leap to increased commercial applications. More Ada compilers will become available; the Ada programming support environment will strengthen just as it has for mini and microcomputers; we can expect to see the development of syntax-directed editors which will assist programmers who write in Ada and databases that will keep track of the times at which various parts of large Ada programs were written. All such development will make it increasingly easy for individuals and companies to make Ada one of their basic tools of operation.

Although Ada is currently at the earliest, most preliminary stage of such developments, stirrings are already evident. The Post Office Department is beginning to request software in Ada, NASA is planning to switch to Ada, and the DoD has recently been the host of several trade missions, sent by other countries to investigate the properties of the new language and study its potential applications. □

"Last spring, as a senior mechanical engineer at a small Boston-area manufacturing firm, I began a search for the best low-cost CAD software for mechanical design and drafting. This article is the result of my research to date. It might well be titled...

Not very long ago, computer-aided design systems were little more than a dream of the future for the typical design engineer. With the exception of a few large corporations that owned mainframe- or minicomputer-based design systems costing hundreds of thousands of dollars, "laying lead" was the name of the game. Then CAD programs for microcomputers appeared on the market. Although they were affordable, these programs lacked the speed and the powerful drawing, editing, and association abilities of minicomputer- and mainframe-based systems. Full-featured CAD programs are big, complex, and computation intensive. They require large, direct-access memory spaces and fast number crunching hardware to handle geometric calculations and to redraw complex, high-resolution screen images without intolerable delays. Central processing units of 16 to 32 bits with floating-point math coprocessors, 1MB random access memories, and hard disks are the minimum requirements for an effective CAD system, at least in mechanical design.

The past year has seen an explosion of microcomputer CAD products. New or upgraded software packages appear almost weekly, and even deal-ers who specialize in CAD systems are hard pressed to keep track of the capabilities and relative merits of each new offering.

The six software packages I have examined in detail are VersaCAD Advanced, AutoCAD 2, CADkey, Anvil 1000MD, Personal Designer, and Design Board Professional. (For information about the vendors see page 34.) All run under MS-DOS and use the 8087 or 80287 floating-point coprocessor chip. All the dealers I spoke with strongly recommended using an IBM PC/AT-compatible computer to ensure adequate processing speed and the ability to eventually upgrade the system. Although an 8086/8087-2-based system running at 8 to 10 Mhz is almost as fast and somewhat cheaper, its limited memory expansion capability is a drawback.

The greatest force driving the CAD market at the moment is probably the plummeting price of high-powered 16- to 32-bit microcomputers. The basis of an effective CAD system can now be purchased for $3000 to $5000. With the addition of a high-resolution display for $1000 to $4500, a digitizer tablet for $700 to $1200, and software for $2000 to $3000, a workstation capable of handling most of the design and drafting tasks typically performed by a prod-

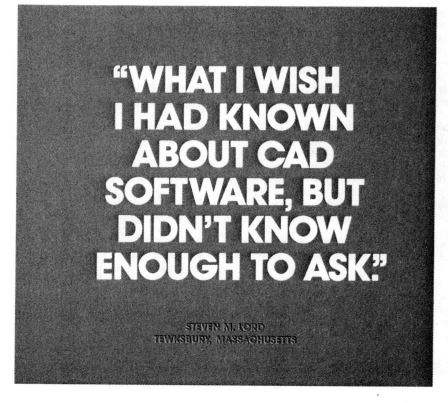

"WHAT I WISH I HAD KNOWN ABOUT CAD SOFTWARE, BUT DIDN'T KNOW ENOUGH TO ASK."

STEVEN M. LORD
TEWKSBURY, MASSACHUSETTS

Reprinted from Mechanical Engineering, November 1985, 24-34, © 1985 American Society of Mechanical Engineers, Inc.

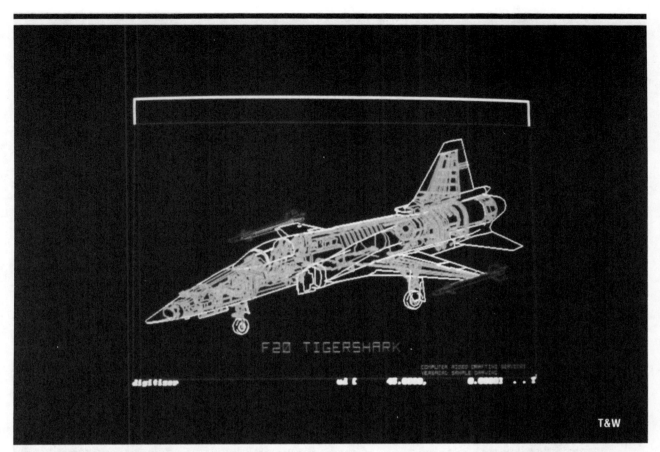

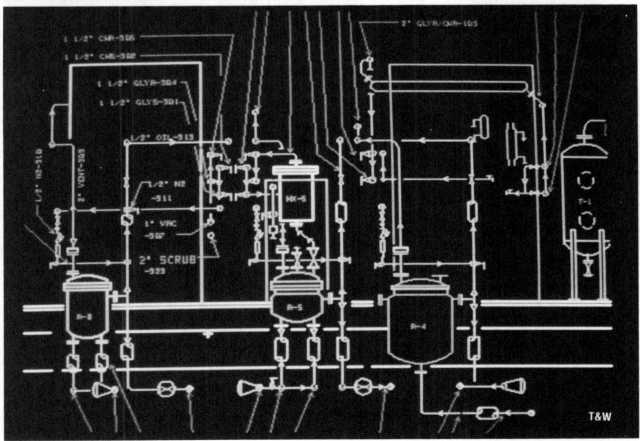

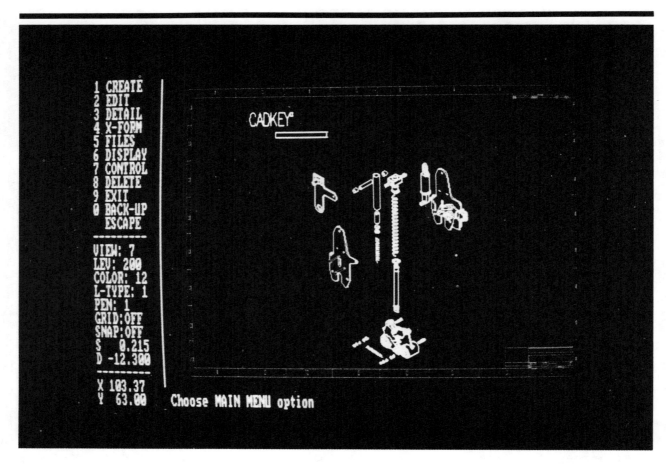

uct development group can be assembled for a total cost of between $7000 and $14,000. Assuming only a very conservative 25 percent increase in engineering efficiency, plus a substantial decrease in drafting and checking time, the investment should pay for itself in less than a year.

The increasing popularity of micro-based CAD systems can also be attributed to the rapidly improving quality of software. A CAD system is only a costly piece of hardware unless the software can significantly accelerate the design and documentation process. Having spent the past six years designing industrial and consumer products manually, I approached the selection of a CAD package with an eye to making my own work as fast, easy, and free of errors and frustration as possible. I found that the software now available is definitely good enough to be of real value. Some of the more sophisticated features of the larger design systems, such as associative dimensioning, are lacking, but the micro-based systems are still unparalleled in cost efficiency.

Because I could not afford the investment of time and money that real fluency with each package required, I conducted this investigation as would any other potential user. I identified the available packages, studied the literature, and asked questions. Demonstrations of each package by local dealers specializing in CAD systems were the real test. In these I provided a typical design drawing and watched an experienced user attempt to re-create and modify my drawing on the screen, demonstrating every feature of the program I thought I might use. In some cases I went back for a second demonstration when experiences with competing programs provoked new questions.

It is important to remember when testing CAD programs that the time it takes to produce a drawing depends on the skill of the user, who must visualize the drawing and work out the best strategy within the program for constructing it on the screen. Another complicating factor is that the hardware used in demonstrations of particular packages may differ. Depending on what the dealer has available, an IBM PC/AT, PC/XT, or compatible may be used, with or without a math coprocessor and various brands of hard disk systems and graphics controller cards. In all the demonstrations performed on PC/AT-based systems, the screen redraw speed and overall system response were reasonably fast, regardless of the software package. Delays in constructing geometry from elements were negligible (the screen followed the digitizer or keyboard input in real time), and the screen redraw time for a drawing of moderate complexity was on the order of a few seconds. Response was noticeably slower with PC/XT-based systems, but no more than one would expect given the difference in hardware. My strong impression was that productivity in day-to-day use is determined primarily by the structure of the software and the skill of the operator, and that comparisons of screen redraw speeds on standard configurations of hardware do not provide a useful measure of software quality. However, unusually complex drawings containing thousands of geometric elements, such as injection molds, are an exception. Here, screen redraw speed might become a significant factor, and speed comparisons are probably in order.

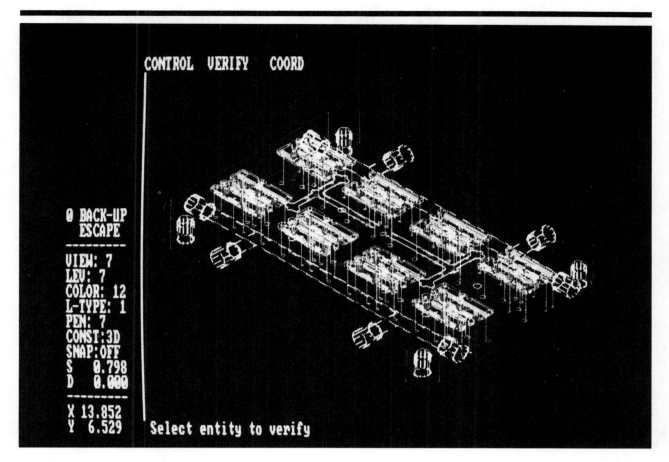

```
CONTROL  VERIFY  COORD

0 BACK-UP
  ESCAPE
---------
VIEW: 7
LEV: 7
COLOR: 12
L-TYPE: 1
PEN: 7
CONST:3D
SNAP:OFF
S    0.798
D    0.000
---------
X  13.852
Y   6.529    Select entity to verify
```

DESIGN AND DOCUMENTATION

After a number of demonstrations, I discovered that my assumptions about the abilities of CAD systems had changed. I found that it was unrealistic to expect to execute original part designs with a CAD program in half the time it takes to sketch and dimension them with a pencil. At best it is possible to go about as fast, and only after a lot of practice. The connection between pencil and brain is much more direct than that between computer screen and brain, at least with current technology. So why spend a lot of money on a CAD system? Because the creation of an original design sketch is only one step in a long process, and it is in the remaining stages that the savings appear.

With a CAD system, it is unnecessary for a draftsman to convert a pencil sketch to a drawing and for the engineer to recheck it. And unlike a sketch, a CAD drawing is always perfectly to scale, making it easier for both engineer and fabricator to detect errors. With the system's overlay and zoom features, a drawing can instantly be checked against other parts in an assembly for correct alignment, clear-

ance, and noninterference. These features also allow a group of engineers to share common layouts and to cross-check part fits by means of overlays. But the greatest gains in productivity are in modifying and revising existing drawings. A few dozen key strokes can save hours of redrawing by a draftsman and, again, exact scaling and overlay checks keep mistakes to a minimum.

A CAD system can also eliminate problems with storing documentation and drawings. Bulky microfilm and paper files are replaced by tapes, floppy disks, or hard-disk cartridges, and backup drawings can be revised on a weekly or even a daily basis instead of just a few times a year. However, the facility with which modifications can be made often leads to a proliferation of similar drawings; careful management of the archives is still essential.

Most CAD programs also allow text to be associated with drawing elements and then to be extracted later to create a bill of materials, manuals, or spreadsheets. Some programs produce three-dimensional isometric views that can be used as assembly drawings or for making exploded

views. All the systems allow drawings to be plotted at reduced or increased scale for customer documentation packages.

GENERAL PROGRAM FEATURES

The programs I examined are of two types: two-dimensional drafting packages (AutoCAD 2, VersaCAD Advanced, Anvil 1000MD) and three-dimensional design packages (Design Board Professional, CADkey, Personal Designer). The 2-D packages are primarily aimed at creating engineering drawings efficiently. They tend to have sophisticated sets of commands for creating and editing geometry and many features for adding dimensions and text to the completed drawings. The 3-D products are focused more on design. They allow the engineer to view a 3-D wireframe representation of a proposed design at exact scale and to examine its interaction with other objects that are represented in geometrically correct relationships. However, since most 2-D software vendors are now working on 3-D extensions to their products, and 3-D vendors are incorporating many of the powerful editing features of the better 2-D packages, the distinction between the two types will probably become obsolete within the next few years.

All the programs support high-resolution color displays. The minimum acceptable display resolution for mechanical design work is about 600-by-400 pixels, and a higher resolution, such as 1024 by 800, is highly desirable, especially for detailed viewing of objects larger than a foot or so across. Anvil 1000MD and Personal Designer do not currently support screen resolutions much over the 600-by-400 range. Some users, however, do not agree with me about the importance of screen resolution, and very high-resolution displays do substantially increase the cost of hardware. All the programs support interfaces to a variety of pen plotters. In a production environment, a high-speed plotter with multiple pens that are changed and capped automatically is almost a necessity.

TWO-DIMENSIONAL DRAFTING PACKAGES

The four essential functions that any CAD system must provide are geometry creation, geometry editing, dimensioning and labeling, and plotter output. Geometry creation is the construction of a shape on the screen out of a set of standard elements which generally include points, lines, arcs, circles, and polygons. Most programs can also construct free-form curves through a series of arbitrary points. The speed with which these basic elements can be placed, scaled, and joined to form a drawing is the first measure of a CAD system's effectiveness. The three 2-D programs tested also allow the user to create a library of special predrawn entities. These can be recalled as primitives, scaled, rotated, or mirror-imaged to quickly create drawings using standard shapes or parts. Another very powerful and important feature is the ability to define any object in a drawing as a unit and then to repeat that object as a rectangular or circular array, such as bricks in a wall or teeth on a gear. Once a block has been defined, it can also be scaled, moved, rotated, mirror-imaged, or deleted as a unit.

Objects are placed on the screen by means of a keyboard, a digitizer pad with stylus or puck, a mouse, or a joy stick. In my opinion, the most effective form of input is the digitizer pad with a combination of command menu and manual cursor control. Speed and control are significantly lower with a mouse or joy stick. The keyboard is used in all systems to set up general drawing parameters and to enter commands not available on the digitizer pad.

Other features common to all the programs are layering, true zoom, on-screen reference grids for geometry construction, automatic cross-hatching, and a menu-based command structure. A drawing can be constructed in layers that can be edited, scaled, deleted, plotted, and turned on or off independently. The layering feature is very useful for checking bolt-hole alignments, interferences, and part fits against a layout. It can also be used to suppress dimensions and text by placing them on a separate layer of the drawing. True zoom allows the designer to close in on a small area for viewing or constructing small details at full resolution. It also permits zooming out for an overall view of an object, such as a building or even a city, that is far larger than the screen.

A grid of evenly spaced dots on the screen speeds up the creation of scaled drawings. It provides a set of reference points for cursor motions at any scale. With the snap-to-grid command, the user can force a line generated by the cursor to follow a point-to-point pattern. The grid system enables the designer to create simple rectangular and isometric shapes at exact scale without having to type in dimensional data. Automatic cross-hatching allows any closed region to be picked out and filled with a selected crosshatch pattern at the touch of a key. The menu-based commands facilitate use by beginners and minimize reliance on reference manuals.

It is, however, important to be able to bypass the menu structure and have direct access to commands with a digitizer pad or user-defined keys for increased speed.

All the two-dimensional programs claim to have automatic dimensioning but none of them actually does. This is not surprising considering the sophistication that would be required of an expert system program that could look at a piece of complex geometry, decide what dimensional information the fabricator needed, and create all the required witness lines, arrows, and dimensions. In practice, the points to be dimensioned and the location of the dimensional data are selected manually by moving the cursor. The witness lines, arrows, and correct dimensions are then produced automatically. At best, this is a semiautomatic process, but it is certainly faster and more accurate than element-by-element dimensioning using the general editing commands. All the programs allow line widths, styles, and colors to be changed for maximum legibility.

AutoCAD 2. Autodesk, Inc. claims that its product is the most widely used of all microcomputer CAD software, with more than 20,000 users. It consists of a basic program plus two advanced drafting extensions. A new extension has some 3-D capability. AutoCAD has tailored versions of its program for a wide variety of IBM PC/XT/AT-compatible machines, and supports many different plotters, digitizers, and displays. French, German, Italian, and Swedish versions are available, and Spanish and Japanese versions are in the works.

Overall, AutoCAD is an excellent program for layout, design, and drafting. Its drawing and editing commands, which are powerful and easy to use, include dynamic drag of objects under cursor control, rubber-banding of lines connected to a moving point, a freehand sketch mode, chamfers, fillets, creation of rectangular and circular arrays, and object snap for drawing tangents and perpendiculars to an existing construction. AutoCAD's cross-hatching capability is extremely flexible and sophisticated. Crosshatch lines will automatically work around any closed object or lettering contained within the crosshatch area, and over forty standard crosshatch patterns are included in the program.

The user macro language allows the command structure to be designed for specific applications and should save the advanced user a great deal of time. I have been told, however, that in its current form the language is rather difficult to work with.

AutoCAD supports both single- and dual-screen displays; color graphics cards provide a main screen resolution of up to 1024 by 1024. In the dual-screen system, the commands and menu are transferred to the second screen, typically a standard 12-inch monitor, reserving the main screen for drawing. Although some users disagree, I find the dual-screen display to be a significant advantage.

Text attributes, both visible and invisible, can be attached to a defined object in a drawing and extracted for later use in a data base such as a bill of materials. The program includes five standard text fonts, which can be scaled, centered, and justified at the left or right. Witness lines, arrows, and dimensions are separate objects in the data base. This is a liability in that it may increase the time required to make a dimensional change, but it is also an asset in that it permits witness lines that cross to be broken at the crossing. It also expedites revision by overwriting dimensions, and it can be used to eliminate multiple overlapping witness lines on pen plots. Whether or not dimensions have tolerances is up to the user.

AutoCAD does lack several features found in other programs. It has no built-in facility, as Anvil and VersaCAD do, for calculating areas (other than simple polygons), volumes, sectional properties, and center of mass. Unlike VersaCAD, which provides for step-by-step undelete, it has only one level of undelete or undo. It does not have an automatic save to protect against work being lost during a power failure. Ordinate dimensioning is not among the semiautomatic dimensioning options, although it can be done manually with the general editing commands. Line styles and colors cannot be mixed within a single drawing layer. The program cannot draw splines or obliquely mirror, except by means of a complex series of commands. An experienced user claimed that the 3-D wireframe program included in the latest extension is quite limited compared with such programs as Design Board Professional and CADkey.

A major advantage of AutoCAD is that it can link up with programs for mainframe-based CAD/CAM, finite element analysis, and machine tool control. An especially powerful combination is the Design Board Professional 3-D design program and AutoCAD's editing features. Specialized component libraries, menus, and supplementary programs are available from AutoCAD users' groups.

AutoCAD is not copy protected, so back-up copies and hard-disk file sharing are not a problem. However, the licensing agreement only grants the right to use the software for up to twenty years, not outright ownership. Finally, the large user base ensures that maintaining and upgrading the software will continue to be profitable and that related software and accessories will continue to be compatible with AutoCAD.

VersaCAD Advanced. This program from T&W Systems is missing a few of AutoCAD's features but it has others that AutoCAD lacks. The VersaCAD command set for building and editing drawings is substantially the same as AutoCAD's. Its features include dynamic drag, rubberbanding, array creation, object snap, the attachment of text attributes to objects, and support for high-resolution displays and dual screens. Drawings are saved on disk as they are created, and a sequential undelete feature makes it easy to withdraw from an unsuccessful construction to an earlier stage of the drawing.

Individual objects can be assigned different line styles and colors within a single layer. VersaCAD's group manipulation commands provide many ways of quickly identifying or selecting a group of objects that share (or do not share) a common attribute and of editing them simultaneously. A construction can be mirror imaged at an angle, which in AutoCAD requires a complicated sequence of commands. The program can be set to keep text reading from left to right even when all the geometrical constructions have been rotated or reversed. This greatly simplifies the drawing of parts that are mirror images of one another.

VersaCAD's line-trim and fillet commands, including a chain fillet that can draw all the corners of a closed polygon with a single command, are more versatile than AutoCAD's. VersaCAD can also automatically calculate the area, perimeter, center of gravity, and sectional moments of an object, although the calculations take a relatively long time. Text can be centered, justified at the left and right, and scaled to fill a predefined area. A recent extension is said to include some important 3-D capabilities, including automatic hidden line removal, perspective views, and shaded surfaces. This extension, if it is well de-

signed, will give VersaCAD a significant advantage over most of its competitors.

Among VersaCAD's drawbacks is its lack of a macro command language or a way of creating user-defined menus. (T&W says it will release its first version of a macro command language this month.) Neither does VersaCAD have a sketch mode or a chamfer command, and it does not provide for library objects with parametric features. It has only one standard crosshatch pattern and does not support ordinate dimensioning as a standard option. VersaCAD does not handle radii and angular dimensions as easily as AutoCAD does, and it does not include optional automatic tolerancing.

T&W has changed VersaCAD's copy protection scheme from hardware to a software lock that will allow a limited number of backup copies. VersaCAD can be used on a number of IBM-compatible micros, including Tandy, Texas Instruments, and AT&T. It also is available for the HP9000 series 200 microcomputer and includes a link program that can pass files to other CAD systems, including mainframes, using an industry standard format. Like AutoCAD, it can be linked to Design Board Professional's 3-D design program. Most popular digitizers and plotters are supported, and special predrawn symbol libraries are available. T&W Systems also offers a less sophisticated design program called CADapple for Apple II series computers.

Anvil 1000MD. This program from MCS Corporation has some excellent features as well as one flaw that rules it out as an acceptable design and drafting package: lack of an undelete or undo function. If you hit a wrong key with Anvil, the only way to recover is through the general editing commands. During a demonstration conducted by an experienced salesman from MCS, the lack of an undelete required some awkward corrections.

Cursor control has been by means of a joy stick, which provides very poor control compared with digitizer pads. (Version 2, released in September, does support a digitizer.) Anvil also lacks a macro command language. It relies on the keyboard much more than the other programs do, and it is limited to a single screen with 640-by-480-pixel resolution containing both menus and drawings.

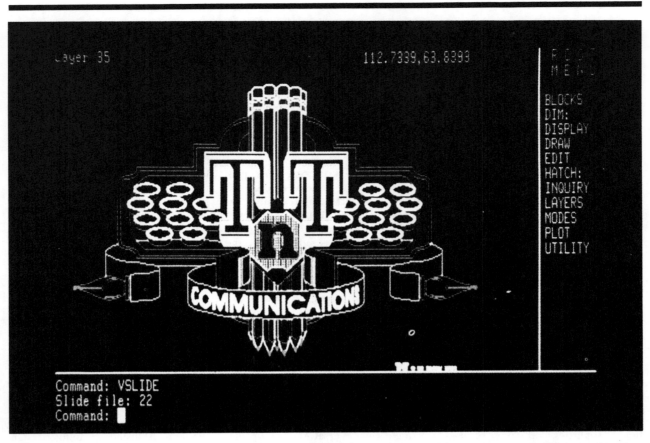

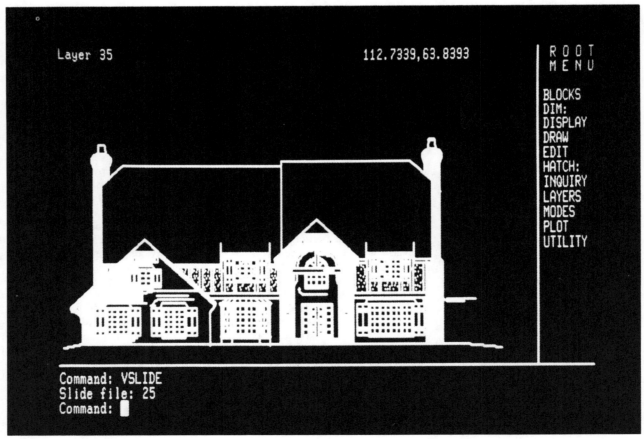

However, Anvil does have assets not enjoyed by AutoCAD or Versa-CAD. In spite of a few missing items, its geometry construction and editing commands are exceptionally complete. Its ability to calculate area, center of mass, and sectional properties is powerful and very fast. It has some useful construction aids for producing isometric wireframe views by extrusion and can even produce elevations of a developed isometric, giving it some minimal 3-D capabilities and accelerating the creation of auxiliary views for simple objects.

Anvil's screen redraw speed is noticeably faster than that of the other programs, a significant consideration when dealing with complex drawings. Anvil has a zoom command that will automatically cause any selected object to fill the screen. The zoom facility is instantly available from within any of the editing and construction commands, as are commands to change line color or font.

Anvil's dimensioning and annotation command set has many features not found in the other programs, such as ordinate and base-line dimensioning, surface finish marks, balloons, and optional dual units and tolerances. Each dimension group, including arrows, witness lines, and leaders, is treated as a unit. Entities in the dimension group do not have to be edited separately, accelerating the revision of dimensions. A nice touch is a command to make all dimensions and associated witness lines in a drawing invisible. This makes it easy to temporarily strip away dimensional information from a finished drawing, which can then be used in a manual or assembly drawing or as an overlay for checking alignments and fits against another drawing. Anvil has an automatic save similar to VersaCAD's to prevent data loss. It also has a chain fillet command and a spline generator for producing a smooth curve through an arbitrary sequence of points.

Several construction and editing commands available in one or both of the other programs are missing in Anvil, including rectangular array duplication, dynamic drag of objects under cursor control, independent X and Y scaling, and a free sketch mode. Anvil also has no facility for defining and saving specific views as windows for future recall or separate plotting, and

no on-screen running feedback of the current cursor position.

Anvil does not at present support a wide variety of hardware or interface options. An IBM PC/XT, /AT, or similar computer is required, and the /AT is strongly recommended. Anvil provides data transfer to Anvil-3000D and Anvil-4000 software running on minicomputers and mainframes via industry standard formats. Presumably these standard format files can be accessed by other programs using the same standard, although this is not mentioned in the MCS literature.

Anvil's user interface features are surprisingly weak, considering that its older siblings, Anvil-3000D and Anvil-4000, are widely used in larger systems. Anvil's strongest points, its speed and powerful dimensioning and annotation features, are suited to the creation of large, complex drawings, such as those found in mold and casting design. If the user interface were better, the hardware options more flexible, and at least one level of undo available, Anvil would be a superior 2-D package.

THREE-DIMENSIONAL DESIGN PACKAGES

The distinguishing feature of a true 3-D program is its three-dimensional data base. This permits an object model to be created as a single 3-D entity from which the standard drafting views are extracted, rather than as a group of only nominally related 2-D drafting views. One, at least potential, advantage of a 3-D system is that design changes can be made by modifying a single entity (the 3-D drawing) and then regenerating the 2-D views, which are thus updated all at once. In practice, because none of the programs discussed here is able to automatically update dimensional data and witness lines along with the geometry, a certain amount of redimensioning of the 2-D views is required. Associative dimensioning is available in only one microcomputer package that I am aware of—SuperCAD—and it is sold only as part of a complete system that costs at least $24,000, about double the cost of the systems being considered here.

Another advantage of 3-D programs is that they permit a designer to view a perfectly scaled wireframe representation of an object from any angle, and in some programs from any per-

spective. This ability, especially in combination with automatic hidden line removal, as in Design Board Professional, is invaluable to designers of consumer products for whom appearance may be as important as function. It can also be useful in producing 3-D drawings for proposals, design presentations, customer manuals, and product documentation. The Personal Designer package goes one step beyond wireframes in permitting wire mesh and shaded surface modeling and, with an optional program, some finite element analysis.

Of the three programs reviewed here, CADkey and Personal Designer attempt to provide a fairly complete 2-D drafting capability along with their 3-D features. Design Board Professional offers an easy link to two powerful 2-D drafting programs, AutoCAD 2 and VersaCAD Advanced. Because the 3-D programs must construct and edit geometric shapes on the screen much as the 2-D programs do, they possess many of the same drawing commands. They also have a set of commands for handling 3-D rotations, translations, and mirror images and for creating 3-D figures by extrusion or revolution of 2-D figures. All these programs can use a variety of digitizer pads as primary input devices and can generate on-screen dot grids at any scale and orientation as construction aids. In addition to the usual drawing elements (points, lines, arcs, polygons, and circles), the 3-D programs can accept three-dimensional as well as two-dimensional objects into their libraries of special shapes defined by the user. This feature can be used in mechanical design, for example, to develop a library of predrawn standard fasteners, sprockets, motors, and valves.

CADkey. The combined 2-D and 3-D features of this package from Micro Control Systems, Inc. may provide a low-cost solution to most design and drafting problems. In CADkey, a part is drawn as a three-dimensional isometric wireframe using direct 3-D coordinates or extrusion or rotation of a 2-D shape. A digitizer pad is the primary input device. The software supports the usual line construction commands, including grid snaps, parallels, tangents, end points, rubberbanding, and keyed-in coordinates. The editing and construction command set is well designed and quite easy to use, although it is not as complete as those in AutoCAD and VersaCAD. For example, version 1.11 does not have independent *X* and *Y* scaling, polygons, ellipses, chamfers, simple mirror-imaging, dynamic drag, or automatic text justification. CADkey does have a multilevel undelete function, dimensional tolerancing, excellent 3-D filleting, and the ability to define a selected area as a window to be saved or printed independently. Zoom, pan, and line-style, color, and layer changes can be implemented immediately from within any command in CADkey.

Version 2.0, introduced in late September, adds ellipses, polygons, 2-D and 3-D chamfers, and mirror-imaging about an arbitrary axis.

A fundamental problem in version 1.11 is the inflexible command structure, which often requires passing through several levels of menu to reach a particular command. Although the program does have a useful set of immediate-mode commands that can be reached directly from within any other command, there is no way to bypass the menu if the required command is not in immediate mode. It is possible to access commands by punching in sequences of integer numbers on the keyboard, but this requires an uncommonly good memory and is not an adequate substitute for direct access to commonly used commands from a tablet menu that other programs provide.

The set of immediate-mode commands has been expanded in version 2.0, but a tablet menu cannot be supported until a macro command language is made available in a future release.

Creating standard drafting views from a wireframe model in CADkey is unnecessarily complicated. A long sequence of commands is required for what should be a standard, single-command operation. In addition, hidden lines are retained when the drafting views are generated and must be removed later by a manual select-and-delete process. CADkey does not have automatic hidden line removal or perspective views, as does Design Board Professional. Dimensions, including witness lines and arrows, are treated as a single entity and the resulting advantages and disadvantages are the the same as in the 2-D packages. Ordinate dimensioning is not supported as a standard dimensioning format. An automatic save facility is also absent. Maximum screen resolution is at present 832 by 624, and there is no dual-screen option. (Pete Smith, president of Micro Control Systems, contends that having to look back and forth at two screens is counter-productive.)

Of the two programs under review that combine reasonably full drafting functions with a true 3-D data base and 3-D wireframe drawing capability, CADkey is the least expensive. For those who require a 3-D program and are willing to sacrifice some of the more sophisticated drafting features to keep costs down, this may be a good choice. CADkey will run on the IBM PC, PC/XT, PC/AT, and many compatibles. Interfaces to several pen plotters and digitizer tablets are supported.

Unfortunately, copy protection is hardware-based; the device plugs in between the computer's parallel port and printer.

Design Board Professional. Two powerful features particularly suit this package from Mega Cadd, Inc. to the needs of designers. These are automatic hidden line removal and perspective views. A finished design can be viewed from any angle and perspective without extraneous lines cluttering up the view, and a structure can be viewed from the inside if the 3-D model is sufficiently detailed.

In two fairly simple demonstrations the automatic hidden line removal was extremely effective, although in more complex drawings some manual touch-up may be required. Compared with manual line-by-line deletion, however, automatic line removal should increase productivity considerably. Another attractive feature of Design Board is a command to create three standard drafting views automatically from a 3-D wireframe view. Watching those views appear at the push of a key was very impressive. Another useful feature is 3-D surface illustrations using rectangular mesh patterns, which, with some effort, can be made to follow complex curved surfaces.

Design Board can display up to four views simultaneously and can take planar sections from a 3-D model to display partial views. When several views are displayed, changes in one view are reflected in the other views as the screen is redrawn. Separately drawn 3-D objects can be merged to form a single entity, although the joint lines are retained unless they are deleted manually. This could be useful when creating assembly drawings from a series of parts. Design Board is well suited for creating exploded views of mechanical assemblies for documentation.

Of the programs under review, Design Board is the only one incapable of producing finished engineering detail drawings on its own. It is intended strictly as a visual rendering tool and has no dimensioning and only limited annotating capability. However, it has hooks to AutoCAD and VersaCAD which permit the drafting views to be easily brought over for completion. Combining Design Board with AutoCAD or VersaCAD results in a very powerful, albeit somewhat expensive, design tool.

One disadvantage of Design Board is that the command menus take up a large part of the screen, leaving only a small area for drawing compared with the other programs. For this reason, a dual-screen option would be a great improvement. Design Board does not provide an easy way of drawing fillets, which limits its ability to create detailed 3-D views of certain types of objects. Fillets can, of course, be added to views that are ported to AutoCAD or VersaCAD for completion. Automatic save, multilevel undelete, cross-hatching, layers, and a macro language are also lacking, and only one line style and color are available unless plots of drawings are overlaid.

It is difficult to evaluate Design Board with respect to the other programs because its purpose is clearly different. Rather than attempting to outdo its competitors in features required for creating detail drawings, Design Board makes use of its competitors' abilities to augment its own. What the program is able to do it does better than any of the others, but by itself it would be a poor choice for an engineering design group or drafting department. In conjunction with AutoCAD or VersaCAD, however, it offers some useful and very powerful features. Design Board runs on many IBM-compatible computers and supports a variety of digitizers, plotters, and graphics cards.

Personal Designer. This package is the first contribution of Computervision Corporation, of Burlington, Massachusetts, to micro-based design. Personal Designer is based on Computervision's Cadds software package,

Comparison of Program Features

	AutoCAD	VersaCAD Advanced	Anvil 1000MD	CADKEY (1.1)	Design Board Professional	Personal Designer
Three-dimensional	limited	new	limited	yes	yes	yes
Auto save-to-disk	no	yes	yes	no	no	yes
Dual screen	yes	yes	no	no	no	no
1000-pixel display	yes	yes	future	no	yes	no
User-defined macros	yes	no	no	no	no	yes
User language	yes	no	no	yes	no	yes
Tablet menu	yes	yes	future	future	no	yes
Digitizer input	yes	yes	yes	yes	yes	yes
Mouse input	yes	yes	no	yes	yes	no
Cursor-coordinated feedback	yes	yes	no	yes	yes	no
Section property calculation	no	slow	fast	no	no	future
Undelete (undo)	yes	yes	no	yes	yes	yes
Sequential undelete	no	yes	no	yes	no	no
Sketch mode	yes	no	no	yes	no	no
Text association	yes	yes	yes	soon	no	yes
Parametric geometry	yes	no	no	no	no	yes
Rectangular array	yes	yes	no	no	no	yes
Circular array	yes	yes	yes	no	no	yes
Save window view	yes	yes	no	yes	yes	yes
Independent XY scale	yes	yes	no	no	yes	yes
Ordinate dimensioning	no	no	yes	no	no	future
Dimensions one entity	no	no	yes	yes	no	yes
Dual-unit dimensions	no	no	yes	no	no	no
Tolerances	yes	no	yes	yes	no	yes
Drafting feature symbols	no	no	yes	no	no	no
Number of layers	no limit	250	1024	256	1	128
Number of text fonts	5	2	4	2	none	1
Dynamic drag	yes	yes	no	no	no	yes
Rubberbanding	yes	yes	no	yes	yes	soon
Conic sections	ellipse	ellipse	ellipse	no	no	yes
Bezier curve or spline	no	yes	yes	no	yes	yes
Chamfer	yes	no	yes	no	no	no
Fillet	yes	yes	yes	yes	no	yes
Chain fillet	no	yes	yes	no	no	yes
Cross-hatch patterns	40	1	10	soon	none	editable
Multi colors/line styles within layer	no	yes	yes	yes	no	yes
Justify text	yes	yes	yes	no	no	yes
Bill of materials	add-on	option	no	future	no	yes
3-D rotate	yes	yes	no	yes	yes	yes
3-D user library	no	no	no	yes	yes	yes
3-D hidden line removal	yes	yes	no	manual	yes	future
Drafting views from 3-D	no	no	no	awkward	easy	future
Perspective views	no	yes	no	no	yes	future
Complex sections	no	no	no	no	no	yes
3-D mesh surface	no	no	no	no	yes	yes
3-D shaded surface	no	yes	no	no	no	yes
3-D finite element link	no	no	no	yes	no	yes
IGES compatible files	soon	extra	yes	extra	yes	yes
Copy Protection	none	software	key disk	hardware	none	key disk

Source: Mechanical Engineering, November 1985, 24-34

which is used on many minicomputer- and mainframe-based design systems throughout the world. At $5800 for the basic package, it is by far the most expensive of the software packages under review. Although it is in many ways a very powerful and effective program, its advantages are probably not worth the price for many users. It lacks some of the 2-D drafting features found in other programs, it is reputed to be relatively difficult to learn, and it does not support dual screens or resolutions over 640 by 480. Nevertheless, it has some special capabilities that make it the best (and perhaps the only) choice for certain applications.

Personal Designer's strongest suit is its ability to model complex 3-D surfaces and surface intersections. In this area it far surpasses any of the other programs. Surfaces can be displayed as wire meshes or shaded solids, and shaded surfaces can be lit from various angles. The model of a 3-D object can be sectioned in almost any way, and links to a powerful 3-D finite element analysis program are available. Where these capabilities are essential, such as in the design of complex cast and molded parts or 3-D assemblies, Personal Designer easily justifies its higher price.

Other attractive features are the program's powerful macro command language, geometry creation and editing commands, built-in documentation, and controllable automatic save. Included in the geometric commands are independent X and Y scaling, crosshatch patterns that can be edited by the user, spline generation, and colors and line fonts that can be changed within a single layer. Personal Designer does not at present support ordinate dimensioning, section property calculations, perspective views, automatic hidden line removal, or automatic generation of drafting views from a 3-D model. Version 2.0, due out in the fourth quarter of this year, is supposed to provide perspective views. Computervision claims that the other features will be included in future revisions of the package. If they are, provided that the price is not raised and that less expensive programs do not come to match its unique features in the meantime, Personal Designer may become the best all-purpose design program available on microcomputers.

Personal Designer runs on IBM PC/XT and /AT computers and supports a limited selection of graphics controller options, including the IBM professional graphics controller, and several major brands of plotters. It has links to the MSC/pal finite element analysis package and to Computervision's larger CAD systems, and it supports networking.

MAKING A CHOICE

A feature that is conspicuously missing from all the programs I have seen, with the exception of Anvil, is ordinate dimensioning. This is widely used and very important in mechanical drafting, especially in the detailing of very complex parts such as injection molds or castings. Another invaluable capability is associative dimensioning. This could save more time than any other feature and it is missing from all the low-cost programs.

Because of fierce competition and the continual release of improved versions, reviewing these programs is something of an exercise in futility. In the course of writing this article I have learned of no fewer than five other programs aimed at the same market. One of these, called Vision 2000, appears to offer a reasonable 2-D drafting capability at less than $500, which could make CAD an option for many individuals and educational institutions that could not otherwise afford it.

None of the programs that have been discussed is clearly superior to all the others in every respect. The best choice depends on the specific design or drafting application, the available budget, and sometimes on the necessity of interfacing with a larger design system or with auxiliary programs such as finite element analysis or CAM packages.

Dealer support is also an important factor. These programs are far more complex and subtle than word processing or spreadsheet programs. Some training is generally needed to use them efficiently, and advice on interfacing to auxiliary software and to hardware such as plotters, digitizers, and printers may also be required. Setting up the new equipment, using it productively, and integrating it into the existing design environment should also be considered. A knowledgeable dealer can help with

these questions and with correcting and upgrading the program in the future. Good CAD dealers tend to be specialists: it is unlikely that a general computer dealer who happens to carry a CAD program or two will command the level of detailed knowledge required to support these programs properly.

The CAD software market is in many ways like the microcomputer market of the past few years. Intense competition leads to rapid product evolution, declining prices, narrow profit margins, and the failure of weaker competitors. Chances are that no matter what choice you make something substantially better will be available for less money six months or so down the line. But the packages currently available are good, with capabilities that were unheard of in this price range less than a year ago. Designing with paper and pencil will soon be as obsolete as manual typewriters and inkwells, and I, for one, won't miss it a bit. ⓂⒺ

For More Information

VersaCAD. T&W Systems, Suite 106, 7372 Prince Drive, Huntington Beach, Calif. 92647; (714) 847-9960. Version 4.0: $1995. Options: Versalist (bill of materials), $495; IGES conversion utility, $500; 3-D to 2-D surface modeling, $495.

CADkey. Micro Control Systems, CADkey Division, 27 Hartford Turnpike, Vernon, Conn. 06066; (203) 647-0220. Version 1.1: $1895; Version 2.0: $2695; IGES translator: $750.

Design Board Professional. Mega CADD, Inc., 401 Second Avenue South, Seattle, Wash. 98104; (800) 223-3175, (206) 623-6245. Version 2.04: $1750.

AutoCAD. Autodesk Inc., 2320 Marinship Way, Sausalito, Calif. 94965; (415) 332-3244. Version 2.1 (including Advanced Drafting Extensions 2 and 3): $2500.

Personal Designer. Computervision Corp., 15 Crosby Drive, Bedford, Mass. 01730; (617) 275-1800. Geometric construction and detailing (main module), Version 1.02: $5800; Version 2.0: $5800. Surface modeling module: $2800.

Anvil 1000MD. Manufacturing and Consulting Services, Inc., 9500 Toledo Way, Irvine, Calif. 92718; (714) 951-8858. $2995.

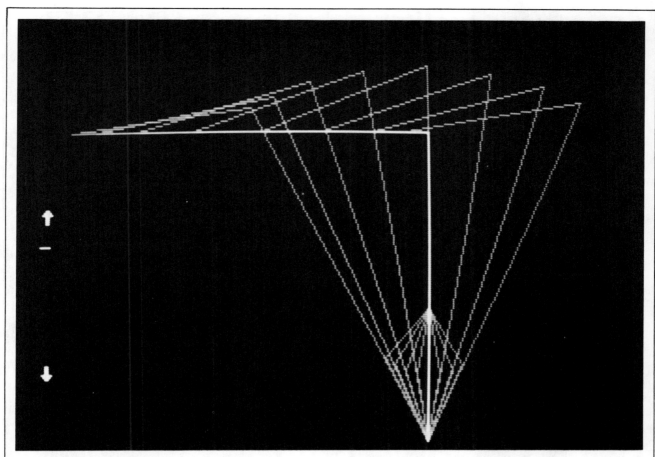

Micro-Mech mechanism analysis. A multiple-exposure photo of positions of a quick return mechanism. *(Ham Lake Software)*

In Search of Component Design And Analysis Software for the IBM PC

There's a lot out there. The best programs not only handle complex "what if" questions, but also allow the novice user to come to a design by only stating elementary criteria.

Donald Smith

Associate Professor
Department of
Mechanical Engineering
University of Wyoming
Laramie, Wyo.

The IBM PC has made itself known in mechanical design. This was amply demonstrated at the trade exhibit of the 1984 ASME Computers in Engineering Conference, where more IBM PCs were called into service than any other micro as software vendors demonstrated their wares. This article will not debate the usefulness of the machine; it will instead direct the reader to the software for component design that runs on it. It is also my hope that the information presented here will stimulate readers to write new software.

This article—essentially an annotated bibliography—stems from my research over the past summer at IBM's

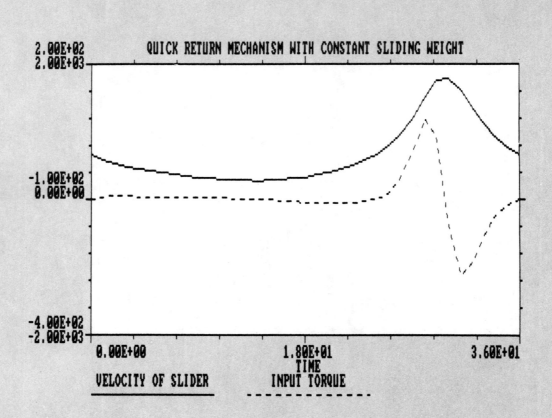

Mechanism analysis. The Micro-Mech program displays the velocity of the sliding link (solid line) and the torque required (dashed line) to drive the mechanism shown on the previous page at a constant angular velocity input. *(Ham Lake Software)*

Information Products Division in Boulder, Colo. I have used the following conventions: when a vendor has software sitting on its shelves that is ready to run on an IBM PC, that company's name is printed in **boldface**. The table on page 56 gives the address and phone number of all the vendors highlighted in this way. Programs from other vendors or from private developers that are not directly aimed at the PC, but are likely candidates, are referenced.

I have also referenced some useful sources of information on methods in different areas of computer-aided design. Some of the works are not addressed specifically to microcomputing; but, as the micro-mini-mainframe "classes" come closer, this distinction becomes less important.

Springs

Before discussing special-purpose spring software, we should look at the often useful spreadsheet. A spreadsheet is not just for accountants. Al-

though not aimed directly at an IBM PC, a good overview of the spreadsheet approach to a variety of mechanical engineering design problems can be found in [1]. Another manual for spreadsheet use, *Engineering Problem Solving with Personal Computers* [2], might also be useful.

Similar to the spreadsheet approach but perhaps easier to use for most engineers is TK!Solver, from **Software Arts.** With this product the equations appropriate to the problem are entered along with the values of the parameters. One of the impressive features of the software is that the solution parameters need not be explicitly defined by equations. For example, in the equations EXP $(x) = 2$-$x*y$ or SIN $(x*y) = 3$-x-y, TK!Solver will solve for either x or y given the other value. So what does all this have to do with springs? A mechanical engineering package that operates with TK! Solver has the helical spring equations set up for both compression and extension springs.

A product from **Interlaken Technology** competing with TK!Solver is Calfex, which also has an applications package for mechanical engineering. For the main Calfex program the user enters the appropriate equations; an algorithm can also be entered as a Basic subroutine. The ME pack has equations for helical springs of round or rectangular wire. (Also see Werner, F.D., "Calfex: An Equation Solver for Engineers," *CIME,* Vol. 2 No. 3, Nov. 1983, pp. 53–59.)

A set of programs—not implemented on a micro—for the design of flat or contact springs would apply to springs used to retain circuit boards while supplying an effective electrical contact [3]. The eventual springs typically are combinations of curved and tapered beams with one or more free ends.

All of these programs have considerable merit but are not on target for our purposes. The spreadsheet approach is excellent for addressing "what if" questions, but it requires that the user be reasonably knowledgeable about acceptable stress levels, available wire sizes, and other design considerations. We would really like to see a program that includes this information and one that would allow the infrequent user to come to an acceptable design by only stating the load and length combinations desired. The software described in the remainder of this section comes closer to this goal.

Engineering Software is selling a program for the design of helical compression and extension springs which has the material properties of several spring alloys internally stored. The user specifies material, design load magnitude and type, coil diameter, free length, end conditions, and, optionally, the wire diameter. The program calculates the number of coils, spring rate, solid height, wire diameter (if not specified with the input), and safety factor.

A compression spring design program from **Practical Engineering Applications Software (PEAS)** has the user select the stress in the coil, spring rate, material, end type, and applied force. With additional input including mean diameter and free height, the output includes total coils, wire diameter, solid height, inside diameter, outside diameter, index, pitch, and shear stress.

Extension spring design is part of a rather comprehensive spring design series by **American Dynamics**, which also includes programs for compression, torsion, torsion bar, and leaf springs. The programs are interactive and largely menu-driven. The approach to design is iterative—with the user doing the iteration. The user supplies appropriate stress levels for the design. End loops are not considered for extension springs; however, recommended minimum radii for leaving the coil body are given by the program.

EMJ Inc.'s program for design of helical compression springs runs on the Texas Instruments TI-59 programmable calculator [4]. Wire diameter, number of coils, solid height, stress at solid height, spring rate, free length, and weight are calculated, and failure

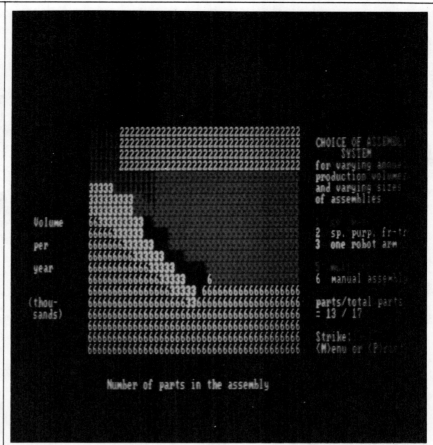

Assembly system economics. The chart shows ranges of batch sizes and number or parts in a product. *(Boothroyd & Dewhurst)*

by fatigue is considered. Material yield strength and spring diameter are input parameters.

From the Lawrence Livermore Laboratories comes a program for the interactive design of extension, compression, and torsion helical springs [5]. The program includes standard wire sizes and materials, and takes fatigue into account. It runs on a Northstar micro, which, except for graphics, is reported to be compatible with the IBM PC.

A Basic program in the open literature could be used for the design of helical compression springs made from round wire [6]. Fatigue considerations are not a part of the program.

A program that designs helical springs for a particular natural frequency [7] has been extended to design for minimum weight, minimum volume, and minimum length [8]. Finally, in [9] one can find along with a discussion of computer-aided design of springs a description of a computer-controlled spring tester.

Cams: To Each His Own

Information on the design and manufacture of cams is abundant. However, I could not find a single general-purpose design package on the market.

One could write a cam design program based on the information in the many articles in the open literature. Any one article tends to be rather specific; the following is a scatter-shot sampling of useful leads.

"Designing Cams on a Microcomputer" [10] is concerned with calculating the profile for a cycloidal motion plate cam and a radial roller follower. Bennet [11] approximates the irregular arcs of the path for an NC cam cutting operation with the circular arcs used by some NC machines. A programmable calculator has been used to generate the paper tape instructions for cutting a cam with an NC milling machine [12].

The functional specifications of cam position, velocity, and follower location have been converted by a

Source: Computers in Mechanical Engineering, January 1985, 53-61

Software on the Shelf

The following companies, which have been indicated by **boldface** in the article, have products available for the IBM PC. Many of the companies have a number of other mechanical engineering programs in categories not covered in my search, or for various subareas of component design. IBM itself has released a useful catalog of software: *Engineering and Scientific Programs for IBM Personal Computers Available from Non-IBM Sources* (Book No. GC34-0588; $12.50). The catalog contains information for developers who want to submit software. Contact a local IBM Product Center or IBM, P.O. Box 1328, Internal Zip 4642, Boca Raton, Fla., 33432.

Addison-Wesley Publishing
Applications Software Div.
Reading, Mass. 01867
(617) 944-3700
Micro Dynamo (continuous simulation)

American Dynamics
12562 Loraleen St.
Garden Grove, Calif. 92641
(714) 539-4498
Spring Design (extension, compression, torsion, leaf springs; torsion bars)
Beam Design

Applied i
200 California Ave., Suite 205
Palo Alto, Calif. 94306
(415) 325-4800
Tutsim (continuous simulation)

Boothroyd & Dewhurst
56 Sherman Lane
Amherst, Mass. 01002
(413) 549-6863
(A package for assembly system economics, design for automatic or manual assembly, design for automatic handling, assembly machine simulation, and "tips and analysis")

c²b² Software Design
763 27th Ave.
San Francisco, Calif. 94121
(415) 841-7175
Beam Analysis

CAD/CAM Systems
28840 Southfield Rd., Suite 142
Lathrup Village, Mich. 48076
(313) 569-0028
Springs
Gear Design, AGMA Standards, and other gear analysis programs.
Shafts

Celestial Software
125 University Ave.
Berkeley, Calif. 94701
(415) 841-7175
Images 2-D and Images 3-D (FE analysis)

Coade
8550 Katy Freeway, Suite 122
Houston, Tex. 77024
(713) 973-9060
Spring (coil spring)
Finite/GP (FE analysis)

Ecom Associates
8634 Brown Deer Rd.
Milwaukee, Wis. 53224
(414) 354-0143
(Plane frame and truss, beam, and column design)

Engineering Software
1405 Porto Bello
Arlington, Tex. 76012
(817) 261-2263
Spring1 (helical compression and extension springs)
HASGAP (helical and spur gears)
Shafts
Shells of Revolution (FE analysis)

Engineering Technologies
P.O. Box 979
Cary, N.C. 27511
(919) 467-8960
Gearmaster (general gear design)

ESDU International
1495 Chain Bridge Rd., Suite 200
McLean, Virg. 22101
(703) 734-7970
Geared Systems
Rolling Element Bearings

Franklin Institute
The Franklin Research Center
Benjamin Franklin Parkway
Philadelphia, Penn. 19103
(215) 448-1566
Genrol (rolling element bearings)

Geartech Software
1017 Pomona Ave.
Albany, Calif. 94706
(415) 524-0668
AGMA218 (life rating for load spectrum, power rating, and Miner's rule)

Ham Lake Software
631 Harriet Ave.
Shoreview, Minn. 55112
(612) 483-0649
Micro-Mech (mechanism analysis)

Intercept Software
3425 S. Bascombe Ave.
Campbell, Calif. 95008
(408) 377-4998
Libra (FE structural and heat transfer analysis)

Interlaken Technology
6535 Cecilia Circle
Minneapolis, Minn. 55435
(612) 944-2627
Calfex and M.E. Pack #1 (equation solver)

MacNeal-Schwendler
815 Colorado Blvd.
Los Angeles, Calif. 90041
(213) 258-9111
MSC/pal (FE analysis)

Moldflow (Aust.) Pty. Ltd.
Commerce Park
4695 Main St.
Bridgeport, Conn. 06606
(203) 374-6181
Moldflow (mold design)

Number Cruncher Micro-systems

1455 Hayes St.
San Francisco, Calif. 94177
(415) 922-9635
SAP86 (FE analysis)

Optimal Software

1445 Crone Ave.
Anaheim, Calif. 92802
(714) 732-6023
Optisolve (equation solver)

Practical Engineering Applications Software (PEAS)

7208 Grand Ave.
Neville Island
Pittsburgh, Penn. 15225
(412) 264-3553
Spring Design (helical compression)
Beam Design

Simulation Software Systems

2470 Lone Oak Drive
San Jose, Calif.
(408) 270-2463
PC Model (assembly line simulator)

Structural Research and Analysis

1661 Lincoln Blvd., Suite 100
Santa Monica, Calif. 90404
(213) 452-2158
Cosmos/M (FE structural analysis)

Sun Soft

P.O. Box 661654
Miami Springs, Fla. 33266
(305) 885-0131
Pack-er (packaging design)

Universal Technical Systems

1220 Rock St.
Rockford, Ill. 61101
(815) 963-2220
Spur and Helical Gear Analysis
AGMA Rating
Shaft Deflection and Stress

computer into machining instructions [13]. Weber [14] reports the continuous-path numerical control of cam manufacturing with a computer. He specifically addresses the problem of matching slopes between curves. A Belgian company has written on the success of its efforts to automate the design and manufacture of cams [15].

Disk cams can be designed from both standard functions and numerical data that describe the follower motion. With interactive graphics, the tasks take shape [16]. The effects of machining errors on cam follower motion have been analyzed with an 11-degree-of-freedom model [17].

A "comprehensive procedure" for the interactive design of plate cams with a DEC minicomputer appeared in 1983 [18]. The magnitude and location of cam profile errors can be determined by a method presented in [19]. Numerical optimization techniques have been used to determine the design parameters of a plate cam with translating or oscillating followers [20].

Principles from optimal control theory are incorporated in the synthesis of high-speed cam follower systems in [21]. Cam profiles suitable for varying high-speed applications are compared in [22]. Maximum allowable contact stresses have also been considered in the computer-aided design of plate cams [23].

The above sampling of topics relating to cam design is from a group of several hundred references resulting from a data base search combining the keywords *cam* or *cams* and *computer* or *computer-aided*. To access more old-fashioned data bases, one can consult the following comprehensive bibliographies and surveys relating to cams.

In 1975 Soni published a cam design survey with 1509 references [24]. That year he also surveyed 82 technical papers that deal with cam dynamics [25]. A similar survey by Chen, with 128 references, reviews the literature on kinematics and dynamics of cam and follower systems [26]. A 1979 survey by Soni cites 365 sources on cam manufacturing methods [27].

Happy hunting!

Spur and Helical Gears

To date, cam design software has been a write-it-yourself affair. The market is much more attractive for the software shopper in gear design.

The Hasgap program of **Engineering Software** is for the analysis of helical and spur gears. The program uses gear type, tooth form, pressure angle, helix angle, horsepower, velocity ratio, material properties, operating factors, and pitch and face width combinations to calculate the pinion and gear bending and surface stresses. Allowable transmitted power is calculated based on user-specified allowable stresses. Contact ratio, transmitted load, and geometry factors are also calculated.

Engineering Technologies offers the Gearmaster programs, in which the user works his way through an analysis or design problem by responding to questions. The input information includes gear ratio, pinion speed, horsepower, basic stress level for the materials, pressure angle, and endurance limit if fatigue is to be considered. There are screen prompts with common values for the basic and endurance limit stresses. The output consists of teeth numbers, diametral pitch, face width, and the geometry, which includes addendum, dedendum, whole depth, clearance, and pitch diameters. Actual stresses and allowable stresses are calculated and the weaker gear is identified. The torque and load on each gear are calculated as well as the shaft reaction forces. The allowable wear load and endurance load are calculated as a function of dynamic tooth load.

CAD/CAM Systems has a number of gear design programs that address the details of manufacturing. The program for spur and helical gear design calculates the distance over or under gage pins. This information is used to set up hob cutters for gear manufacture. The vendor's programs also implement a number of American Gear Manufacturers Association (AGMA) standards. Most of the programs appear to be tutorial.

AGMA218, from **Geartech Software**, can rate a gearset in three ways: a life rating for single horsepower and

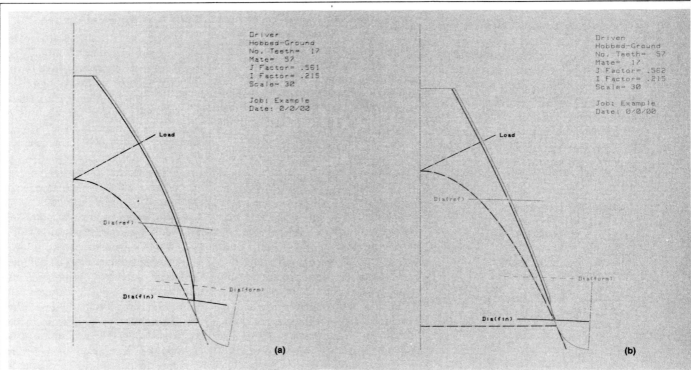

Driver
Hobbed-Ground
No. Teeth= 17
Mate= 57
J Factor= .561
I Factor= .215
Scale= 30

Job: Example
Date: 0/0/00

Driven
Hobbed-Ground
No. Teeth= 57
Mate= 17
J Factor= .562
I Factor= .215
Scale= 30

Job: Example
Date: 0/0/00

Load

Dia(ref)

Dia(form)

Dia(fin)

(a)

(b)

Helical gear analysis. Drawing of driver **(a)** and driven gear **(b)** showing rough and finished machine profiles. **(c)** The sliding ratios give a measure of the noise and

in-out speed, a power rating for a given pinion speed and gearset life, and a Miner's rule rating for a spectrum of loads.

The software of **Universal Technical Systems** is oriented toward manufacture of gears with hobs, shaper cutters, and shaving cutters. For example, one of their programs calculates the load rating for spur or helical gear sets according to appropriate AGMA standards. One program calculates flash temperature to determine the probability of gear tooth scoring; another analyzes the contact conditions with a hone or shaving cutter. As mentioned, the problems addressed seem to be those of the gear manufacturer rather than the designer.

In the library. The open literature abounds with material relating to gear design. In fact, at the ASME-sponsored quadrennial meeting on power transmission the predominate topic is gearing; see, for example, [28]. The proceedings of each of the ten annual meetings of the National Conference on Power Transmission also hold much for the interested reader. Here is a sampling of a few articles from other sources that apply to the computer-aided design of gears.

In [29] a program is presented that determines the gear module based on noninterference and satisfactory dynamic performance, wear, and endurance. As shown in [30], the analytical aspects of gear design are resolved by computer, and the engineer selects the best design from several generated by the computer. How CAD and graphics can be used to determine the geometry strength factors for helical gears is shown in [31].

A computer can be used to calculate an efficient and economical transmission [32]. Fatigue is considered in a program for the design of gears in automobiles, trucks, and tractors [33]. The program will design gears for equal fatigue life between the pinion and the gear and will balance the resistance to bending fatigue with the resistance to surface damage. Finally, a description of a rather comprehensive set of gear programs can be found in [34].

Shafts and Rolling Element Bearings

A program from **Universal Technical Systems** calculates the stress and deflection for two bearing shafts with one end overhung. Input includes

loads applied at any angle around the shaft. Bearing reactions and shaft slope are also calculated.

In the shaft stress and deflection program of **Engineering Software** a shaft can be loaded in three dimensions with bending, torsional, and axial loads or specified support displacements. Supports can be determinate or indeterminate; stress concentration effects are included at points where the geometry changes.

CAD/CAM Systems offers a shaft design tutorial section in their collection of machine design programs. The vendor's approach to mechanical design is from a slightly different direction than others. They provide what might be considered a machine design textbook on a disk. The intent is to teach the user enough on any particular topic so that he can then focus on his specific problem.

An interactive shaft design program has been reported as part of a project to implement micro software for the academic community [35]. The program, which runs on a Northstar, includes stress concentration and fatigue effects, and accepts three-dimensional loading.

The market thins out for rolling ele-

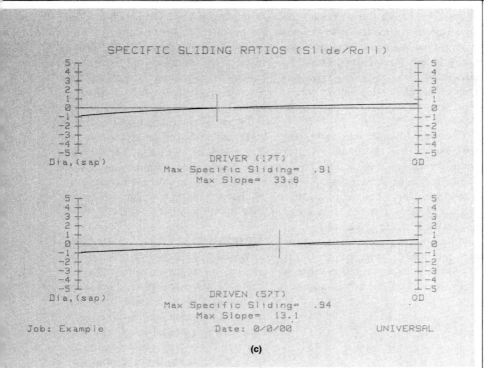

SPECIFIC SLIDING RATIOS (Slide/Roll)

DRIVER (17T)
Max Specific Sliding= .91
Max Slope= 33.8

Dia.(sap)   OD

DRIVEN (57T)
Max Specific Sliding= .94
Max Slope= 13.1

Dia.(sap)   OD

Job: Example   Date: 0/0/00   UNIVERSAL

(c)

heat in the gearset. *(Universal Technical Systems)*

ment bearings. I found only one program, from **ESDU International,** for the IBM PC, but could not get details on its scope.

The Genrol program, by the **Franklin Institute,** is being adapted for the PC. It uses as input the number of bearings in the system, applied loads or displacements, RPM, and bearing type. In addition, internal geometry information, materials characteristics, and preload information are required. The output consists of the system B-10 life, bearing reactions and displacements, bearing spring rates, internal element load and stress, contact area size, and geometry.

Finite Element and Structural Programs

At the 1984 ASME Computers in Engineering Conference a speaker commented that every engineer will have Nastran on his desktop in two to five years. What every engineer would *do* with Nastran or its equivalent wasn't made clear—but certainly more mechanical engineers will be using finite element method codes in their design tasks.

Finite element analysis and structural analysis programs were not specifi-cally included in the initial objectives of my search for design-oriented software. However, several interesting packages turned up. (Also see Falk and Beardsley, "Finite Element Analysis Packages for Personal Computers," *Mechanical Engineering,* Vol. 107, No. 1, Jan. 1985.)

Engineering Software markets a program for the FEM analysis of shells of revolution. The program includes mesh generation and is applicable to deflection, strain, and stress analysis of pressure vessels, formed heads, and piping. Analysis of general beams, 2-D frames and trusses and axisymmetric vessels, shells, and plates can be done with other programs from the company.

PEAS offers several programs that will perform the analysis of single span or continuous beams, selection of structural sections for beams or columns, and other functions relating to structural applications.

A beam analysis and design program that has built-in tables of common structural shapes for steel, timber, and concrete beams is offered by **c²b² Software Design.** Programs for the design and analysis of frames and columns are also in the works.

A set of programs for 2- and 3-D FE modeling—static or dynamic—is offered by **Celestial Software.** Images 2-D and Images 3-D programs have color graphics for modeling and to augment the tabulated results. An internal library of six element types allows the user to simulate a variety of systems. Interactive graphics routines assist in the model verification; dynamic mode shapes are animated in the display.

Intercept Software has the Libra package for 2- and 3-D static analysis. The software contains an interactive model builder and a mesh generator. Nearly 20 element types can be used. (The company has an FE heat transfer program also called Libra that works with its namesake.)

A similar package is currently offered by **Structural Research and Analysis.** One particularly attractive feature of this software is that its preprocessor will accept data in Ansys format. Ansys is a powerful system that resides on minis or mainframes. Thus, the user could move from a PC-based system to a larger system without having to learn a new input language.

MacNeal-Schwendler has acquired a license to the Mace 3-D FE program by Mohango Inc. [36], and is marketing this program under the name MSC/pal. MacNeal-Schwendler has considerable experience in support, based on their version of Nastran for mainframes. The MSC/pal program provides screen graphics and preprocessors to specify the geometry. If a problem becomes too large for MSC-Pal, the program assists in translating the input to MSC/Nastran.

Coade offers Finite/GP for the modeling of 2-D and axisymmetric shapes. The software addresses the problems of structural integrity, heat transfer, plate bending, and potential flow. Other programs by Coade are for the analysis and design of pressure vessels and 2-D frame analysis.

Ecom Associates offers a number of plane frame and truss programs which include both static and dynamic capabilities. Also available from the company are several beam and column design programs.

Number Cruncher Microsystems has taken the mainframe SAPIV program, squeezed it down to the microcomputer, and called it SAP86. SAPIV has been used for a number of years, and those experienced with it may find the new micro version (one of the many "Saplings") easy to learn. However, SAP86 runs in batch mode only, and has no graphics on its own. For the program to display models and deformations, it must be linked up with third-part CAD software.

Mechanisms and Linkages

The Micro-Mech program of **Ham Lake Software** is for the analysis of single or multi-loop, multiple degree-of-freedom mechanisms. Barker has developed an attractive IBM-PC program for the kinematic analysis of planar four-bar mechanisms; it is discussed in detail in [37]. Both programs display an animation of the device being modeled in addition to the tabulated results.

One must be quite careful in using the results of a kinematic analysis when the device may in fact be dynamic for some reason, e.g., a real motor that cannot maintain a constant angular velocity. (One can find a good discussion of these problems in [38].) The leading mainframe-mini software for modeling dynamic systems, DRAM, for planar machinery systems, and Adams, for 3-D configurations, is from Mechanical Dynamics Inc., Ann Arbor, Mich. Portions of this software may be implemented on a PC in the future.

Also on the market for micros are Micro-Kinsyn and Micro-Linkages, for kinematic synthesis of four-bar and some six-bar linkages [39]. Micro-Kinsyn runs on an Apple and requires special hardware. Micro-Linkages, marketed by Terak, was developed by Erdman at the University of Minnesota. It will probably be on the IBM PC sometime this year.

Simulation

Applied i markets a continuous simulation language called Tutsim. Tutsim will accept mathematical model input as a block diagram or a bond graph representation. The direct writing of the differential equations, as is typically done with other continuous system simulators, is not a valid way of entering the model. The block diagram approach is very similar to the hooking up of an analog computer. A good review of Tutsim is contained in [40].

Micro Dynamo, available from **Addison-Wesley,** models continuous systems with up to 300 active variables. Accuracy may be a problem with this package because only Euler integration is available.

Miscellaneous

The following are packages that relate to mechanical design but are not oriented toward a particular component. Also included is software for design topics that do not conveniently fit in any of the previous categories.

Boothroyd and Dewhurst Inc. has developed a set of programs to address problems associated with the assembly and manufacture of mechanical systems. This software is for use at the design level to assist the engineer in arriving at a system that is economical to manufacture. This work is well reported and significant cost savings have been predicted if these programs are used. Xerox, for one, estimated a potential annual savings of $150 million from their use [41]. Another assembly line simulator, PC Model, is from **Simulation Software Systems.**

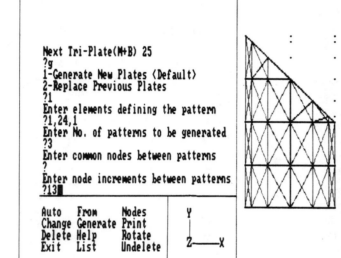

```
Next Tri-Plate(M+B) 25
?g
1-Generate New Plates (Default)
2-Replace Previous Plates
?1
Enter elements defining the pattern
?1,24,1
Enter No. of patterns to be generated
?3
Enter common nodes between patterns
?
Enter node increments between patterns
?13
```

Auto	From	Nodes
Change	Generate	Print
Delete	Help	Rotate
Exit	List	Undelete

(a)

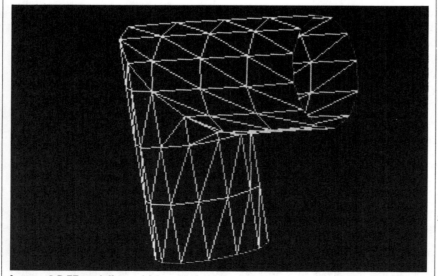

Images 3-D FE modeling. **(a)** A model of a piping elbow is created and **(b)** the complete geometry is displayed with hidden lines removed. *(Celestial Software)*

The program can be used to predict the mean time between failures.

Sun Soft has put out a program for the design of packaging for electronic and consumer equipment. The program contains data on the most common cushion materials. The program evelutes different materials, examines different cushioning techniques, and compares material performance. It gives cushion dimensions, finds peak accelerations, and calculates dimensions.

Moldflow has a series of programs to improve the design and function of injection-molded parts. The programs predict the flow of plastic into a mold, allowing design of parts for improved strength with sized and balanced runners and gates.

Optisolve, from **Optimal Software,** solves a single equation in one unknown or a system of simultaneous equations in several variables, and performs maximization or minimization of a function with constraints. Another optimization program is available from **Engineering Software.** The reader should also look into—or become a member of—the Design Optimization Laboratory, directed by K.M. Ragsdell. The Lab has issued a number of important mainframe and mini Fortran codes on aspects of optimization, nonlinear programming, and CAD. Dr. Ragsdell is at the Dept. of Mechanical and Aerospace Engineering, Univ. of Missouri, Columbia, Mo., 65211. (314) 882-2785. ∎

Acknowledgments:
I would like to thank the IBM Information Products Division for the opportunity to participate in this project, and for their willingness to share the results with the mechanical engineering design community.

References

1 Keene, B., "Solving Mechanical Problems with Personal Computers," *Machine Design* (hereafter *MD*), Feb. 10, 1983, pp. 121–125.

2 *Engineering Problem-Solving with Personal Computers*, Binary Engineering Assoc. Inc., P.O. Box 528, Holden, Mass., 01520, (617) 829-4362.

3 Wolberg, J.R. and Nickerson, E.H., "SPREAD—A Family of Programs for Computer-Aided Spring Design," in *ASME Advances in Computer Technology-1980*, Aug. 1980, pp. 439–445.

4 EMJ Inc., Product Design and Development, 849 Ruth Ave., Erie, Penn., 16509, (617) 232–1194.

5 Comfort, W.J. III, (personal communication), Lawrence Livermore National Laboratories, Microcomputer-Aided Engineering Project, P.O. Box 808, Livermore, Calif., 94550, (415) 422–4908.

6 Hundal, M.S., "Designing Helical Springs on a Minicomputer," *MD*, Vol. 55, No. 27, Nov. 23, 1983, pp. 118–123.

7 Anand, S.K., et al., "Computer Program for the Design of a Helical Spring for Optimum Frequency," *J. of the Inst. of Engineers* (India), Mechanical Engineering Div., Vol. 59, No. ME-4, Jan. 1979, pp. 171–176.

8 Raghavacharyulu, E. and Asnani, N.T., "Computer Program for the Design of a Helical Spring for Minimum Weight, Minimum Volume, and Minimum Length," *J. of the Inst. of Engineers* (India), *op. cit.*, No. ME-6, May 1979, pp. 274–281.

9 Mahler, D., "Spring Design Problems Reduced by Computer-Controlled Machines," *Design Engineering* (U.K.), Vol. 35, Jan. 1980, pp. 37–38.

10 Hundal, M.S., "Designing Cams on a Microcomputer," *MD*, Vol. 55, No. 18, Aug. 11, 1983, pp. 94–97.

11 Bennet, K.W., "Simple Program for Cutting Cams," *MD*, Vol. 55, No. 17, April 7, 1983, pp. 92–94.

12 Davidson, J.K., "Calculating Cam Profiles Quickly," *MD*, Vol. 50, No. 27, Dec. 7, 1978, pp. 151–155.

13 "Precise Cam Profiles are Generated if Computer Interpolates Equations," *Product Engineering*, Vol. 48, No. 5, May 1977, pp. 29–31.

14 Weber, T. Jr., "Simplifying Complex Cam Design," *MD*, Vol. 51, No. 7, March 22, 1979, pp. 115–119.

15 deFraine, J., "Integration of Computer-Aided Design and Computer-Aided Manufacturing for Cams Driving Mechanisms," *Proceedings*, 5th World Congress on the Theory of Mechanisms and Machines (Published by ASME), July 1979, Vol. 1, pp. 122–125.

16 Russon, V.K. and Chase, K.W., "Computer-Aided Design of Disk Cams from Numerical Follower Motion," *ASME Paper 8-WA/DSC-9*, Nov. 1980.

17 Kim, H.R. and Newcombe, W.R., "Effect of Manufacturing Tolerances on Cam Follower Motion," *Proceedings*, 8th Canadian Congress on Applied Mechanics, 1981, Vol. 2, pp. 455–456.

18 Marek, S. and Kinzel, G.L., "Comprehensive Procedure for the Interactive Computer-Aided Design of Plate Cams," in Computers in Engineering 1983—*Proceedings*, 1983 International Computers in Engineering Conference, ASME.

19 Smith, D.L. and Soni, A.H., "Simplified Design and Evaluation of Cam Surfaces," *Forsch Ingenieurwes* (W. Germ.), Vol. 45, No. 1, 1979, pp. 11–15.

20 Fenton, R.G. and Lo, C.H., "Computer-Aided Design of Disk Cams," *ASME Paper 76-DET-62*, Sept. 1976.

21 Chew, M., et al., "Application of Optimal Control Theory to the Synthesis of High-Speed Follower Systems," *ASME Paper 82-DET-100* and *82-DET-101*, Sept. 1982.

22 Chen, F.Y., "Assessment of the Dynamic Quality of a Class of Dwell-Rise-Dwell Cams," *ASME J. of Mechanical Design*, Vol. 103, No. 4, Oct. 1981, pp. 793–802.

23 Terauchi, Y. and El-Shakery, S.A., "Computer-Aided Method for Optimal Design of Plate Cam Size Avoiding Undercutting and Separation Phenomena," *Mechanism and Machine Theory*, Vol. 18, No. 2, 1983, pp. 157–163.

24 Grant, B. and Soni, A.H., "Cam Design Survey," *Proceedings*, 1st Technology Transfer Conference, ASME, Oct. 1974, pp. 177–184.

25 Lee, I.P. and Soni, A.H., "Survey of Techniques in Cam Dynamics," *Proceedings*, 4th Applied Mechanisms Conference, Oklahoma State Univ., Stillwater, Nov. 1975, Univ. Paper No. 32.

26 Chen, F.Y., "A Survey of the State of the Art of Cam System Dynamics," *Mechanism and Machine Theory*, Vol. 12, No. 3, 1977, pp. 201–244.

27 Grant, B. and Soni, A.H., "A Survey of Cam Manufacture Methods," *ASME J. of Mechanical Design*, Vol. 101, No. 3, July 1979, pp. 455–464.

28 4th International Power Transmission and Gearing Conference, ASME, Cambridge, Mass., Oct. 1984.

29 Manjunatha Prasad, M.K., "Computer-Aided Design of Spur Gears," *J. of the Inst. of Engineers* (India), Mechanical Engineering Div., Vol. 63, No. ME-6, May 1983, pp. 232–236.

30 "Computers in Gear Design," *Engineering Materials and Design* (U.K.), Vol. 27, No. 10, Oct. 1983, pp. 28–32.

31 Davey, R.J., et al., "The Use of CAD and Computer Graphics in the Determination of Geometry Strength Factors for Helical Gears," *Proceedings*, Conference on Computers and Engineering, Australian Inst. of Engineering, Sept. 1983.

32 Wadlington, R.P. and Hirschfeld, F., "Computer-Designed Gearing," *Mechanical Engineering*, Vol. 101, No. 6, June 1979, pp. 32–33.

33 Hughson, D., "GODA5—Gear Optimization and Design Analysis 5," *SAE Paper 801026*, 1980.

34 Walton D. and Taylor, S., "Computer-Aided Gear Design," *Desktop Computing for Engineers*, May 5, 1982, pp. 24–29.

35 Comfort, W.J. III, et al., "Technology Transfer and Development of Computer-Aided Engineering with the University Community," *ASME Paper 81-WA/TS-3*, Nov. 1981.

36 Mohango Inc., Mechanical Design Center, Floydada, Tex., 79235, (806) 741-8652.

37 Barker, C.R., "Analyzing Mechanisms with the IBM PC," *Computers in Mechanical Engineering*, Vol. 2, No. 1, July 1983, pp. 37–44.

38 Chace, M.A. and Dawson, G.A., "An End to the Guesstimate in Mechanical Design," *Computer-Aided Engineering*, July-Aug. 1983, pp. 22–30.

39 Krouse, J.K., "Designing Mechanisms on a Personal Computer," *MD*, March 24, 1983, pp. 94–99.

40 Murray, W.H., "Mathematical Modeling with Tutsim," *PC Tech Journal*, July 1984, pp. 41–45.

41 Boothroyd, G. and Dewhurst, P., "Design for Assembly," a four-part series in *Machine Design*: "Selecting the Right Method," *MD*, Nov. 10, 1983; "Manual Assembly," *MD*, Dec. 8, 1983; "Automatic Assembly," *MD*, Jan. 26, 1984; "Robots," *MD*, Feb. 1984.

Solid Models Define Finite-Element Geometry

Solid models increase the accuracy of finite-element analysis and bridge the gap between drafting and manufacturing by providing a common geometric database.

There are two schools of thought on the integration of solid modeling and finite-element analysis. One views this merger as the key to integrating all aspects of factory automation. The other is skeptical about the need for such sophisticated design tools and backs up this observation with the assertion that the software is neither perfected nor affordable. Although each camp has valid points, firms that have tied solids to FEA may well have a competitive advantage over those companies that have not made this link.

The experience of the firms that have integrated solids and FEA has proved that product quality and design productivity are improved. And increasing user sophistication has pushed vendors to provide comprehensive design systems for industry groups such as automotive, aerospace, and other heavy users of CAD/CAM.

The choice of software relies on integration of single function programs for tasks such as analysis and drafting. To link these packages a substantial amount of software development effort is still required. However, many of the packages such as separate drafting and preprocessing programs are already working and providing productivity and accuracy improvements. And better software is rapidly being developed.

Primary Benefits

Three principle reasons for employing solids modeling to drive analysis codes are higher accuracy, improved productivity, and reduced errors. Any assessment of part's physical properties can only be as accurate as its geometric definition. For example, in shape-sensitive applications such as vibration analysis only precise geometric descriptions can yield reliable results. Many users report that by increasing the precision of part descriptions, errors can often be reduced by 15% or more.

Defining this part geometry to a FEA system is one of the most time-consuming aspects of a structural analyst's

Shared Geometry Links Software

Computerized product development uses a series of programs operating on a common geometric database. In developing a clevis, for example, designers begin with an initial concept, displayed as a solid model, and then proceed through analysis. With software from firms such as Catronix Corp. each task, from solids modeling through postprocessing, is performed with a different program. Because geometry must be passed between program segments, any change to the database should be transparent to the user and not affect later processing. Refining a FEA mesh, for example, should have no effect on the layout of a part.

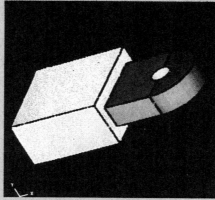

Initial design concept of a clevis was built with a series of color-coded primitives.

A coarse FEA mesh derived from the solid model provides a rough estimate of stresses at the hole.

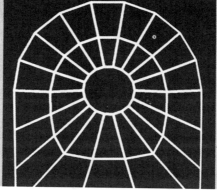

Fine mesh imposed on the entire clevis can yield reliable results for shape sensitive analyses such as vibration response.

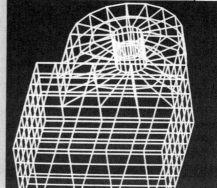

Refined mesh of the same hole follows actual part contours and provides an accurate estimate of stress response.

Reprinted with permission from CAE Computer Aided Engineering, August 1985, 34-35, 38-39, 42, © 1985 Penton/IPC

job. Older alphanumeric interfaces to analysis codes contribute significantly to this barrier. Although newer, solids-driven interfaces are easing this task, geometric definitions are often locked in the CAD system and are inaccessable to the analyst.

Many analysis departments are isolated from design, with information transfered between the two on paper drawings. This isolation can lead to several problems. One hazard is that prints often present ambiguous information which can lead to errors. And because data required by each department is so different, there is often a substantial duplication of efforts in recreating the geometry that the drawings represent.

An outgrowth of the duplication of efforts is generation of a variety of errors from rekeying data. For example, handwritten memos from the design group to the engineering department often results in differences of opinion on part size. Error size from these communications can range from percentage errors in digitizing positions to orders of magnitude blunders from mis-keying. Without a graphic interface to FEA, the most frustrating error may be the miskeying of a single digit, which then goes on to ruin entire FEA runs.

Front End Functions

The terms pre and postprocessing describe the software used before and after runs of the analysis codes. The terms by themselves are not descriptive and considerable variance exist in their interpretation. Because of differences in opinion over the definitions of the terms, many buyers have been disappointed to find that their particluar package does not include functions they expected.

Generally, the so-called front-end functions fall into three categories, geometry definition (best done with solids modeling), geometry documentation (best done with drafting), and preprocessing. Preprocessing, in turn, consists of three main properties: mesh generation, boundary value definition, and material properties definition.

Some less sophisticated preprocessors handle only simple mesh generation on 2D objects. With this limitation, it is not surprising that other functions such as drafting and geometry definition are not included in the software. Of those package with mesh generation, the FEA meshes are composed of nodes, (the coordinates of the mesh), elements (the geometric configuration of the nodes), and type of elements (both geometric and stress response are included in the definition of element type.)

From the platform of mesh generation, the importance of geometry definition can be seen. If models can be developed from geometry residing a computer generated database, the defi-nition of node and element information is simplified, and accurate models can be developed with little effort. On the other hand, if structural engineers must build node and coordinate data from scratch in the form of x, v, and z coordinates, much time and effort is spent recreating data that already exists.

Geometry has the potential of unifying FEA with drafting and mechanical engineering. In a larger perspective, geometry is the basis of nearly all computer-aided design and manufacturing functions. Primarily for historical reasons, computer-aided analysis has been an island unto itself and was not integrated with the rest of the design and manufacturing operations.

Modeling Methods

The trend in geomtric modeling has been toward increasingly comprehensive part definitions. It has been discovered that any attempts at geometric shortcuts create loopholes for future errors. For example, inadequate model definitions may not permit accurate machining or analysis. Therefore, modeling has progressed from 2D, to 2 1/2D, and now to 3D. Three-dimensional representations model complex shapes such as aircraft wing sections and ergonomic hand tools.

Performance of many industrial products has been enhanced by the use of complex 3D sculptured surfaces. For example, streamlined cars give better

Smooth Contours Produce Accurate Results

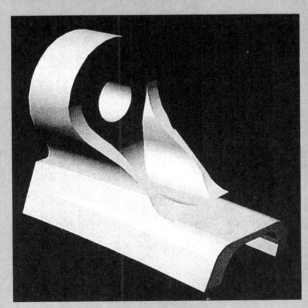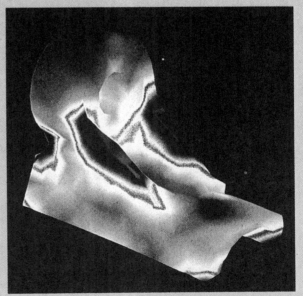

On complex surfaces, accurate solid models can increase the reliability of stress analysis. For example, if this jet engine clevis was modeled with a mesh based on rough contours, the stress concentrations at the corners may be incorrect.

Source: CAE Computer Aided Engineering, August 1985, 34-35, 38-39, 42

gas mileage. Sculptured handles, knobs, keyboards, and switches can be easier to work with and may also prevent certain accidents. One possible drawback of 3D is that once it has been used for some features on a project, invariably, the en-tire design must be done in 3D to ensure database compatibility.

Similarly, representations of the geo-metric models have become increasingly sophisticated. For 3D design, there are three approaches to representing and displaying computerized geometry: wire-frame, surface, and solid modeling.

Historically, the least sophisticated method came first, and progress has moved toward more comprehensive so-lutions. For example, the most sophis-

Comparing Programs

Software systems can be easily com-pared by developing a system config-uration template. This template contains all possible functions, even though few users require every possi-ble feature. Overlaps and gaps in cov-erage can be determined by super-imposing one vendor's template over the other.

This partial listing of products shows relative coverage for eight of the major software vendors. The de-gree of shading indicates coverage, with darker representing more com-plete coverage. Evaluating suit-ability for any one application also involves looking at specific features inside each functional circle.

Looking at this selection of eight of the hundred or so products, it is clear that conciseness is not yet common. No single vendor currently has all the bases covered. However, most are pursuing aggressive expansion plans. SDRC, the developer of Su-pertab, for example, plans to expand the product's FEA functions to match Ansys and MSC/Nastran. Swanson Analysis has just intro-duced solids modeling. And Auto-trol is working on a full-function in-tegration of solids, drafting, and pre-processing.

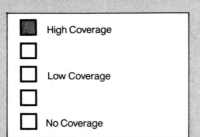

High Coverage

Low Coverage

No Coverage

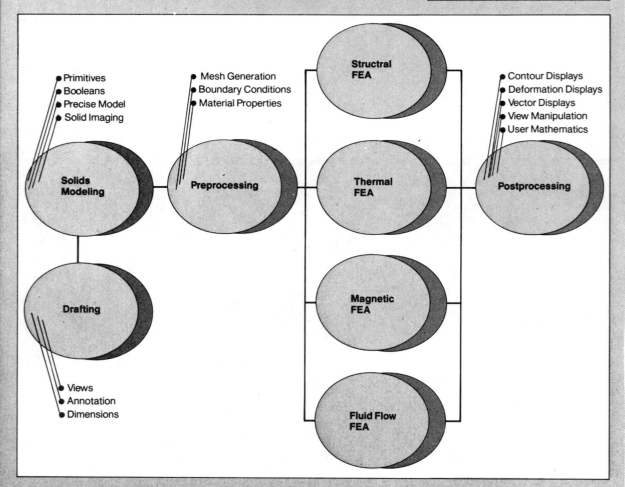

Few users require all aspects of solids modeling and FEA. Ideally, systems integrators could purchase only the portions of programs that they need.

ticated model type, solids, can be used to indicate whether a particular point lies inside, outside, or on the surface of a part. Surface modelers can only tell whether or not a point exists on the defined surface. And wire-frame modelers, given an x, y, z point are not usually able to tell whether or not this is a given point, even when it is on the surface.

Solids have many advantages over wire-frame models so it may be difficult to understand why wire-frame is still the most widely used method of geometry representation used for FEA. Solid models do take almost twice the disk space to store, but this disadvantage is diminishing as disk prices are halving every year. The three principle advantages of solids over wire frames for analysis are, removal of ambiguity, avoidance of impossibility, and detection of interference. For example, two engineers looking at the one wire frame can perceive different parts. This can be a problem if the drafter defined a hole running in one direction and the structural engineer interpreted the hole in the other direction. Also, parts that could not exist in the real world can be created using wire frames. Finally, wire-frame modelers are often unable to detect interference when curved surfaces run near other parts. In finite-element heat transfer problems, for example, it is imperative to know the existance of gaps or connectivity as these factors could have a major effect on results.

One important distinction between wire-frame and FE models that must be made is that the meshes for each are different. For example, if a wire-frame modeler is used, that mesh is seen first. Then, if necessary, the FE mesh is generated. Although the FE mesh follows the confines of the original geometry, this new mesh may bear little relationship to positions seen in the wire-frame.

Systems that have hidden-line removal, for example, contain more geometric information than wire-frame data and consequently are usually better suited to FEA. Full-surface modelers, however, should also produce shaded color images of solid models of contoured surfaces. There are many such surface packages for pre and post processing including Femgen and Femview from Greatwest Technology Transfer in Minneapolis, MN and Strudl from Georgia Institute Technology in Atlanta. Sperry Corp., Blue Bell, PA has a fully integrated CAD/CAM system based on a solid modeler from Baustein of the University of Berlin. This system can feed surface geometry into Femgen, its FEA preprocessor.

Solid Ingredients

Although there are over 25 products on the market which claim to be complete solid modelers, not all are full-function packages. For example, a restrictive definition of a solid with interior, exterior, and surface points may rule out many so-called advanced packages. Complete solids systems should also display 3D color shaded images.

Another feature demanded by the user community is the use of primitives and Boolean operations to create complex part geometries. Primitives are

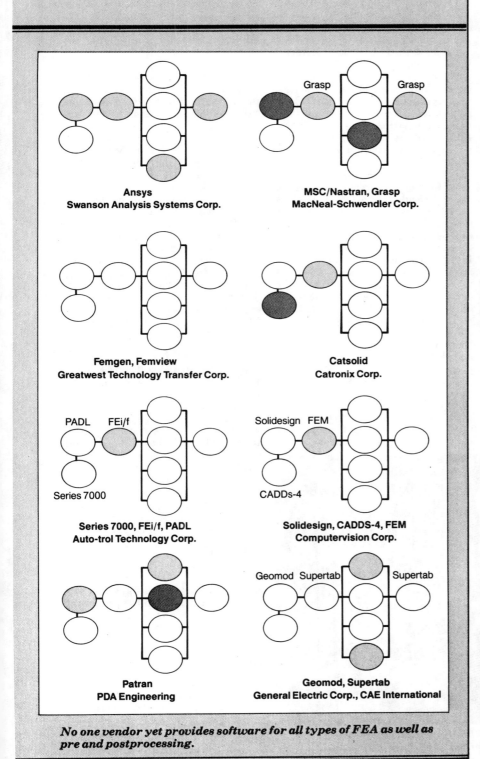

Ansys
Swanson Analysis Systems Corp.

Grasp Grasp
MSC/Nastran, Grasp
MacNeal-Schwendler Corp.

Femgen, Femview
Greatwest Technology Transfer Corp.

Catsolid
Catronix Corp.

PADL FEi/f
Series 7000
Series 7000, FEi/f, PADL
Auto-trol Technology Corp.

Solidesign FEM
CADDs-4
Solidesign, CADDS-4, FEM
Computervision Corp.

Patran
PDA Engineering

Geomod Supertab Supertab
Geomod, Supertab
General Electric Corp., CAE International

No one vendor yet provides software for all types of FEA as well as pre and postprocessing.

Source: CAE Computer Aided Engineering, August 1985, 34-35, 38-39, 42

simple shapes such as cubes, cylinders, and spheres and are defined within the solids modeler. Boolean operations stem from set theory and include operators such as union, intersection, and difference, which combine the primitives to build up detailed part geometries. Comprehensive modelers, such as Geomod from Structural Dynamics Research Corp. of Milford, OH, has eight primitives and six Boolean operations.

Modelers are not restricted to Boolean operations and primitves for geometry creation. Solids of rotation and translation, for example, are excellent for certain geometries such as slabs of constant thickness or objects with circular symmetry. Another useful manipulation is translation of a 2D template through a more complex path, such as a spiral, to create springs or light bulb filaments. Simply because a modeler has complex translation techniques does not mean that Boolean operations and primitives are left out. Synthavision from Mathematical Applications Group Inc. of Elmsford, NY, for example, has a complete repertoire of techniques to build almost any geometry imaginable.

There are a number of other properties of solids modelers which should be investigated before a purchase. These include integration with other parts of computer-aided design, analysis, and manufacturing. Within the modeler itself, it is important that an accurate representation of geometry is used. Some earlier modelers used a faceted approximation of part geometry which could lead to errors in calculations such as determination of mass from a volume. Faceted packages currently marketed include, GMS from Interactive Computer Modeling of Reston, VA, Sol-

ids Modeling from Applicon of Burlington, MA, and CS5 from Cubicomp Corp. of Berkely, CA. Some accurate software programs are Romulus from Evans & Sutherland, Geomod from SDRC, and Catsolid from Catronix.

Support Products

Assembling an FEA system with solids modeling support requires a view of the big picture of systems integration as well as an elaboration of the specific functions of each candidate ingredient. When assembling comprehensive systems for all computer-aided functions, buying different packages from several vendors is unavoidable. When CAE implementation was small-scale, it was practical to have a turnkey vendor provide all software. However, now that corporations are moving out of the experimental and evaluation phase of CAE, the scope of activities tends to grow beyond turnkey approaches. Because of the multivendor approach, conciseness has become an issue in configuring CAE systems. Ideally users would prefer that each vendor choose one particular function and perform that one task well. Instead, it appears that many vendors are trying to cover too many bases and failing to cover them properly. In the concise approach, there are nine building blocks for full function CAE systems: solids, drafting, preprocessing, static strucutral, dynamic strucutral, thermal, fluid flow, electromagnetic, and plastic molding FEA, as well as post processing.

Midwest Avionics, a manufacturer of radio housings, for example, illustrates the need for conciseness. Designing their product requires thermal, strucutral, dynamic, and fluid flow FEA. Because the geometries of the parts were

complex, a solids modeler with close links to drafting was chosen so that proper documentation of designs was possible. Software selection for these tasks included Romulus for solids modeling and drafting, Patran and Femgen for preprocessing, Ansys and Nisa for FEA. As Ansys could not perform arbitrary geometry fluid flow analysis, EMRC Fluids was used for these computations. And for dynamic analysis of vibrations, Systan from SDRC was used.

Running all these packages may well provide a solution to the design problem, but all this software creates is own procedural difficulties as well as being expensive. This system is also unavoidably redundant. For example, both Patran and Ansys have solids modeling, but the need for primitives, Boolean operations, and integral drafting makes Romulus a necessity. And Patran also contains routines for static, dynamic, and thermal FEA, but nonlinear analysis requires codes such as Ansys. For fluid flow, both Ansys and Nisa-Fluids can perform the computations, but Ansys is limited to pipe geometries. Also, while both Ansys and Patran have preprocessing, Ansys has no interface with Romulus so Patran must be used. The net result is that Midwest purchased three solids modelers, three FEA codes, and two preprocessors to solve one problem.

Mix and Match

It is estimated that 70% of the new CAD/CAM systems are integrated by the purchaser. This trend is supported by looking at the reduced market share of turnkey vendors such as Computervision Corp. Clearly, systems integrators would like to have more support from the software developers by not having to buy redundant modules. Because software selection tends to be a mix and match proposition with detailed charts listing technical specifications, purchases would be easier if users could choose only the desired software.

Today, few software houses offer users concise, unbundled, and integrated solutions. Some products which excel at integration lack conciseness and vice versa. Nevertheless, vendors are making substantial progress towards meeting the wishes of the users.

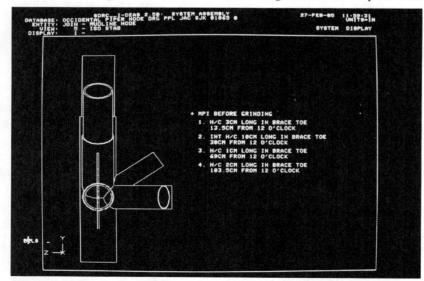

When a single solid model created with software such as Geomod serves as a basis for assembly drawings, drafting, and FEA, little time is wasted on error-prone recreation of part data.

By ROBERT BOWERMAN, *President, Heliosoft, Fairfield, IA.*

Mr. Bowerman and an associate, Robert T. Fertig, are authors of a 200-page research report covering workstation trends and a possible shakeout in the CAD/CAM industry. The report is available from Enterprise Information Systems Inc., 600 Summer St., Stamford CT 06901-1154. Tel. 203-323-0075

The Intelligent Computer–Part II

NATURAL LANGUAGE
THE DIALOG HAS BEGUN

With natural-language systems, users direct their computers in commonplace English, not cryptic commands. **by Henry Fersko-Weiss**

Using personal computers has never been as easy as we thought it was going to be, and certainly not as easy as it should be. The more powerful a program is, it seems, the more arcane is the command language used to operate it. Even pull-down menus and icons, touchscreens and mice do not make computing universally approachable. But a new set of program techniques, called natural-language systems, are giving users the most intelligible way to interact with computers: their own language. Users who want to search corporate data bases, hunt through decades of case law or check on the competition, are beginning to converse with computers much the same way they would with a peer.

Natural-language systems serve as front ends to applications. They take plain English (or French, or German, etc.) input from a user and refract it, like a prism, into application commands, so the computer can understand it. Because these systems allow people to carry on a dialog with the computer in everyday language, they open up personal computing to novice and occasional users. But they also give experienced users a fallback approach to complex application features or a faster, more natural way to accomplish a particular task.

For the moment, there are relatively few natural-language systems being marketed; but over the next six months, several more will appear in the form of data base interfaces or as front-end additions to other applications, such as spreadsheets or word processors. Natural-language facilities will eventually be built into virtually every type of software, from operating systems to accounting packages, from order-entry systems to expert systems.

The most common use of natural language today is for data base inquiry. Clout, from Microrim, in Bellevue, Wash., was among the first bridges between personal computers and their users. It was built for Microrim's R:Base 4000 relational data base. "It [Clout] makes life easier," says Stephen Lowens, a senior transportation planner at BKS Associates, a transportation and engineering design firm, in Oakland, California.

Lowens set up a system of 1,600 records for BKS. "The whole reason for using Clout was to let an engineer or marketing assistant, who doesn't know how to use the data base, sit down and type in a request in English—and be able to get the information he needs," Lowens says. The key to making Clout effective was setting it up—that is, "teaching" it the key words related to BKS's business; a process Lowens admits was time-consuming.

The first screen you see in Clout is the main menu. It consists of six choices, the first of which is: "Examine the data using conversational requests." The other main menu selections have to do with examining the structure of the data base, or dictionary, opening a different data base, refining a request, or exiting Clout. If you choose the first option, you get a nearly blank screen that reads, at the very top: CURRENT SERIES OF REQUESTS, followed by two more lines: ENTER REQUEST (or [ESC] for main menu), and R>. A user at BKS might then type in a request like, SHOW ME ALL THE PROJECTS IN MARIN COUNTY, or SHOW ME ALL THE PROPOSALS CONTAINING TRANSIT PLANNING. Clout then searches the data base to comply with the request—a process Lowens says takes only seconds on the company's IBM PC/XT. When the answer is displayed on the screen it is fol-

How Natural-Language Systems Work

While there are two different styles of natural-language systems—one in which you type English sentences free-form, so to speak, and another in which you pick words and phrases out of menu windows—the inner workings are basically the same. In simple terms, a natural-language system deciphers the parts of speech and action in a common sentence and then translates it into application commands, or compiled program language, so the computer can respond.

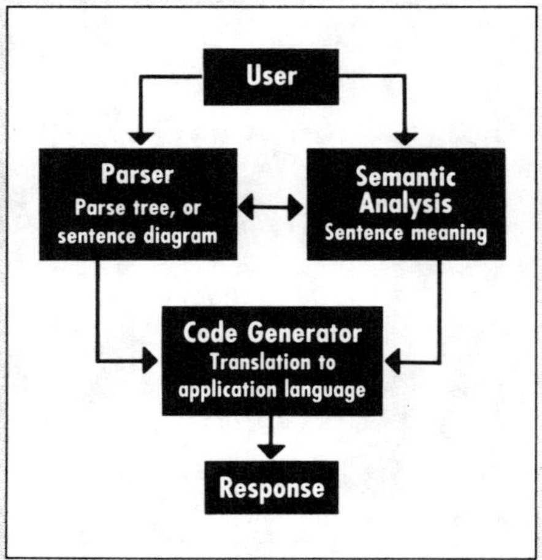

The first part of the process, mapping the sentence, or syntactic analysis, is accomplished by an inner part of the program, called a parser. "The parser figures out how a sentence is pasted together," explains Robert Berwick, an assistant professor at MIT who works in the Artificial Intelligence Laboratory. "For example, if you have the sentence: 'Mary ate ice cream,' the parser would designate 'Mary' as a noun phrase and 'ate ice cream' as a verb phrase. The verb phrase would be further broken down into the verb, 'ate,' and a noun phrase, 'ice cream,' modifying it. The net result would be a parse tree that looks very much like the sentence diagrams we used to make in our notebooks in school."

There are many different kinds of parsers. Some work by first breaking the sentence into large grammatical chunks—like noun phrase and verb phrase—and some go right to the word level to identify parts of speech. Other parsers work in parallel, jumping back and forth between words and phrases until the parse tree has been completed.

Once the syntactic analysis is finished, the natural-language system does a semantic analysis of the sentence. The heart of this process is a dictionary that contains information about words. For most business uses of natural language, this dictionary is nothing more than a list of command synonyms for key English words. The semantic analysis usually proceeds by translating the verb first because that's the guide to the main action in the sentence. Then the modifiers of that action are translated and, finally, the nouns. Once the whole sentence has been translated, the personal computer can then respond to the user's request.

Although it is logically necessary to think of the syntactic and semantic analyses as two separate stages in the process of understanding, natural-language systems may do both at the same time. "The dictionary look-up and parsing are typically interleaved processes," says Berwick. "But here, again, there are a variety of approaches. Some systems start at one end of the process, some at the other end, and still others accomplish both jobs simultaneously."

As natural-language systems become more refined, this basic two-part procedure will probably include a third part: an expert system facility that will help clarify ambiguous expressions. When that happens we will be able to carry on much more complex dialogs with our computers.

—H.F-W.

[Diagram: User → Parser (Parse tree, or sentence diagram) ↔ Semantic Analysis (Sentence meaning) → Code Generator (Translation to application language) → Response]

lowed by another R> for the next request.

In the past, Lowens had tried to set up an easily accessible data base using Ashton-Tate's dBase II program, only to find that it was too difficult for most non-power users. R:Base 4000, by itself, was hardly any better. But the combination of Clout and R:Base 4000 has worked "because anyone can sit down and type English—that's the key," Lowens says.

While Clout brings a great deal of ease to data base inquiry, it does not facilitate direct access to the data base. In order to update the data base, you have to exit Clout. A new program that solves this problem, and also has a more powerful facility very much like Clout, is Q&A from Symantec. (See the story on page 71 in the October 1985 issue of *Personal Computing*.) Q&A is actually an integrated package that offers word processing, data base facilities and a report generator. In addition, there is a natural-language mode that users can jump into at the touch of a key. This mode, called Your Intelligent Assistant, comes with a built-in vocabulary of several hundred words, and users can add to this vocabulary by filling in blanks as they are prompted by the system.

The basic approach to data base querying in Q&A is the same as in Clout, except the options are couched in friendlier terms: "Get acquainted," "Teach me about your data base" and "Ask me to do something." As you type an English request, the words appear at the top of the screen in a rectangular box. When you're done, the sentence is redisplayed in a form much closer to a traditional data base query so you are sure the Assistant has understood.

Because users of Clout and Q&A do not have to learn complex command routines to get at the information they want, they can concentrate on their work instead of on the computer. They don't have to waste time trying to recall a very unnatural sequence of keystrokes or go thumbing through a 200-page manual. However, some users find that staring at a blank screen—with nothing much more than a flashing cursor—leaves them blank as well. They become too intimidated to start. And there is another problem with free-form systems like Clout and Q&A. If the program doesn't understand a word or phrase you've typed in, then you have to try using a synonym it *may* understand, or take the time to enter the unknown expression in the system's dictionary.

Texas Instruments has taken an approach to natural language that avoids the problems of unknown words. TI's NaturalLink program already contains all the acceptable words and phrases that can be used in a valid query. NaturalLink starts off with the beginning of a sentence request displayed at the top of the screen: I WANT TO... Underneath this fledgling sentence are several windows, filled with acceptable words and phrases that can be used to complete a request. You simply use the cursor keys or a mouse to jump from one window to

the next, highlighting your choices as you go. As you build a query, your request is written at the top of the screen.

For example, if you were using the NaturalLink-based front end to the Westlaw legal data base—called LawSearch, from Direct-Aid, in Boulder, Colo.—your first window choice might be SEARCH FOR CASE LAW DOCUMENTS... In the next window you might pick CONTAINING THE TERM ... At this point a pop-up window would appear in the middle of the screen and you would have to type in the first term you wanted to search on, such as "accident." You would continue to make choices until the request was complete.

Texas Instruments offers NaturalLink as a complementary front end to a number of popular applications programs. These include Lotus 1-2-3, dBase II, WordStar, MultiMate, Multiplan, BPS Business Graphics, MS-DOS and the Dow Jones News/Retrieval Service. Texas Instruments also offers tool kits for building NaturalLink facilities into other programs. Recently, for example, IntelliSource, of Dallas, Texas, introduced a series of accounting modules that uses the NaturalLink query-building technique. In October, another company, Brodie Associates, in Boston, Mass., adopted this same approach to natural language. It offers a tool kit called NatPack. One of the main differences between NaturalLink and NatPack is that NatPack allows you to build more words and phrases into the windows as you work with the program.

NatPack is used in conjunction with Golden Common Lisp, an artificial-intelligence programming language and development system from Gold Hill Computers, in Cambridge, Mass. NatPack starts by showing the user a menu of valid phrases for building a request. The user points to a phrase using a

Wally Rhines, president of the data systems group at Texas Instruments, says natural-language systems can make a corporate or public data base accessible to managers on an ad hoc basis without them being specialists in the structure and access method of the data base. As for expert systems, Rhines says such programs can "make factory experts accessible to the local dealer's technicians, diagnosing even a complex, infrequent problem."

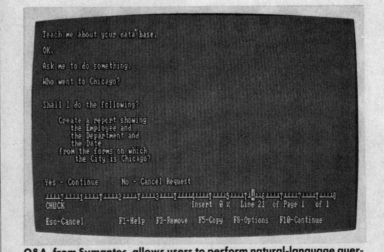

Q&A, from Symantec, allows users to perform natural-language queries of the program's data base. Users may add words to those not already included in Q&A's vocabulary of several hundred words.

Source: *Personal Computing*, November 1985, 93-96

129

mouse or the cursor positioning keys. Scrolling facilities allow the user to move quickly to the phrase he wants. The system then shows the user a menu of phrases which can follow his first choice, and the process is repeated until a complete sentence has been formed. There are six units in the tool kit: a grammar builder, vocabulary builder, a screen builder, an input module for the run-time version, a parser and a set of semantic actions for determining the meaning of each word or phrase.

The advantage c˚ window-based systems is that you can never go wrong. You can't type a sentence that the system doesn't understand. With a system like Clout, if it doesn't understand a word or phrase typed in you have to enter it into the system and define it. This can be an annoying and time-consuming interruption to a search for information you want right away.

While it is clear that natural-language systems offer a way into computing for people who don't, for various reasons, learn application languages, they are, as a result, slower and more cumbersome to use than conventional programs. Furthermore, natural-language systems are currently limited in their ability to understand language. Therefore, natural-language techniques are arguably unsuited for the experienced computerist. And some application categories, such as spreadsheets, seem to be better off without the added baggage of a natural-language interface.

Bill Gross, the president of GNP Development Corp., in Pasadena, Calif., disagrees. "If you could type 'total all columns' in Lotus 1-2-3, instead of having to enter the command for total and then drag the cursor across the whole range of what you want totaled, it would be much faster and easier, no matter

how skilled a Lotus user you were," says Gross. He backed up his belief by introducing a natural-language front end to 1-2-3 in October. In beta tests of the product, called English, GNP found that experienced users can accomplish some tasks 66 percent faster with natural language, defying those who say how slow and inappropriate natural language is for power users.

English comes with more than 1,600 words in its dictionary, about half for power users and half for novices. Users can add as many words to the dictionary as they like. What promises to make this product especially useful is the fact that users can jump in and out of the English mode. They can use regular Lotus commands when they want to and can switch to English when that seems faster or lets them feel more comfortable with some of the complex program functions. English will even allow users to write macros in all natural English sentences or in a combination of Lotus commands and English phrasing.

The ability to fall back on natural language whenever you need, or want it, could make many applications much easier to use —regardless of a user's experience level. For one thing, points out Eric Frey, president of Frey Associates, in Amherst, N.H.: "Users often don't know what information is available to them in corporate or external data bases. With natural-language front ends, the user can ask the system what subjects it knows about and what the data fields are." Frey Associates has a natural-language system that does just this on Digital Equipment Corp.'s VAX computers. Themis, as the system is called, will also be available for personal computers, sometime in early 1986. Themis comes with 900 or so words in its dictionary, but Frey says it can be added to without functional limit. In fact, his company has built a system with a 10,000-word vocabulary, and he claims its size didn't slow down the system appreciably.

Frey sees the eventual use of natural-language facilities in many programs where the universe of information or capabilities available to users is quite large. Users generally utilize only a small portion of such systems on a consistent basis, so they are rusty when it comes to less-used portions of the program. This is where natural language would be useful even to users experienced with a product.

In the coming months and years, more and more natural-language tools will be used as front ends to personal computer applications or will be embedded within them. This will let users in every industry and in every type of job carry on a dialog, in natural English, with the personal computer. And perhaps the most exciting prospect for natural language comes in conjunction with expert systems. When these two parts of artificial intelligence are united, personal computing will be much smarter, as well as smoother.

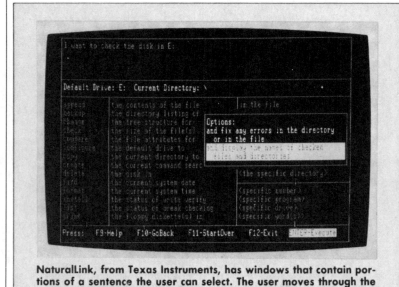

NaturalLink, from Texas Instruments, has windows that contain portions of a sentence the user can select. The user moves through the sentence, choosing clauses, until the request is completed.

ROBERT HOLMGREN

PROGRAMMING MAKES SENSE FOR BUSINESS

COBOL

BASIC

PASCAL

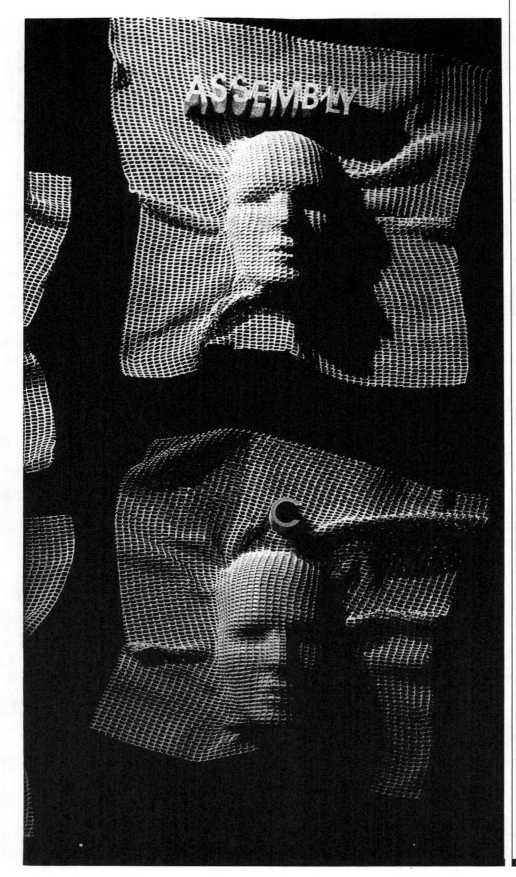

You may not be a professional programmer, but you spend a lot of time programming your PC— for practical reasons and for fun. But do you use the right language products? Should you change? Here's help in deciding.

PC Magazine's reader surveys tell us that over 60 percent of you are programming your own PCs these days. The overflowing mailbox backs up this statistic daily with letters requesting more programming and technical columns. And our Interactive Reader Service bulletin board is kept so busy sending you source code for our published programs that we're not sure our expansion plans for it are ambitious enough.

Those same surveys say that only 4 percent of you are professional programmers, so the obvious question is, Why are so many of you programming your own PCs? With thousands of commercially available software packages that let you do everything from simple typing to professional word processing, and from financial analysis to grocery-coupon-analysis, why do you choose to put up with the vagaries of a cryptic, arcane computer language to make your computer do something for you? Why are you learning new languages and complex techniques that were, until recently, solely the domain of professional programmers? Why don't you just buy a package that already does what you want your program to do?

Several good reasons come to mind. One major reason is that no prepackaged commercial program, including the PC's own DOS, can satisfy all of you in all ways at all times. You always have one more thing you want it to do or one more way you want to do it. But with so many of you using their products, the commercial software vendors simply can't keep up with your demands, so you "roll your own," and write a program to do the job.

Another reason to program your PC is

to build bridges that transport data between the various software packages and hardware products that you regularly use but that don't get along with each other well enough to communicate effectively. We know because you regularly fill our User-to-User, Power User, and Spreadsheet Clinic coffers with tips on how to transfer those awkwardly formatted *WordStar* files to other programs or use printers in some way that *1-2-3* or *Microsoft Word* don't know about.

The Game

But perhaps the most compelling reason of all is that you enjoy the challenge of programming your machine and the sheer joy of winning the "game." The visceral experience of computer programming closely resembles that of competitive sports or, more realistically, a good stiff game of penny-ante poker or *Trivial Pursuit*.

Unlike many things that challenge you in life, computer programming is a game that can be won, lost, and won again without damage to your ego or physical abilities. It's not unusual for a PC user to spend all night getting a program to work right and then, without a second thought, spend the next night meeting the programming challenge all over again by adding features to it. The only real damage may show up in a divorce court, but it's difficult to imagine a judge granting separation papers on grounds of "cruel and unusual programming."

No matter what problem you're trying to solve, programming a PC is safe, rewarding, and fun—that's all there is to it.

But is it easy? Programming a PC may seem difficult the first time you try it, but in the long run it is easier, and infinitely more satisfying, than programming a typical microwave oven. How easy it is depends on what type of program you're trying to write and what language you've chosen to write it in—and those issues are not unrelated.

A Marvelous Continuity

An interesting thing about computer languages is that they tend to survive longer than the computers they were designed for. When you look at the history of lan-

guages in contrast to the history of computer hardware, there is a marvelous continuity rivaled only by the history of human languages.

That continuity is partly fallout from the fact that hardware development has advanced at a much faster pace than software development. But it is more significantly the result of a good business decision that dictates a stable human environment for using and programming computers. Corporate America has a seemingly insatiable need for better, faster computers of all classes from micros to mainframes, and it would be a financial and human disaster if every business application had to be re-

> **I**t's not unusual for a PC programmer to spend all night getting a program to work right and then, without a second thought, spend the next night adding features to it.

written at the rate new hardware is usually introduced.

Roots

Just as history and geography are filled with a variety of human tongues, computerdom is filled with a variety of languages. Each computer language has its history, purpose, and place, and all have proponents who are willing to challenge would-be interlopers on either technical or purely emotional grounds.

Computers were around long before programming languages, but not before people had to program them. The first computers had to be programmed by flipping switches and wiring "bread boards" that connected various instructions and locations of the computer. When computer

memory was invented, programs could finally be stored inside the computer, but the only way to write programs was in binary and, later on, hexadecimal or octal codes. It was difficult, but it beat switching and wiring by a country mile.

Programmers finally wrote themselves a program that translated short computer instruction mnemonics into the computer's binary codes. It was called an "assembler" because each mnemonic was assembled directly into one computer instruction. Later enhancements included symbolic (named) variables, which replaced direct memory references with simple names. These improvements made programming enormously easier. The gods of computing looked at assemblers and declared that they were good.

But Not Easy Enough

Forward-thinking programmers were still not satisfied, however. A few years later, higher-level computer languages were designed that allowed a programmer to think directly about the problem at hand rather than about how to instruct the computer to solve it. The resulting program ran more slowly than one written in assembly language but took so much less time to develop that the performance penalty was well worth it.

The first such language was FORTRAN (FORmula TRANslation), which was designed to solve scientific and mathematical problems. The next was COBOL (COmmon Business Oriented Language), which was designed to handle large-scale accounting and recordkeeping problems.

Both FORTRAN and COBOL, originally developed by IBM, have survived for nearly 30 years in the fast-changing computer market and are available today for your PC. But they have not survived without serious attacks. One offensive came from people who challenged the strategy of program development itself.

FORTRAN and COBOL, and most computer languages in use today, are known as compilers. Compilers use a multistep process to translate source language statements entered by the programmer into machine language. First the code is entered, then it is compiled, and after an intermediate step called link editing, it is

Source: PC Magazine, October 29, 1985, 108-114, 116-122, 124-138, 140-145, 150, 152-155, 158, 160-161

BENCHMARKING THE LANGUAGES

For PC Magazine's programming language roundup, our reviewers tested each product against a series of six benchmark tests. Here's a description of each.

A benchmark is a standard against which you can measure or judge something. Since each computer language has its own strengths and weaknesses, no single, global measurement will do the job. And time flies so fast in a computer that you can't just measure something once—you have to measure it thousands of times.

The language benchmarks used here are the same ones used in *PC Magazine*'s last language roundup (Volume 2 Number 4). Rather than attempt global measurements of language performance, these benchmarks demonstrate each language's ability and speed in handling six simple tasks that are usually repeated millions of times when a typical computer program is run. You'll find the execution speeds for each version of each language in the performance table accompanying each language article.

If you are selecting a language, how hard it is to program the benchmark test may be more important to you than how fast it runs. Because it is your time, not your computer's, that is spent learning the language and writing programs with it, succinct code should be important to you. We've listed the source code of each test for the editor's choice product in all five languages (except for the floating point test in assembler).

The Empty Loop

Doing anything only once with a computer program is almost not worth doing at all. The reason is that most programs spend most of their time looping through a sequence of instructions (iterating). The "empty loop" benchmark, which does nothing 10,000 times, is meant to show how well and quickly each language controls the iteration process.

Integer Addition

After looping, adding a series of integer numbers together is the most frequent function in computer programs. While integer addition is often the sole purpose of a program loop, a series of integer additions more often controls exit and entry conditions for a series of loops. In short, it is an important application of a computer language, and *PC Magazine*'s benchmark is designed to demonstrate each language's integer addition skills.

Floating-Point Arithmetic

Floating-point arithmetic is a difficult test for a PC's programming language because, unlike integer arithmetic, the PC's 8088 CPU does not know how to do it. The language product designer's skills come into play here more than in any other benchmark because the design of the floating-point emulation technique directly determines its performance. If you have a mathematical, scientific, or finan-

> **I**n a computer, you can't just measure something once—you have to measure it thousands of times.

cial application in mind, the floating-point benchmark should be critical to you when you select both a language and a product.

Character String Concatenation

Character manipulation is obviously important for word and other text-processing applications, but it also has wide applicability in general-purpose computer memory management. Other string operations can be good determinants of a language's string-handling skills, but concatenation is by far the most commonly used. While the time results are interesting, you may be more fascinated by the variety of language designs for the simple task of pasting two sentences together.

Table Lookup

Storing a series of numbers in a table and then later looking for them is a very common function in computer programs. Languages vary greatly in how they handle the problem; the benchmark speeds as well as the source code styles vary significantly. The reason is that some languages (called "strongly typed") are more concerned about numeric types; they have more rules about using them and converting them to other numeric types than do weakly typed languages, which play fast and loose with numbers.

File Access

Reading and writing data from disks is obviously critical to any good PC application. How fast each language does it is probably not too important, however, because they invariably use DOS function calls to handle the boring chores of manipulating the disk drive. We tried to make the test a bit unfair to both DOS and the languages by using 132-byte records, which is absolutely not an optimum or standard size and forces either the language or DOS to do extra work. Once again, it's probably more important to consider how fast you'll be able to write a program in a given language to manipulate files than how fast the language actually handles them.—**John Dickinson**

run. If the results are wrong, the errant source statement (or statements) has to be corrected, and then, compiled, linked, and run again.

Subject to Interpretation

Compilers are relatively efficient at their job of composing computer programs, but inefficient when it comes to making the best use of today's most precious computer resource: a programmer's time. Compilers are particularly inefficient tools for program development when the programmer is new to the game and trying to learn. To improve the odds in the programmer's favor, two Dartmouth College professors, John Kemeny and Thomas Kurtz, and an IBM employee named Kenneth Iverson, developed alternatives to compiled languages known as interpreted languages—Kemeny and Kurtz's creation was BASIC (Beginner's All-purpose Symbolic Instruction Code) devised in 1964, and Iverson's was APL (A Programming Language), created in 1962.

An interpreter looks at each source statement and then immediately composes a binary computer program to execute the task. The nature of interpreting makes an interpreted-language program slower to run than a compiled one. But the interpreter's ability to enable you to interactively enter a program, run it and see the results, then modify the program and run it again without stopping to compile and link edit each time, makes interpreters attractive and quick as learning and program-development tools. Like FORTRAN and CO-BOL, BASIC and APL are long-haul survivors and can be used today on a PC.

Structure

A more recent trend is to write programs that follow a somewhat vague set of concepts and rules that has become known in the trade as "structured programming." Structured programming, which has evolved almost into an art form all its own, results in programs that are simple, often elegant, statements of both the problem and the solution, which makes them a great deal easier to develop and infinitely easier to maintain than unstructured programs. The concepts may be vague, but the benefits of the structured approach are appreciated in the commercial programming community, where it is generally accepted that structured programming makes better use of a programmer's time.

While some structuring of programs can be done in any language, several languages have emerged in recent years that are specifically designed to produce well-structured programs. They almost force you to use the tools of program structuring, which include functions, subroutines, and other program block structures, and data structures (also called records or sets) that allow meaningful names to be used for memory locations. These languages include Europe's Algol, which never caught

---

Structured programs, which are simple, often elegant, statements of both the problem and the solution, are infinitely easier to maintain than unstructured programs.

---

on in this country, IBM's PL/I, and, more recently, Knuth's Pascal and Bell Laboratories' C.

You Are What You Program In

A gaggle of languages are available for your PC, but you use only a few of them widely. For this issue, *PC Magazine*'s editors have selected only the most significant ones and have asked reviewers to evaluate representative products that assemble, compile, or interpret each selected language. We couldn't review the entire range of products available for each language, so we've tried instead to give you some insights into the products that are at the top of their category, along with those that offer the most flexible programming environments for your PC.

Here then are the languages covered in this issue, ranked according to what percentage of *PC Magazine* readers who program use them.

BASIC

It's no big surprise to us that most of you (87 percent of our programming readers) use BASIC for programming your PCs, even if you use other languages as well. Microsoft's by-now-legendary BASIC interpreter comes installed in an IBM PC's ROM chips and comes on disk with the operating system for most compatibles. In either case, it's free for the asking—all you have to do is boot it up and start coding in BASIC (you don't even need to have a disk installed on an IBM PC, although it's preferable).

Perhaps in trying to explain the language's popularity, it's more important to note that BASIC is marvelously easy to use and one of the most powerful languages available for the PC when it comes to using your computer's resources, especially screen formatting and graphics. Most other languages require you to buy additional library routines or write your own assembly language programs to get the sort of full-screen user interfaces you're used to from professionally written programs.

BASIC is an interpreter, so it's relatively slow. But BASIC's interactive, interpretive environment does just what its original developers, Kemeny and Kurtz, wanted—it gives you a place where it's easy to learn how to write, run, and debug a computer program. You can apply many of those lessons to more-advanced, faster languages, or you can buy a BASIC compiler to make the programs you debugged using the interpreter run faster. The choice is up to you, and perhaps this issue of *PC Magazine* will help you decide.

8086/8 Assembly Language

It is something of a surprise that so many of you (38 percent of readers who program) program in assembly language. No language is more arcane or difficult to learn and use; the Intel version's mnemonics and memory formats for the 8086/8 and DOS's program interface are no joys.

But after we reread our mail, we began

THE AUTHORS

A who's who of PC Magazine's language reviewers plus some personal insights into their specialty areas.

To bring you the latest word on PC programming languages, *PC Magazine* naturally sought out an expert for each language. Each of our experts is highly qualified and has his own strong opinions on that language. Here's some background on each author along with his comments on his language of choice.

Eric Bank teamed up with *PC Magazine* associate editor Stephanie Stallings to review CO-BOL. Bank, an independent consultant who specializes in designing COBOL-based business applications, learned COBOL as a student while working his way through college as a data processor. He has continued to expand his fluency in the language throughout the past 12 years. While he programs in other languages as well, Bank thinks CO-BOL is the ultimate business language because "it is quick, easy to learn and to read, and handles large data files with precision." Bank has recently been experimenting with programs that extend COBOL beyond purely business applications, including one that uses a complex algorithm to plot graphs of imaginary number sets.

Jeff Duntemann is technical editor for *PC Tech Journal* and author of *The Complete Guide to Turbo Pascal* (Scott, Foresman, 1985), which he describes as "the first book written specifically for Turbo Pascal." He learned Pascal in 1979 and immediately fell in love with the language for its elegance, read-ability, and modular structuring, which has allowed him to stockpile a considerable library of useful subroutines. In fact, one of the reasons he continues in his commitment to Pascal is his time investment in these subroutines. "In Pascal," he says, "I can write a piece of code, pick it up a year from now, and know at a glance what it does." Duntemann also favors Pascal because of its "generalist" orientation. "I expect my programming language to follow me into anything I want to explore," he says. True to his word, Duntemann has written Pascal programs that run stepper motors, plot star charts (look for a public-domain program called KEPPLER), and keep addresses, as well as countless other utilities and short applications.

Kaare Christian puts together computer systems for vision research at the neurobiology lab of New York City's Rockefeller University. He is also the author of *The Unix Operating System* (John Wiley, 1983) and a soon-to-be-released book on the Modula 2 programming language. Christian initially learned C in 1976 while programming graphics and UNIX applications at the New York Institute of Technology. He continues to use C today primarily because it is extremely portable. His work often involves hooking up exotic hardware systems, and he says C is "one thread that runs through all systems." He also favors C because of the direct control it gives the programmer. "Programming in C is like driving a sports car," says Christian. "Everything is manual and rudimentary, but control is extreme."

Richard Aarons is a contributing editor of *PC Magazine* and a senior editor at *Business and Commercial Aviation* magazine, specializing in aviation-related software. He is also president of RNA Associates, a Connecticut-based firm that develops law-enforcement applications (almost exclusively in BASIC). He has been a BASIC programmer for 8 years, starting out on a Radio Shack Model 1 in the days when if you wanted software for your microcomputer, BASIC was your only option.

Aarons believes that BASIC's popularity, its ability to access machine-specific functions, and its ongoing evolution into the modular structures popularized by C and Pascal make it truly the language of the future. He describes BASIC as the language for "the guys in the trenches. I've tried to move into the fancier languages, but when you gotta make a buck and the deadline's coming, you do it in BASIC."

Charles Petzold is a programmer and freelance writer. *PC Magazine* readers know him as the PC Tutor editor and as a Programming column author and frequent editor of the Power User column. He found his way into assembly language programming in 1979 when he built and programmed a computer-controlled music synthesizer around a Z-80 microprocessor. In his former post as office-automation coordinator with New York Life, he wrote lengthy mainframe assembly language programs to calculate premium payments quickly from personal data variables, but he now uses assembly language mostly to create utility software for the IBM PC. Why does Petzold program in what he admits is the most difficult language to master? "Power," he replies. "You can do the most with it, and virtually anything you write will execute very quickly."—**Paul M. Stafford**

to understand. Assembly language is in some ways the only suitable alternative to BASIC for getting the most programming bang for the buck out of your PC. Most of the compilers available don't get you to as many PC resources as easily as the assembly language (or BASIC) can. If something can't be done using the PC's assembly language, it can't be done at all, and many of you have discovered how great that can be when writing your own programs.

Assembly language programming is arduous and slow work, but the programs are incredibly fast when they're finally running, and in terms of performance and your own sense of achievement, the end result is particularly gratifying. In fact, it's more like winning a high-stakes poker game in Las Vegas than a penny-ante one in the den. If you haven't tried assembly language programming, give it a whirl—you might just fall in love.

Pascal

Pascal's popularity (used by 30 percent of our programming readers) came as no surprise to *PC Magazine*'s editors. It would have 2 years ago, but since then, Borland International invaded the languages market with a $69.95 high-performance, RAM-resident compiler called Turbo Pascal that changed the language's image forever.

Pascal was invented in 1971 by Niklaus Wirth, the Swedish computer science professor who, like Kemeny and Kurtz, wanted to make learning how to program easier. He differed from his predecessors, however, in that he wanted to make it easier to learn how to write well-structured programs. As a result, Pascal (named for the 17th-century French mathematician and philosopher, Blaise Pascal) was designed to force good structure on programmers using the language.

It has succeeded quite well as a teaching language, but some of Pascal's limitations have led to serious problems when it has been used in the business world. Like most block-structured languages, Pascal's formal definition is limited to a kernel of fundamental instructions. This restriction isn't serious in an academic environment, but business applications require a richer variety of language facilities. When adding extensions, implementers of Pascal compilers have gone their separate ways, and the result is a hodgepodge of incompatible Pascals.

The compatibility problem had made Pascal less than suitable for business applications because transporting a program between machines with different Pascal compilers was next to impossible. Then Borland stepped in by making its inexpensive, fast compiler available on a number of micros besides the PC.

Since you're a PC user, you get the benefit of a standardized compiler for a terrific language. With any luck at all, Borland's

> T he oldest business programming language survives not because it is such a great language, but because the installed base of applications written in COBOL is so large.

standard will become even more accepted on a wider variety of computers.

C

Looking at C (used by 21 percent of our programming readers) source code might make you think its name stands for "compact" because it's such an economical, yet powerful and well-structured language. I'm sorry to have to tell you the very unromantic story of its name: *C* is the next letter in the alphabet after *B*, and B was C's predecessor as Bell Laboratories' internal systems-development language.

C was developed by Brian W. Kernighan and Dennis M. Ritchie as a systems programming language for the UNIX operating system. It is a block-structured language that contains some of the best features from other languages, including PL/I, Algol, and Pascal. UNIX itself was originally written in C for DEC PDP-11 minicomputers. Several variations of both C and UNIX are around for many computers, including the PC, but Bell Laboratories and AT&T have made substantial efforts to maintain standards for them.

C is fast becoming the language of choice among professional applications and systems developers for the PC, other micros, and most classes of computers. A fifth of you are already using it, and I suspect that number will grow in the next year or two because C's general portability, simplicity, and power can only increase its popularity.

COBOL

This statistic may come as a shock to you, but the majority of all currently operating computer programs (about 80 percent) are written in COBOL (which is used by 8 percent of our readers who program). The oldest business language survives not because it's a great language (although plenty of programmers love it), but because the installed base of COBOL applications is so large. Rewriting those programs would be uneconomical; a large base of COBOL programmers is needed to maintain them. Those same programmers usually favor COBOL when they develop new applications.

You probably don't use COBOL because it's best suited to applications that don't usually run on PCs, or at least not ones that you write for yourselves. COBOL does the mundane chores of accounting and recordkeeping better than just about any language invented before or since. Many computer scientists would like to see it replaced with something more in line with today's philosophies of program design and structure. Many have even tried to replace COBOL, but they have all failed because COBOL is a survivor in the world of business computing. If you want to write accounting programs for your PC, COBOL may be your language of choice. ∎

John Dickinson was recently appointed special projects editor of PC Magazine. *He has programmed in over 20 languages.*

BASIC: If it can be done, BASIC can do it, and what's more, there's a BASIC for everyone. One user makes a strong case for the world's most widely used computer language and looks at some of the current crop.

Years ago, a friend of mine returned home from two stints as a chopper pilot in Vietnam. He spent a month "getting civilized" and then went around to the airlines looking for a job.

Shortly thereafter, I ran into him at a local watering hole. He was more down than I'd ever seen him. "What's wrong?" I asked.

"Well," he sighed, "I thought 2,000 hours flying experience was more than enough to get an airline job, but it turns out that having helicopter time in your logbook is like having a social disease in your medical records." Airline people apparently don't believe helicopters are *real* aircraft.

A similar situation exists with programming languages. Veteran data processing types just don't believe that BASIC is a *real* computer language.

In fact, even among hobbyists, it's getting so that "if you don't speak 'C' you ain't a *real* man—or woman." The implication here is that BASIC programmers eat spinach quiche, own small, yippy dogs, and suffer many nonspecific disorders of the psyche.

Well, at the risk of losing credibility with my friends and readers in the microcomputing world, I'll admit it right here: I program in BASIC.

Let me tell you why.

When microcomputers first came along, the only language you could program in was BASIC—good old linear, two-character-variable, everything's-double-precision BASIC. Of course, you could have programmed in assembly or machine language—and some did. But the few that chose those roads either went crazy or invented software like *VisiCalc*. In either event, no one ever whipped up a workable user application in a day or two in hopes of scratching out a living.

It is true that BASIC hasn't been very pretty in the classical programming sense until recently. But it's just as true that until

recently BASIC was the only high-level language that could get a desktop computer to exercise all of the machine's capabilities. That's simply because each machine had its own custom-made BASIC to optimize its features.

True, customizing the language did lead to portability problems. But what good is a highly portable language that can't begin to touch the special features offered by the various hardware vendors? If

To us, Summit Software Technology's **Better-BASIC** *system stands out from the crowd because it combines the best elements of interpreters, compilers, and language structures in a single environment for the programmer of typical talents. Its authors have taken those extra steps to make the language usable in day-to-day programming tasks—such as including an interface with the well-known* Btrieve/ISAM *utility.*

all the industry wants is portability, it might as well develop one bland machine with one beautifully structured programming language and call it quits.

So, much to its credit, BASIC is a high-level language that can do all sorts of low-level things on its host machine.

Four BASIC Flavors

Some new versions of BASIC can run much more quickly than the original interpreted versions while retaining the language's interactive qualities. In fact, BASIC now comes in four different flavors: the traditional interpreted and compiled/linked versions, plus two hybrids. One of these hybrid forms, represented by Better-BASIC, is what I call an "interactive compiler." As with an interpreted language, you enter all your source code and run it from within the language's own internal

editor, and you can get an almost immediate response to your program upon typing RUN. But an interactive compiler runs that code much faster than an interpreter: It compiles your code into memory rather than interpreting it a line at a time or compiling it from DOS in the traditional compiler fashion. However, like a traditional compiler, an interactive compiler can generate .EXE and .COM files.

The other hybrid, represented here by True BASIC, performs as a traditional interpreted language. But you can also ask it to "compile" pseudocode, which in theory makes your program portable to any machine that supports the language.

And it's also become pretty. Many of these new interpreters and compilers now qualify as fully structured languages. In fact, you can now write a complicated BASIC application without ever saying GOTO or GOSUB. The pretty BASIC has multiline functions, procedures with arguments both local and global, and nifty structures like DO...UNTIL and CASE... ENDCASE. It even has b-trees and ISAM support, both integrated and offered by outside vendors such as SoftCraft and Computer Control Systems, as well as fancy screen generators like *Screen Sculptor* from The Software Bottling Company of Maspeth, New York.

In short, BASIC programmers now have all the goodies that those who program in other languages have. In addition, they have, for the most part, the good old comfortable keywords and syntax they learned years ago.

Sure, Turbo Pascal is pretty and it generates nifty code in an interactive compiler environment. But so does BetterBASIC, and you can bet Microsoft isn't too far away from releasing a low-cost, interactive compiler BASIC.

So why bother putting a semicolon after each line and driving yourself nuts with nested BEGIN...END blocks? If it can be done, BASIC can do it. In fact, the only

Source: PC Magazine, October 29, 1985, 108-114, 116-122, 124-138, 140-145, 150, 152-155, 158, 160-161

139

BASIC

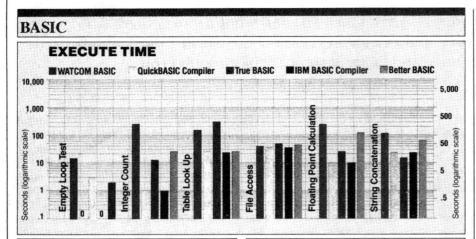

EXECUTE TIME

■ WATCOM BASIC   QuickBASIC Compiler   ■ True BASIC   ■ IBM BASIC Compiler   Better BASIC

Seconds (logarithmic scale): 10,000 / 1,000 / 100 / 10 / 1 / .1

Seconds (logarithmic scale): 5,000 / 500 / 50 / 5 / .5

Empty Loop Test, Integer Count, Table Look Up, File Access, Floating Point Calculation, String Concatenation

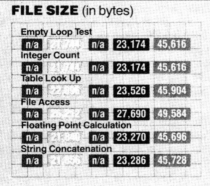

FILE SIZE (in bytes)

Empty Loop Test				
n/a		n/a	23,174	45,616
Integer Count				
n/a		n/a	23,174	45,616
Table Look Up				
n/a		n/a	23,526	45,904
File Access				
n/a		n/a	27,690	49,584
Floating Point Calculation				
n/a		n/a	23,270	45,696
String Concatenation				
n/a		n/a	23,286	45,728

COMPILE TIME (in seconds)

Empty Loop Test				
n/a		n/a	99	n/a
Integer Count				
n/a		n/a	99	n/a
Table Look Up				
n/a		n/a	99	n/a
File Access				
n/a		n/a	99	n/a
Floating Point Calculation				
n/a		n/a	99	n/a
String Concatenation				
n/a		n/a	99	n/a

Benchmark test results for WATCOM BASIC, Microsoft QuickBASIC Compiler, True BASIC, IBM BASIC Compiler, and BetterBASIC. All tests were run on an IBM PC with two floppy disk drives. All file sizes are listed in bytes, all times listed in seconds. Note: True BASIC and WATCOM BASIC are interpreters and thus do not produce object code or have compile times. BetterBASIC compiles in the interpreter, so compile times are instantaneous. Compile times for Microsoft QuickBASIC and IBM BASIC were 2 seconds for each test. Times shown include standard link time of 97 seconds. BetterBASIC, Microsoft QuickBASIC, and IBM BASIC also compile run-time object code that can be executed only with additional run-time modules. File sizes for run-time code and modules are not listed here.

problem BASIC programmers have today (besides admitting they're BASIC programmers) is picking the appropriate BASIC for their applications.

My guess is that Microsoft BASICs (including the Microsoft-authored IBM BASICs) will always be around in steadily improving versions. BetterBASIC, a relative newcomer, shows what is possible with a fully structured language in an interactive compiler environment. MTBASIC (not reviewed here) is a highly specialized BASIC for those interested in asynchronous multitasking on a single-processor CPU. True BASIC, offered by the original au-

thors of this language, attempts to bring some degree of standardization to the language in its ANSI (draft) garb. WATCOM BASIC is a language designed to be used in a multisystem environment where applications must be moved from machine to machine, vendor to vendor.

In short, there is a BASIC for everyone, somewhere.

My advice to my fellow "closet" BASIC programmers is to take heart. Don't be afraid to speak up for your language. You can do all the things the big boys can do, and you can often do it faster, cleaner, and better.

BetterBASIC

BetterBASIC may be the best of all BASIC programming worlds. It combines all the best features of other microcomputer BASIC environments.

However, it seems that designing modern interactive compilers is an art of compromise. The cost of a marvelous programming environment with Better-BASIC is relatively fat code. For example, our simple table-lookup benchmark requires 45,904 bytes as a standalone, BetterBASIC .EXE program.

The same source code compiled with Microsoft's QuickBASIC compiler generates a standalone .EXE program that comprises 22,096 bytes. Interestingly enough, the BetterBASIC file completes the table benchmark test in 88 percent of the time required by the QuickBASIC version. It would seem, then, that BetterBASIC is like a small college fullback—fat and fast.

The BetterBASIC interactive compiler environment does offer some delights, though. First, it is syntactically identical to the later versions of Microsoft BASIC. Admittedly, Microsoft BASIC may not be the best in the world, but it is the BASIC that most of you cut your teeth on and continue to use today under the guise of the various PC BASICs. In addition, Better-BASIC supports structured programming with strong subprograms and user-defined, multiline functions. It also lets you create your own libraries of functions and subroutines. Support is available for the 8087 math coprocessor (as well as the 80287), and a run-time module lets the user generate standalone .EXE-type programs. The program also supports chaining and uses all available memory.

One of the best features of BetterBASIC is its interactive programming environment. It works like this: You call up BetterBASIC just as you would any interpreter BASIC, and it comes up with its own screen that is very similar to the PC BASIC screen. Here you begin to write your main program in the traditional way, using line numbers. At any point, you can type procedure (name) or function (name). The screen will then change to a new work space that belongs to the procedure or function. Here you write the function as a standalone BASIC program with line numbers. (You needn't worry about conflicts with line numbers or variables in other sections of the program; everything is local unless defined as an argument or argument variable for passing data among the routines.)

You can use Microsoft-type declarations such as ($) for string, (#) for double precision, a Pascal-type declaration system such as REAL: A,B,C; STRING: X[89], and so forth.

Such user-defined structures as records and sets are also supported as well as simple assembly language interfaces, windows, and graphics. Although you use line numbers, a BetterBASIC program can be totally structured.

One way to think of BetterBASIC is as Pascal using Microsoft BASIC keywords, commands, and built-in functions and procedures.

As your program develops with each function and procedure in its own work space, debugging is ongoing and fully interactive. Everything must work together as the program is built up.

You can save all of the procedures and functions (or a subset of them) into a library for use in other programs. Thus you can add permanent elements to the language.

The program offers full support for windows and graphics as well as the usual Microsoft file types. Full support for SoftCraft Inc.'s *Btrieve*, the popular independent b-tree, ISAM file management system, is available as an option.

Unwilling to trust my own judgment, I asked two other language reviewers what they thought of BetterBASIC. The vote was unanimous: BetterBASIC is the best if you can put up with fat code. So good, in fact, that other language purveyors, most probably Borland and Microsoft, are expected to bring out similar products by year's end.

BetterBASIC is available on a modular basis, so you have the option of purchasing only what you need. The BASIC programming system costs $199. The 8087 support module costs $99. Add $49 for binary math support, $99 for *Btrieve* support, and $250 for a run-time converter, and you've got the works.

IBM BASIC Compiler and Microsoft QuickBASIC Compiler

Microsoft probably knows more about microcomputer BASIC than any other company. In 1975, Microsoft developed a BASIC interpreter for the MITS Altair, and it's been supplying increasingly more-advanced interpreter BASICs for microcomputers ever since. In fact, 9 out of 10 microcomputers use some version of Microsoft BASIC, with over 3.5 million users worldwide.

The three versions of IBM interpreter

```
SOURCE
PROCS=0
INTEGER: X,L
STRING: START$[8],STOP$[8]
INTEGER ARRAY(25): A
     1  ' Table lookup test
     7  START$= TIME$
    10  FOR X = 1 TO 1000
    15    RESTORE
    20    FOR L = 0 TO 24
    30      READ A(L)
    40    NEXT L
    50  NEXT X
    60  STOP$=TIME$:PRINT START$,STOP$
   200  DATA 1,2,3,4,5
   210  DATA 6,7,8,9,10
   220  DATA 11,12,13,14,15
   230  DATA 16,17,18,19,20
   240  DATA 21,22,23,24,25
   250  END
ENDFILE
```

BetterBASIC: Table lookup test.

```
SOURCE
PROCS=0
INTEGER: X
STRING: START$[16],STOP$[16]
     1  'Empty loop in BASIC
     7  LET START$ = TIME$
    30  FOR X = 1 TO 10000
    40  NEXT X
    45  STOP$ = TIME$
    50  PRINT START$,STOP$
    60  PRINT "START"," STOP"
ENDFILE
```

BetterBASIC: Empty loop test.

```
SOURCE
PROCS=0
INTEGER: X
STRING: START$[8],STOP$[8]
REAL: A,B,C
     1  'Floating Point
     7  START$ = TIME$
    10  FOR X = 1 TO 10000
    20    A=0:B=1234.56:C=78.9
    30    A=B*C
    40    A=B/C
    50  NEXT X
    55  STOP$ = TIME$
    60  PRINT START$,STOP$
ENDFILE
```

BetterBASIC: Floating point test.

BetterBASIC
Summit Software
Technology Inc.
40 Grove Street
Wellesley, MA 02181
(617) 235-0729
List Price: $199; 8087
support module, $99; binary
math support, $49; *Btrieve*,
$99; run-time converter, $250
Requires: 192K RAM, one disk drive, DOS 1.1
or later.

QuickBASIC Compiler
Microsoft Corp.
10700 Northup Way
Bellevue, WA 98009
(206) 828-8080
List Price: $99
Requires: 128K RAM,
one disk drive, DOS 2.*x*.

BASIC Compiler
IBM Entry Systems
5201 South Congress Ave.
Boca Raton, FL 33431
(305) 998-2000
List Price: $495
Requires: 128K RAM,
one disk drive, DOS 2.1 or later.

True BASIC
Addison-Wesley
Publishing Co. Inc.
One Jacob Way
Reading, MA 01867
(617) 944-3700
List Price: $149.90
Requires: 192K RAM, one disk drive, DOS 2.*x*.

WATCOM BASIC
415 Phillip St.
Waterloo, Ontario N2L 3X2
(519) 886-3700
List Price: $250
Requires: 192K RAM, one
disk drive, DOS 2.*x*.

BASIC (cassette, disk, and advanced) are all Microsoft products. So, too, are Versions 1.0 and 2.0 of the IBM Personal Computer BASIC Compiler. And, not to be outdone by itself, Microsoft also markets a BASIC compiler under its own label. It's called the Microsoft QuickBASIC Compiler.

For this review, *PC Magazine* combines the Microsoft QuickBASIC compiler and the IBM Personal Computer BASIC compiler because of their many similarities and the fact that both were born in the Microsoft shop.

These two compilers were introduced within a week of each other in late July, and, despite their similarities, they arrived on the market at opposite ends of the price spectrum. QuickBASIC lists for $99 and was thus priced "to put a fast BASIC compiler within the reach of all BASIC users—hobbyists and recreational programmers as well as professional software developers," according to Microsoft. The IBM compiler, with a $495 price tag, is targeted at the business professional. So, besides general pricing philosophy, what accounts for the $400 price difference? The IBM package includes b-tree, ISAM file management utilities, and a library utility. Other than that, the features of the two compilers are identical.

The QuickBASIC compiler is actually Version 2.0 of Microsoft's original BASIC compiler. And the IBM compiler is Version 2.0 of the original IBM release. Both compilers have many new and improved features over their predecessors.

Most important of these features is the fact that traditional linear BASIC is taking the long-awaited turn toward facilities that let you use it to develop fully structured and modular programs.

For example, both QuickBASIC and the IBM compiler now support multiline subprograms and parameter passing with global or local variables. You can separately compile subprograms and then link them with the main program before you run it. You can also create libraries of separately compiled modules for other uses.

Another important change that both compilers implement makes line numbers optional while permitting alphanumeric labels—a giant step for programming purists. When subprograms and new multiline functions are used in an environment without line numbers, you can create BASIC programs that are just as structured as any Pascal program. This should end much of the grumbling among the ivory-tower types that BASIC is not a *real* language because of its lack of structure.

Other improvements to both QuickBASIC and IBM's compiler include new support for all of the sound, color, and graphics features of BASICA Version 3.0, (PLAY, SOUND, DRAW, GET, PUT, LINE, CIRCLE, COLOR, and so forth).

Programs can be significantly larger with these new BASIC products that allow

Traditional linear BASIC is taking the long-awaited turn toward facilities that let you use it to develop fully-structured and modular programs.

a full 64K bytes of RAM for the program and another 64K bytes for data.

Another important change in both versions is the support of dynamic arrays—that is, arrays whose dimensions the program can control.

The $99 Microsoft QuickBASIC compiler has more than enough features to handle any but the most file-dependent applications. However, when you have a full database management project you want to solve with BASIC, the ISAM support in the IBM product (or ISAM support from an outside vendor such as Computer Control Systems' *FABS* or *Btrieve*) is a must.

IBM's b-tree, ISAM, is highly polished for operation with the IBM system; therefore, interface is simpler with the supplied ISAM than it is with ISAM from an outside vendor.

Both QuickBASIC and the IBM compiler are well documented, but as is often the case, IBM's documentation effort is best. The IBM package includes two manuals. The first is a tutorial and reference on BASIC in general, the compilation process, and ISAM theory. The other is a syntax reference with coverage of all BASIC keywords and commands.

Microsoft, on the other hand, offers a single manual. The major difference is that Microsoft documents only those BASIC keywords and functions that differ from similar interpreter BASIC keywords and

```
SOURCE
PROCS=0
STRUCTURE: REC
    STRING: RECORD$[132]
END STRUCTURE

REC: R1
INTEGER: RECNUM
STRING: START$[8],STOP$[8]
    1 ' File access test
    5 START$ = TIME$
   10 OPEN "TEST.DAT" AS #1 LEN = SIZE (R1)
   30 FOR RECNUM = 1 TO 100
   40    WRITE RECORD #1 RECNUM R1
   50 NEXT RECNUM
   60 'read records back.
   70 FOR RECNUM = 1 TO 100
   80    READ RECORD #1 RECNUM R1
   90 NEXT RECNUM
  100 'modify and rewrite records.
  110 FOR RECNUM = 1 TO 100
  120    READ RECORD #1 RECNUM R1
  125    R1.RECORD$ = "MODIFIED"
  130    WRITE RECORD #1 RECNUM R1
  150 NEXT RECNUM
  155 CLOSE
  160 STOP$ = TIME$
  165 PRINT START$,STOP$
  170 END
ENDFILE
```

BetterBASIC: File access test.

functions. You are expected to use interpreter BASIC documentation for those language elements that are used identically in both environments.

Both of these products offer a run-time module (56K bytes for IBM; 58K bytes for Microsoft) that can be loaded and used by a series of compiled programs. Standard routines are kept in the run-time module; thus each of the compiled programs with access to the run-time module need not contain them. In a system with a number of program modules, this method can save significant disk space. If these run-time modules are used in commercial applications, they must be licensed with IBM or Microsoft.

Alternatively, you can opt to create standalone .EXE files that do not need run-time support. These files contain all the routines they'll need (and then some) and thus occupy more disk space than a run-time unit. Interestingly, these standalone programs seem to run about 12 percent faster after loading than those that use the run-time module.

Both the Microsoft and the IBM compilers are excellent products, truly the bedrocks of BASIC language compilers. Even if you're a professional BASIC programmer who may opt for some other system—BetterBASIC, perhaps—you may still want at least one of these Microsoft products (QuickBASIC or the IBM compiler, Version 2.0) on your shelf.

```
SOURCE
PROCS=0
INTEGER: X
STRING: START$[8],STOP$[8]
    1 'Integer count in BASIC
    5 START$ = TIME$
   10 WHILE X <32767 DO
   20    X = X + 1
   25    PRINT X
   30 REPEAT
   35 STOP$ = TIME$
   40 PRINT START$, STOP$
ENDFILE
```

BetterBASIC: Integer count test.

```
SOURCE
PROCS=0
REAL: X
STRING: A$[16],B$[49],C$[65],START$[16],STOP$[16]
    1 ' String concatenation test
    7 START$ = TIME$
   10 FOR X = 1 TO 10000
   20    A$ = "This is a string"
   30    B$ = "This is a longer string with lots of words in it."
   40    C$ = A$+B$
   50 NEXT X
   55 STOP$ = TIME$
   60 PRINT START$,STOP$
ENDFILE
```

BetterBASIC: String concatenation test.

True BASIC

Did you ever have one of those horrible moments in life where you slap yourself on the forehead and say, "Damn, I had that idea first, but somebody else made a million bucks out of it—I just gotta do something."

True BASIC is the result of one of those moments. Back in 1963, Dartmouth College professors John G. Kemeny and Thomas E. Kurtz responded to a college commitment that computers would become easily available to all students by designing a new language that would be easy to learn, yet useful for any programming task. The language, of course, is BASIC, and it is currently the most widely used computer language in the world—alas, largely due to the foresighted folks at Microsoft, not Kemeny and Kurtz.

According to Kemeny and Kurtz in the BASIC manual, the fate of BASIC in the world outside academia was not too bright. "When microcomputers first appeared, BASIC was the most popular language for them because it was a clean and simple language. The first microcomputers had very limited memories, so that implementors had to make a number of compromises—some of which were most unfortunate. Many of these compromises became features of the language and were kept when the original reasons for compromising had disappeared. Today's personal computers are large and powerful machines that allow the implementation of a full modern BASIC. Yet the versions most widely used are what the trade calls Street BASIC—a horrible dialect of a beautiful

> When microcomputers first appeared on the market, BASIC was the most popular programming language because it was so simple.

language."

Kemeny and Kurtz both agree that because "the authors of these languages violated the fundamental design principles of BASIC. Street BASIC is heavily dependent upon the particular hardware that is being used. The same BASIC program will not run on different personal computers; indeed, it typically cannot be run on two different models from the same manufacturer."

Last year Kemeny and Kurtz decided it was time to replace that obsolete and often ugly implementation of BASIC with a well-designed, modern version. True BASIC, obviously, is the Kemeny-Kurtz idea of a well-designed, modern version. It conforms closely to the latest drafts of the still-unsettled ANSI BASIC. And I'll have to admit that it's both modern and pretty, enabling you to attack your problems while adhering to the best principles of structured programming.

However, True BASIC is also bound to its interactive compiler and shows up erratically on the benchmark tests. Its execution time of our string concatenation benchmark was about half the time for the Microsoft, IBM, and BetterBASIC compilers. However, True BASIC's table-lookup benchmark was several orders of magnitude greater than the others—minutes for True BASIC versus seconds for Microsoft, IBM, and BetterBASIC. True BASIC's control structures include IF-THEN-ELSE, SELECT-CASE, DO-WHILE, and DO-UNTIL.

Most of the True BASIC statements, commands, and built-in functions and procedures are similar enough to the more familiar Microsoft BASIC that they can be read easily. However, many have slight differences in syntax, and, therefore, writing a True BASIC program will take a bit of extra study. Some matters, such as file reading, writing, and maintenance, are entirely different from the more familiar forms of microcomputer BASIC. However, first time users will probably find it easier to grasp the file routines used by True BASIC.

Part of the Kemeny-Kurtz idea was to develop a highly transportable BASIC, a machine-independent language that closely adheres to the ANSI draft standard. At the same time, they wanted to keep the language interactive for programming ease. To meet these requirements, they went to an interactive compiler. Errors show up immediately, and a built-in help system usually pinpoints the error and suggests an appropriate correction.

True BASIC's compiler generates an intermediate corrected code that is then run against a machine-dependent module. The language stays portable because only the machine-dependent module must change.

True BASIC lets you use all the installed memory for your BASIC programs.

Documentation for True BASIC is nicely presented in two spiral-bound, softcover binders. One is a reference manual and the other is a user's guide that has instructions for the IBM PC.

It seems unlikely that True BASIC will replace the Microsoft and Microsoft lookalike BASICS—at least in this generation. But it is sure to gather a following, I suspect, particularly in teaching centers.

WATCOM BASIC

The WATCOM BASIC interpreter is a relatively new member of a family of languages and support systems developed and distributed by the University of Waterloo's computer systems group.

This package deserves consideration if you are thinking of developing applications on a microcomputer system that will ultimately be used in the minicomputer

BASIC is the language for the guys in the trenches— when you gotta make a buck and the deadline's coming, you do it in BASIC.

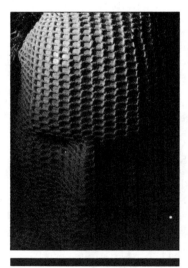

world. WATCOM BASIC can produce executable programs for a number of different computers, including the Commodore SuperPET (MC6809 processor), the IBM PC, IBM 370, Digital VAX, Digital PRO series (pdp 11), and MC68000-based systems.

WATCOM BASIC is not only highly portable, it permits highly structured programming. Subprograms and multiline functions are supported as well as recursive functions.

WATCOM BASIC's syntax is quite similar to the customary Microsoft BASIC key words and statements. The only things that are missing from this BASIC are those highly machine-dependent functions that would destroy the portability of developed applications.

WATCOM BASIC comes packaged with a powerful standalone programming line editor called WEDIT. The editor, typical of mini and mainframe text editors, has fast search and replace features and can simultaneously manipulate multiple file segments.

Although WEDIT is nice to use, it is not necessary. You can develop WATCOM BASIC programs on any text editor or in the program's interactive interpreter environment. In fact, the interpreter manages its own file director, making loads, saves, and other file management chores easy from within either the interpreter or applications programs.

WATCOM BASIC allows you to use long names for variables and other program entities. Procedures can be called by these names and variables can be passed back and forth.

This implementation supports true integer arithmetic and bit logical operations using integers, and MAT statements support operations on entire matrices. WATCOM BASIC is relatively slow when compared with the other interpreters that are reviewed here, but that's the compromise that was made in order to gain a high degree of portability.

WATCOM BASIC is probably most valuable in educational or engineering environments that simultaneously use various machines and languages.. The Waterloo group also has interpreter versions of APL, COBOL, FORTRAN, and Pascal, which all use the same full-screen editor that WATCOM BASIC does.

In addition, WATCOM produces several versions of FORTRAN, C, COBOL, BASIC, and Pascal for the IBM 370 and DEC VAX minicomputers. The company offers networking and terminal emulation capabilities for all machines supported by its languages.

The bottom line is that WATCOM BASIC is the ideal BASIC if you have to move applications programs all over the shop. In this sense, it seems to come closer than any other BASIC to bridging the micro-mini-mainframe canyons. ∎

Photograph: Craig Cutler

COBOL: This language is widely used for programming business applications, but at age 25, is it fast enough to compete with other languages for the PC?

In the late 1950s, you couldn't do computer programming in low-level languages, such as assembly language, unless you had an engineer's intimate knowledge of the particular computer on which you wanted to run your programs. At best, it was an unsatisfactory situation. If computers were to become a useful business tool, an easier way had to be found to convey requirements to the computer. To this end, the Department of Defense convened the Conference on Data Systems Languages in 1959 to define a programming language for business data processing applications.

The conference determined that a language had to be machine independent, so that you could easily transport programs from one computer to another without having to rewrite them. The language also had to be easy to understand, so that it could be taught to and maintained by business people rather than engineers. But it also had to produce efficient object code because the programs had to run quickly enough not to bog down the computer or frustrate users.

The outcome of the conference was the COmmon Business Oriented Language, or COBOL. Since its inception, COBOL has been approved by and is periodically revised by the American National Standards Institute and is often called ANSI COBOL. The current published standard for COBOL is ANSI x3.23-1974.

Business data processing problems are very file-oriented because they entail a great deal of repetitive information, such as customer records or product types and quantities. These files need to be input using a language with procedures that perform a few mathematical or logical operations. The output is the modified files. COBOL facilitates file handling by recognizing a variety of file types, sorting and merging files, and generating printed reports from the file data. It has powerful verbs that carry out these functions without your having to code them in detail.

The tradeoff is that COBOL is not an ideal language for structured programming, although revisions of the ANSI standard, which are now under way, include several new features that will make COBOL more structured. Unlike in languages such as C and Logo, you cannot make up new commands (verbs) in COBOL. Nevertheless, its combination of features have made COBOL the most widely implemented programming language for busi-

Among the COBOL compilers we reviewed, Micro Focus's **VS COBOL Workbench** *is the clear winner, offering a panoply of programming tools. The forms facility paints screens and creates data descriptions. You can write your program with the editor, then check its syntax. The animator debug facility lets you halt execution at any point and inspect data. You can convert your finished program into a .EXE file with the build facility.*

ness applications.

Until recently, nearly all COBOL programming has taken place on mainframes. However, several manufacturers are now marketing COBOL for the PC, and *PC Magazine* tested all four available PC COBOL compilers (see the benchmark tests). But it's really not enough to determine if these new COBOL compilers are bug-free, the additional productivity tools are well-designed, and the manual is clear. A more basic question is: Even though COBOL has been successful on mainframes, is it in fact a good programming language for the PC?

What really counts is speed. A language's speed is not determined by the wordiness of a program but to what extent the compiler can turn that source code into lean, fast machine code—a real challenge for any COBOL compiler for the PC.

A compiler can also obtain a speed edge at run time. Most compilers continue to do their input and output through DOS, an approach that makes the compiler more transportable but adds a layer of code that slows down processing. Alternatively, a compiler can bypass DOS and deal directly with the BIOS. This method significantly increases speed, but the company—and you—take on the responsibility of having to install revisions for each new model of your machine.

Speed and size are both affected by the way a COBOL implementation handles libraries. If the libraries are copied into the loading module, that module will be larger and will require more memory, but it will run faster because the library members are right there. The library members can also be dynamically called up at run time, creating a smaller load module that uses less RAM, but the libraries must be waiting in storage whenever you run the program. The processing time of this setup will be longer.

The size of COBOL source code or loading modules on PCs is a minor issue because both storage and memory are becoming cheaper and more plentiful.

Custom Versus Store Bought

Besides speed, your type of application, such as whether it's a one-of-a-kind program or a generalized package, is another factor to consider before you buy a COBOL compiler for your PC. A major difference between mainframes and PCs is that the majority of mainframe systems are custom-written, while PCs usually run off-the-shelf packages. It's cost-efficient for large companies to invest in large-scale data processing environments and to develop systems from scratch. On the other hand, PCs are generally used on a comparatively limited scale, either by a smaller company in which it isn't economically feasible or for that matter necessary to write custom-tailored systems or in large

COBOL compiler, the Workbench offers other tools to help you enter and edit your program, check syntax, debug your program, and create a standalone, executable module.

While Workbench is based on the ANSI standard, it also includes elements from IBM VS compilers and innovations of its own. These differences make this COBOL easier to use on the PC, but they are likely to cause errors when you run your PC-developed program on a mainframe.

Workbench's environment division includes a lock clause. This clause is useful in a multiuser environment because it lets you specify exclusive or shared use of a file.

An unfortunate decision of Micro Focus's was to omit the report section of the data division. This section supports the report writer utility that greatly speeds up the generation of reports and is widely used on mainframes. The company has retained all internal data formats in the data division except floating point.

The procedure division has several augmentations. Explicit scope delimiters of the type END-ADD and END-IF make it easier to write structured programs. Negated exceptions, such as NOT ON SIZE ERROR and NOT AT END, help you isolate conditions that fail to occur. The Workbench also lets you put COPY statements into your copy libraries, which nest COPY statements.

Additional verbs in the procedure division include CONTINUE, a useful no-operation statement; EVALUATE, which compares subjects of the evaluation with one or more objects and helps you detect a complex state and then take an appropriate action; and INITIALIZE, which sets a list of data items to zeros, spaces, or a user-defined character.

Micro Focus has turned COBOL's ACCEPT and DISPLAY verbs into very useful PC-oriented screen-painting tools. Once you use the Workbench's forms utility to place data fields on-screen, the Workbench creates two copy library members. The first of these contains the standard COBOL data descriptions for the display screen data fields, and the second includes procedure division code to accept

```
ENVIRONMENT DIVISION.
CONFIGURATION SECTION.
SOURCE-COMPUTER.  IBM-PC.
OBJECT-COMPUTER.  IBM-PC.
DATA DIVISION.
WORKING-STORAGE SECTION.
77    THE-X                        PIC 9(5) VALUE ZERO.
01    THE-INTEGERS.
      05    THE-A                  PIC 9(5)V99.
      05    THE-B                  PIC 9(5)V99.
      05    THE-C                  PIC 9(5)V99.
01    START-TIME                   PIC 9(8).
01    END-TIME                     PIC 9(8).
PROCEDURE DIVISION.
A000-START.
      ACCEPT START-TIME            FROM TIME.
      PERFORM A010-CALC VARYING THE-X FROM 1 BY 1
          UNTIL THE-X GREATER THAN 10000.
      ACCEPT END-TIME              FROM TIME.
      CALL "TIMER" USING START-TIME END-TIME.
      STOP RUN.
A010-CALC.
      MOVE ZERO                          TO THE-A.
      MOVE 1234.56                       TO THE-B.
      MOVE 78.9                          TO THE-C.
      COMPUTE THE-A = THE-B * THE-C.
      COMPUTE THE-A = THE-B / THE-C.
```

VS COBOL Workbench: Floating-point test.

```
ENVIRONMENT DIVISION.
CONFIGURATION SECTION.
SOURCE-COMPUTER.  IBM-PC.
OBJECT-COMPUTER.  IBM-PC.
DATA DIVISION.
WORKING-STORAGE SECTION.
77    THE-X                        PIC 9(5) VALUE ZERO.
01    THE-STRINGS.
      05    THE-A                  PIC X(20).
      05    THE-B                  PIC X(60).
      05    THE-C                  PIC X(80).
01    START-TIME                   PIC 9(8).
01    END-TIME                     PIC 9(8).
PROCEDURE DIVISION.
A000-START.
      ACCEPT START-TIME            FROM TIME.
      PERFORM A010-STRING VARYING THE-X FROM 1 BY 1
          UNTIL THE-X GREATER THAN 10000.
      ACCEPT END-TIME              FROM TIME.
      CALL "TIMER" USING START-TIME END-TIME.
      STOP RUN.
A010-STRING.
      MOVE "THIS IS A STRING"      TO THE-A.
      MOVE "THIS IS A LONGER STRING WITH LOTS OF WORDS IN IT."
                                   TO THE-B.
      STRING THE-A THE-B           DELIMITED BY SIZE
                                   INTO THE-C.
```

VS COBOL Workbench: String concatenation test.

```
ENVIRONMENT DIVISION.
CONFIGURATION SECTION.
SOURCE-COMPUTER.  IBM-PC.
OBJECT-COMPUTER.  IBM-PC.
DATA DIVISION.
WORKING-STORAGE SECTION.
77  THE-X                        PIC 9(5) VALUE ZERO.
77  LOOP                         PIC 9(5) VALUE ZERO.
01  THE-A-TABLE.
    05  THE-A  OCCURS 25 TIMES   PIC 99.
01  THE-BZ-TABLE.
    05  FILLER                   PIC X(10)  VALUE "0102030405".
    05  FILLER                   PIC X(10)  VALUE "0607080910".
    05  FILLER                   PIC X(10)  VALUE "1112131415".
    05  FILLER                   PIC X(10)  VALUE "1617181820".
    05  FILLER                   PIC X(10)  VALUE "2122232425".
01  THE-BZ-TBL REDEFINES THE-BZ-TABLE.
    05  THE-BZ OCCURS 25 TIMES   PIC 99.
01  START-TIME                   PIC 9(8).
01  END-TIME                     PIC 9(8).
PROCEDURE DIVISION.
A000-START.
    ACCEPT START-TIME            FROM TIME.
    PERFORM A010-MOVE VARYING THE-X FROM 1 BY 1
        UNTIL THE-X              GREATER THAN 1000
            AFTER LOOP           FROM 1 BY 1
                UNTIL LOOP       GREATER THAN 25.
    ACCEPT END-TIME              FROM TIME.
    CALL "TIMER" USING START-TIME END-TIME.
    STOP RUN.
A010-MOVE.
    MOVE THE-BZ (LOOP)           TO THE-A (LOOP).
```

VS COBOL Workbench: Table lookup test.

```
ENVIRONMENT DIVISION.
CONFIGURATION SECTION.
SOURCE-COMPUTER.  IBM-PC.
OBJECT-COMPUTER.  IBM-PC.
INPUT-OUTPUT SECTION.
FILE-CONTROL.
    SELECT TEST-FILE ASSIGN      TO EXT-FILENAME
        ORGANIZATION IS          RELATIVE
        ACCESS MODE IS           RANDOM
        RELATIVE KEY IS          FIELD1.
DATA DIVISION.
FILE SECTION.
FD  TEST-FILE
    LABEL RECORDS STANDARD
    RECORD CONTAINS 132 CHARACTERS
    DATA RECORD TEST-REC.
01  TEST-REC.
    05  RECORD-DOLLAR            PIC X(132).
WORKING-STORAGE SECTION.
77  FIELD1                       PIC 9(5).
77  THE-X                        PIC 9(5) VALUE ZERO.
77  EXT-FILENAME                 PIC X(10) VALUE "B:TEST.DAT".
01  START-TIME                   PIC 9(8).
01  END-TIME                     PIC 9(8).
PROCEDURE DIVISION.
A000-START.
    ACCEPT START-TIME            FROM TIME.
    OPEN I-O                     TEST-FILE.
    PERFORM A010-WRITE VARYING THE-X FROM 1 BY 1
        UNTIL THE-X              GREATER THAN 100.
    MOVE 1                       TO THE-X.
    PERFORM A020-READ  VARYING THE-X FROM 1 BY 1
```
(continues)

VS COBOL Workbench: File access test.

and display the screen. The FORMS facility also puts the accompanying COPY statements into your program. Despite the success of the FORMS utility on the PC, it can't be ported to mainframes; the monitors are different from those used with PCs, and Micro Focus's enhanced ACCEPT and DISPLAY verbs aren't standard COBOL.

In addition to the compiler, the Workbench offers a program editor. For some reason, this editor does not do line numbering, but it satisfactorily performs most other editing functions.

The syntax checker receives COBOL source code, finds language errors, and delivers intermediate-level code. You can directly compile this code into machine code or test it further with the "animator," the highlight of the Workbench. The animator is an excellent debugging tool that allows you to set breakpoints, change data items, and trace execution through subprograms.

Once you have compiled your program, you execute it with the RUN facility. If other people are going to use the program, you wouldn't want them to have to buy the Workbench just to run it, so the Workbench includes a BUILD facility that converts your compiler output to .COM files and a COM2EXE utility to convert files over 64K bytes to .EXE files.

The Workbench also includes a session recorder. You can use it to keep track of all your keystrokes, which is useful for training users and locating problems.

The Micro Focus VS COBOL Workbench offers a lot more than just a COBOL compiler. The components have been tailored to take advantage of microcomputers, so while the programs you produce with the Workbench can certainly be ported to a mainframe, they will run best on a PC.

Realia COBOL

Realia Inc.'s COBOL takes IBM's VS COBOL rather than ANSI standard COBOL as its guide. Like its mainframe counterpart, Realia COBOL is command-driven rather than menu-driven. Stepping right into the footprints of its mainframe leader, the package doesn't even include a language manual but suggests that you purchase an IBM VS COBOL (main-

```
(File access test continued)
          UNTIL THE-X             GREATER THAN 100.
     MOVE 1                       TO THE-X.
     PERFORM A030-MODIFY VARYING  THE-X FROM 1 BY 1
          UNTIL THE-X             GREATER THAN 100.
     CLOSE                        TEST-FILE.
     ACCEPT END-TIME              FROM TIME.
     CALL "TIMER" USING START-TIME END-TIME.
     STOP RUN.
A010-WRITE.
     MOVE THE-X                   TO FIELD1.
     WRITE TEST-REC               INVALID KEY STOP RUN.
A020-READ.
     READ TEST-FILE               INVALID KEY STOP RUN.
A030-MODIFY.
     PERFORM A020-READ.
     MOVE "MODIFIED"              TO RECORD-DOLLAR.
     PERFORM A032-REWRITE.
A032-REWRITE.
     REWRITE TEST-REC             INVALID KEY STOP RUN.
```

frame) manual to use in conjunction with Realia's user manual. You can purchase the IBM manual from Realia for $25 plus shipping or, better yet, from IBM for $17 plus tax.

Realia COBOL is missing some features from both ANSI and VS COBOL standards, but it has several extensions, some of which are based on the proposed changes to the ANSI standard and some of which improve efficiency and flexibility on the PC. For instance, in the environment division, Realia has added several means for associating internal filenames with DOS files. You can assign a file directly to a literal containing a DOS filename or you can assign it to a data item and associate the internal and external filenames with the DOS SET command. The compiler does not support the EBCDIC collating sequence, which means you can't upload your Realia program to a mainframe.

The data division is missing both the report writer and the communications program. A nice enhancement, based on the proposed ANSI COBOL revision, is that it allows condition names or 88s to be set to TRUE. This would allow 88s to be set as well as tested.

Although the procedure division does not support the SORT/MERGE command, it does include 16 explicit scope delimiters, including END-IF and END-COMPUTE. The in-line PERFORM statement allows you to place the code to be performed immediately following the PERFORM command. This procedure speeds up execution by eliminating branching. You can use the TEST BEFORE and TEST AFTER clauses to indicate whether you should do limit testing before or after the UNTIL phrase of the PERFORM statement. The DELETE FILE statement lets you delete an entire file, rather than just a record, from within a COBOL program.

In addition to the language extensions, Realia COBOL includes an interface that lets you call DOS facilities from within COBOL programs and a program for interfacing with Lattice C programs via calls.

The Realia Editor (RED) is completely command-driven, though most of the commands are not letters or mnemonics but combinations of the nonalphanumeric keys. These combinations are difficult to remember unless you frequently use the editor. To use any of the nine commands activated with alpha keys, you must first press the Esc key. RED does have some nice features, such as line numbering and a recovery mode in case your system crashes during an edit. However, an SPF-like, menu-driven program editor is much easier to use.

Realia's COBOL compiler, REAL-COB, can accept compile parameters either from switches appended to the command line or from directives included in the source file. You can direct the compiler to do such things as check for data exceptions and subscript ranges. You can also ask it to treat PERFORM statements such as CALL, which increases the efficiency of the compiler. Most parameters can be put into effect by either a switch or a directive but not by both. This means that you will usually have to look in both places to confirm your parameters.

REALCOB can produce a full cross-reference listing, similar to that on IBM mainframes. This listing is especially useful for long or multiple programs.

The Realia compiler relies on the DOS linker to create load modules. You can use the linker to combine COBOL programs, or you can wait until run time to call subprograms. With so many options and steps, it will be worthwhile for you to set up a batch file to run the compile and link process. It would have been nice of Realia to set up menus to guide you through these steps.

Realia does not support the ANSI standard COBOL DEBUG, but it offers its own debugger, FOLLOW THE SOURCE, as a replacement. The Realia debugger has an extensive set of commands, including ones to modify subscripts and set a range of lines to be executed. You can debug COBOL programs with the CALL command from an initial program and can display on-line help.

A mainframe COBOL programmer will feel right at home with Realia, but a PC user new to COBOL will want more guidance than Realia offers.

mbp COBOL

Mbp COBOL is a European entry in the COBOL sweepstakes, produced by mbp Germany in Dortmund, West Germany, and marketed here by mbp Software in Alameda, California. Mbp has been a software consulting firm for nearly 30 years and has recently entered the retail arena. The mbp COBOL is entitled "ANSI '74 COBOL compiler," and the language elements closely follow the standard. Mbp has added some utilities that give mbp COBOL a distinctive personality.

In the environment division, mbp has made a small but necessary extension to the SELECT-ASSIGN statement so that the compiler will recognize DOS-format filenames. The data division change is equally small; hexadecimal constants are permissible.

Mbp's screen management system

(SMS) is a utility that helps you paint menus or data entry screens. Most commonly, a screen painter eases the handling of screen I/O by generating much of the working-storage code, often placing it in COBOL copy libraries. You use ACCEPT, DISPLAY, or CALL commands in the procedure division to pass control to the screen painter, which does the screen I/O for you. Mbp COBOL uses CALLs. The SMS screen painter generates I/O masks, which it places in SMS mask libraries—not COBOL copy libraries. Therefore, you must still write your own working-storage screen descriptions. This means that to change a screen, you must change both the I/O mask and the program. SMS could reduce this effort by generating the working-storage code for you.

Mbp has a special utility for dynamically calling COBOL programs at run time. Usually COBOL requires you to put calls to the subprograms in the procedure division of the main program and pass arguments to the subprograms' linkage sections. Mbp's technique bypasses the linkage section. Instead, you must execute the main program with the CHAINC command. The called program then picks up the passed arguments by calling CHAIN, rather than by using the linkage section. Using this technique in the COBOL benchmark tests would have required changing the program code, so we used static linking instead. Unfortunately, this meant that the TIMER module had to be linked into the test programs, increasing

COBOL is the ultimate business language—it is quick, easy to learn and to read, and handles large data files with precision.

the size of each .EXE module.

This COBOL package has an especially strong emphasis on segmentation and overlays. This means of dividing a program into segments that stay in memory and those that can be overlayed by new segments is standard COBOL, and it used to be helpful when computers had limited memory as compared with the program size. It's rarely needed now and hardly deserves the emphasis mbp gives it.

Mbp's ISAM maintenance utility is helpful for reorganizing ISAM files,

while network versions of mbp COBOL offer record locking to support several PC networks.

The mbp's compiler works fine once you get rid of your syntax errors. A missing hyphen in the phrase FILE-CONTROL caused 45 errors; fixing it reduced the errors to zero. Since FILE-CONTROL and similar phrases are required in CO-BOL, most compilers will, in this instance, put out a message saying "FILE-CONTROL not found; assumed present" and let it pass. The compiler listing spreads a lot of extraneous information over five pages and doesn't give you a way to reduce the amount of printed material. The listing also prints 132 characters across, requiring you to set your printer accordingly.

You use the DOS linker to link your programs and mbp's run-time module to execute them. The cost of mbp includes 25 copies of the run-time module, allowing that many people to use your programs without having to buy the full package.

Some of mbp's COBOL utilities seem unnecessarily complex, like the screen management system, dynamic calls through chaining, and program segmentation techniques. The relative inaccessibility of these utilities is compounded by the manual, which makes simple concepts hard to understand. The mbp compiler can be recommended for its leanness and because it adheres so closely to the ANSI standard, but the company could have improved its COBOL package by streamlining the auxiliary facilities as well. ■

ASSEMBLY

ASSEMBLER: If you master this devilishly difficult language, it will reward you with the tightest code, fastest execution, and the most gratifying experience on the PC.

It's two o'clock in the morning, and the first part of your new assembly language program almost works. The listing is four pages long and full of instructions like MOV, JMP, SHR, ADC, MUL, DIV, PUSH, and POP. None of the individual instructions gives the slightest clue as to what they're supposed to do in tandem. You feel as if you're trying to construct a suspension bridge out of toothpicks. What does the program do so far? Well, it does something equivalent to the BASIC statement

```
INPUT "Enter 3 numbers: ",A,B,C
```

It still needs work. If one of the numbers entered has two decimal points, the program crashes and fills the screen with ampersands. It fills the screen instantaneously, so it's fun to watch, but you still have to get rid of that bug. Do it tomorrow? No, you're so close now that another hour or two won't hurt.

One thing's for sure, though. Getting an assembly program to run properly has to be one of personal computing's most gratifying experiences: You've accomplished one of the most difficult things you can do on the PC.

An assembly language program translates directly into machine code that the computer's microprocessor reads from memory and executes. Higher-level languages are also translated into machine code by a compiler, but the translation is done in a generalized and roundabout manner that leaves the resultant program bloated and sluggish. Until compilers become smarter than people, assembly language will continue to be the one that produces the tightest code and the fastest execution. In the hands of a master, assembly language can create power-programs like Lotus's *1-2-3*, speed-demons like *Xy-Write*, and the unbelievably tiny Turbo Pascal compiler and editor.

Why aren't all programs written in assembly language? There are some very

good reasons why not. Every computer has its own assembly language, and each is different because the language is dependent upon the architecture of the machine. The big problem with assembly language, particularly from the viewpoint of a company developing commercial software, is the lack of portability. Programs written for one machine must often be almost completely rewritten to run on another machine that uses a different microprocessor.

IBM's **Macro Assembler** *is the top choice in its field for one reason: The documentation is outstanding. The IBM manual is simply the best reference source around on 8086/8088 assembler instructions. Version 2.0 corrects most of the bugs that were found in its predecessor and incorporates some additional features that make it a solid product that should be part of every serious PC user's software library.*

Byte the Devil

The assembly language used on the IBM PC is called 8086/8088 Assembly Language, named after the microprocessor that runs the PC. It is a devilishly difficult language to learn, master, and maintain. Since each line of source code translates into 1 to 6 bytes of machine code, assembly language source programs are generally huge compared with the files they create. The language has no inherent structure: Within 8086/8088 assembly, for instance, there are 32 variations of the JUMP instruction. Unlike in most higher-level languages, you can't program in assembly unless you jump around a lot.

8086/8088 Assembly Language has no PRINT statement, no INPUT statement, and can do arithmetic only on integers. You say you need to do floating-point calculations? Would you like to program your own floating-point routines? For reading

the keyboard, printing to the screen, and accessing files, an assembly language programmer must not only learn the assembly language mnemonics but must be intimately familiar with the PC-DOS function calls and BIOS services.

Even the simple act of converting a number stored in hexadecimal into readable decimal ASCII (which most programmers do without thinking by using a PRINT statement or the equivalent) becomes a large subroutine, with logic dedicated to rounding, suppressing leading zeros, and putting commas and decimal points in the right places. Having total control means also that you have to take care of everything.

The program that translates your assembly language source into machine code is the assembler. The assembler won't tell you if your PUSHes and POPs aren't paired up right or if you've forgotten a RET statement. You'll find out soon enough, though—a system crash is the normal result of running an assembly language program in the early stages of development.

At times, a veteran assembly language programmer will pull his or her bleary eyes from the display and say, "It's just not worth it. Maybe I should do this in Pascal or C." It's certainly a temptation, and many programmers and companies are going that route. Yet assembly language still retains a certain mystique and functionality. You can create a useful program (a rudimentary *WordStar*-to-ASCII converter, for instance) with a .COM file of less than 20 bytes. Some programs, such as utilities that remain resident in memory, can realistically be programmed only in assembly language.

Assembly Hall of Fame

PC Magazine restricted the survey to three professional, full-featured macro assemblers from IBM, Microsoft, and a new entry from Phoenix. A macro assembler

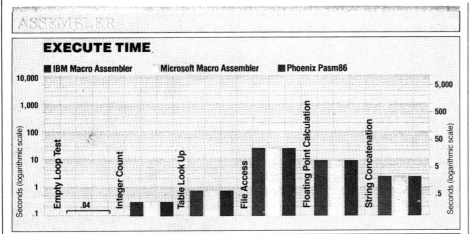

EXECUTE TIME

■ IBM Macro Assembler ■ Microsoft Macro Assembler ■ Phoenix Pasm86

Empty Loop Test .04; Integer Count; Table Look Up; File Access; Floating Point Calculation; String Concatenation (Seconds, logarithmic scale)

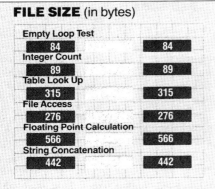

FILE SIZE (in bytes)

	IBM	Microsoft
Empty Loop Test	84	84
Integer Count	89	89
Table Look Up	315	315
File Access	276	276
Floating Point Calculation	566	566
String Concatenation	442	442

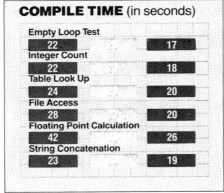

COMPILE TIME (in seconds)

	IBM	Microsoft
Empty Loop Test	22	17
Integer Count	22	18
Table Look Up	24	20
File Access	28	20
Floating Point Calculation	42	26
String Concatenation	23	19

Benchmark test results for IBM Macro Assembler, Microsoft Macro Assembler, and Phoenix Pasm86 Macro Assembler. Tests were run on an IBM PC with two floppy disk drives, without an 8087 coprocessor. All file sizes are listed in bytes, all times listed in seconds. Note: File sizes and execute times are the same for each package because assembly language code does not vary among different assemblers. The file access test when performed with a hard disk yielded a significantly faster result for all products.

lets you save commonly used code with replaceable parameters in separate files and, in turn, use this code by just specifying the name of the macro.

In coding the six test programs, we tried to be fair to high-level language readers. The string concatenation test is somewhat generalized, for instance, and the table-lookup test avoids use of the fast REP MOVSW instruction.

The floating-point subroutines used were developed by Robert Gray and published in the series "Significant Figures" (*PC Tech Journal,* Volume 2 Numbers 4 and 5). The .ASM listing of these routines to multiply and divide single-precision numbers were five pages long.

Using an 8087 floating-point coprocessor speeds up and simplifies floating-point calculations immensely since the 8087 instructions can be used directly in the .ASM file. (All three assemblers tested support the 8087.) The complete program using the 8087 was a 136-byte .COM file, of which over half was devoted just to print the elapsed time. The 10,000 floating-point multiplications and divisions took the 8087 just under 1.5 seconds. Amazing? No, just assembly language.

IBM Macro Assembler

The IBM Macro Assembler, Version 2.0, is the obvious top choice in the field. This package includes an attractive two-volume manual and one program disk. Al-

though the MASM.EXE program shares a copyright between IBM and Microsoft (and is very similar to the new Microsoft assembler), the IBM manual clearly makes this the superior package. It is simply the best reference source I've seen on 8086/8088 Assembly Language instructions.

The biggest change from Version 1.0 of IBM's Macro Assembler is the addition of instructions for the 8087/80287 coprocessor and the 80286 in real-address nonprotected mode.

Thankfully, most of the glaring bugs in Version 1.0 of IBM's assembler have been exterminated. The SHR and SHL pseudo-ops now work as they should, and the type checking has been improved by a significant margin. Amazingly for usually taciturn IBM, the manual actually admits that the previous version did not always work as documented.

As a real improvement over Version 1.0 of the assembler, Version 2.0's manual (and sample skeleton files on disk) actually show you what's needed to create an executable assembly language program. It may sound trivial, but the process really stumped a lot of people who started with Version 1.0. There's even a good solid discussion of .COM- and .EXE-file structures and the advantages and disadvantages of each.

Programmers who worked in assembly language before the IBM PC appeared on the scene often bemoaned the absence of a library utility in the earlier version of the assembler. A library is a collection of often-used subroutines stored in OBJ format that can be pulled into a program in the linking process.

Version 2.0 of the Macro Assembler includes LIB, a plain-vanilla library manager also sharing a Microsoft copyright. Unfortunately, the IBM manual uncharacteristically falls down in its discussion of the LIB facility, devoting a mere ten pages to the program and covering only command-line syntax. The principles of constructing a good library (such as using consistent segment names and providing for near and far versions) are not discussed at all.

Thirty-eight pages of the manual, however, are devoted to an IBM-developed

```
CSEG            Segment Public 'CODE'
                Assume CS:CSEG, DS:CSEG, ES:CSEG, SS:CSEG

                Extrn   StartTime:Near, PrintTime:Near

                Org     0100h

Entry:          Call    StartTime               ; External Subroutine

                Mov     CX,10000                ; 10,000 repetitions
Test1Loop:      Loop    Test1Loop               ; Same line loop

                Call    PrintTime               ; External Subroutine
                Int     20h                     ; Exit Program

CSEG            EndS

                End     Entry
```

IBM Macro Assembler: Empty loop test.

```
CSEG            Segment Public 'CODE'
                Assume CS:CSEG, DS:CSEG, ES:CSEG, SS:CSEG

                Extrn   StartTime:Near, PrintTime:Near

                Org     0100h

Entry:          Call    StartTime               ; External Subroutine

                Sub     AX,AX                   ; AX starts at zero
Test2Loop:      Cmp     AX,32767                ; Count until 32767
                Je      Test2End
                Inc     AX                      ; Otherwise increment
                Jmp     Test2Loop               ; And do it again

Test2End:       Call    PrintTime               ; External Subroutine
                Int     20h                     ; Exit

CSEG            EndS

                End     Entry
```

IBM Macro Assembler: Integer count test.

```
CSEG            Segment Public 'CODE'
                Assume CS:CSEG, DS:CSEG, ES:CSEG, SS:CSEG

                Extrn   StartTime:Near, PrintTime:Near

                Org     0100h

Entry:          Jmp     Test4                           ; Skip over data
ShortString     db      (ShortStringEnd - ShortString - 1)      ; Length
                db      'This is a string'                      ; String
ShortStringEnd  Label   Byte

LongString      db      (LongStringEnd - LongString - 1)
                db      'This is a string with lots of words in it.'
LongStringEnd   Label   Byte

CatString       db      0,255 dup (0)

Test4:          Call    StartTime       ; External Subroutine

                Cld                     ; String Moves forward
                Mov     CX,10000        ; Number of Repetitions

Test4Loop:      Push    CX

                Mov     SI,Offset ShortString
                Lodsb                           ; Get length
                Mov     DL,AL                   ; Save length of 1st string
                Sub     AH,AH                   ; Zero out type byte
                Mov     CX,AX                   ; CX = length
                Mov     DI,1 + Offset CatString ; Destination
                Rep     Movsb                   ; Move it in
```

(continues)

IBM Macro Assembler: String concatenation test.

preprocessor called SALUT (Structured
Assembly Language Utilities). To use it,
you write programs with the .SAL extension, which includes specially structured
statements for controlling program flow.
You then run SALUT, which translates the
listing into an .ASM file to be assembled
by MASM.

It's an interesting and even worthwhile
idea to try to add some structure to assembly language programs, but the first time
you run SALUT and you hear your printer
reset (a good sign that SALUT was programmed in BASIC), it really turns you
off. Using a BASIC program to preprocess
structured utilities in an assembler program is just too surreal for me.

A good library system is much more
important than a wimpy preprocessor like
SALUT, and I'm sorry IBM did not take
the opportunity to discuss library management with the detail it deserves.

IBM's package includes a linking program, a cross-reference utility, and a
smaller version of the assembler called
ASM. ASM does not support macros or

8087 instructions, and it lists errors by number only. It's only useful if you're really hurting for memory.

The IBM Macro Assembler is a good, solid, full-featured program with a manual that opens the package up for both beginners and experienced programmers. It should be a part of every serious PC user's software library.

Microsoft Macro Assembler

The Microsoft Macro Assembler, Version 3.0, is nearly identical to the IBM Macro Assembler Version 2.0, but with a few added features. While it also includes 8087, 80287, and 80286 instructions in nonprotected mode, Microsoft's assembler adds the protected-mode 80286 instructions. (However, these instructions have very limited value to most programmers.) While the package does not have anything like the IBM SALUT program (which is no great loss), Microsoft includes LIB, MAKE (a program maintainer), and SYMDEB, a nifty symbolic debugger. SYMDEB alone is worth the price of the package and can be used with other Microsoft language products.

The Microsoft manual, however, is a disaster. It is badly printed and difficult to use. If you buy this package, you will need some other reference source for the assembly language instructions because each instruction is documented by exactly one line of text, like "JCXZ label—Jump on CX zero" or "Loop label—Loop."

Like IBM, Microsoft includes a LIB program with a 14-page explanatory discussion in the manual. Microsoft's information is more helpful than that supplied by IBM, but again doesn't come close to what is really required. A program unique to the Microsoft package is MAKE, a program maintainer. When set up with the proper description file, this program checks file-change dates on your .ASM, .OBJ, .MAP, and .EXE files and does any necessary assembling or linking to bring everything up to date.

SYMDEB

The real jewel of the Microsoft package is SYMDEB, a symbolic debugger. It looks and acts like a souped-up version of

```
(String concatenation test continued)

                   Mov      SI,Offset LongString
                   Lodsb                          ; Length of second string
                   Mov      CX,AX                 ; Set to CX also
                   Add      AL,DL                 ; AL = Length of both strings
                   Jnc      StringLengthOK        ; If under 255, no problem
                   Mov      AL,255                ; Truncated length of total string
                   Mov      CL,AL                 ; Now CX = 255
                   Sub      CL,DL                 ; Now CX = truncated length of 2nd
StringLengthOK:    Rep      Movsb                 ; Move in 2nd string after first
                   Mov      [CatString],AL        ; Put in the total length
                   Pop      CX                    ; Get back repetition counter
                   Loop     Test4Loop             ; Do it CX times
                   Call     PrintTime             ; External Subroutine
                   Int      20h                   ; Exit
CSEG               EndS
                   End      Entry
```

```
CSEG            Segment  Public 'CODE'
                Assume   CS:CSEG, DS:CSEG, ES:CSEG, SS:CSEG

                Extrn    StartTime:Near, PrintTime:Near

                Org      0100h

Entry:          Jmp      Test5                      ; Skip over data

Array1          dd       1.0 , 2.0 , 3.0 , 4.0 , 5.0
                dd       6.0 , 7.0 , 8.0 , 9.0 , 10.0
                dd       11.0 , 12.0 , 13.0 , 14.0 , 15.0
                dd       16.0 , 17.0 , 18.0 , 19.0 , 20.0
                dd       21.0 , 22.0 , 23.0 , 24.0 , 25.0

Array2          dd       25 dup (?)

Test5:          Call     StartTime                  ; External Subroutine

                Mov      CX,1000                    ; Number of trials

Test5Loop:      Push     CX

                Sub      BX,BX                      ; Initial Index
                Mov      CX,25                      ; Number to move

Test5Move:      Mov      AX,Word Ptr Array1[BX]     ; Get from first
                Mov      DX,Word Ptr Array1[BX + 2]
                Mov      Word Ptr Array2[BX],AX     ; Put into second
                Mov      Word Ptr Array2[BX + 2],DX
                Add      BX,4                       ; Push up BX
                Loop     Test5Move

                Pop      CX
                Loop     Test5Loop                  ; Do it 1000 times

                Call     PrintTime                  ; External Subroutine
                Int      20h                        ; Exit
CSEG            EndS

                End      Entry
```

IBM Macro Asembler: Table lookup test.

the DOS DEBUG utility, with some important differences. Devoid of all nicely descriptive labels and address names, a normal DEBUG listing is difficult for most people to follow. The code may look familiar, but a CALL 0E87 command doesn't tell you that your FATAL__ER-

ROR subroutine is about to be executed.

To fully make use of the SYMDEB feature, you have to link with the map option, then run the MAPSYM utility to convert the map file to a symbol file that SYMDEB reads in with your program. It will then display procedures, groups, and labels

```
;       TEST6.ASM -- Assembly Language Test Program 6 -- File Input/Output
;       --------------------------------------------------------------------
;;      Charles Petzold, August 3, 1985


CSEG            Segment Public 'CODE'
                Assume  CS:CSEG, DS:CSEG, ES:CSEG, SS:CSEG
                Extrn   StartTime:Near, PrintTime:Near
                Org     005Ch
FileBuffer      Label   Byte                         ; Put buffer in PSP
                Org     0100h
Entry:          Jmp     Test6                        ; Skip over data
FileAsciiz      db      'TEST.DAT',0                   ; Name of File
FileHandle      dw      ?                             ; File Handle
RecordSize      dw      132                           ; Record Size
ModifiedFile    db      'Modified',0                  ; New File contents
ErrorMessage    db      '-- File I/O Error -- $'       ; All-purpose message
Test6:          Call    StartTime    ; External Subroutine
                Cld                                  ; String Moves Forward
                Mov     DX,Offset FileAsciiz         ; File name
                Mov     AL,2                         ; Read / Write Access
                Mov     AH,3Ch                       ; CREATE file Call
                Int     21h                          ; Do it
                Jc      Test6Error                   ; Error Exit
                Mov     [FileHandle],AX              ; Otherwise save handle
                Sub     AX,AX                        ; Record Count
WriteLoop:      Push    AX                           ; Save current record
                Call    LSEEK                        ; Points file to record offset
                Jc      Test6Error                   ; Error exit
                Call    WRITE                        ; Writes a 132 byte record
                Jc      Test6Error                   ; Error exit
                Pop     AX                           ; Get back current record
                Inc     AX                           ; Up it by 1
                Cmp     AX,100                       ; See if reached 100 yet
                Jb      WriteLoop                    ; If not, keep going
                Sub     AX,AX                        ; Phase 2 Record Count
ReadWriteLoop:  Push    AX                           ; Save record count
                Call    LSEEK                        ; Move pointer to record
                Jc      Test6Error
                Call    READ                         ; Read 132 byte record
                Jc      Test6Error
                Mov     DI,Offset ModifiedFile       ; Beginning of string to write
                Mov     SI,DI                        ; Set SI to it also
                Mov     CX,[RecordSize]              ; Max characters in record
                Sub     AL,AL                        ; Search for terminating zero
                Repnz   Scasb                        ; Find it
                Sub     CX,[RecordSize]              ; Convert to string length
                Not     CX
                Mov     DI,Offset FileBuffer         ; Destination is buffer
                Rep     Movsb                        ; Move it in
                Pop     AX                           ; Get record number again
                Push    AX
                Call    LSEEK                        ; Set pointer to record
                Jc      Test6Error
                Call    WRITE                        ; Write the buffer
                Jc      Test6Error
                Pop     AX                           ; Record number
                Inc     AX                           ; Increment it




                Cmp     AX,100                       ; Continue if less than 100
                Jb      ReadWriteLoop
                Mov     BX,[FileHandle]              ; Handle of file
                Mov     AH,3Eh                       ; CLOSE File Call
                Int     21h
                Jnc     Test6Exit
Test6Error:     Mov     DX,Offset ErrorMessage       ; Error exit
                Mov     AH,9
                Int     21h
Test6Exit:      Call    PrintTime                    ; External Subroutine
                Int     20h                          ; Exit
```

(continues)

IBM Macro Assembler: File access test.

with their names as well as the normal address. Of course, in a single-module program, your map file is not going to have much information, but you can make all the important variables and routines public in the source file to get them listed and usable by SYMDEB.

SYMDEB has lots of other nice features, like dumping real variables in short, long, and 10-byte fashion. If you happen to have a dump terminal connected to the serial port of your computer, you can redirect SYMDEB input and output to the terminal so it doesn't interfere with the keyboard and screen workings of your program.

SYMDEB can also be used with other Microsoft language products, such as the C, FORTRAN, and Pascal compilers, for debugging at the source-code level. Any public address in linked-object modules or libraries is picked up by SYMDEB so you can trace through the inner workings of the compiled program.

SYMDEB is a nice little program, and the Microsoft Macro Assembler package is the only place you can get it. I bought this package for SYMDEB alone, and I'm glad I did.

If you go out and buy both the IBM and Microsoft assemblers, you'll probably use IBM's manual and Microsoft's program. The MASM.EXE programs are just about indistinguishable in operation, but the Microsoft program has a creation date 4 months later than IBM's. I use the Microsoft assembler under the assumption that something must have been fixed in those 4 months, even though I'm not quite sure what it is.

Pasm86

Pasm86 is a recent entry in the professional macro assembler market from Phoenix Computer Products, the folks who write IBM-compatible ROM BIOS programs. Pasm86 is purportedly IBM- and Microsoft-compatible (except in that it has fewer bugs) and twice as fast. Is it? Well, almost and yes.

As advertised, Pasm86 is about twice as fast as the IBM and Microsoft assemblers for most programs. If you spend a lot of time staring at your screen while

```
(File access test continued)
;         Subroutines for File I/O
;         ------------------------
LSEEK:          Mul     [RecordSize]         ; DX:AX = offset in bytes
                Mov     CX,DX
                Mov     DX,AX                ; CX:DX = offset in bytes
                Mov     BX,[FileHandle]
                Mov     AX,4200h             ; Move from beginning of file
                Int     21h
                Ret
READ:           Mov     AH,3Fh               ; Read Call
                Jmp     Short READWRITE
WRITE:          Mov     AH,40h               ; Write Call
READWRITE:      Mov     DX,Offset FileBuffer ; Buffer
                Mov     CX,[RecordSize]      ; Number of bytes
                Mov     BX,[FileHandle]      ; Handle
                Int     21h                  ; Call DOS
                Jc      ReadWriteExit        ; Carry set if error
                Cmp     AX,CX                ; Also set if AX < CX
ReadWriteExit:  Ret
CSEG            EndS
                End     Entry
```

MASM digests your program, you may want to look into this alternative. (You may also want to divide your program into multiple-object modules and start using libraries.) This increase in speed was not so evident in the standard test programs because they were all rather short and disk-access time predominates in assembling short programs. But for assembly language listings of five pages and up, the two-fold speed increase is real and impressive.

Compatibility?

Compatibility with IBM and Microsoft, however, is a problem. One of the first differences you'll notice is right on the command line. The .ASM filename cannot be followed by a semicolon, and all the parameter flags must be specified in UNIX fashion with a preceding dash instead of the normal DOS slash.

Another incompatibility showed up in the floating-point benchmark test. The floating-point multiply and divide test routines assumed that the numbers were stored in DD statements in BASIC (rather than 8087) format. The IBM and Microsoft assemblers can generate either format, with the BASIC format as the default. The Phoenix assembler used 8087 format and could not be switched to BASIC format. Thus, the floating-point routines had to be altered somewhat to work correctly when assembled by Pasm86.

For most programs, assembly under Pasm86 and linking under IBM's LINK program created files identical to those of the IBM or Microsoft assemblers. The intermediate .OBJ files were different, but this is of little consequence if LINK correctly translates them.

However, the Phoenix assembler had problems with several programs that I had written over the past few years and assembled under the IBM and Microsoft assemblers without difficulty. These problems

Why program in assembly language? Power. You can do the most with it, and virtually anything you write will execute very quickly.

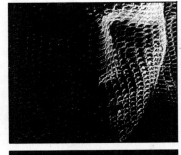

involved assembly language directives rather than instruction code. The Phoenix assembler is not compatible with IBM/Microsoft in some uses of the EXTRN command in multimodule programs and with certain constructions using the ampersand (&) and percent sign (%) macro directives. Sometimes I found I could change something in order to make the program work under the Phoenix assembler that would make it impossible for the IBM and Microsoft assemblers to use. I'm not going to say that one version is wrong and one version is right, because programmers are often forced to discover by trial and error what will and will not work under a certain assembler or compiler if the manual is wrong or somewhat obscure on an issue. Somebody else who writes in assembly language may have a different interpretation. But just be aware that the Pasm86 assembler is not an IBM clone. It will probably be fine if you are beginning to program in assembly language but don't expect it to work flawlessly on all the programs you have already written.

Sometimes a program assembled without errors under the Pasm86 assembler but linked by LINK, Version 2.2 (delivered with DOS 3.1), said "DUP record too complex", a LINK message I had never gotten before. I suspect the message is a subtle advertisement to buy Phoenix's Plink86 program.

Phoenix's manual contains more information than the Microsoft manual, but it is not as complete or well printed as the IBM manual. The manual devotes at least one page each to all of the 8086/8088/80286 instructions and assembly language directives, but the 8087 instructions are not adequately documented.

If a faster, full-featured macro assembler is what you need, Phoenix Software has it. It's not fully compatible with IBM and Microsoft assemblers, but you will eventually adjust to the differences.

Conclusion

All three of these macro assemblers do just about the same thing. Despite the increased speed of the Phoenix assembler and some nice utilities included in the Microsoft package, the best of the three is the IBM Macro Assembler, Version 2.0. It is the best simply by virtue of its including something that even very experienced assembly language programmers must keep within arm's reach—an excellent reference manual. ∎

160

PASCAL: This block-oriented, structured language is tailor-made for business programming, where getting the job done is more important than computer calisthenics.

There are quite a number of reasons why I do all of my programming in Pascal, and one of them is APL.

Back in early 1976, when a useful personal computer was still prohibitively expensive, I had an account on an APL mainframe system. I had a canned application that I ran from time to time, but I also wanted to do some text processing for an in-house newsletter. I bought a good book on APL and spent some long evenings in front of the Diablo daisywheel terminal and soon found myself writing a text formatter.

It took several weeks to complete and ran to about 300 lines of APL code. APL, if you aren't familiar with it, uses Greek letters and unique symbols in an extremely terse notation. The formatter worked, after a fashion, and I used it for several months. Then, abruptly, my company sold its mainframe and started buying time on an APL time-sharing service. The new APL interpreter was similar, but not identical, to the old one, and my formatter didn't run on it. The new APL interpreter kept printing error messages and tossing me out on my ear. I suspected a little rewriting would get it running again.

I printed it out, looked at it, and realized with welling horror that I hadn't the slightest idea how my program worked. The code was literally Greek to me. No matter how hard I tried, I could not dope out the algorithm by looking at the code. I ended up abandoning the old formatter and writing a new one completely from scratch.

In hindsight, there were three reasons why I had to abandon my formatter: It was completely unstructured and existed as one block of 300 lines of code. APL permits nested functions, but my reference book did not emphasize them and explained them poorly.

Another reason was that the terseness of the APL notation encouraged me to cram a great deal of function onto a single line of code. My reference book said this ap-

proach would make my program run faster. Each line was a dense conglomeration of symbols that required laborious examination to understand.

To top it off, I had not had the discipline to plan and properly annotate my work, and nothing in the APL language encouraged me to do so. To the contrary, everything in APL encourages spontaneous, unstructured patching-together of operators and lines without forethought.

For small-to-middling projects, there is only one compiler to choose: Borland International's **Turbo Pascal.** *It is simple, inexpensive, lightning fast in operation, and bursts at the seams with its vast array of features and extensions. Programming will always take a certain amount of time, but in Turbo Pascal it doesn't take as much time as it used to—and that has made a world of difference.*

All in all, APL made for rapid program development and for programs that set like concrete into stone as soon as I put them aside for a week or two. When I bought my first real personal computer a few years later, I sniffed around for a language that would prevent this disaster from ever happening again. The language, of course, is Pascal.

The Secret Is Structure

Pascal imposes discipline and readability on programmers because it is a block-oriented, structured language. A block is a series of statements between the key words BEGIN and END. A statement is a single program action, and while statements are often written one to a line, you can put several statements on one program line or spread a complicated statement across many program lines.

Pascal is called a structured language

because it is particular about where its different parts are placed. Constants must be declared first, followed by variables, followed by functions and procedures, followed by the block of statements that accomplish the program's work. The program must have a name, followed by constant and variable declarations (if any), followed by the actual block of program statements between the BEGIN and END words, followed by a period. Pascal's characteristic indented style is a typographical convention used only to improve program readability. A Pascal compiler is completely indifferent to the way the program is distributed across source-file lines.

Within a Pascal program there may be subprograms, called functions or procedures. Subprograms are miniature programs, identical to programs in nearly all respects. Subprograms can have their own private constants and variables or they can use those of the main program. Furthermore, subprograms can have their own private subprograms declared within them, and so on, like nested Chinese boxes.

Straitjacket Effect

Apart from where you put certain parts of a program, Pascal has definite feelings about how different types of data are treated within the program. Integers and characters, for example, mean very different things, regardless of how they are stored in the computer. Adding a character to an integer makes no sense in Pascal, so the compiler simply won't allow it. It has well-defined ways of moving values from one data type to another in those specific cases where such transfers make logical sense. These are called "transfer functions," and they are a major force in keeping nonsense out of programs.

Open-ended concepts are verboten in Pascal. If you need an array of data items, you must tell the compiler how big it is as soon as you define it, and it can never grow larger or smaller than your definition

Source: PC Magazine, October 29, 1985, 108-114, 116-122, 124-138, 140-145, 150, 152-155, 158, 160-161

states. This restriction can keep you out of certain subtle and not-so-subtle trouble, like arrays accidentally overlapping in memory or arrays growing too large for memory to hold. (I should point out that most modern Pascal compilers have plenty of sneaky tricks to get around this and other restrictions. You don't have to use them—it is often difficult to use them—but they are there to get you out of bad spots if you need them.)

Many critics of Pascal cite these very points as its flaws: It is verbose, it puts the programmer into a straitjacket, and so on. Like so many things in life, it depends entirely on what you consider your priorities to be. If you mostly want to have a good time programming, Pascal takes a backseat to C and FORTH. But if retaining your investment in a piece of code over a long period of time is your primary goal, then what others call restrictions are only the enforcement of a necessary discipline—and the preservation of your investment of time.

Pascal's Uneasy Evolution

Like most distinctive languages, Pascal originated as the product of one mind: Niklaus Wirth, a Swiss professor of computer science. Pascal had some roots in an earlier language called Algol 60, but Pascal's characteristic features—its rigid program structures and emphasis on strict segregation of data by type—were largely original with Wirth. Wirth developed Pascal in the late sixties as a tool for teaching good programming practices to computer science students. The first Pascal compiler was made to work on a large mainframe computer in 1970.

The language as Wirth defined it (which later became standard Pascal) was severely limited in many ways. It was not suitable for any kind of extensive interactive programming, and it certainly was not written with CRT displays in mind. CRT terminals were rare creatures indeed in 1970. It had very little file I/O and no provision for making calls to the operating system or otherwise operating computer peripherals. Wirth intended it to be used by students in batch mode on mainframes, and he strongly resisted the suggestion that the scope of the language be broadened to

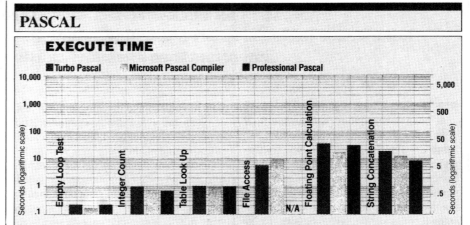

PASCAL

EXECUTE TIME
■ Turbo Pascal ▨ Microsoft Pascal Compiler ■ Professional Pascal

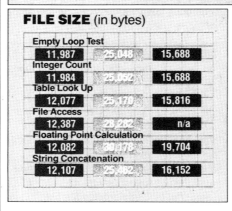

FILE SIZE (in bytes)

Empty Loop Test		
11,987	25,048	15,688
Integer Count		
11,984	25,052	15,688
Table Look Up		
12,077	25,170	15,816
File Access		
12,387	28,282	n/a
Floating Point Calculation		
12,082	30,178	19,704
String Concatenation		
12,107	25,482	16,152

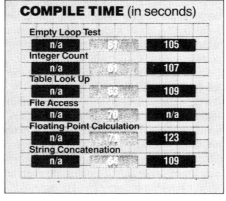

COMPILE TIME (in seconds)

Empty Loop Test		
n/a	51	105
Integer Count		
n/a	51	107
Table Look Up		
n/a	53	109
File Access		
n/a	78	n/a
Floating Point Calculation		
n/a	71	123
String Concatenation		
n/a	64	109

Benchmark test results for Turbo Pascal, Microsoft Pascal, and Professional Pascal. All tests were run on an IBM PC with a hard disk drive. All file sizes are listed in bytes; all times listed in seconds. Note: Pascal stores data tables in arrays only, so the table lookup test is included to show the speed with which Pascal reads from an array. Professional Pascal contains no built-in library function to perform file update-in-place, as required by the file access test. Turbo Pascal compiles instantaneously in the interpreter.

make it commercially useful.

Pascal, however, became very popular, first with university programmers and then with programmers in the business world. The inevitable happened, and the vendors who sold Pascal compilers commercially began to expand the features of the compilers they sold to please commercial program developers.

By the late seventies, there were microcomputer Pascals with random file I/O, graphics, operating system interfaces, and full-screen control for developing interactive applications. Wirth disavowed all of this, implied it was ruining the language, and promised a better solution. That solution, Wirth's new Modula 2 language, was announced in 1981. Modula 2 may, in

fact, be a better way to program than Pascal, but the language is in its infancy, and no significant Modula 2 compilers exist for the IBM PC.

Wirth's fears that "extending" the standard Pascal definition would create a myriad of different dialects of the language and destroy any hopes for portability have largely come true. On the other hand, the choice between a useful language and a portable one is an easy one to make, and the situation is certainly no worse than it is for any other high-level computer programming language. Finally, Pascal's readability (a feature burned into its structure by Wirth) makes "porting" programs from dialect to dialect easier in Pascal than in any other language.

Pascal Compilers

Pascal compilers are difficult to write, harder still to document, and nearly impossible to market. Compilers have come and gone over the years, and most have met with little commercial success. In my view, there are only three Pascal compilers available today for the IBM PC that merit attention from business programmers: Microsoft's MS-Pascal, MetaWare's Professional Pascal, and Borland International's Turbo Pascal. IBM offers a Pascal compiler that is a private-label version of MS-Pascal, so I will not be dealing with IBM Pascal separately in this review. As IBM's private version is always several releases behind Microsoft's, there is little, if any, reason to buy it.

There is also a version of UCSD Pascal that is available for the IBM PC that can be used under Softech's P-System operating system, and if you are using the P-System, in all probability you simply have no other choice. While reasonably well documented, USCD Pascal is a slow, primitive version of the language, and it is nowhere near as portable as the vendor claims it to be.

MS-Pascal

Microsoft's MS-Pascal is a reasonably priced ($300) and tremendously powerful Pascal compiler. It has some distant roots

> **I**f you already know something about programming and are willing to probe, MS-Pascal could be your compiler of choice.

in UCSD Pascal, but in general, it is highly original and very well thought out. The code it produces is extraordinarily fast, if somewhat bulky, and the current version of the compiler (3.3) compiles very quickly when compared with its progenitors.

MS-Pascal supports overlays, the 8087/80287, and the 8086 large-memory model. It has the WORD and LONGINT data types, DOS call primitives, and a large number of extensions to the Pascal language. One interesting addition to the current release is file- and record-locking, a process useful only in multiuser or multitasking (two or more programs running at once) environments. Since multitasking environments such as TopView and Windows are beginning to appear, it is good to see that tools are available to make use of this power.

The most recent releases of all Microsoft languages are link-compatible with one another. In other words, you can take routines compiled with MS-FORTRAN and link them into programs compiled with MS-Pascal. Many scientific and numeric algorithms have been published in FORTRAN, and this provides one way of making use of this material without writing exlusively in FORTRAN.

The fundamental difficulty with MS-Pascal is that its awesome power tends to get in the way. This is especially true for people who are new to Pascal and attempting to learn the language by experimentation. The compiler has its own internal command set, called a *metalanguage* (meaning, literally, a "language about a language"), which is badly explained and often difficult for beginners to grasp.

Much of the difficulty stems from the documentation. While fairly complete, MS-Pascal's twin slipcase manuals are poorly organized and difficult to scan for needed information. Many necessary features (the DOS-call library function, for example) are not mentioned in the typeset documentation and are covered in an addendum tacked on to the end of the manual. Several evenings of intense study will be necessary before you can begin to do any serious work in MS-Pascal. You will probably need to read the manual for a good hour before compiling even the simplest program.

```
PROGRAM INTCOUNT;

VAR X : INTEGER;

{$I SHOWTIME.SRC}
{$I ZEROTIME.SRC}

BEGIN
  ZERO_TIME;
  X := 0;
  WHILE X < MAXINT DO X := X+1;
  SHOW_TIME;
END.
```

Turbo Pascal: Integer count test.

```
    PROGRAM FLOATER;

VAR X : INTEGER;
    A,B,C : REAL;

{$I SHOWTIME.SRC}
{$I ZEROTIME.SRC}

BEGIN
  ZERO_TIME;
  FOR X := 0 TO 10000 DO
    BEGIN
      A:=0; B:=1234; C:=78.9;
      A:=B*C;
      A:=B/C;
    END;
  SHOW_TIME;
END.
```

Turbo Pascal: Floating point test.

On the other hand, if you already know something about programming, and you are willing to probe and experiment and spend some considerable time and energy getting to know the compiler's labyrinthine ways, MS-Pascal could be your compiler of choice.

Professional Pascal

MetaWare's Professional Pascal is by far the newest of the three compilers mentioned here, and, at $595, it's certainly the most expensive. On the other hand, it could well be the most powerful Pascal compiler ever implemented on a microcomputer.

Professional Pascal is part of a larger

> **U**sing Professional Pascal for small projects is shooting mosquitos with a howitzer, but if you must, the mosquitos don't have a chance.

family of assemblers, cross-assemblers, and cross-compilers (which allow a single version of a program to be developed on one machine and then cross-compiled to many other computers). Professional Pascal code generators are available for both DOS and the VAX/VMS operating environments, with other environments, including UNIX, still in development. Professional is one way to beat Pascal's portability problem, but it comes at a fairly high price: having to buy a separate $595 compiler for each target machine you wish your program to run on.

The compiler has all the necessary features: 8087 support, overlays, interrupt support, and strings. There is no support for sound or graphics or any other IBM PC peripheral devices, but this is true of MS-Pascal and most versions of Turbo Pascal.

```
      PROGRAM STRINGER;

VAR X     : INTEGER;
    A,B,C : STRING[80];

{$I SHOWTIME.SRC}
{$I ZEROTIME.SRC}

BEGIN
  ZERO_TIME;
  FOR X := 0 TO 10000 DO
    BEGIN
      A := 'This is a string';
      B := 'This is a longer string with lots of words in it.';
      C := CONCAT(A,B)
    END;
  SHOW_TIME
END.
```

Turbo Pascal: String concatenation test.

```
      PROGRAM EMTYLOOP;

VAR I : INTEGER;

{$I SHOWTIME.SRC}
{$I ZEROTIME.SRC}

BEGIN
  ZERO_TIME;
  FOR I := 1 TO 10000 DO;
  SHOW_TIME;
END.
```

Turbo Pascal: Empty loop test.

In preparing the current release, MetaWare went through MS-Pascal feature by feature, making sure that anything MS-Pascal could do, Professional Pascal could do. As far as I could determine, it was successful in meeting this standard.

Professional Pascal incorporates some of the philosophy of the Ada language specification, as well as some of its jargon: It uses *packages* to contain libraries of separately compiled modules and *pragmas* to issue commands to the compiler. (MS-Pascal calls such commands *metacommands*. What a pity we don't all speak a common tongue. . . .)

All 8086 memory models are supported. Memory models are ways of allocating machine memory to program data and code. Each model requires a separately compiled copy of all utility libraries, and, consequently, the Professional Pascal compiler is shipped on nine diskettes. The compiler and libraries for a single memory model require about 1.5 megabytes of disk storage. This compiler will not run on a floppy-based PC.

The compiler itself, in fact, occupies 566K bytes, and is too large to fit on a single disk. An installation program reads portions of it from two diskettes and assembles it into a single enormous program file on your hard disk. The installation program is excellent. It builds whatever subdirectories it needs and loads whatever files are necessary from the nine distribution diskettes, placing them in the appropriate subdirectories.

Professional Pascal is quite difficult to learn and use. The documentation, while better organized than MS-Pascal's, is in smaller, fuzzier print, so it is quite difficult to read. It is not complete; some libraries are documented only in the source files and are not mentioned in the manual at all. However, the documentation is Professional Pascal's only real weakness. The software itself performed beautifully, never once behaving in an unexpected fashion. The code it produces is tighter than MS-Pascal's and about as fast.

In a way, the difficulties I discovered working with Professional Pascal were not really the compiler's fault. It is an enormously complicated product, and it is not intended for the beginning or casual programmer, as you might expect from the rather stiff price tag. Professional Pascal's target buyers are people who develop software to be sold on the commercial market. Such people are generally familiar with

```
        PROGRAM FILER;

TYPE STRING132 = STRING[132];

VAR X          : INTEGER;
    TESTER     : FILE OF STRING132;
    DATA_STRING : STRING132;
    A_STRING   : STRING132;

{$I SHOWTIME.SRC}
{$I ZEROTIME.SRC}

BEGIN
  ZERO_TIME;
  A_STRING :=
  '
  DATA_STRING := A_STRING;
  DATA_STRING := CONCAT(DATA_STRING,A_STRING);
  ASSIGN(TESTER,'TEST.DAT');    { Create/open new file }
  REWRITE(TESTER);
  FOR X := 1 TO 100 DO   { Write records to file }
    BEGIN
      SEEK(TESTER,X);
      WRITE(TESTER,DATA_STRING)
    END;
  CLOSE(TESTER);              { Close file to flush buffer }
  RESET(TESTER);             { Re-open file for read }
  FOR X := 1 TO 100 DO   { Read records back from disk }
    BEGIN
      SEEK(TESTER,X);
      READ(TESTER,DATA_STRING);
    END;
  FOR X := 1 TO 100 DO    { Read records, modify them, }
    BEGIN                 {   and write them back out again }
      SEEK(TESTER,X);
      READ(TESTER,DATA_STRING);
      DATA_STRING := 'Modified!';
      SEEK(TESTER,X);
      WRITE(TESTER,DATA_STRING)
    END;
  CLOSE(TESTER);
  SHOW_TIME
END.
```

Turbo Pascal: File access test.

```
        PROGRAM TABLER;

VAR X,A,LOOP  : INTEGER;
    INT_ARRAY : ARRAY[1..25] OF INTEGER;

{$I SHOWTIME.SRC}
{$I ZEROTIME.SRC}

BEGIN
  ZERO_TIME;
  FOR X := 1 TO 25 DO INT_ARRAY[X] := X;   {Fill the array}
  FOR X := 1 TO 1000 DO
    FOR LOOP := 1 TO 25 DO A := INT_ARRAY[LOOP];
  SHOW_TIME;
END.
```

Turbo Pascal: Table lookup test.

compilers, separate compilation, and all the other myriad details that must be understood to bring a commercial package to market. Using this program for small projects is definitely shooting mosquitos with a howitzer, but if you must, well, the mosquitos don't have a chance.

Turbo Pascal

MS-Pascal and Professional Pascal are similar in the way they operate: You edit a source-code file with a separate text editor, compile the source-code file to a linkable object-code file with the compiler itself, and then link the object-code file together with one or more utility libraries to pro-

> **T**urbo Pascal bundles a good text editor, fast compiler, and certain debugging facilities into a single package occupying only 36K bytes of disk space.

duce a runnable .EXE file. This process involves a good deal of reading from and writing to disks, and a whole edit/compile/link cycle rarely takes less than 5 to 7 minutes.

In 1983, Borland International released a new kind of Pascal compiler that allowed a complete edit/compile cycle (it eliminated the link step) to take as little as 15 to 20 seconds for short programs. This remarkable product, Turbo Pascal, bundled a good text editor, fast compiler, and certain debugging facilities into a single package occupying only 36K bytes of disk space. Professional Pascal, by contrast, requires more than 1.5 megabytes of hard disk storage to operate.

Turbo Pascal obtains much of its speed edge by setting up an "environment" in memory that contains the compiler, the editor, plus both your source program and object code. Nothing must be read in from disk to edit, compile, or run a program. Everything remains in memory until you explicitly choose to save something on disk. Editing scrolls the screen without delay and compiling takes place in seconds. Movement from one feature to another is instantaneous, since all features exist concurrently in memory.

Loaded with Features

In richness of features, Turbo Pascal is fairly bursting at the seams. It includes full-screen control, elementary graphics, sound, DOS calls, interrupt calls, full pathname support, strings, and a host of minor but very useful extensions to Standard Pascal. Borland offers even more features as "toolbox" utility libraries. Turbo Pascal was originally offered for $49.95, and in its third major release, the price has increased only $20.

A special version of the program is available that uses the 8087/80287 math coprocessor chip, and yet another version offers high-precision BCD or "financial" real numbers. Turbo has always been available on a large number of operating systems and computers, including Z80, CP/M-80, CP/M-86, and PC/MS-DOS. More than 300,000 copies have been sold, far more than all other Pascal compilers put together.

Turbo Pascal's major virtue is simplicity. It has a strong visual orientation and a very clean design. Commands are single keystrokes; the several menus are uncluttered and self-explanatory. Its reference manual is an order of magnitude better than that of MS or Professional Pascal's, at least in part because there is less complication to explain.

The Great Compromise

To achieve its simplicity, the designers of Turbo Pascal made an important compromise: They optimized Turbo Pascal as a fast development tool for short-to-middling programs 5,000 lines of code and smaller. Longer programs will not fit in memory and must be read from disk as they are compiled. Still-longer programs will not compile to less than Turbo's maximum file size of 64K bytes and must be cut up into overlays.

Since Turbo Pascal does not do separate compilation, it doesn't need a linker or a link step to produce executable programs. The drawback is that without a linker to bring in already-compiled subprograms from a library, Turbo must recompile the entire program each time you compile it to test a new piece. Even if 3,000 lines' worth of subprograms are completely solid and debugged, you must

I n Pascal, you can write a piece of code, pick it up a year from now, and know at a glance what it does.

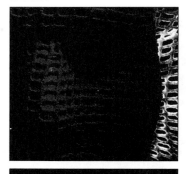

still wait for them to compile while you work on the remaining code. In other Pascal compilers that can link library subprograms into the final executable program file, the link operation does take time. The break-even point between time lost to linking under ordinary compilers and time lost to compiling fully developed code under Turbo Pascal is about 6,000 to 7,000 lines. After that point, you come out ahead with separate compilation.

But for programs smaller than about 5,000 lines there is nothing to equal Turbo Pascal. Nothing even comes close. There are minor lapses in the language, the worst of which is probably the lack of a 32-bit "long integer." That and the inability to compile from batch mode and the fact that the compiler pauses after each error encountered in your source file are the major

complaints against Turbo.

These problems once again point to Turbo Pascal's specialty: developing short and fairly simple programs in a hurry. The machine code produced by Turbo Pascal is at least as good as that produced by the $595 Professional Pascal and better in many ways than the code generated by MS-Pascal. Turbo's executable files are virtually always smaller than those produced by MS-Pascal using similar source code. Experts in compilers point to an awkward code generator in MS-Pascal as the reason.

The ultimate advantage of Turbo Pascal lies in sheer numbers. Lots of people are using the program: Magazines publish articles about it, there is a vigorous user group devoted to it, and third-party vendors are bringing out numerous add-in products for it. Borland's support for the product (by telephone and CompuServe) is legendary, belying the contention of some vendors that low-priced products cannot be supported at a profit.

Conclusion

There are languages that require less "waiting around" than Pascal, like BASIC. Certainly, once you get beyond Turbo Pascal, Pascal compilers are not inexpensive. But when you balance speed of development against the cost of losing code to unreadability, Pascal clearly comes out ahead.

For small-to-middling projects, there is only one compiler to choose: Turbo Pascal. You will, if you work in Pascal long enough, grow beyond Turbo's limitations, in which case MS-Pascal will serve you well until you have grown well beyond the capabilities of the IBM PC. Professional Pascal is really for professional software developers rather than businesspeople; its price and complexity make it impractical for moderate, in-house projects.

Overall, Pascal is tailor-made for business programming, where getting the job done well is more important than computer calisthenics. Its restrictions and structure should be looked upon as guides rather than barriers, guides toward the goal of keeping your software creations available and malleable as long as you require them. ∎

Source: PC Magazine, October 29, 1985, 108-114, 116-122, 124-138, 140-145, 150, 152-155, 158, 160-161

C: This language produces such tight code that it could have been named for the word "compact." Here's a look at four versions of high-performance, low-overhead C.

If it weren't for C, many of the best DOS programs might never have been written. Although some personal computer software is written in other languages, more and more high-performance software is written in C because it is efficient—for both the machine and the programmer.

C is an ideal language for a small computer like the IBM PC because it was developed as a low-overhead, high-yield language for "power-programmers." C gives the programmer complete freedom, and that freedom can produce responsive, useful software or it can be abused, bearing buggy, hard-to-understand software sludge.

C was developed in the early 1970s by Dennis Ritchie, a noted computer scientist at Bell Laboratories. As you might expect from its name, C is an extension of the computer language B, which itself evolved from a once-popular language called BCPL. C was developed so that Ritchie and his coworker, Ken Thompson (the original developer of UNIX), could more easily support and maintain UNIX on a variety of host computers. C, then, was designed to allow gifted programmers complete control of the machine. The C philosophy has proven to be so popular that the language is now widely used throughout the world on a wide variety of computers.

There are more versions of C for the IBM PC than any other language. At last count, there were more than a dozen compilers, at least four interpreters, and dozens of add-on packages. Yet all of this availability is for a relatively small market, at least compared with the size of the market for spreadsheets or word processors. Actually, C's great availability is one of the greatest strengths of the PC. A low-budget software entrepreneur can afford the same programming tool that is used by established firms such as Microsoft or Ashton-Tate.

In the past, C's popularity has been dampened by its steep learning curve. Doing simple things in C is easy, but mastery of the language is very difficult. C programs can be very obscure, and there are many ways to create subtle bugs that would normally be prevented in "safer" languages. Although C was primarily designed for professional programmers, it can be used productively by the less sophisticated if it is approached cautiously. C demands a greater investment on the part

of the programmer than easier languages such as BASIC.

The C Standard

Currently, the C standard is set forth in the classic C reference *The C Programming Language*, by Brian Kernighan and Dennis Ritchie, C's inventor. My favorite introductory C text is *Learning to Program in C* by Thomas Plum, and a good advanced book on C is *C: A Reference Manual*, by Harbison and Steele.

There are several notable C programming systems for the PC. First you must identify your needs, and then you can choose the system that is best for you. For example, the Aztec system runs on several machines, including the IBM PC and the Apple Macintosh, and it includes several useful utilities. It might be a good choice if

you want one C vendor for projects on several different machines. Another unique compiler is Mark Williams's MWC86, which comes with a useful interactive debugger named CSD.

Choosing a few products to review from the large group of good C products was very difficult indeed. In the end I decided to review an old champion, a fresh contender, and two coaches (interpreters). There are many excellent products that are not reviewed here, and you should make a careful study before making a sizable investment in C.

Traditionally, C software development has been a three-step process: The source program was modified, then it was compiled (translated from source to machine language), and then it was tested. This traditional approach is exemplified by Lattice C, the current champion in the C arena. Until recently, Lattice was sold by Microsoft under the Microsoft name as well as under its own name. Now Microsoft is selling its own compiler, and the performance and completeness of the new Microsoft C makes it a strong contender for the honors (and rewards) of being the best C compiler for the PC.

Unfortunately, the modify-compile-test cycle can be painfully slow on a small computer because large programs (text editors and compilers) have to be loaded in from disk. So just as the business world is turning to integrated packages to make it easier to jump from spreadsheet to word processor, the programming world is realizing the advantages of "integrated" program-development systems. The two revolutionary interpreters reviewed here are Instant-C from Rational Systems and Run/C from Lifeboat Associates. Each of these products consists of a simple editor and a means of executing C programs. Both of these systems reduce the modify-compile time lags to zero, which allows the programmer to concentrate instead on testing.

C

■

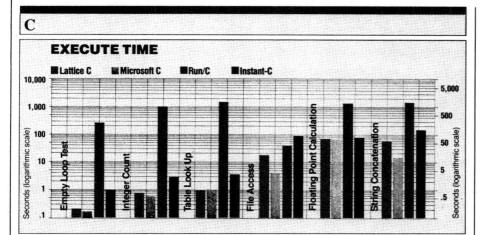

EXECUTE TIME

■ Lattice C   ■ Microsoft C   ■ Run/C   ■ Instant-C

Empty Loop Test / Integer Count / Table Look Up / File Access / Floating Point Calculation / String Concatenation

Seconds (logarithmic scale) — 10,000 / 1,000 / 100 / 10 / 1 / .1

Seconds (logarithmic scale) — 5,000 / 500 / 50 / 5 / .5

FILE SIZE (in bytes)

Empty Loop Test			
10,344	7,022	n/a	n/a
Integer Count			
10,342	7,006	n/a	n/a
Table Look Up			
10,408	7,072	n/a	n/a
File Access			
11,246	9,946	n/a	n/a
Floating Point Calculation			
12,158	19,754	n/a	n/a
String Concatenation			
10,714	7,202	n/a	n/a

COMPILE TIME (in seconds)

Empty Loop Test			
39	41	n/a	n/a
Integer Count			
41	38	n/a	n/a
Table Look Up			
43	40	n/a	n/a
File Access			
51	52	n/a	n/a
Floating Point Calculation			
48	49	n/a	n/a
String Concatenation			
45	42	n/a	n/a

Benchmark test results for Lattice C, Microsoft C, Run/C, and Instant-C. All tests were run on an IBM PC with two floppy disk drives. All file sizes are listed in bytes, all times listed in seconds. Note: Run/C and Instant-C are interpreters and thus do not produce object code and do not have compile times.

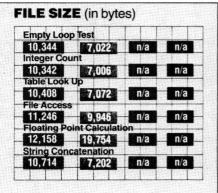

```
#include <time.h>

main()
{
long t;

time(&t);
puts(ctime(&t));

for (i=1;i<=10000;i++)
        ;

time(&t);
puts(ctime(&t));
}
```

Microsoft C: Empty loop test.

```
main()
{
int i;
long t;

time(&t);
puts(ctime(&t));

i = 0;
while (i < 32767)
        i++;

time(&t);
puts(ctime(&t));
}
```

Microsoft C: Integer count test.

Microsoft C

I am impressed by this product. Microsoft C is a professional C implementation that conforms to extremely high standards. It produces compact code, it implements the complete, latest version of C, it compiles quickly, it is easy to use, it has the most extensive library of any C compiler for the PC, its library is fast and compact, and the documention is marvelous. The only major improvement would be if Microsoft actually wrote and debugged your C programs.

The Microsoft C package consists of three manuals (in two binders), three disks, and a quick-reference guide. Your system must have 256K bytes of memory and DOS 2.0 or later, and you must use the latest version of LINK which is supplied with the package. Microsoft C is not copy-protected.

Microsoft is faced with the difficult task of trying to wean their old Lattice C customers to their new product, and one of their inducements is a liberal upgrade offer. Current owners of either Lattice C or the previous version of Microsoft C can upgrade to Microsoft C Version 3.0 for $100. Although both Lattice C and Microsoft C are extremely close to the C standard, there are minor differences between the packages. Microsoft devotes an extensive section of the manual to pointing out problem areas to ease the transition. Like Lattice C, Microsoft C supports all of the PC's memory models and supports and emulates the 8087 numeric coprocessor.

Another goal of Microsoft's compiler is to provide some degree of compatibility between DOS applications and XENIX applications. The documentation clearly indicates which subroutines are common to the two operating systems, and it identifies the differences in usage of the subroutines on the two systems. This is clearly the only compiler to consider if your software development is for both DOS and the UNIX/XENIX environment.

Microsoft provides three different ways to run the compiler. Method 1 is a program called MSC that is similar to Lattice's LC compiler interface. Compiling with MSC is a two-step process. First you run MSC to translate C programs into machine language files, and then you run LINK to combine your machine code with the library machine code to produce an executable program. Method 2 is a program called CL that automates both the compilation and the linking steps, making it easier to use the compiler for programs that don't require unusual linkages. Method 3 is a version of the CL program that uses a different set of option flags, making CL closely resemble the UNIX C compiler usage. I used this third method because it is most natural to me. Unfortunately, the convenient CL program is documented only in a manual appendix, and it is completely omitted from Microsoft's otherwise-handy quick-reference guide.

Exemplary Manual

The manuals contain a total of 900 pages, divided into a *User's Guide,* a *C Language Reference,* and a *Run-Time Library Reference.* The *User's Guide* and the *Run-Time Library Reference* present the standard material, but they are unusually thorough and extremely easy to use as a reference. The indexes are good. The *C Language Reference* is a well-written description of C, with special attention paid to providing examples. It is organized somewhat like a language-standards document, but provides more insight and examples than a formal standards document. The *C Language Reference* wouldn't be a good manual for learning C but is definitely a good place to look for help when troubles arise.

As indicated by the benchmark results, Microsoft C produces fast programs. This is especially true for programs that rely on library functions, because Microsoft C's library appears to be highly optimized. Examination of the machine language code is easy because the compiler has an option to intersperse the original C code with the generated machine code. In its default operation Microsoft C uses registers somewhat more sparingly than Lattice C, but this allows saving two registers for use as C-register variables. Thus, Lattice C automatically attempts to optimize register usage, whereas Microsoft C gives more power to the programmer to optimize register usage. Code generation for integer expressions is similar; the Lattice subrou-

```
int table[25] = { 1, 2, 3, 4, 5,
                  6, 7, 8, 9, 10,
                  11, 12, 13, 14, 15,
                  16, 17, 18, 19, 20,
                  21, 22, 23, 24, 25 };
main()
{
int x, loop;
int a[25];
long t;

time(&t);
puts(ctime(&t));

for (x=0;x<=1000;x++)
        for (loop=0;loop<25;loop++)
                a[loop] = table[loop];

time(&t);
puts(ctime(&t));
}
```

Microsoft C: Table lookup test.

```
#include <stdio.h>
char record[132];
main()
{
int i;
FILE *f;
long t;

time(&t);
puts(ctime(&t));
f = fopen("test.dat","wb+"); /* read/write binary */
for (i=0;i<100;i++) {
        fseek(f,(long)(i * sizeof(record)),0);
        fwrite(record,sizeof(record),1,f);
        }
for (i=0;i<100;i++) {
        fseek(f,(long)(i*sizeof(record)),0);
        fread(record,sizeof(record),1,f);
        }
for (i=0;i<100;i++) {
        fseek(f,(long)(i*sizeof(record)),0);
        fread(record,sizeof(record),1,f);
        strcpy(record,"Modified");
        fseek(f,(long)(i*sizeof(record)),0);
        fwrite(record,sizeof(record),1,f);
        }
fclose(f);
time(&t);
puts(ctime(&t));
}
```

Microsoft C: File access test.

PC FACT FILE

Lattice C
Lattice Inc.
P.O. Box 3072
Glen Ellyn, IL 60138
(312) 858-9750
List Price: $500
Requires: 256K RAM, two disk drives, DOS 1.1 or later.

Instant-C
Rational Systems Inc.
P.O. Box 480
Natick, MA 01760
(617) 653-6194
List Price: $495
Requires: 320K RAM, one disk drive, DOS 1.1 or later.

Run/C
Lifeboat Associates
1651 3d Avenue
New York, NY 10028
(212) 860-0300
List Price: $150
Requires: 256K RAM, one disk drive, DOS 2.x.

Microsoft C
Microsoft Corp.
10700 Northup Way
Bellevue, WA 98009
(206) 828-8080
List Price: $395
Requires: 256K RAM, one disk drive, DOS 2.x, LINK 3.0.

tine interface is somewhat faster, while Microsoft's is better at arranging in-line code for long and unsigned variables.

Lattice C

Lattice C is currently the most talked about and the most imitated C compiler for the PC. Until recently, it was sold both under its own name and also as the Microsoft C compiler. Lattice's popularity stems from many features. It was one of the first compilers for the PC with a solid, complete implementation of C. It was also one of the first C compilers for the PC to provide multiple memory models so that programmers could choose between compact, efficient programs or slightly bulkier but potentially much larger and more complex programs.

The Lattice C compiler comes on three disks. Because it supports four memory models, there are four copies of the compiler, four copies of the library, and so on. Installation is easy on a hard disk. Lattice supplies an installation batch file that creates all the necessary directories and moves everything into them. Installation onto a floppy is somewhat harder. The compiler is not copy-protected, and it requires at least 128K bytes of memory.

Lattice C is harder to use than some other compilers because running the separate pieces of the language system is not completely automated. Unfortunately, the documentation emphasizes the most manual methods and provides less information on a half-way automation aid called LC. A separate OMD (Object Module Disassembler) program must be run to produce assembly language listings of programs.

The Lattice subroutine library is reasonably complete, but it certainly could be more inclusive. The program has the standard C portable I/O library plus some of the lower-level I/O calls first popularized on UNIX systems. The library also contains UNIX-style memory allocation procedures, SETJMP/LONGJMP error-handling routines, standard math functions, and common string-handling functions. In the manual, the discussions of the library functions are organized topically, which makes a good introduction but a poor reference.

DOS the Problem

I'm surprised a compiler that has enjoyed such long and overwhelming success on the PC has so few DOS-specific facilities. Like most other PC C compilers, Lattice supplies UNIX-style I/O facilities, while neglecting many things that are required in virtually every program written in the visual, interactive style of the PC. For example, you would have to call DOS directly to find out the time, date, size of a file, or free space on a disk. Forget about reading a directory in one of your programs unless you want to learn the DOS system calls to do it yourself. There is no graphics or screen-oriented text output support. Most of these facilities are available as separate extra-cost packages from other vendors. Today, however, the Lat-

tice C enhancement business is a small industry.

Lattice C has always done well in benchmarks, so it is not surprising that in this simple series of tests it demonstrated excellent performance. Examination of the object code reveals the secret to its success. The program uses the meager register set of the 8088 CPU well and avoids the code fluff common to many other compilers. You might note that two of the benchmarks, file access and string concatenation, measure the speed of the Lattice subroutine library rather than the speed of the generated code. In these two benchmarks, Lattice did surprisingly poorly, indicating that the program should attempt to further optimize these important facilities.

Lattice C is a very complete implementation of the C language. The manual lists ten areas where Lattice deviates from the C standard, but many of these deviations are minor language extensions that can easily be avoided. However, two deviations are potentially serious: Lattice disallows structures or unions as function arguments and allows application of the "address-of" operator to an array name. These are serious problems for anyone concerned with moving C software to different computers.

Although the Lattice manual is clearly written, I disliked its organization. The manual is probably good for people just learning C, but I found it difficult to use as a product reference. One of the better parts of the manual is the section on Lattice's assembly language interface. Many C programs will need to have small sections coded in assembly language, and it is admirable of Lattice to provide instructions—many other compilers don't.

Run/C

Run/C is a simple, easy-to-use C interpreter. Using Run/C is much like using BASIC with all of the C data types and control structures. Ostensibly, Run/C's goal is to give you the convenience of BASIC coupled with the power of C. On the first point it succeeds, on the second, it falls a little short.

The Run/C manual and advertising copy claim that the program interprets "pure, unadulterated Kernighan and Ritchie C.'' This is one of those advertising claims that sounds too good to believe, and it is. In Section 5 of Run/C's manual there is a more honest discussion of the areas where the program falls short of being a complete C implementation. Run/C is complete enough to teach you C, but it is not complete enough to run many example programs. Fortunately, the package is a proper subset of full C, so it is easy to move a Run/C program to almost any other C system. The converse operation, however, is not likely to be easy.

A more serious problem with Run/C is its speed. Interpreters need not execute programs as quickly as compilers, because most programmers are willing to trade some execution speed for ease of use. However, only so much execution speed can be traded before ease of use becomes irrelevant. As the simple benchmark results indicate, it takes Run/C about 15 minutes to count from 1 to 32,000. That's faster than you or me, but it's five times slower than interpretive BASIC. Run/C's performance may be adequate for some programs. However, many applications could not be tested thoroughly with Run/C because it typically executes 1,000 times slower than a good C compiler.

The Run/C interpreter and about 100 extremely short example programs are supplied on the distribution disk. The software is not copy-protected. Run/C requires at least 256K bytes of memory, one double-sided disk drive, and DOS 2.0 or later. For serious programming, you will also need a true compiler to produce finished versions of your programs.

Basically BASIC

Using Run/C is easy, especially if you've ever used BASIC on the PC. The Run/C display looks like BASIC, its simple editor works like BASIC's, and it automatically numbers lines for easy reference. In addition, it uses familiar commands like NEW, LOAD, LIST, and SAVE—just like BASIC. For editing longer programs, you can automatically chain to your favorite text editor and then automatically return to Run/C.

As promised, there is no "compilation" delay once you have finished editing your program text. Just enter the RUN

```
main()
{
int i;
double a, b, c;
long t;

time(&t);
puts(ctime(&t));

for(i=1;i<=10000;i++) {
        a = 0.; b = 1234.; c = 78.9;
        a = b * c;
        a = b / c;
        }

time(&t);
puts(ctime(&t));
}
```

Microsoft C: Floating point test.

```
main()
{
int i;
char a[80], b[80], c[100];
long t;

time(&t);
puts(ctime(&t));

for(i=1;i<=10000;i++) {
        strcpy(a,"This is a string");
        strcpy(b,"This is a longer string with lots of words in it.");
        strcpy(c,a);
        strcat(c,b);
        }

time(&t);
puts(ctime(&t));
}
```

Microsoft C: String concatenation test.

command, and your program starts. If your program has bugs, you can activate TRACE (printouts of changing variables), TRON (printouts of executing statements), or PRON (a simple execution analyzer). These operations are as simple as they are in BASIC.

What is missing from Run/C? Since the interpreter is organized by lines, Run/C disallows multiline comments and multiline strings. Run/C also disallows data declarations in nested blocks—a relatively unused feature of C. Two important omissions center on the preprocessor; the program disallows defines with parameters,

and it doesn't handle conditional compilation. But the worst drawback is that Run/C doesn't fully support the C data types. Use of the following key words will lead to an error message: register, auto, enum, typedef, and extern.

The manual is generally excellent. All features are listed in an alphabetic reference section. This makes it easy to look things up. There is a good index and a reasonable tutorial to help the beginner get started quickly. Occasionally the manual writers were a bit sloppy. For example, on page 5-4 a table of octal codes for control characters was identified as being in '' decimal except where otherwise specified." An experienced C programmer would know that the "\015" is an octal specifier, but most people would be confused. In the description of the DOSTIME and DOSDATE functions, the documentation muddles a discussion involving pointers and arrays. Similar descriptions for the STRCAT and STRCPY functions are much better.

In summary, Run/C gets excellent marks for ease of use, excellent marks for the manual, good marks for completeness (although they should advertise more honestly), and poor marks for speed. If the package gets faster and is made more complete, it could be a winner.

Instant-C

Instant-C from Rational Systems may change your conceptions about interpreters. Instant-C combines the convenience of an interpreter with the execution-speed advantage of a compiler. The subset of C supported by Instant-C is almost complete, and it should be truly complete with the upcoming shipment of Version 2. The price of Instant-C is a clear indication that this is a professional productivity tool, not a hacker's toy.

The first thing I noticed about Instant-C was its speed. This program produces fat native-object code. Examination of the simple benchmarks shows that Instant-C is in the same ballpark as the Lattice and Microsoft compilers in terms of speed. The technique used in Instant-C is conceptually simple. Each time you change a function the program automatically recompiles it to true PC-native executable code. This process is fast even with a large program because only one function at a time needs recompilation. Since everything is memory resident, the automatic function recompilation occurs without any disk access.

Instant-C requires 320K bytes of memory and DOS 1.25 or later. The program is supplied on two disks; one contains the interpreter and the attached full-screen editor, and the other contains the C language subset, including files and the sources for the library. The package is not copy-protected. Keyboard- and screen-reconfiguration files are included in the package so that Instant-C can be adapted to MS-DOS

machines that aren't fully compatible with the PC.

The style of Instant-C should attract people who are fluent in C because the interactive environment is that of C itself. This program does not present an "imitation BASIC" environment. You can run a program by typing in the command

```
main()
```

Programming in C is like driving a sports car: Everything is manual and rudimentary, but control is extreme.

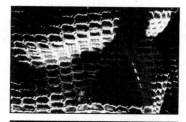

instead of the customary BASIC-style RUN command (although RUN is also available). The advantage here is enormous. You can type many valid C constructs for execution by the interpreter. For example, the command

```
0xff − 010 + 3
```

displays the result 250. (The C parlance above is an expression that starts with hexadecimal ff, subtracts an octal 10, and then adds a decimal 3.) Those of you who don't understand C pointers can interactively enter pointer-arithmetic expressions to see just what is happening. Even more important is the program's ability to call any function interactively and display the results.

Standing Tall

Another important feature of Instant-C is its primitive ability to produce standalone programs. As the documention free

ly admits, for professional software development you will also need a professional compiler for producing finished versions, but for some less-demanding applications, Instant-C's standalone execution facility can be helpful.

Most real C programs are stored in several files so that related functions are grouped together. Unlike Run/C's primitive "one-file-one-program" view of the world, Instant-C lets you work on applications stored in multiple files, all of which are resident simultaneously in the program.

This feature is much less confusing than it sounds, although it is not well described in the manual. Future versions of this product will allow you to link your Instant-C programs to external-object module libraries as well as work on programs 8,000 to 10,000 lines long.

The editing portion of Instant-C is well designed. There is a simple full-screen programmer's-style editor for editing one function (or external declaration) at a time. Instant-C produces and accepts plain ASCII text files, so you can do extensive file editing using an external editor if you wish.

Unfortunately, I found several bugs in my version of the product. However, a call to Rational Systems produced instant help for Instant-C. Two bugs were known and had been fixed for the next release, and one was something they were aware of and were trying to fix. The product support was excellent.

The weakest part of this generally excellent package is the documentation. Portions of the manual are good, especially the first three chapters. However, when you get to Chapter 9, a description of Instant-C's function library, it is outrageously terse. Over 100 functions are detailed in 25 pages. For example the description of the FOPEN command doesn't mention the "+" that you can append to the mode argument to open a file in a read/write mode. The index could be more complete, and several of the appendixes appear to be lacking contents. The manual is clearly written, it just needs more beef. But, overall, the best thing about Instant-C is its speed. Instant-C means instant gratification. ■

Relational Data Base Managers

This is a guide with all the facts and information you need to choose relational data base manager software. These programs can help you keep vital information current by letting you easily add to, subtract from or transform information into lists, tables, ledgers, forms, invoices or labels.

The overriding rationale of the Computer Age seems to be that information is power. And, it stands to reason that if information is power, the more efficiently you can collect, manipulate and use information, the more power you have. The tool that can help do this most effectively, according to experienced users, is a computerized data base system.

When a non-user thinks of a computer or computer system, it's a safe bet that one of the first associations—however foggy or ill-defined—is of a vast storehouse of information that can be tapped for pertinent details on any subject. The non-user may see the computer as omniscient— even sinister—in its ability to create a vast library of facts, then store and sift them, and effortlessly report their significant relationships.

In reality, data base systems that allow truly infinite and random access to information without a rigid, pre-imposed data structure, are as yet unknown. But limited variations on this grand theme abound, especially for personal computer users.

Data base managers are computerized systems for maintaining records of information or searching and gathering that data. Generally, regardless of what type—relational, hierarchical or networking (more on that later)—they have the ability to create, maintain and manipulate a collection of information. It's misleading to compare a data base manager to a file cabinet—with files analogous to drawers, records to pieces of paper, and fields to line items on the paper. Why? Because a good data base manager allows you to manipulate data in a file more efficiently than a cadre of clerks could ever do, and much faster, too. For instance, it lets you instantly transform the information from its raw state into lists, tables, ledgers, forms, invoices or mailing labels—things that could take days by hand. It may also let you do calculations, and other applications-like functions.

How should you buy a data base management system? What should you think about before you walk into a computer store? Plan carefully. If you have a fuzzy idea of what you want, you will wind up with a fuzzy system. For example, if you plan to create a custom system for handling the inventory of a small auto parts business, you'll probably need a program that allows you to tailor the features to suit your needs. This is possible with a program that offers a command language. If you plan to do only a listing of customers, compiling information from various files and updating it frequently, you can probably get by with a program that lays everything out for you, that is, gives you an inflexible range of options from which you choose to progress through the program. This is commonly called a menu-driven program. But remember, the functional life of the system will be short if you buy a package that runs out of features before you run out of uses you can devise for it.

The job of designing and administering a large data base is a giant task. But it's almost as hard to identify a data base management system by type, at least in today's market. It's not unlike trying to find every piece of music that qualifies as a love song. There are really

The non-user sees the computer as omniscient—even sinister—in its ability to create a library of facts.

three categories of data base managers: file and report size programs (single file, single report products); high-level programmer's tools; and products in the middle—with these broken down into packages that may lean more toward one end of the spectrum than the other.

There are further divisions in the more elaborate packages. These programs are conventionally categorized as networking, hierarchical, or relational. With networking and hierarchical systems, the user must know the links between the information stored in the data base—that is, the relationship between one entry and another, and the conceptual path to each one. With a relational data base manager, ideally at least, the user need not know the links between the data—though they exist in the program—in order to use the system. Conceptually, a hierarchical structure is like an inverted tree—you start at one data item, and the relationship between it and another in the data base is defined by branches or pathways from one to the other. A true relational

structure has a data field that has a defined pool of possible values. It can be likened to a table of rows and columns, from which is constructed smaller tables of the desired information. Again, all of this is invisible to the user, the important thing is that with a relational data base structure you should be able to select, join, add and subtract data items easily.

The chart which accompanies this article lists products that are positioned by their makers as relational data base managers.

Most data base systems require you to systematically design the manner in which you plan to use the program. You have to supply the definitions of your data, sometimes from a menu and sometimes with program commands. In any case, a useful data base management system should allow you to:

1. Create a data base structure.
2. Add new information.
3. Sort the information.
4. Search for the information.
5. Report the data in the form you want, including as a file that is ready for use by a variety of applications programs, such as word processors, spreadsheets, etc.
6. Correct the data or make editing changes.
7. Remove data when necessary.

Price (though not an overwhelming factor for experienced users who know precisely what they want, or who are in a business situation where the cost of inputing data will amount to more than the actual cost of the software) may be

What Can A Data Base Manager Do?

Hayes Microcomputer Products, Inc., of Atlanta, Ga., gave their data base manager 'Please' a test drive before it was released this spring. The experience of the Georgia Dental Association in using the product helps show the impact that a data base manager can have on an organization.

The dental group used the product for a year. They had had no computerized system before that time. They found that it saved them a tremendous amount of time and enabled them to do things for their membership that they were never able to do before.

The association used the program to maintain information on some 3079 dentists. The data base contained 96 pieces of information on each member, including the specialties they practiced, the papers they had published, and the status of membership.

The program allowed group administrators to pass out telephone referrals to patients who needed emergency treatment or who wanted a particular dentist who, for example, offered senior citizen discounts or free screenings. If, for instance, a caller wanted to see a doctor specializing in oral surgery who had graduated from a certain school, the software produced a list of dentists who matched the specification.

The program enabled the association to make up special reports that would have been difficult, if not impossible, to do before. Please has generated such items as lists of dentists eligible for life membership, dues owed from the wives of the dentists for their auxiliary, and reports on third party insurance payments. The program has also been used to send out bills stating exactly how much has been paid and how much is owed.

critical for first-time users. Take a look at the package and what you're getting for the price. Are there expensive add on options? Report generators, which let you get reports out of the data base—through menus or simple commands—without having to write a lot of programming code, should be included with the package.

The buying process is going to depend in part on whether you have any data processing experience. Business and professional people new to the world of personal computers should avoid programmer or technically-oriented data base systems. Take, as an example, a data base manager that uses a query language. Typically this kind of program comes up with a blank screen and a prompt, and you type something to get it to work. This may be confusing for a beginner. Once again, a novice should probably look for a menu-driven program, which operates much like filling in a form—you choose what you want to do from a range of options, a list of multiple choices. There are front end programs that add menu features for some of the command-language programs, but they also add to the cost of the basic system.

Try this approach. Define your business problem. Devise possible solutions. Then make a first pass at some of the packages that could meet your needs. Look at what your current system does—whether it is automated or manually-operated. Try to determine how your data must relate to itself (each piece of information), and think about its end use. What do you do most frequently? Do you take requests for information over the telephone? Make up summaries of activities on a periodic basis? Try to analyze how the bits of information you use most frequently relate to one another. Use a chart if necessary, drawing maps of how information flows—broken down by type and frequency—from one point to another, like on a wall chart. Check with subordinates or the people who currently manipulate the data.

Take a look at the hardware you have on hand and the hardware you'll need to buy and try to determine the kind of training you'll need for yourself and others. Think about how you can make the system grow. Will the software let this happen gracefully?

Ask your dealer for a list of people who use the software you're considering. Contact as many of them as possible and ask about the pros and cons of the package, and their experience with the company's support—or lack it.

You should be able to query the data base easily. You might typically use a few keywords to find the information you want and devise macros, or command sequences that you store in the

Data Base Vs. File Managers

There are almost as many opinions about what a data base management program is as there are programs on the market. A large part of the problem centers around the distinctions between so-called file-manager programs and those that are termed data base managers. Here are some opinions on the matter.

Rob Hershfield, product support manager at Concentric Data Systems, Inc., Westboro, Mass., maker of the Concentric Information Processor, says, "Everybody has their own definition, but I think of a file manager as a program that will work with one file at a time, while a data base manager will transparently handle all the interactions between a bunch of files. The user doesn't have to worry about what is in which file. You might have a list of customers and a list of invoices. The computer may keep each in different files in the data base. But it will do it without causing you to think about it. With a file manager you would have to make up two different files and worry whether they are in sync."

At Executec (Dallas, Tex.), manufacturer of the Series-OnePlus integrated data base manager, Jeff Anderson, marketing services representative, says, "Generally a file manager is not as complete as a data base manager. A file manager cannot break down individual areas of data separately, but with a data base manager you can pull up part of the information and search and extract from that rather than the whole file. The difference is the way the information is handled."

Greg Bruner, customer service representative at Desktop Software Corp., Princeton, N.J., maker of the NPL data management system, says, "Basically a file manager is more limited than a data base management system. In general, this is because it doesn't offer the reporting capabilities. With a data base manager you have an extensive reporting capability, along with a powerful data entry facility."

"With a file management program, you enter data, store that in the data base, and then you get reports on key fields in that file," says Fred Sconberg, marketing manager at Advanced Data Systems, Sacramento, Calif., maker of Aladin, a data base management system. "It's similar to a file cabinet—you pull out a file, look for the information you want, and then put it back into the cabinet. With a data base management program, you create your file and put your information in it; then you can generate the reports that you want from that file. And in generating that report you can take information from a second, third, fourth file—or more. You can do calculations on the information, and then analysis. If you want to know the percentage of total sales the revenue from a certain item represents, you can find out and use it in the form of a graph. A file manager is a storage and retrieval system only. A data base manager allows you to manipulate data in such a way that the information can be used to make a business decision."

"It really has to do with the power of the two systems—how much they can hold and also how many files they can access at once," said Katy Johns, product manager at Information Unlimited Software, Sausalito, Calif., maker of EasyFiler, a file management system. "With a file management system, you input your data, set up your data base, and then you generate reports from that informa-

program, for frequently-used searches. You should also look for a way to customize your input screen, especially if you have to frequently and rapidly enter a lot of data.

Try to determine whether you can change reports and the relationships between data after you have created a data base—you may otherwise be stuck with bits of information that you can't get out of the data base. That is, don't get locked into an rigid format. You may have information in the data base, but you may only want part of it. Some programs lock you into a certain format—you put it in and that's the way you have to take it out. In others, you can take out pieces of information and join that with information from other files—and it doesn't matter what sequence the information is in. You can take it out, even search and sort, on any part of it.

The point is, you should be able to maintain the integrity of the data base. If you add to the information, or subtract outdated information, you should still be able to access the remaining data without having to rewrite the input routines or report formats. The query structure should stay the same and programs that use the data should not have to be altered. If you make minor changes in the information, even after entering all the data, you should not have to re-enter everything in order.

What are some special features? With an advanced program, you should be able to lock unauthorized users out of certain files—payroll and other confidential files, for instance. You can use this feature to prevent two different people from updating the same file in different ways. Integrating with the corporation's existing data system is another useful feature.

An accountant would want a system that lets him check entries to reduce the chance of an error later on. A businessman may need a little more flexibility, such as a system that lets him keep track of inventory.

The documentation is often one item which is overlooked. Is it understandable? Is there a comprehensive index? Are examples clear and well-illustrated? Many data base programs are accompanied by documentation that is ponderous, complicated and nearly unintelligible. Try to estimate how long it will take you to learn the program. In many cases this could involve a considerable investment of time. If the program allows special programming, will you be able to learn the language quickly? Remember—the disk will wear out, the program won't. You'll waste time and effort finding out the hard way that a program is the wrong one.

—Charles Miller

tion. In a data base management system, you input your information and access that information from many different kinds of fields—you don't have to set up a separate data base every time you need to use the information. You have a greater capacity with a data base management system—it lets you continually update the data. Most of the file managers are menu-driven; that is, you are presented with options and the program carries on the processes for you."

Jennifer Godward, product marketing manager for Micropro's data base manager InfoStar in San Rafael, Calif., says, "There are several different ways to look at it. If you can work on more than one file at a time, then you can set one up as a transaction file and the other as a reference file that will do updating. With a file manager you can never do that because you're only processing one file at time—you're limited if you're trying to build a system, such as an inventory. File managers tend to be more limited in capacity, in terms of the number of characters per field, fields per record, records per file. A data manager tends to be more powerful in sorting. The more keys fields you are allowed, for instance, the more powerful a system is—it leads to more things to sort on. Any product that can only access one file at a time is really a file manager."

Jeff Garbers, vice-president of product development at Microstuf (Atlanta, Ga.), says, "Those kinds of labels are used more by marketing departments than anyone else. The distinction is kind of arbitrary. I think a data base management system is distinguished by programmability, whereas a file manager is fairly self-contained—everything is there. You don't have to write a program, you just punch numbers from a menu to make things happen. The question should be: Is the product more a program unto itself or is it an environment for describing or manipulating data bases? The term data base itself is very broad. A telephone book is a data base when you think about it. Anything that manipulates information in any way is, to some extent, a data base management system."

Ellen Rony, manager of public relations at Stoneware, Inc. (San Rafael, Calif.), maker of the DBMaster data base manager, says, "File management systems tend to use single-purpose files, they work only with one file at a time. They are generally menu-driven, easy to use, learned very quickly and have limited functionality. They are also less expensive. Data base management systems have more complex procedures, can be complicated to use, have more comprehensive information manipulation capabilities, tend to be expensive, and take longer to learn."

Says Ellen Smith, Software Manager at Readers Digest Services, maker of the ListMaker file manager, "File managers simply manage linear files—a data base management system is more dynamic and you can manipulate data in more elegant ways. Static versus dynamic is the way I would differentiate the two. It's a function of the type of files you're dealing with and the ways you can deal with them."

"A file management system usually does not give you control of data or as many possibilities of control as a data manager," says John Peterson, technical support representative at VisiCorp, San Jose, Calif., maker of VisiFile, a data base manager. "With a data base system you can use different files. A file management system gives you the ability to file customer names and that sort of thing. With a data base manager, you can manipulate the data within files. You have a lot more control with a data base management system. It's harder to learn, but it gives you a lot more flexibility and control."

SECTION V
Going On-Line

SECTION V
Databases: Going On-Line

A database, in contemporary terminology, is a repository of organized units of information. The term is broad enough to encompass a file folder, a personal address book, the contents of filing cabinets, and so on, up to and including the Library of Congress. The term also implies that the information contained in the database is stored in accordance with a special system or code of access, so that a unit or item of information may be readily withdrawn from the database to be reviewed, used, changed, or otherwise operated on prior to its deletion or return.

The articles in this section cover a variety of database types and applications. Commercial databases have evolved into a growth business, by which, for a fee usually based on time, the user can, with the proper computer communication hardware, access a commercial database such as SOURCE, or the American Society for Metals METADEX library, to obtain items or article abstracts on any of the subjects stored in the database.

A "do-it-yourself" database is described for those readers who want to develop their own. Data management for company databases — both theory and practice — is discussed in two of the articles. There is also an article on factory databases, an area receiving increased attention as the "factory automation" concept matures into the actual practice stage.

Getting The Best From Data Banks

Data-base retrieval systems can be a valuable
tool for any growing business

by Marvin Grosswirth

Among the myriad acronyms and
initials plaguing the world of
computers, DBRS may be somewhat
less memorable than some. It may be
easier to remember, however, if one
thinks of it as DBR$.

The initials stand for "data-base
retrieval system," a computerized
service that's giving the competitive
edge to thousands of executives, managers, entrepreneurs, and professionals by providing valuable information
at low cost in a matter of minutes—
information that would otherwise be
available only through dogged, time-
consuming, and expensive research if
at all.

That information is contained in
data bases, or data "banks," electronic libraries of millions of facts that
can be selected, arrayed, arranged,
saved, analyzed, and updated. Some
of the information is so time-sensitive
that many data bases are updated
daily or even hourly. (DBRSs are
sometimes called "on-line data
bases" or "on-line data banks," but
those terms are inaccurate, because
many data bases are "on-line" only to
authorized personnel of the organizations that own them.)

Law, medicine, corporate profiles,
agribusiness, technical and professional journals in 40 languages, con-

*Marvin Grosswirth is a New York-
based free-lance writer who specializes in computers.*

sumer buying habits, solar energy, financial management and planning,
coal, oil, and forestry resources, commodities, pharmacology, aquatic sciences, alcoholism, Latin America,
the Middle East, money markets,
stock markets, supermarkets, and
even the fertilizer market—these,
and some 600 more fat files of "live"
data are available. And all that is required to obtain the specific information is access to a personal computer
or a terminal, a modem, communications software, and a telephone.
Subscribers to DBRSs can "log on"
to the services' mainframe computers
and use the facilities of the big machines in a number of ways, not only
in the handling of information, but, in
some instances, for electronic mail
and for computing power above and
beyond the capabilities of their personal computers. Frequently, the
DBRS's computer serves as a
"front-end communicator," linking
the subscriber's computer with the
data-base owner's.

The data bases are usually independently owned—by an individual,
a corporation, a news service, a government agency—and made available, over public or private communications carriers, by "vendors" who
provide the DBRS. The vendor
charges the user (known as a subscriber) a fee, and pays the data-base
owner a royalty for each use.

This is no small-time business.

Three of the best-known vendors are
subsidiaries of large corporations
who have years of experience in selling information. The Source is owned
by The Reader's Digest Association,
CompuServe is a subsidiary of
H. & R. Block, and DIALOG is the
property of Lockheed Corp. Boeing,
General Electric, McGraw-Hill,
Chase Manhattan Bank, Dow Jones,
TRW, and Dun & Bradstreet are just
a few of the major corporations sharing the profits to be gained by providing information services to other
business operations, both great and
small.

A better barrister

Chicago attorney Steve Bashaw has
been practicing law for seven years,
and has been a regular DBRS user
for the last two years. He claims that
using a DBRS has made him a better
lawyer and has brought him more clients. Bashaw estimates that 85 to 90
percent of his practice is devoted to
representing "mortgage houses—
savings and loan associations, mortgage bankers, insurance companies
that invest in mortgages .... Mortgage law, mortgage litigation—real
estate mortgage foreclosures—are
my specialty," he explains. "These
are very heavily impacted by federal
regulations.

"We do a newsletter for our clients

A LISTING OF SOME ON-LINE DATA BASES

ACCOUNTANTS' INDEX
Provider: American Institute of Certified Public Accountants
On-line through: SDC Search Service
Subject: Information for accountants

ADTRACK
Provider: Corporate Intelligence, Inc.
On-line through: Dialog Info. Services, Inc.
Subject: Ads in print

AGRICOLA
Provider: U.S. Department of Agriculture
On-line through: BRS Inc.; Dialog Information Services, Inc.
Subject: Literature on agriculture

AGRICULTURE
Provider: Data Resources, Inc.
On-line through: Data Resources, Inc.
Subject: Agricultural economic data

AGRICULTURE FORECAST
Provider: Chase Econometrics/Interactive Data Corp.
On-line through: Chase Econometrics/Interactive Data Corp.
Subject: Agricultural economic data

AMERICAN PROFILES
Provider: Donnelley Marketing
On-line through: National CSS
Subject: 1980 Census demographic data

AUERBACH COMPAR
Provider: Auerbach, The Information Company
On-line through: BRS Inc.
Subject: Data-processing products

BALANCE OF PAYMENTS
Provider: International Monetary Fund
On-line through: Chase Econometrics/Interactive Data Corp.; Data Resources, Inc.; Rapidata, Inc.
Subject: International economic transactions

BILLBOARD INFORMATION NETWORK
Provider: Billboard Information network
On-line through: Billboard Information Network
Subject: Play listings from 400 radio stations

BOOK REVIEW INDEX
Provider: Gale Research Company
On-line through: Dialog Info. Services, Inc.
Subject: Periodical and book reviews

BOOKS IN PRINT
Provider: R.R. Bowker Company
On-line through: BRS Inc.
Subject: Books currently in print

CALL/SINKING FUND PROVISIONS
Provider: Chase Econometrics/Interactive Data Corp.
On-line through: Chase Econometrics/Interactive Data Corp.
Subject: Call and sinking fund characteristics for bonds

CHASE ECONOMETRICS/INTERACTIVE DATA CORP.
Provider: Chase Econometrics/Interactive Data Corp.
On-line through: Chase Econometrics/Interactive Data Corp.
Subject: Wide range of financial and economic data bases

CSS/QUOTES+
Provider: National CSS
On-line through: National CSS
Subject: Stock history data base

COMMODITIES FUTURES
Provider: Market Data Systems, Inc.
On-line through: General Electric Information Services Co.
Subject: Commodity futures contracts

COMPUSERVE INFORMATION SERVICES
Provider: CompuServe Inc.
On-line through: CompuServe Inc.
Subject: Multifaceted information service providing many data bases

DOW JONES NEWS/RETRIEVAL SERVICE & STOCK QUOTE REPORTER
Provider: Dow Jones & Co.
On-line through: Dow Jones & Co.
Subject: Business news and stock prices

DRI CAPSULE
Provider: Data Resources, Inc.
On-line through: I.P. Sharp Associates; United Information Service
Subject: Time series on economic indicators

FOUNDATIONS
Provider: The Foundation Center
On-line through: Dialog Info. Services, Inc.
Subject: Non-profit foundations

HARVARD BUSINESS REVIEW
Provider: Harvard Business Review
On-line through: BRS Inc.; Dialog Information Services, Inc.
Subject: Abstracts and comprehensive indexing of all articles in HBR from 1971 to present, plus some older articles; full texts available for articles from 1976

HORSE
Provider: Bloodstock Research Statistical Bureau
On-line through: Bloodstock Research Statistical Bureau
Subject: Thoroughbred pedigrees, breeding records, race records

ICC
Provider: Interstate Commerce Commission and Federal Railroad
On-line through: The Computer Company
Subject: Traffic and financial data

INTERNATIONAL SOFTWARE DATA BASE
Provider: Imprint Software Ltd.
On-line through: Dialog Information Services, Inc.; Knowledge Index
Subject: Microcomputer software

LEGAL RESOURCE INDEX
Provider: Information Access Corp.
On-line through: Dialog Info. Services, Inc.
Subject: Citations to law and law-related literature

LEGI-SLATE
Provider: Legi-Slate
On-line through: I.P. Sharp Associates
Subject: Contains complete history of all bills and resolutions introduced during regular and special sessions of U.S. Congress

MEDLINE
Provider: National Library of Medicine
On-line through: Dialog Information Services, Inc.; BRS Inc.; National Library of Medicine
Subject: Worldwide biomedical literature

MICROCOMPUTER INDEX
Provider: Microcomputer Index
On-line through: Dialog Information Services, Inc.; Knowledge Index
Subject: Microcomputer magazines

NATIONAL TECHNICAL INFORMATION SERVICE
Provider: National Technical Information Service
On-line through: BRS Inc.; Dialog Information Services, Inc.
Subject: Unclassified reports on government and non-government sponsored engineering and scientific research

SOCIOLOGICAL ABSTRACTS
Provider: Sociological Abstracts, Inc.
On-line through: Dialog Info. Services, Inc.
Subject: Abstracts of sociological literature

THE SOURCE
Provider: Source Telecomputing Corporation
On-line through: The Source, Source Telecomputing Corp.
Subject: Contains many information services

STANDARD AND POOR'S INDUSTRY FINANCIAL DATA BANK
Provider: Data Resources, Inc.; Standard and Poor's Corp.
On-line through: Data Resources, Inc.
Subject: Time series on industry groups

This listing is, in part, based on information from the *Directory of Online Databases* published by Cuadra Assoc., Inc., 2001 Wilshire Blvd., Suite 305, Santa Monica, CA 90403, (213) 829-9972

Source: Personal Computing, May 1983, 111-113, 115, 117, 166

ON-LINE SERVICES

BILLBOARD INFORMATION NETWORK
1515 Broadway
New York, NY 10036
(212) 764-7300
Connect charge: $200, incl. training (deferred for home use)
Report charge: 25¢ to $5 each
Time charge: 40¢/minute

BLOODSTOCK RESEARCH STATISTICAL BUREAU
801 Corporate Drive
P.O. Box 4097
Lexington, KY 40544
(606) 278-0411
Charge: $1.75 to $16.50 per pedigree

BRS INC.
1200 Route 7
Latham, NY 12110
(800) 833-4707
Subscription: Charge depends on contract commitment
Telecommunications fee: $6 to $11/hr, depending on network
Connect charge: $16 to $30/hr plus royalties and telecommunications
Royalties: Charge determined by producer
Open access plan: $35/hr plus royalties and telecommunications; $50 start-up fee, credited to account

CHASE ECONOMETRICS/INTERACTIVE DATA CORPORATION
486 Totten Pond Road
Waltham, MA 02154
(617) 890-1234
Subscription: Depends on individual data base
Connect charge: $19.75/hr
Time charge: 16¢ per unit on-line, 8¢ overnight

COMPUSERVE INCORPORATED
5000 Arlington Centre Blvd.
Columbus, OH 43220
(800) 848-8990
Subscription: $39.95, includes five introductory hours
Connect charge: $5/hr, no monthly minimum; $22.50/hr prime time, two-hour monthly minimum; includes communication network

THE COMPUTER COMPANY
1905 Westmoreland St.
Richmond, VA 23230
(804) 358-2171
Connect charge: $1/hr
Other charges: 55¢ per 1000 charcters; 25¢ per CRU

DATA RESOURCES, INC.
24 Hartwell Ave.
Lexington, MA 02173
(617) 863-5100
Subscription: Depends on individual data base
Connect charge: $22.75
Time charge: 7.2¢ per CRU

DIALOG INFORMATION SERVICES INC.
3460 Hillview Ave.
Palo Alto, CA 94304
(800) 227-1927; (800) 982-5838 in CA
Time charge: $15 to $100/hr

DOW JONES & CO.
P.O. Box 300
Princeton, NJ 08540
(800) 257-5114
Subscription: Standard—$1.20/minute prime time, 20¢/minute non-prime, for news; 90¢/minute prime time, 15¢/minute non-prime, for quotes. Blue Chip—for non-prime-time users; lower rates. Executive—$50/month; lower prime-time rates.
Connect charge: $50

DUN & BRADSTREET, INC.
99 Church St.
New York, NY 10007
(212) 285-7669
Subscription: Contract required. Charge depends on number of units (one per report); minimum contract depends on type of service.
Access charge: Depends on type of contract purchased

GENERAL ELECTRIC INFORMATION SERVICES
401 N. Washington St.
Rockville, MD 20850
(301) 340-4000
Subscription: $40 to $50/month
Connect charge: $3 to $45/hr, depending on type of terminal device used and baud rate
Time charge: 16¢ per CRU; up to 10¢ per 1000 characters

I.P. SHARP ASSOCIATES
2 First Canadian Place
Exchange Tower, Suite 1900
Toronto, Ont., Canada M5X 1E3
Connect charge: 25¢ to 45¢ CPU charge; 70¢ per 1000 characters
Time charge: $1/hr

KNOWLEDGE INDEX
3460 Hillview Ave.
Palo Alto, CA 94304
(800) 227-5510; (415) 858-3796
Subscription: $35
Connect charge: $24/hr, including network fees

NATIONAL CSS
187 Danbury Road
Wilton, CT 06897
(203) 762-2511
Minimum charge: $300/month
Connect charge: $17/hr including telecommunications
Time charge: $27\frac{1}{2}$¢ per application resource unit (ARU)

NATIONAL LIBRARY OF MEDICINE
8600 Rockville Pike
Bethesda, MD 20209
(301) 496-6661
Minimum charge: $15/month
Connect charge: $22/hr prime time, $15/hr non-prime time; telecommunication charges included

RAPIDATA, INC.
20 New Dutch Lane
Fairfield, NJ 07006
(201) 227-0035
Minimum charge: $100/month
Connect charge: 16.7¢/minute; national toll-free access
Time charge: 30¢/minute, 300 baud; 50¢/minute, 1200 baud

SOURCE TELECOMPUTING CORP.
1616 Anderson Road
McLean, VA 22102
(800) 336-3366
Subscription: $100, includes user's manual, lifetime subscription to newsletter and other publications
Connect charge: None in 480 cities; toll-call charge for outlying regions
Time charge: $20.75/hr prime time; $7.75/hr non-prime time; $5.75/hr economy rates.

UNITED INFORMATION SERVICE
5454 West 110th St.
Overland Park, KA 66211
(913) 341-9161
Subscription: $200 to start, $300/month minimum; catalog maintenance fee
Connect charge: Option A—10 to 30 characters per second (cps), $10.50/hr plus 21¢ per kilocharacter (kch) input or output; 120 cps, $20.50/hr plus 21¢ per kch. Option B—10 to 30 cps, $15.50/hr plus 27¢ per kch; 120 cps, $25.50/hr plus 27¢ per kch.

*CRU designates computer resource unit

> *He believes his use of the DBRS is responsible for a significant growth in business.*

DATA BANKS

(continued from page 111)
about every 45 days," Bashaw continues. "When I'm getting ready to do the newsletter, I review all the news that has an impact. I excerpt from that and make comments about it, and advise my clients about what's going on." In the old days, before DBRS, Bashaw tried to be diligent about checking and clipping the daily newspapers, with varying success. When it came time to prepare his newsletter, he was confronted with file folders filled with ragged, rapidly yellowing clippings. Then he discovered personal computers and, shortly after, DBRS.

After a typical 12-hour day at the office, Bashaw goes home, does whatever a mortgage lawyer does when he gets home, and around midnight, sits down at his Apple and "logs on" to his DBRS. "I keyword-search UPI (United Press International) under the word 'mortgage' and pull in any and all articles that mention the word . . . ." If he misses a night, the data base fills him in on what happened the day before. First, he gets a headline synopsis of available articles so that he can select the ones he wants to read. "If they're good, I capture them in a buffer," he explains, for printing out hard copies after he "logs off." Actually, he doesn't have to do that; the DBRS will mail him hard copies if he requests it (for a fee, of course, but a relatively nominal one). He also searches other news data bases for relevant national, business, regional, and congressional events that could affect his clients.

Bashaw believes his use of the DBRS is responsible for a significant growth in business. The quality of his newsletter, which is, after all, generated by his data-base searches, has enhanced his reputation as an expert. As a result, there has been a marked increase in the number of speaking invitations he receives. That, he claims, has increased his exposure and, therefore, his clientele. His firm,

Pierce & Bashaw, currently oversees the Illinois holdings of national clients, but he plans to expand his activities into other states. He can do this because of his expertise on federal and state mortgage laws—expertise he has developed by keeping abreast of the field via his DBRS.

Portfolios with DBRS

Bashaw also uses the DBRS to take care of his personal business. "About a year ago," he says, "I chose to invest my IRA funds in stocks." Now he uses the DBRS to stay abreast of his portfolio as well as the stock market in general. "I save my stock prices on my buffer," he explains. "When I log off, I bring up my VisiCalc [spreadsheet program] and plug in the current prices. I press a button and change the value of the current price on my matrix, and it lets me know how I'm doing." Bashaw is on-line almost every weeknight. His average time: 24 minutes.

The researching capabilities of a DBRS have had a powerful impact on Tom Trone's business, too. Trone is co-owner (with John Kiser) of Lox, Stock & Bagel, a chain of three restaurants in Champaign and Normal, Ill., with ambitious plans for future expansion.

The restaurants are sort of eat-in gourmet delicatessens, and while they would appear to be quite similar, there are differences among them. One is located on a university campus, another in a downtown business district, and a third in a regional shopping mall. (Five more are now "on the drawing board," according to Trone; ultimately, there will be 14.) "We have three different markets and three different customer profiles," Trone says. So while the menus may look the same, the rates of consumption of the various items are likely to be different. That, in turn, affects pricing and profit. Pricing and profit are carefully monitored by Lox, Stock & Bagel—mostly through a DBRS. About a year ago, when

---

HOW TO FIND ON-LINE DATA BASES

Although finding facts through a DBRS is easy, finding the right data base can sometimes be a problem. Some suggestions to aid you in your search:

Ask the executive director or secretary of your trade association or professional society. Many such organizations have their own DBRSs or can recommend ones that are directly related to your work. Also ask around among your colleagues. Most people who use DBRSs like to brag about it.

Write or telephone the large DBRSs and ask about data bases that serve your specific needs. Chances are you'll find what you're looking for among one of these three: CompuServe Inc., 5000 Arlington Centre Blvd., Columbus, OH 43220, (800) 848-8990; Source Telecomputing Corp., 1616 Anderson Rd., McLean, VA 22102, (703) 734-7500 or (800) 336-3366; and Dialog Information Services, Inc., 3460 Hillview Ave., Palo Alto, CA 94304, (800) 982-5838 (CA) or (800) 227-1927.

Pick up a copy of *Data Bases for Business* by Van Mayros and D. Michael Werner (Chilton Book Co., Radnor, PA, 1982; $19.95 paperback, $27.50 hardbound), an excellent guide to using DBRSs. In addition to a full explanation of what DBRSs are and how to use them, the book contains a directory of about 400 DBRSs with details such as names, addresses, telephone numbers, the vendors through which the services are available, and cross-referenced indexes so that you can find the data bases by name, by vendor, or by type of business or service.

There are other, similar data-base directories that tend to be larger and more expensive. Check the business section of your local public library, or, if you work for a large organization, talk to your company librarian.

Source: Personal Computing, May 1983, 111-113, 115, 117, 166

The DBRS's most important competitive advantage is its communications feature—electronic mail.

Trone first began using The Source, he explains, "we were in the process of making up new pricing scheduling, looking at our purchasing decisions and prices that we could lock in. We tied in to the Commodity News Network. From that, we got information about grain price movements, meat price movements, and so on. We actually make decisions on future purchasing and pricing based on information that we get through The Source."

Trone also uses his DBRS for what he calls "basic research." "In the process of expansion," he says, "we need a lot of information. In a town like Champaign, or in any community, it's difficult to get a broad base of information to make decisions with. The Source brings that information right here into the office. We use their research bases, where we can go in and research a topic of very broad scale, over a wide range of information bases, right here." Trone would not discuss which broad-scale topics he reviews, but certainly, if he plans to open restaurants, he will be looking at economic trends, consumer eating habits, area demographics, and other considerations necessary for determining whether, where, and how to establish a restaurant. All that information is available on DBRSs. "We can get information that deals specifically with what we're doing and can save us the old trial-and-error process. At least when we do a trial now," he says, "it's going to be an educated one."

Getting market updates

Michael Merker also uses a DBRS to monitor economic trends that relate to his business. Merker is the manager of a Xerox retail store in Great Neck, N.Y. Although the store, one of a chain of 52 outlets, is owned by Xerox, Merker is expected to run it as though it were his own business. Thus, while the company dictates the line of products to be carried—which consists of a full line of office ma-

chines, including the Xerox copiers and the 820-II personal computers—it's Merker who's responsible for sales, advertising budgets, promotions, and special offers.

"I use the business update on a regular basis," Merker says, "just to keep up with what's happening in the market. I can't always get to read the entire *Wall Street Journal* every morning—the DBRS allows me to look around and monitor what's going on in retail and the business community as a whole. If everything indicates, for example, that there's a recession, I know I should cut back on my inventory levels and perhaps stress my sale items a bit more—alter my advertising strategies. It allows me to change my 'head set.'"

But for Merker, the DBRS's most important competitive advantage is its communications feature—electronic mail. A "sender" can transmit a message to the central DBRS computer which the "receiver" can access whenever it's convenient. Inasmuch as all 52 Xerox stores are on the DBRS, Merker can play all sorts of inventory games. "There are times," he admits, "when I'll take an order here and I just physically don't have the merchandise. If I go through my normal ordering system (through a mainframe computer in Dallas) . . . it takes a week and a half for the merchandise to arrive . . . . But if I use my distribution list, I can send a 'letter' to all 52 stores and say, 'I need 50 Toshiba calculators . . . .' Within an hour and a half, I can usually get my entire order shipped . . . ." The accommodating sibling store ships by one of the express services and delivery to the customer is in 24 to 36 hours.

Merker has even used the DBRS electronic mail to beat out one of Xerox's field representatives who called on a potential customer at his office after the prospect had shopped in the store. The field rep promised delivery of a copier and computer in

(continued on page 166)

DATA-BASE DIVERSIONS

All is not strictly business on data base retrieval systems (DBRS). Among the data bases offered by The Source and CompuServe are software programs that can be downloaded to personal computers so technophiles can experiment at their leisure. And there are, of course, games.

Some subscribers use the electronic mail facility to vent their political passions. Mike Sidoric, of Informedia, likes to keep tabs on the troubles plaguing the social security system. "As a sole proprietor," he says, "I'm afraid that the social security system's demise is going to increase our tax burden considerably. They're going to look to anybody who's making money to help bail it out." So he scans the DBRS wire services for social security news, especially in Congress. "I can find out what's going on in the hearings and who's saying what, and use electronic mail to send him a tacky mailgram."

If he used CompuServe, Sidoric wouldn't have to write his own letters. He could call on the services of Lobby Letters of America which, upon receipt of the barebones facts and an indication of your mood, will compose and dispatch an appropriate letter.

Also on CompuServe are special interest groups, among them a CB group whose members emulate CBers on their computers, software and hardware users groups, the National Satirists, a work-at-home group for the self-employed, and World Wide Exchanges, for people who like to travel and enjoy exchanging facilities. Other special interest groups abound on both The Source and CompuServe, and it's easy to form new ones via the "bulletin board" facility that both services provide. Computer pen pals are accessible through the bulletin boards, as are advertising opportunities and requests for anything.

two weeks; Merker promised it in 24 hours, and came through.

Texas entrepreneur Michael Sidoric attributes noticeable increases in his business to using the electronic mail facility of his DBRS. His company, Informedia, was founded 13 years ago as a partnership with his wife, June, when both were still students at the University of Texas in Austin. Today, the company, which produces audiovisual presentations, is still in Austin, and has added one employee. Their list of some 60 clients include IBM's Austin division, Texas Tourist Development Agency, Southwest Bell, and Southland Corp. of 7-11 Stores' fame.

Sidoric says the main advantage to using a DBRS is the time savings. Before he became a subscriber, a presentation to a client in Dallas or Houston cost him a full day, not to mention air fares and car rental fees. "Now," he says, "we're able to send things via electronic mail . . . . I can send the entire file to the client in a matter of minutes. When he gets in the next morning, he can get that file from his electronic mail box, print it out, react to it, get other people to look at it and react to it, and send me ·comments and criticisms. He can even feed it into his own word processor and make factual corrections, and send it back to me. All that time,

I'm still working on projects that require attention here in Austin."

Although Sidoric has two personal computers—including a modified Apple that specializes in audiovisual work—he also uses his DBRS's computing power. "I can do word processing and scores of things that I couldn't afford the software for," he says. "It gives the small-computer owner tremendous power for very small cost."

"It's the same with financial planning," he says. "I can do our day-to-day stuff here, but when I'm working on really involved things, it's a lot easier to use their on-line system, which is virtually unlimited."

DBRSs can offer measurable benefits to most business people and professionals—if they select their DBRS with care and use it judiciously. The first step is finding the DBRS most likely to provide the required information.

Costs considered

On-line costs vary with the data base, starting at $5 an hour (at night or on weekends) and going up from there. Some services require a one-time subscription fee of $100; others are as high as $50,000 and more. The more popular DBRSs, such as The Source, CompuServe, and DIALOG (which

is heavily oriented toward business, science, and technology) offer special discounts. In addition to the usage charge, there are often communications network charges and local telephone call charges. (Many DBRSs use private communications carriers readily accessed by a local call in most U.S. and Canadian cities.) According to DIALOG, "A typical search on a single data base averages about 10 minutes for experienced searchers and 15 minutes for beginners. Depending on the data base, a 10-minute search could range from $5 to approximately $15 (or somewhat more in some of the specialized data bases)"—hardly a clearcut estimate. Although charges are quoted on an hourly basis, they are always calculated to the nearest whole minute of actual use. Most DBRSs provide a cost breakdown at the conclusion of the search.

In July 1980, when Eastman Kodak raised its film prices by 30 percent, Mike Sidoric raised his by 10 percent. He has not raised them since, and, at this writing, had no intention of doing so. "It's got to stop somewhere," he says of the inflationary spiral. "Instead of working harder, we need to work smarter."

Plugging in to a DBRS may be one way of doing just that.

A Buyer's Guide To Communications Software

This is a guide with all the facts and information you need to purchase communications software. There are many packages that do many things and throughout this guide the functions are explained to help you decide what you need. A glossary of the terms is included to help guide you.

If you're thinking of using your computer for communications, you may also be thinking you only need to add a modem and you'll be online with the world. Well, it's not quite as simple as that. Communications, like any other personal computing application, depends on a successful combination of hardware and software. And while some modems come bundled with software, others don't, so it's up to you to find a program that supports the modem you're using and has the features you're likely to need. To do that, you need to know a little about the communications process, and more particularly, the modem's place in that process.

Modems work by taking data from a computer and converting it into a string of audio signals which can be transmitted over telephone lines. The modem connected at the other end of the line reconverts the data into digital form, making it acceptable to the receiving computer. The name modem comes from the combination of the words "modulation," the conversion process, and "demodulation," the reconversion process.

In the absence of software, the data must be entered into the keyboard as it is sent. At the receiving end, data appears on the computer screen, scrolls past, and is gone. The limitations of this approach are obvious: data can't be sent from a file in the computer's memory, and once received, it can't be stored or manipulated in any useful way. In this kind of process, the computers are known as dumb terminals—they send, they receive, and that's all they do.

To turn a personal computer into a smart terminal, you need software that will let the computer store and manipulate the data as well as transmit and receive it. Those capabilities in themselves open up a lot of possibilities for computer-to-computer communications, but good software can do even more than that. It can set up a computer to handle different forms of data, change transmission speeds to suit the hardware, store phone numbers and provide automatic routines so you don't have to repeat the connect process every time you communicate.

Of course, not all packages perform all functions. Even if they did, you might not need every feature, and it

A COMMUNICATIONS GLOSSARY

ASCII Code—American Standard Code for Information Interchange. A seven-bit code used widely in data communications to transmit the letters of the alphabet, plus the standard punctuation marks and certain control characters.

Acoustic coupler—A device that allows modem-generated audio signals to be transmitted into and received from a telephone handset. The handset is placed into rubber cups on top of the coupler. A small transducer at the bottom of the cups produces the actual signal.

Auto-answer modem—A modem that can answer an incoming call by generating a carrier tone that signals the originating modem its call has been received.

Auto-dial modem—A modem that can simulate a telephone dialer using either pulse or touch-tone dialing signals.

Baud—A unit of signaling speed, usually given as "baud rate." For most personal computer communications applications, baud rates of either 300 or 1200, which approximate 10 and 40 characters per second, respectively, are used.

Binary Number—Any number in a "base two" system, where the digits have the value of 1 or 0.

Bit—Short for binary digit (either 1 or 0), the elemental unit of digital information.

Buffer—Often called capture buffer. In general, a temporary storage place for data. A capture buffer is temporary storage for data "captured" from a communications link.

Byte—A unit of eight bits. Characters are often referred to as bytes, but each character in the ASCII set can be represented by only seven bits. Thus, a byte can be thought of as equivalent to a character for approximation purposes only.

Carrier—A steady signal that can be changed in tone (modulated) to transmit data.

Checksum—A method of totaling bits received by a computer for the purposes of checking errors in transmission.

Control Characters—ASCII characters that do not print out, but are used to control communications. Control characters can, for example, signal a sender to stop transmitting information when the receiver is busy.

Data—Information in code, text or numerical form, generally represented in ASCII code for digital communications.

Duplex—Refers to the two-way nature of modem communications. In full-duplex communication, both terminals can send and receive simultaneously. In half-duplex operation, both ends can send and receive, but not at the same time. With full-duplex, echo-back communications, a transmitted character is not displayed until it has been verified by the receiver.

Frequency—The number of cycles of an oscillating waveform that occur each second.

Modem—A device that modulates audio tones to carry digital signals and also demodulates the signals at the receiver so they can be understood by a computer.

Noise—Random disturbances that degrade or disrupt data communications, present to some degree in all transmission links.

Originate/Answer Modem—A modem that can either start a telephone call or receive one automatically. Some modems automatically assume originate or answer status, others require manual switching to the proper state.

Parity—A means of checking for errors by adding an extra bit to each ASCII character transmitted.

Protocol—A set of rules for the transmission of data. Protocols describe when transmission will start and stop, what error checking system is in effect and the like.

RS-232—A standard for transmission of serial data covering both hardware configurations and transmission parameters. Different manufacturers may implement some or all of the RS-232 standard in their communications products.

Serial data—Data sent one bit at a time, as opposed to parallel data sent several bits at a time. Modems operate on serial data.

Terminal—A device that receives or transmits digital information. Communications software is designed to control computers during terminal mode operation.

makes no sense to pay for capabilities you won't use. There are other considerations as well—you need a package that will support your computer, and more important, your modem. And you need one you can afford. As with any software, there's a lot to choose from. So what should you look for?

At its simplest, communications software should be able to transmit and receive data represented by standard ASCII code characters. ASCII, which stands for American Standard Code for Information Interchange, is a seven-bit code that represents all the letters in the alphabet, both upper and lower case, punctuation marks, and some control codes. Files represented in this code are often called simply "text files." Most information utilities, like The Source, Dow Jones and CompuServe, send their data to their subscribers in ASCII code. Personal computers can store their data in ASCII code, too. So

communications software should be able to understand it. It should also be able to save data to, and transmit data from, an external storage device (disk drive, cassette tape etc.), and it should provide a screen display of the data that's going out or coming in.

A communications package should also be able to control the operation of the modem. For example, if the modem has an auto-dial feature, it should be activated from the software. If it can

transfer at different speeds, then the software should be able to make the changes as needed.

Another feature of some modems is the provision for both full- and half-duplex transmission. In full-duplex communications, both the receiving and the transmitting computer can send and receive at the same time. In half-duplex mode, both ends can send and receive, but not at the same time. If the modem has the ability to switch between full- and half-duplex modes, the communications package you choose should be able to make the switch.

That's a basic shopping list for communications packages, and most of the software available will meet these requirements. But most of them will also do more.

One of the most valuable additional features is the capability to transfer data which is not in ASCII, or text form. Other types of files include binary,

ANATOMY OF A MODEM

Throughout the communications software buyer's guide, constant reference has been made to the modem. This has made it clear that a modem is an essential piece of equipment; here's what a modem is, and what it does, in more technical terms.

The plain and simple fact is that unless you're a Ham radio operator or will spend many thousands of dollars for data communications, you're going to communicate over the in-place telephone network. A modem must then provide the connection to that network.

While you can transmit computer-compatible signals over wires, over long distances it gets very expensive. So most personal computer communications take place over the phone voice network. Computer-compatible signals are not voice signals, so must be converted for transmission.

A modem accomplishes the conversion by taking the computer's output, in the form of electrical impulses that represent ones and zeros, or "ons and offs," and changing them into audio signals that can be transmitted over voice lines. At the other end of the connection, another modem reads the tones from the telephone line and converts them back into binary signals.

In actual practice, a continuous audio signal, called a carrier, is changed abruptly in pitch, so that the changes represent the presence or absence of data. The changing of this carrier is called modulation. Demodulation, at the other end, is the removal of the audio signal, which exposes the original digital impulses that the computer can understand. The term, modem, comes from combining the first few letters of modulation and demodulation.

Modulation and demodulation operate under standards that were developed by the telephone company. The most popular modems today follow the Bell 103 standard which sets the method of modulation and the speed of transmission, along with other parameters. Bell 103-compatible modems can communicate at 110 or 300 baud. Baud rate is a measure of communication speed: Three-hundred baud equates roughly to 10 characters per second. As the cost of electronics decreases, more modems are becoming available that use the Bell 202 or 212 standard. These modems communicate at speeds of up to 1200 baud, or at about 40 characters per second.

There are a number of modem manufacturers serving the small-computer communications market. Their products can be classified in a number of ways, which introduces some confusion. There are direct-connect and acoustically-coupled modems. The former connect directly to the telephone line, as the name implies, while the latter connect by contact with a standard telephone handset.

Direct-connect modems can have automatic dialing and answering capability. That is, they can initiate the tones or pulses required to dial another telephone and recognize when that phone has answered (auto-dial modems); auto-answer modems can detect an incoming call on the telephone line to which they are connected, and complete the telephone circuit automatically.

Many modems available today have additional capabilities. For example, some can store lists of telephone numbers or complex log-on sequences. Some feature an automatic redial of busy numbers. To most effectively control these features, you need communications software.

A modem may not be the only piece of hardware you need for data communications, however. Some modems come complete with an RS-232 interface, the most common form of small-computer connection, while others don't. You may need to purchase such an interface, also called a serial port, so your modem can connect to your computer. If you do need to get a serial port, make sure to also get the cable that connects the port to the modem. Such cables follow an interconnect standard only loosely; getting the interface and the cable at the same time and place is one way to reduce the possibility of interconnect problems.

which is used by many word processors, and files written in programming language. The ability of the software to transmit and receive these kinds of files saves you the work of converting them into text files.

Another capability which can save you time and frustration is error checking. Transmission of data can be affected by noise on the telephone line, atmospheric conditions, and other types of interference. Weather disturbances can garble the data, resulting in unintelligible files filled with random characters. When this happens, it can be very difficult, and sometimes impossible to correct the files and put them into a form you can use. An error checking feature in a communications software package will check the accuracy and flag problems before transmitting the data.

The simplest method of checking for errors is by adding a parity bit (binary digit) to the byte that's about to be transmitted. The bit is added so that the sum of all the bits in the word will be either even (even parity) or odd (odd parity). The receiving computer checks the bytes coming in to determine whether their parity is correct. If it isn't, the receiver sends a request for retransmission to the transmitting computer.

Parity checking has some drawbacks. It's rather slow, and it isn't terribly accurate, because it's limited to finding mistakes caused when one bit in a character gets changed in transmission. If two bits are affected, the parity check will still work, but the character will be wrong. So alternative error checking protocols (rules for data transmission) are commonly used. One of the most widely used of these is the Ward Christiansen protocol, which computes a number based on the bits in the data being sent. When this protocol is used, whole disk sectors are sent at a time, and a total sum, or checksum, is computed for the sector. The receiving computer compares the checksum it computes with the checksum that came in over the line. If the checksums match, the sector is fine, and the receiver tells the transmitter to send the next sector. If not, the receiver asks for a retransmission of the sector that didn't transmit correctly.

It's important to remember that the computer you're communicating with has to be running a program with the same error checking protocols as the one you're using, otherwise, the error checking won't work.

There's one basic question you should ask yourself beyond whether a program has or doesn't have certain capabilities. You really should know how big the program's capture buffer is. If you want to transfer large files from one computer to another, the communications program you use has to have a large amount of memory set aside as a capture buffer. If the file is larger than the buffer, data can be lost in transmission.

Actually, this problem can occur even before you try to transmit data. If, for example, your word processor's buffer is larger than the one on your communications program, you'll have to limit the size of the file you want to

> *If you want to transfer large files, you really should know how big the program's capture buffer is.*

send, or send the file directly from the word processor to the modem. Not all programs will do this.

Some programs get around this problem by including a feature called auto buffer save. When the buffer is full, the program will automatically issue a stop character to the transmitter and save the full buffer. Then it will continue to receive data. This is a valuable feature if you're planning to work with large files.

Any communications program that incorporates the elements outlined thus far—storing received files on disk, transmitting a file from disk, controlling a modem, checking for errors in transmission, and automatically saving the buffer when it's full, will probably meet most of your communications needs. But there are some features which, while not essential, are convenient.

Some programs will store log-on routines (the sequence of characters that opens the communications line between two computers) and activate them automatically. Automatic log-on saves a lot of time for people who call several numbers, because each log-on sequence is different.

Another convenient extra is the ability to edit files while still in the communications program. This allows you to make changes to the file before and after transmission, without downloading to a word processing program.

A LISTING OF SOME ON-LINE DATA BASES

ACCOUNTANTS' INDEX
Provider: American Institute of
Certified Public Accountants
On-line through: SDC Search Service
Subject: Accounting

ADTRACK
Provider: Corporate Intelligence, Inc.
On-line through: Dialog International
Subject: Advertising

AGRICOLA
Provider: U.S. Department of
Agriculture
On-line through: BRS, Dialog
Information Services, Inc.
Subject: Literature on agriculture

AGRICULTURE BANK
Provider: Data Resources, Inc.
On-line through: Data Resources, Inc.
Subject: Agricultural economic data

AMERICAN PROFILES
Provider: Donnelley Marketing
Information Services
On-line through: Dunn & Bradstreet,
Control Data Corp./Business
Information Systems
Subject: Demographics and
Population, Demographics in U.S.

BALANCE OF PAYMENTS
Provider: International Monetary
Fund
On-line through: Chase
Econometrics/Interactive Data
Corp.; Data Resources, Inc.
Subject: International finance

BILLBOARD INFORMATION NETWORK
Provider: Billboard Publications, Inc.
On-line through: Billboard
Publications, Inc.
Subject: Music and music industry

BOOK REVIEW INDEX
Provider: Gale Research Company
On-line through: Dialog Information
Services, Inc.
Subject: Social science and
humanities

BOOKS IN PRINT
Provider: R.R. Bowker Company
On-line through: BRS, Dialog
Information Services, Inc.
Subject: Books and
periodicals—catalogs

CSS/QUOTES+
Provider: Dunn & Bradstreet
Computer Services, Inc.
On-line through: Dunn & Bradstreet
Computer Services, Inc.

Subject: Securities-Canada,
Securities-U.S.

COMMODITIES FUTURES
Provider: Market Data Systems, Inc.
On-line through: General Electric
Information Services Co.
Subject: Commodities U.S.

COMMODITIES/FUTURES
Provider: Call Computer, Inc.
On-line through: Call Computer, Inc.
Subject: Commodities U.S.

COMPUSERVE CONSUMER
INFORMATION SERVICES
Provider: CompuServe, Inc.
On-line through: CompuServe, Inc.
Subject: Multifaceted information
services providing many data bases

COMPUSERVE EXECUTIVE
INFORMATION SERVICES
Provider: CompuServe, Inc.
On-line through: CompuServe, Inc.
Subject: Multifaceted information
services providing many data bases

DOW JONES NEWS AND DOW JONES
FREE-TEXT
Provider: Dow Jones & Co.
On-line through: Dow Jones & Co.
Subject: Business and industry
corporations-finance,
news-economics and finance

EEI CAPSULE
Provider: Evans Economics, Inc.
On-line through: Control Data Corp./
Business Information Services/Boeing
Computer Services Co.
Subject: Economics—U.S.

FOUNDATIONS
Provider: The Foundation Center
On-line through: Dialog Information
Services, Inc.
Subject: Directory, funding sources
and awards

HARVARD BUSINESS REVIEW
Provider: HBR/ONLINE
On-line through: John Wiley & Sons,
Inc. under agreement with Harvard
Business Review
Subject: Business management

HORSE
Provider: Bloodstock Research
Information Services, Inc.
On-line through: Bloodstock Research
Information Services, Inc.
Subject: Horses

INTERNATIONAL SOFTWARE DATA
BASE
Provider: Imprint Software, Ltd.

On-line through: Dialog Information
Services, Inc.
Subject: Computers and computer
industry

LEGAL RESOURCE INDEX
Provider: Information Access Corp.
On-line through: Dialog Information
Services, Inc.
Subject: Law

LEGI-SLATE
Provider: Legi-Slate, Inc.
On-line through: Legi-Slate, Inc.
Subject: Government, U.S.-Federal,
U.S.-State

MEDLINE
Provider: Australian Medline Network;
BLAISE-LINK; DRS; Dialog Information
On-line through: Services, Inc.;
DATA-STAR; DIMDI; MIC-KIBIC;
National Library of Medicine Japan
Information Center of Science &
Information Technology
Subject: Biomedicine

MICROCOMPUTER INDEX
Provider: Microcomputer Information
Services
On-line through: Dialog Information
Services, Inc.
Subject: Computers

NATIONAL TECHNICAL INFORMATION
SERVICE
Provider: National Technical
Information Services
On-line through: BRS; CISTI;
DATA-STAR; Dialog Information
Services; ESA-IRS;
INKA Karlsruche; SDC Information
Services; CEDOCAR
Subject: Science & Technology

SOCIOLOGICAL ABSTRACTS
Provider: Sociological Abstracts, Inc.
On-line through: BRS; DATA-STAR;
Dialog Information Services, Inc.
Subject: Sociology

THE SOURCE
Provider: Source Telecomputing
Corp.
On-line through: The Source
Subject: Contains many information
services

STANDARD AND POOR'S INDUSTRY
FINANCIAL DATA BANK
Provider: Data Resources, Inc.
Standard and Poor's Corp.
On-line through: Data Resources
Corp.
Subject: Business and
industry—Finance, U.S. finance

This listing is based on information from Cuadra Assoc. Inc.'s Directory of On-line Data Bases.

While all these features can be useful, they're not essential, and so we haven't included them in the comparison charts of communications software that start on page 130. If you want these extras, make sure to ask about them before you shell out for a communications program. You should be aware, though, that the more features a program has, the more expensive it's likely to be. And adding features adds complexity, so multifeatured programs tend to be more difficult to use than those with more limited capabilities. Some software publishers attempt to get around this problem by making their programs completely menu-driven, which is nice, but too many menus can spoil easy interaction with a program.

The only real way to know whether a program will suit you is to try it out. This sounds easier than it is, because it's entirely possible that your computer dealer, or software dealer, won't

Functions added to communications programs can make them hard to use—try them before you buy.

have a modem-equipped computer on which you can try the software. But you should be able to go through some simple configuration steps, move into terminal mode, which will probably get you a message telling you to establish a phone connection, and so forth. Do enough to get an idea of how this program works, and what it can do.

So far this has been a discussion of asynchronous communications software. That's the kind of data communications most common with personal

computers, because it's the easiest to implement. Asynchronous communications proceed without sender and receiver being aware of what the other is doing, to some extent. They know they're connected, because each modem sends a carrier signal at an appropriate frequency to inform the other of its existence. But they don't know when the other is going to do something. Their communications are not synchronized.

Synchronous communications, on the other hand, proceed with sender and receiver locked in step with one another. Since sender and receiver know what each other is doing, their communications can proceed at a much faster rate, because there is less figuring out what's going on in the communications channel. Synchronous communications systems can take advantage of this by allowing more error checking than is possible with asynchronous communications.

CONNECTING TO ON-LINE DATA BASES

Many people who use data communications with a personal computer do so to access on-line data bases. There are a number of these data bases, covering subjects as diverse as stocks and bonds and thoroughbred

bloodlines. The chart below shows subscription charges, storage charges and connect-time charges for a cross-section of on-line services.

SERVICE	SUBSCRIPTION	CONNECT TIME	STORAGE	MONTHLY MIN.
The Source	$100	$20.75/hr prime 7.75/hr non—prime 300 baud 1200 baud—$5/hr surcharge prime, $3/hr non—prime	$.50/rec first ten (1 rec = 2k bytes)	$1 storage $9 connect
CompuServe	$39.95	$6/hr non—prime 12.50/hr prime	First 128k free $4/wk per add 64k	None
Dow Jones News/Retrieval	$75	$1.20/min prime .90/min non—prime	None	
Dialog Information Services	None	Typical searches of the data bases cost $25 to $125.		None
BRS After Dark	$75	$6/hr—to—$20/hr depending on info researched. 6 p.m. local time to 4 a.m. Eastern time.		2 hr/month

Many mainframe and minicomputers support synchronous communications but not asynchronous communications. This isn't true for all of these large computers. The computers of the public information services, for example, support asynch. That's how the large number of subscribers they have with personal computers are able to communicate with them. But many computers inside corporations are only communicating with terminals and smaller computers synchronously, because such communications are faster, and because they implement more sophisticated error checking and correcting protocols than are possible in the case of asynch.

If you want a communications package that will let you communicate with other personal computers and with the mainframes that make up the public bulletin boards, then an asynchronous package is for you. If, on the other hand, you need to communicate with the corporate mainframe, assuming the data processing types allow such communications, and that the mainframe only supports synchronous communications, then you'll wind up using one of the latter packages. It isn't likely that you'll be buying such a package yourself—corporate DP types get very parochial about their large computers. Nevertheless, we've included some of the synchronous packages that run on personal computers in our charts for your information.

Such packages normally emulate one of the IBM communications terminals. The most common of these are the 3270 family of terminals that run a sophisticated communications protocol allowing for high speed and extensive error checking and correcting.

Just getting to a mainframe computer using the proper protocol, which is what synchronous software and hardware terminal emulators are all about, isn't enough. Whether you're talking to a large computer synchronously or not, there's still the problem of transferring a file from the big machine and getting it into some kind of form that a small computer can understand. Spreadsheets, for example, require that the data they use be in a particular format, and data are not normally stored on a mainframe computer in a way that they will load easily into a spreadsheet. With

COMMUNICATIONS SOFTWARE PUBLISHERS

ADVANCED MICRO TECHNIQUES
1291 E. Hillsdale Blvd.
Suite 209
Foster City, CA 94404
(415) 349-9336

ALPHA SOFTWARE CONNECTION
30 B St.
Burlington, MA 01803
(617) 229-2924

APPARAT, INC.
4401 S. Tamarac Pkwy.
Denver, CO 80237
(303) 741-1778

APPLE COMPUTER
20525 Mariani Ave.
Cupertino, CA 95014
(408) 996-1010

ARROW MICRO SOFTWARE
11 Kingsford
Kanata, Canada, K2K 1T5
(613) 592-4609

CAWTHON SCIENTIFIC GROUP
24224 Michigan Ave.
Dearborn, MI 48124
(313) 565-4000

COMPUTER APPLICATIONS, INC.
13300 S.W. 108 St. Circle
Miami, FL 33186
(305) 385-4277

CONTEXT MANAGEMENT SYSTEMS
23868 Hawthorne Blvd.
Torrance, CA 90505
(213) 378-8277

CORPORATION FOR DISTRIBUTION SYSTEMS
17440 Dallas N. Pkwy.
Dallas, TX 75252
(214) 380-0671

CYBERAN SOFTWARE, INC.
11222 Richmond
Suite 140
Houston, TX 77082
(713) 558-8090

DATASOFT, INC.
19808 Wordhoff Pl.
Chatsworth, CA 91311
(213) 701-5161

DATAMARK BUSINESS SYSTEM
279 S. McKnight Rd.
St. Paul, MN 55119
(612) 738-9111

DIGITAL MARKETING CORP.
2363 Boulevard Circle
Walnut Creek, CA 94595
(415) 938-2880
(800) 826-2222

DIRECT·AID
P.O. Box 4420
Boulder, CO 80306
(303) 442-8080

DYNAMIC MICROPROCESSOR ASSOCIATES
545 Fifth Ave.
Suite 1103
New York, NY 10017
(212) 687-7115

EBERT PERSONAL COMPUTERS
4122 S. Perker Rd.
Aurora, CO 80014
(303) 693-8400

FEROX MICROSYSTEMS
1701 N. Fort Meyer Dr.
Suite 611
Arlington, VA 22209
(703) 841-0800

FIRST SOFTWARE COMPANY
5622 E. Presidio
Scottsdale, AZ 85254
(602) 953-1208

HAWKEYE GRAFIX
23914 Mobile St.
Canoga Park, CA 91307
(213) 348-7909

HAYES MICROCOMPUTER PRODUCTS, INC.
5923 Peachtree Industrial Blvd.
Norcross, GA 30092
(404) 449-8791

HEWLETT-PACKARD COMPANY
11000 Wolfe Rd.
Cupertino, CA 95014
(800) 367-4772

HOWE SOFTWARE
14 Lexington Rd.
New City, NY 10956
(914) 634-1821

IBM
1000 N.W. 51st St.
Boca Raton, FL 33432
(800) 447-4700

LINDBERGH SYSTEMS
49 Beechmont
Worcester, MA 01609
(617) 852-0233

LIFEBOAT ASSOCIATES
1651 Third Ave.
New York, NY 10028
(212) 860-0300

LINK SYSTEMS
1655 26th St.
Santa Monica, CA 90404
(213) 394-3664

LOVELLS
4205 Biltmore
Corpus Christi, TX 78413
(512) 852-3096

MADISON COMPUTER
1825 Monroe
Madison, WI 53711
(608) 255-5552

MARK OF THE UNICORN, INC.
P.O. Box 423
Arlington, MA 02174
(617) 576-2760

MICROCOM, INC.
1400A Providence Hwy.
Norwood, MA 02062
(617) 762-9310

MICROCORP
913 Walnut St.
Philadelphia, PA 19107
(215) 627-7997

VisiCalc, for example, each data element needs additional characters tacked onto it that tell the program where the element is to go in the spreadsheet.

Publishers are becoming aware of needs like taking a mainframe file and loading it into a personal computer program, and file-transfer programs to accomplish just this kind of action are starting to appear. The ones we know about are listed in our charts. With these programs, the selection chore is considerably simpler, because the choice is determined by the kind of mainframe computer you are trying to talk to, and the program into which you're trying to load the data. These programs aren't general data communications programs, like most of those listed in the charts. They are special-purpose programs intended for data-file transfer only.

In the case of the general programs, though, the selection process is pretty clear. Any such program should be able to send and receive ASCII text files, and to store received files on a floppy disk. It should also be able to read a file from disk and send it over the telephone line. Its buffer must be large enough that you can comfortably load the largest file you expect to get, and it should have some provision for automatically saving your capture buffer when it becomes full so data aren't lost.

If you have an auto-dial modem, or an auto-answer modem, or one with switchable baud rates, or one that can operate in half or full duplex, then the communications software should be able to control the modem's parameters. That means the software has to be compatible not only with your computer, but with your modem as well.

Finally, there are the things that are nice to have in communications software—the ability to automatically log on to time-sharing services, for example, or the ability to edit files prior to transmission or after reception. Remember that the more functions the program has, the more it is likely to cost. And at the same time, adding more functions to a program often reduces its ease of use. So the last thing you need to know about buying communications software is: Try it before you buy it.

—**David Gabel**

MICRO–SYSTEMS SOFTWARE
4301–18 Oak Circle
Boca Raton, FL 33431
(305) 983–3390

MING–TELECOMPUTING, INC.
P.O. Box 101
Lincoln Center, MA 01773
(617) 259–0391

MYCROFT LABS, INC.
P.O. Box 6045
Tallahassee, FL 32314
(904) 385–1141

MICROLOG
222 Rt. 59
Suffern, NY 10901
(914) 368–0353

PEACHTREE SOFTWARE, INC.
3445 Peachtree Rd. N.E.
8th Floor
Atlanta, GA 30326
(800) 554–8900

PERSOFT, INC.
2740 Ski La.
Madison, WI 53713
(608) 273–6000

RADIO SHACK
1800 One Tandy Center
Fort Worth, TX 76102
(817) 390–3011

SAMS SOFTWARE
4300 W. 62nd St.
Indianapolis, IN 46268
(317) 298–5400

SANYO BUSINESS SYSTEMS CORP.
51 Joseph St.
Moonachie, NJ 07074
(201) 440–9300
(800) 526–7043

SHARED SYSTEMS TECHNOLOGIES, INC.
Route 7 South
Box 163
Bennington, VT 05201
(802) 442–8008

SMALL BUSINESS SYSTEMS GROUP
6 Carlisle Rd.
Westford, MA 01886
(617) 692–3800

SOFTWARE CONNECTIONS
2041 Mission College Blvd.
Santa Clara, CA 95054
(408) 988–0300

SOFTWARE SORCERY, INC.
7927 Jones Branch Dr.
Suite 400
McLean, VA 22102
(703) 471–0610

SOLUTION SOFTWARE SYSTEMS
3930 Wispering Trail
Hoffman Estates, IL 60195
(312) 259–4800

SOUTHEASTERN SOFTWARE
6414 Derbyshire Dr.
New Orleans, LA 70126
(504) 246–8438

SUPERSOFT TECHNOLOGY
P.O. Box 1628
Champaign, IL 61820
(217) 359–2691

SYSTEMS & SOFTWARE, INC.
1315 Butterfield Rd.
Suite 230
Downers Grove, IL 60515
(312) 960–1181

TELEPHONE SOFTWARE CONNECTION, INC.
P.O. Box 6548
Torrance, CA 90504
(213) 516–9430

TELEVIDEO SYSTEMS, INC.
P.O. Box 3568
Sunnyvale, CA 94088
(408) 745–7760

TELEXPRESS, INC.
P.O. Box 217
Willingboro, NJ 08046
(609) 877–4900

THE MICROPERIPHERAL CORP.
2565 152nd Ave. N.E.
Redmond, WA 98052
(206) 881–7544

THE MICROSTUF COMPANY
P.O. Box 33337
Decatur, GA 30033
(404) 491–3787

TNW CORP.
3444 Hancock St.
San Diego, CA 92110
(619) 296–2115

TRANSEND CORPORATION
2190 Paragon Dr.
San Jose, CA 95131
(408) 946–7400

UNIQUE AUTOMATION PRODUCTS
15401 Redhill Ave.
Suite G
Tustin, CA 92680
(714) 730–1012

UNITED SOFTWARE INDUSTRIES
1880 Century Park East
Suite 311
Los Angeles, CA 90067
(213) 556–2211

U.S. ROBOTICS
1123 W. Washington
Chicago, IL 60607
(312) 733–0497

VECTOR GRAPHIC
500 N. Ventu Park Rd.
Thousand Oaks, CA 91320
(805) 499–5831

VEN–TEL, INC.
2342 Walsh Ave.
Santa Clara, CA 95051
(408) 727–5721

VISICORP
2895 Zanker Rd.
San Jose, CA 95134
(408) 946–9000

VOLKSMICRO COMPUTER SYSTEMS, INC.
202 Packets Court
Suite C
Williamsburg, VA 23185
(804) 220–0005

WESTICO
25 Van Zant St.
Norwalk, CT 06855
(203) 853–6880

WINDMILL SOFTWARE, INC.
Box 1008
Burlington, Ontario,
Canada L7P 3S9
(416) 336–3353

WOOLF SOFTWARE SYSTEMS
6734 Eton Ave.
Canoga Park, CA 91303
(213) 703–8112

Inside Modems

Modems that slip into your PC are crowding the market, but problems of incompatibility and unreliable high-speed transmissions plague some entrants.

Larry Jordan and Jan van der Eijk

That first move into data communications takes you directly into the thick of the modem marketplace, where you'll be assaulted with choices about everything from speed to cost. However, your most crucial decision lurks in a simple guise: Should you buy an external modem or an internal modem? The characteristics of the former are well known (see "The Modem Market," *PCW*, Vol. 1, No. 8). Internal modems, however, bring with them an entourage of possibilities and pitfalls not associated with their desktop cousins.

In the early days of personal computing, only 300-bps modems were available. Back in those olden times, modems were relatively expensive external devices that only a few computer-savvy individuals dared to use. Today, data transmission rates for PC modems have skyrocketed, usable communications programs have arrived, prices have fallen, and board-level products abound. The seven 1200-bps modem boards tested here represent only a sample of the devices that fit nicely into an IBM PC, XT, AT, or compatible.

Like desktop modems, modem boards translate a computer's digital data signals into analog frequencies that can be transmitted across normal telephone lines. Another modem at the receiving end converts the analog signals back to digital form for use by the receiving computer. Modem speeds vary, but 300 bps and 1200 bps predominate—although 2400-bps units are becoming more available.

The 1200-bps modems discussed here have some common characteristics. All PC modems, for example, require special communications software to provide easy access to other PCs and the huge number of computer bulletin boards, on-line data services, and company mainframes. Since Hayes modems dominate the market (more than 50 percent according to market analysts at Future Computing of Dallas, Texas), PC modems usually employ the Hayes command set and are often referred to as "Hayes compatible." These commands let users change certain modem functions such as waiting time for dial tones. As a result of this de facto standard, most communications software houses use the Hayes command set in their products.

For that same reason, the Hayes 1200B internal modem is used in this article as a standard against which the other products are measured. But be careful. Timing differences in a manufacturer's modem chip can corrupt compatibility. A program that runs beautifully on one modem may not work at all on an-

Table 1: Modem characteristics

Modem Feature	Qubie' PC212A 1200	Cermetek Info-Mate 212PC	Microcom Era 2	Bizcomp Intelli-modem XL	IDEA-Comm 1200	Novation Smart-Cat Plus	AST Reach!	Hayes 1200B
Command capability:								
Manufacturer-specific		●		●				
Hayes command set	●		●	●	●	●	●	●
Modem type:								
Short expansion board		●					●	
Long expansion board	●		●	●	●	●		●
Line monitors speaker:								
Uses PC speaker						●		
Provides a speaker	●		●	●	●		●	●
None		●						
Telephone:								
Telephone set	●	●	●	●	●	●	●	●
Handset				●				
Switch control:								
Manual switch	●			●	●		●	●
Software switch	●	●	●	●	●	●	●	●
Dial types available:								
Pulse (rotary)	●	●	●	●	●	●	●	●
Touch-tone	●	●	●	●	●	●	●	●
Adaptive		●				●		
Dialing modes:								
Keyboard input	●	●	●	●	●	●	●	●
Modem directory		●						
Automatic dial design:								
Redials busy number		●						
Redials last number	●	●	●	●	●	●	●	●
Links other numbers		●				●		
Tone recognition:								
First dial tone		●		●	●	●	●	
Second dial tone		●		●	●	●	●	
Busy signal		●		●	●	●	●	
Dead line		●			●	●		
Blind dial	●	●	●	●	●	●	●	●

other. For example, with many software packages you can press a function key to disconnect the modem from an on-line service, but the timing sequence in the modem chip might not register the command properly. Thus, you may think you've logged off a system or closed a phone connection when the link actually remains open.

Table 1 illustrates the specific features of the internal modems discussed in this article.

Each of the seven modems requires an expansion slot and a standard RJ11 telephone wall jack for connection to a telephone system; each includes a communications program.

All modems reviewed comply with the Bell 103, 113, and 212A protocols to send asynchronous data. That is, the modems transmit characters in an irregular pattern, using start and stop bits to define the length of each burst of data.

We put these seven modems through a series of tests that included connecting them to an electronic mail program on a minicomputer, sending files to a remote PC, and linking them to on-line services such as MCI Mail and The Source. The modems were tested at both 300 bps and 1200 bps. Ease of installation and quality of documentation were also factors in our evaluation. But before making any selection, consider the words of a wise old race driver: "One man's Ferrari is another man's jalopy." The same maxim applies to modems.

AST Reach!

Noted for making multifunction boards, AST Research generates a significant amount of its revenues from PC communications products. Its Reach! half-board modem is a welcome solution to users of the XT and the Portable PC, who need all the full-length expansion slots they can get. This modem is compatible with most features of the Hayes Smartmodem 1200B, and it fits in the IBM Portable PC slots. The latest version of *Crosstalk XVI* is bundled with it.

One appealing feature of the Reach! is the accessibility of its DIP switches at the PC's backplane. Many communications programs differ in the switch settings they require; the Reach!'s switch design eliminates the loathsome task of opening and closing the PC's enclosure each time you change communications programs. While the Reach!'s documentation won't win a Pulitzer, it is adequate. Only 65 pages focus on the modem itself. Most of the manual is dedicated to explaining the use of *Crosstalk*. This emphasis seems unnecessary, since Microstuf, *Crosstalk*'s manufac-

> *The Era 2 responds to all of the Hayes Smartmodem commands. However, Microcom fully expects that you will use its own software.*

turer, provides ample documentation, and users may choose other communications programs instead. During our tests, the Reach! modem did just fine with *PC-Talk III*.

The board appears well made with no loose wires or uneven soldering. The modem responded well to all Hayes commands and provided excellent line noise filtering. The Reach! modem does, however, have two shortcomings. Its speaker volume control must be adjusted with a screwdriver—a tedious task if the back of your system unit is not easily accessible. An outside volume-control knob would have added a crowning touch to an otherwise excellent modem. A less serious flaw is that the user manual comes on loose-leaf paper without a binder.

Era 2

Microcom is making a name for itself in the arena of communications protocols. Its MNP (Microcom Network Protocol) has attracted considerable interest among microcomputer manufacturers and long-haul data networks such as Telenet (see "The Microcom Agreement," *PCW*, Vol. 2, No. 5). But Microcom also has a bread-and-butter modem, the Era 2—a combination hardware/software package whose software works only with the accompanying hardware. The modem is one of the best Hayes Smartmodem 1200B clones tested. Because the software is written exclusively for the Era 2, it takes full advantage of the modem, using menus and simple commands to set parameters. Another Era 2 advantage is the size of its line monitor speaker. Microcom uses a thin metallic speaker on the Era 2, which fits better in an XT than does the thicker, standard speaker Hayes uses in the 1200B.

With both *PC-Talk III* and *Telios* communications software, the Era 2 communicates easily with remote systems. During tests, the modem transmitted files and sent mail with no problems or need to retransmit. The Era 2 responds to all Hayes Smartmodem commands. The only problem in using these commands is the lack of relevant documentation in Microcom's manual. Microcom fully expects that you will use its own software even though Hayes command recognition is available. The only communications package that refused to work with the Era 2 was the RBBS-PC bulletin board system, a popular public domain program used on many community and private computer bulletin board services (BBSs). The Era 2 would not answer incoming telephone calls when operating under the RBBS program.

The Era 2's typeset manual is easy to read and contains excellent modem installation advice. The program documentation will appear familiar to users of the *IBM Personal Communication Manager*—Microcom also wrote that software.

IDEAComm 1200

Like AST, IDEAssociates sells a wide variety of enhancement products for IBM PCs and compatibles. Communications, however, is the company's strength; IDEAssociates supplies local area networks and terminal emulation boards as well as modems. The IDEAComm internal modem will warble a tune to indicate when it is properly installed and ready to go. Our model ran through the standard tests and had no problem making connections at 300 bps. However, at 1200 bps, while sending a file to a remote PC and while attempting to send electronic mail to a remote host computer, the IDEAComm sent 8-bit graphics characters to the screen so fast that the tests had to be stopped. The screen filled with gibberish because (at least on the unit we tested) the IDEAComm modem does an inadequate job of filtering telephone noise at high speeds. At higher speeds, noise on telephone lines has traditionally created havoc with modem output because the bandwidth on the copper wires is not designed to handle high-speed data without sophisticated filtering.

Included with this modem is the IDEAComm communications software. It supports the standard XMODEM file transfer protocol and has a dialing directory to store 70 characters in the logon sequence. In auto-answer mode, the IDEAComm modem allows remote users to upload and download files and leave up to 15 messages.

Info-Mate 212PC

Cermetek has made its name in the communications market quietly by providing modems for other manufacturers to resell. These OEM relationships have served the company well in its venture into the PC internal modem market. Cermetek's Info-Mate is well designed, expertly fabricated, and nicely packaged. Installation instructions are readable and easy to follow, with the exception of some indistinct photographs that proved of no help at all.

The screen filled with gibberish because the IDEAComm modem does an inadequate job of filtering telephone noise at high speeds.

The modem design employs very large scale integration (VLSI) chips. VLSI lets Cermetek pack the power of 1200 bps into a half-length modem board. The Info-Mate contains two modular telephone cable connections and an auxiliary RS-232C port. Modem jumpers are provided to switch between telephone communications and RS-232C digital communications. A second telephone connection port allows you to monitor the progress of a session or place a phone call manually. The modem comes configured as COM1: but is easily changed to COM2: using small cap-type jumpers.

The Info-Mate is one of the few modems on the market that does not use the Hayes command set. It comes with its own software, Modem-Mate. In our tests, the software did not respond properly when used to communicate with *PC-Talk III* on the other end of a transmission. The Info-Mate can be used with a modified version of *Crosstalk XVI;* that seems to be the software of choice for this modem, especially since Modem-Mate provides only a crude directory for telephone numbers, lacks the ability to use batch mode operations, and has no facility for script files such as those available in *Crosstalk* and other popular communications programs.

A couple of minor problems arose with the Cermetek modem. It has no line monitor speaker, and the dialing response result codes are cryptic. Without an aural monitor a user has to be cognizant of exactly what is happening on the screen. It takes some practice to make calls and to understand what the on-screen signals mean once a call is initiated. In addition, the modem's auxiliary RS-232C port does not support a ring-detect signal, not a problem if the port is used only for direct connection to a host computer. But operating an external modem from the port with bulletin board software, for example, requires the ring-detect signal.

Intellimodem XL
Bizcomp is the company that started the intelligent modem industry. Mike Eaton, Bizcomp's president, holds the patent for intelligent modems, and it is his technique that has been licensed by Hayes. The Intellimodem XL is a full-featured board that allows easy switching between voice and data communications. The voice-call capabilities are an asset to anyone who makes frequent telephone calls to a small group of people. Three telephone connection ports are provided on the back of the board. After connecting the modem to the PC, you can connect a telephone handset to the modem and make or receive voice calls independent of the computer when data calls are not in progress. If you don't have a dedicated line for your modem, this capability spares you from constantly having to plug in and unplug the phone. A special command mutes the modem so it won't transmit its high-pitched signal during voice calls.

The Intellimodem XL package includes *Intelli-Soft Plus* communications software, which takes full advantage of the modem's voice call capabilities but does not measure up to *PC-Talk III* and *Crosstalk* in file-transfer and data communications features.

The tests uncovered problems with the Intellimodem. First, the board's default setting is for COM2:. Most modem manufacturers use COM1:, and the majority of communications software packages expect COM1:. Second, the Intellimodem XL

> *Without an aural monitor a user has to be cognizant of exactly what is happening on the screen.*

locked up several times when used with a Hayes-compatible terminal emulation software package. The hang-up turned out to be a modem timing problem during initial connection with the remote host.

On the plus side, the manual is first-class, except for slight disorganization in the sections that cover linking to the major nationwide data networks. This deficiency could prove a problem for users not familiar with the complexities of data communications.

Qubie' PC212A/1200
Qubie's plug-in modem was one of the first internal 1200-bps units containing ROM-based emulation of the popular Hayes Smartmodem command set. Hayes compatibility combined with the modem's low list price (approximately two-thirds of the Hayes Smartmodem price) created an immediate demand for the Qubie'. Many initial buyers were disappointed, however, when they found that the Qubie' emulated the Hayes standard but not its high-quality construction and components.

Early Qubie' buyers suffered through a variety of problems ranging from improperly manufactured expansion-slot plugs to inadequate telephone-line noise filters. Users had to file down the modem's bus connector before it would slide into a PC slot. Some Qubie' buyers found that file transfers at 1200 bps using the XMODEM file transfer protocol were almost impossible due to inadequate noise filtering in the modem hardware.

Qubie' has corrected most of the problems that surfaced in early versions of its modem while keeping the list price low. The modem uses all the Hayes 1200B commands and interacts easily with *PC-Talk III*, which is included in the package. The Qubie' works nicely with a wide variety of remote data bases and PC-to-PC file transfer programs.

Despite these improvements, the Qubie' modem has a few lingering flaws. First, it sends 8-bit graphics characters (square root signs, brackets, and other symbols not in the transmitted text) to the screen when receiving data from a remote system. Second, the modem's default switches are set for auto-answer, an irritant to people who call while the modem is connected to the voice telephone line. Third, the modem's retaining bracket does not slide properly into an XT and must be twisted in order to fit. Finally, the Qubie' speaker's volume control is inaccessible from outside the PC system unit.

Smart-Cat Plus

Novation provides a full line of modem products for the IBM PC. The Smart-Cat Plus package contains an excellent installation and operations manual with ample illustrations for the hardware novice.

The modem comes with *Mite* communications software but operates equally well with several other programs. *Mite* supports both the XMODEM and CLINK file transfer protocols. In addition to a full Hayes Smartmodem command set, the Smart-Cat includes additional commands for such feats as dialing an alternate number when the first number is busy and determining whether pulse or touch-tone dialing is required.

Unlike the other modems reviewed, the Novation unit relies on the PC speaker for sound feedback during telephone dialing. The pair of jumper wires that connect the modem board to the PC speaker are too short, requiring that the modem be placed in an expansion slot close to the speaker in the PC or the XT. To connect the Smart-Cat Plus to an AT speaker,

Qubie' has corrected most of the problems that surfaced in early versions of its modem while keeping the list price low.

you must remove the disk controller to access the speaker connections on the motherboard. In the Compaq the speaker is so far from the expansion slots that the line monitor cannot be used at all.

Making the Connection

After taking these modems for a ride, we can recommend no single device as a dream modem. If price is the absolute criterion, the Qubie' has to be the answer. If sophisticated file-transfer capabilities are a must, the Era 2 is the standout. But if you're looking for a modem that best mimics the popular Hayes modem, go with the Bizcomp or AST products.

Other relevant issues lie beyond the scope of this review. Two of the modems—the AST and IDEAssociates products—are designed to work in tandem with the companies' other PC enhancements. IDEAssociates, for example, uses its modem with its terminal emulation board. In a corporate environment, that compatibility could be of greater value than low price or Hayes compatibility.

One more point needs considering before you buy an internal modem for your PC: Do you really want to fill an expansion slot with a modem? Desktop modems, although usually more expensive, are less hassle and won't interfere later if you want to add another enhancement to your PC's bus. Study your environment and consider your requirements carefully. ◉

Larry Jordan and Jan van der Eijk are freelance writers and communications and networking consultants with NUS Corporation of Gaithersburg, Maryland, and are the coauthors of the TCOMM unattended communications software package. Larry Jordan coauthored the book Communications and Networking for the IBM PC *(Robert J. Brady Company, Bowie, MD, 1983).*

AST Reach!
AST Research, Inc.
2121 Alton Ave.
Irvine, CA 92714
714/863-1333
List price: $549

Era 2
Microcom, Inc.
1400A Providence Hwy.
Norwood, MA 02062
617/762-9310
List price: $499

IDEAComm 1200
IDEAssociates, Inc.
35 Dunham Rd.
Billerica, MA 01821
617/663-6878
List price: $495

Info-Mate 212PC
Cermetek Microelectronics, Inc.
1308 Borregas Ave.
Sunnyvale, CA 94088
408/752-5000
List price: $149

Intellimodem XL
Bizcomp
532 Weddell Dr.
Sunnyvale, CA 94089
408/733-7800
List price: $549

PC212A/1200
Qubie'
4809 Calle Alto
Camarillo, CA 93010
805/482-9829
List price: $399

Smart-Cat Plus
Novation, Inc.
20409 Prairie St.
Box 2875
Chatsworth, CA 91311
800/423-5419, 818/996-5060
* California*
List price: $499

Selected Databases Available Through DIALOG Information Services, Inc.

The DIALOG Information Retrieval Service, from DIALOG Information Services, Inc., has been operating since 1972. There are 160 databases available on the system.

The databases on the DIALOG system contain in excess of 50 000 000 records. Records, or units of information, can range from a directory-type listing of specific manufacturing plants to a citation with bibliographic information and an abstract referencing a journal, conference paper, or other original source. For more information on DIALOG write:

DIALOG Information Service, Inc.
Marketing Department
3460 Hillview Avenue
Palo Alto, California 94304

The DIALOG service may also be called at these numbers:

- (800) 227-1927, marketing and training
- (800) 227-1960, customer services Toll-free in the continental U.S., except California
- (800) 982-5838, California
- (415) 858-3785, all other locations

The following databases related to metals and the metals industry are available on the DIALOG system:

BHRA Fluid Engineering
British Hydromechanics Research Association
Cranfield, Bedford MK43 OAJ
England

1974 to present, 63 460 records, indexing and abstracting of worldwide information of fluid engineering, including theoretical research and the latest technology and applications on the following subjects: civil engineering hydraulics, industrial aerodynamics, dredging, fluid flow, fluid power, fluid sealing, fluidics feedback, and tribology

Chemical Abstracts
Chemical Abstracts Service
Columbus, Ohio

Maintains five separate databases covering information relating to chemicals and the chemical-processing industry:

- **Chemical Industry Notes (CIN).** 1974 to present, 368 300 records. Extracts 78 worldwide business-oriented periodicals which cover the chemical processing industry
- **Chemical Regulations and Guidelines System (CRGS).** Regulatory materials in effect. Index to U.S. federal regulatory material relating to the control of chemical substances, covering federal statutes, promulgated regulations, and available federal guidelines, standards, and support documents
- **Chemname.** 1 179 796 chemical substances derived from Chemical Abstracts Service (CAS) Registry Nomenclature File, quarterly updates. Contains a listing of chemical substances in a dictionary-type non-bibliographic file
- **Chemsearch.** 183 876 chemical substances, derived from Chemical Abstracts (CA) Search. Companion file to Chemname listing most recently cited substances
- **Chemsis.** 1967 to 1971, 850 000 records, closed file: 1972 to 1976, 1 178 410 records, closed file; 1977 to present, 1 320 880 records, irregular updates. Dictionary, non-bibliographic file containing those chemical substances cited once during a collective index period of chemical abstracts

Claims
Plenum Data Company
Arlington, VA

Seven separate files on patent information:

- **Claims/Chem.** 265 000 U.S. chemical and chemically related patents issued from 1950 to 1962
- **Claims/Citation.** 1 757 000 records. Includes all patent numbers cited in U.S. patents from 1947 to 1981
- **Claims/Class.** 100 743 records. Classification code and title dictionary for all classes and selected subclasses of the U.S. Patent Classification System
- **Claims/U.S. Patents.** 1963 to 1970, 483 000 records. Contains all patents listed in the general, chemical, electrical, and mechanical sections of the "Official Gazette" of the U.S. Patent Office
- **Claims/U.S. Patent Abstracts.** 1971 to present, 745 200 records. Citations and abstracts for all patents classified by the U.S. Patent Office in the areas of aerospace and aeronautical engineering, agricultural engineering, chemical engineering, chemistry, civil engineering, electrical and electronics engineering, electromagnetic technology, mechanical engineering, nuclear science, and general science and technology
- **Claims/U.S. Patent Abstracts Weekly.** Companion to previous file; includes most current weekly update and records from the current month
- **Claims/Uniterm.** 1950 to present, 497 000 records. Access to chemical and chemically related patents. Subject indexing to facilitate retrieval of chemical structures and polymers

Compendex
Engineering Information, Inc.
New York, NY

January 1970 to present, 1 001 000 records. Machine-readable version of the "Engineering Index". Worldwide coverage of approximately 3500 journals, publications of engineering societies and organizations, papers from the proceedings of conferences, and selected government reports and books

DOE Energy
U.S. Department of Energy
Washington, D.C.

1974 to present, 640 000 records. Covers all aspects of energy and related topics from journal articles, report literature, conference papers, books, patents, dissertations, and translations. Topics include: nuclear, wind, fossil, geothermal, tidal, solar, as well as the environment, energy policy, and conservation

Metadex
American Society for Metals
Metals Park, Ohio 44073

Two separate databases on metals and related materials:

- **Metals Abstracts/Alloys Index.** 1966 to present, 300 000 citations produced by the American Society for Metals (ASM) and the Metals Society (London). Provides coverage of international literature on the science and practice of metallurgy. Included are *Review of Metal Literature* 1966 to 1967, *Metals Abstracts* 1968 to present, and since 1974, *Alloys Index*. *Metals Abstracts* include about

Reprinted from ASM Metals Reference Book, 2nd Edition, 539-540, © 1983 American Society for Metals

40 000 citations each year from about 1200 primary journal sources. *Alloys Index* supplements *Metals Abstracts* by providing access to the citations through commercial, numerical, and compositional alloy designations; specific metallic systems; and intermetallic compounds found within these systems. In addition to specialized topics (including specific alloy designations, intermetallic compounds, and metallurgical systems), six basic categories of metallurgy are covered: materials, processes, properties, products, forms, and influencing factors. Each month about 3500 new documents related to metals technology are scanned and abstracted, with intensive coverage of appropriate conference papers, reviews, technical reports, and books. These sources are international in scope, including the U.S.S.R. and Eastern European nations among the 43 countries covered

MIDAS, Metals Information Datafile on Alloys and Specifications. Available only through System Development Corporation (SDC), Santa Monica, CA (not available through DIALOG). Commencing Fall, 1982. A numerical data bank on the mechanical and physical properties, composition, and specifications of ferrous and nonferrous alloys. Alloys can be located by specification/designation number, trade name, or by composition. When available, data on product form (such as bar or plate) and condition (heat treatment) are also provided. The mechanical and physical properties covered include: tensile, shear, impact, fatigue, crack propagation, hardness, fracture toughness,

specific heat, thermal conductivity, thermal expansion, melting temperature, electrical resistivity, electrical conductivity, and magnetic permeability. The data is obtained from journal articles, conference proceedings, compilations, and technical reports. The source of the data is listed in each record

Non-Ferrous Metals Abstracts
British Non-Ferrous Metals
  Technology Center
Wantage, Oxfordshire, England

1961 to present, 60 000 records. Covers nonferrous metallurgy and technology. Sources include journals, monographs, British patents, reports, standards, and conference papers

NTIS
National Technical Information
  Service
U.S. Department of Commerce
Springfield, VA

1964 to present, 863 500 citations. Consists of government-sponsored research, development, and engineering, plus analyses prepared by federal agencies, their contractors, or grantees

Standards and Specifications
National Standards Association, Inc.
Bethesda, MD

1950 to present, 72 000 records. Access to all government and industry standards, specifications, and related documents which specify terminology, performance testing, safety, materials, products or other requirement, and characteristics of interest to a particular technology or industry (Relevant standards-issuing organizations included in this database are listed in the

following article on selected standards-issuing organizations)

Surface Coatings Abstracts
Paint Research Association of
  Great Britain
Middlesex, England

1976 to present, 50 200 citations. Contains references to research literature on all aspects of paints and surface coatings, including pigments, dyestuffs, resins, solvents, plasticisers, printing inks, testing pollution, and marketing

Weldasearch
The Welding Institute
Cambridge, England

1967 to present, 55 400 records. Primary coverage of the international literature on all aspects of the joining of metals and plastics, and related areas, such as metals spraying and thermal cutting

World Aluminum Abstracts
American Society for Metals
Metals Park, Ohio 44073

1968 to present, 81 600 citations, monthly updates. Provides coverage of the world's technical literature on aluminum, ranging from ore processing (exclusive of mining) through end uses. Includes information abstracted from approximately 1600 scientific and technical patents, government reports, conference proceedings, dissertations, books, and journals. All aspects of the aluminum industry, aside from mining, are covered, including the following major subject areas: aluminum industry in general, and foundry, metalworking, fabrication, finishing, physical and mechanical metallurgy, engineering properties and tests, quality control and tests, and end uses

Selected Standards-Issuing Organizations Related to Metals and Metalworking

Standards issued by these organizations may be searched using the Standards and Specifications database described in the previous article on DIALOG information services.

Aerospace Industries Association
National Aerospace Standards
  Committee
1725 DeSales St., N.W.
Washington, DC 20036

American Bureau of Shipping
65 Broadway
New York, NY 10006

American Chemical Society
1155 16th St. NW
Washington, DC 20036

American Foundrymen's Society
Golf and Wolf Rd.
Des Plaines, IL 60016

American Gear Manufacturers
  Association
1901 N. Fort Myer Dr.
Arlington, VA 22209

American Institute of Steel
  Construction, Inc.
400 N. Michigan Ave., 8th Floor
Chicago, IL 60611

American Iron & Steel Institute
1000 16th St. N.W.
Washington, DC 20036

American National Standards
  Institute
1430 Broadway
New York, NY 10018

American Nuclear Society, Inc.
555 N. Kensington Ave.
LaGrange Park, IL 60525

American Society for Nondestructive
  Testing, Inc.
4153 Arlingate Plaza
Columbus, OH 43228

American Society for Quality
  Control, Inc.
230 W. Wells St.
Milwaukee, WI 53203

American Society for Testing and
  Materials
1916 Race St.
Philadelphia, PA 19103

American Society of Lubrication
  Engineers
838 Busse Hwy.
Park Ridge, IL 60068

American Society of Mechanical
  Engineers, Inc.
345 E. 47th St.
New York, NY 10017

American Vacuum Society
335 E. 45th St.
New York, NY 10017

American Welding Society
550 N.W. LeJeune Road
P.O. Box 351040
Miami, FL 33125

Anti-Friction Bearing Manufacturers
  Association, Inc.
Century Building, Suite 704
1235 Jefferson Davis Hwy.
Arlington, VA 22202

Architectural Aluminum
  Manufacturers Association
35 E. Wacker Dr.
Chicago, IL 60601

Association of Iron and Steel Engineers
Three Gateway Center, Suite 2350
Pittsburgh, PA 15222

Can Manufacturers Institute
1625 Massachusetts Ave., N.W.
Washington, DC 20036

Cast Bronze Bearing Institute, Inc.
221 North LaSalle, Suite 2026
Chicago, IL 60601

Cast Iron Soil Pipe Institute
1499 Chain Bridge Rd., Suite 203
McLean, VA 22101

Composite Can and Tube Institute
1800 M St. N.W.
Washington, DC 20036

Computer Aided Manufacturing
  International, Inc.
611 Ryan Plaza Dr., Suite 1107
Arlington, TX 76011

Concrete Reinforcing Steel Institute
180 N. LaSalle St., Room 2110
Chicago, IL 60601

Copper Development Association, Inc.
405 Lexington Ave., 57th Floor
New York, NY 10017

Diamond Core Drill Manufacturers
  Association
59 E. Main St.
Moorestown, NJ 08057

Diesel Engine Manufacturers
  Association
c/o A.P. Wherry & Associates
712 Lakewood Center North
Cleveland, OH 44107

Equipment and Tool Institute
1545 Waukegan Road
Glenview, IL 60025

Expansion Joint Manufacturers
  Association
25 N. Broadway
Tarrytown, NY 10591

Fluid Controls Institute, Inc.
Plaza 222, U.S. Highway One
P.O. Box 3854
Tequesta, FL 33458

Forging Industry Association
1121 Illuminating Bldg.
55 Public Square
Cleveland, OH 44113

General Aviation Manufacturers
  Association
1025 Connecticut Ave. N.W.
Washington, DC 20036

Heat Exchange Institute
c/o Thomas Associates, Inc.
1230 Keith Building
1621 Euclid Ave.
Cleveland, OH 44115

Hoist Manufacturers Institute
1326 Freeport Rd.
Pittsburgh, PA 15238

Hydraulic Institute
c/o A.P. Wherry and Associates, Inc.
712 Lakewood Center North
Cleveland, OH 44107

Industrial Fasteners Institute
1505 E. Ohio Building
1717 E. 9th St.
Cleveland, OH 44114

Institute of Electrical and
  Electronics Engineers, Inc.
345 E. 47th St.
New York, NY 10017

Reprinted from ASM Metals Reference Book, 2nd Edition, 541-542, © 1983 American Society for Metals

Institute of Scrap Iron and Steel, Inc.
1627 K St. N.W., Suite 700
Washington, DC 20006

Instruments Society of America
67 Alexander Dr.
P.O. Box 12277
Research Triangle Park, NC 27709

Investment Casting Institute
8521 Clover Meadow Dr.
Dallas, TX 75243

Lead Industries Association, Inc.
292 Madison Ave.
New York, NY 10017

Magnetic Materials Producers
  Association
c/o H.P. Dolan & Associates, Mgrs.
3451 W. Church St.
Evanston, IL 60603

Manufacturers Standardization
  Society of Valve and Fittings
  Industry
5203 Leesburg Pike, Suite 502
Falls Church, VA 22041

Material Handling Institute, Inc.
1326 Freeport Rd.
Pittsburgh, PA 15238

Metal Powder Industries Federation
105 College Rd. E.
Princeton, NJ 08540

Metal Treating Institute, Inc.
1311 Executive Center, Suite 200
Tallahassee, FL 32301

National Association of Architectural
  Metal Manufacturers
221 N. LaSalle, Suite 2026
Chicago, IL 60601

National Association of Corrosion
  Engineers
P.O. Box 218340
Houston, TX 77218

National Association of Pipe Coating
  Applicators
717 Commercial National Bank
  Building
Shreveport, LA 71101

National Association of Recycling
  Industries, Inc.
330 Madison Ave.
New York, NY 10017

National Board of Boiler and
  Pressure Vessel Inspectors
1055 Crupper Ave.
Columbus, OH 43229

National Coil Coaters Association
1900 Arch St.
Philadelphia, PA 19103

National Solid Wastes Management
  Association
1120 Connecticut Ave. N.W., Suite 930
Washington, DC 20036

Pipe Fabrication Institute
1326 Freeport Rd.
Pittsburgh, PA 15238

Plastics Pipe Institute
369 Lexington Avenue, 10th Floor
New York, NY 10017

Refractories Institute
1102 One Oliver Plaza, No. 3760
Pittsburgh, PA 15222

Resistance Welder Manufacturers
  Association
1900 Arch St.
Philadelphia, PA 19103

Semiconductor Equipment &
  Materials Institute
625 Ellis St., Suite 212
Mountain View, CA 94043

Society for Technical Communication
815 15th St. N.W., Suite 506
Washington, DC 20005

Society of Automotive Engineers, Inc.
400 Commonwealth Dr.
Warrendale, PA 15096

Society of Die Casting Engineers, Inc.
Triton College Campus
2000 N. 5th Ave.
P.O. Box 3002
River Grove, IL 60171

Society of Naval Architects and
  Marine Engineers
One World Trade Center, Suite 1369
Washington, DC 20005

Society of the Plastics Industry, Inc.
355 Lexington Ave.
New York, NY 10017

Specialty Wire Association
1101 Connecticut Ave., No. 700
Washington, DC 20036

Spring Manufacturers Institute, Inc.
1211 W. 22nd St.
Oak Brook, IL 60521

Steel Bar Mills Association
1125 W. Lake St.
Oak Park, IL 60301

Steel Founders' Society of America
20611 Center Ridge Rd.
Cast Metals Federation Building
Rocky River, OH 44116

Steel Joist Institute
1703 Parham Rd., Suite 204
Richmond, VA 23229

Steel Structures Painting Council
4400 Fifth Ave.
Pittsburgh, PA 15213

Steel Tank Institute
666 Dundee Rd., Suite 705
Northbrook, IL 60062

Ultrasonic Industry Association, Inc.
c/o Michael Management
P.O. Drawer F
Jamesburg, NJ 08831

Zinc Institute, Inc.
292 Madison Ave.
New York, NY 10017

SECTION VI
Programmable Controllers

SECTION VI
Programmable Controllers

One of the fastest-growing areas of industrial microcomputer application is that of programmable logic controllers, often called PLC's, or simply PC's. Since the reader has already seen the term PC applied to personal computers in Section III, it is important to recognize that the two terms apply to quite different breeds of microcomputers. Once the differences are established, the reader should have no trouble in separating the two.

The PC is a perfect example of the use of microprocessor technology applied to meet a specific need. The first PC's were developed in 1968 by General Motors Corporation in an attempt to eliminate, or at least minimize, the costly scrapping of assembly line relay panels during model changeovers. The first PC's were manufactured in 1969.

The programmable controller, according to NEMA (National Electrical Manufacturers Association) Standard ICS3-1978, Part ICS3-304, is " . . . a digital electronic device . . . that uses a programmable memory to store instructions for implementing specific functions . . . such as logic, sequencing, timing, counting, and arithmetic . . . to control machines and processes . . ." As a working description, NEMA's definition has satisfied practically no one involved with the selection and application of PC's.

More broadly defined, a programmable controller is a flexible, easily reprogrammed industrial computer that performs a chain of commands similar to a group of hard-wired relays. PC's are general-purpose microprocessor controls that accept inputs, evaluate them, and generate appropriate outputs in response.

Unlike digital computers, which execute several tasks in any order, PC's scan a series of tasks in an orderly and sequential manner. PC's interface directly with machines and processes and control them in harsh manufacturing environments. These controllers are generally rated for 0 to 60°C (32 to 140°F) operation at up to 95% relative humidity without error generation or damage to internal electronics.

There are three basic characteristics that help classify an industrial control device as a programmable controller. First, a PC's basic internal operation is to solve logic from the beginning of memory to some specified stopping point, such as end of memory or end of program. Once the end is reached, the operation starts again at the beginning of memory. This scanning process continues from the time power is turned on until it is shut off.

Second, the programming language is a form of relay ladder diagramming (see Fig. 1). Normally open and normally closed contacts and relay coils are used within a format utilizing a left and right vertical rail. Power flow (symbolic positive electron flow) is used to determine which coils or outputs are energized or de-energized.

Third, a PC is designed for the industrial environment from its basic concept. That protection is not added on later. This harsh environment includes unreliable ac power, high temperatures, humidity, vibrations, radio frequency "noise," and other conditions deleterious to electronic devices designed mainly for use in the protected control room or office environment. Figure 2 shows a modern programmable controller system.

A brief chronology of PC developments since 1969 indicates the fast pace of change for these popular devices:

1971 . . . First application of PC's outside the automotive industry

1973 . . . Introduction of "smart" PC's for arithmetic operations, printer control, movement of data, matrix operations, CRT interface, etc.

1975 . . . Introduction of analog PID (proportional-integral-derivative) control that made possible the accessing of thermocouples, pressure sensors, etc.

1976 . . . First use of PC's in hierarchical configurations as part of an integrated manufacturing system

1977 . . . Introduction of very small (fewer I/O) PC's based on microprocessor technology

1980 and since . . . Major advances in ease of programming, access to data networks, options in programming languages, easier PID configuration, etc.

Current progress in PC's includes increases in memory capacity, system functionality, and the ability to communicate with other PC's as well as other components of control systems. There has also been an explosion of PC suppliers. One of the most recent "roundups" of the PC marketplace, for example, listed 158 different models of PC's available from 48 manufacturers.

The articles at the end of this section illustrate the variety of ways the PC is being applied in metalworking operations.

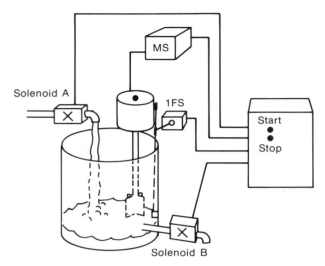

Solenoid A

MS

1FS

Start
Stop

Solenoid B

A — Automatic mixer

Figure A is a sketch of an automatic mixing system. It shows a tank that is filled with a fluid, agitated, and then emptied. Figure B graphically illustrates the complete system cycle, while Fig. C is the ladder diagram for this simple system. It would be used to program a programmable controller.

Let's look at the series of events for the full control cycle by referring first to Fig. A. To start the operation, the Start pushbutton (1PB) is pressed. This energizes a control relay (1CR) located in the Start/Stop box (not shown).

Referring to Fig. C, when the float switch (1FS) is in the empty position and contact 1CR is closed, solenoid A is energized, allowing fluid to fill the tank. When the tank has filled, float switch (1FS) changes to the filled position, de-energizing solenoid A, starting timer relay (1TR) in Start/Stop box, and operating the mixer (MS). After the allotted time, relay 1TR switches off mixer and energizes solenoid B, emptying the tank. When the tank is empty, float switch (1FS) shuts off solenoid B and places the system in Ready for the next manual start.

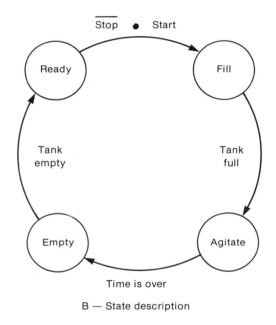

Stop • Start

Ready

Fill

Tank empty

Tank full

Empty

Agitate

Time is over

B — State description

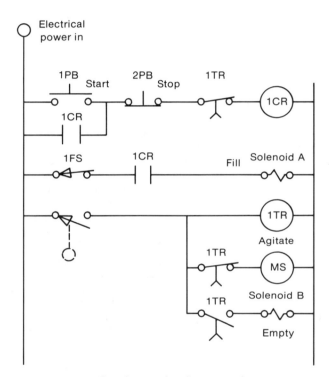

C — Automatic mixer control

212

Typical Sequence for a Two-Heat-Cycle Carburizing Operation
(O's are open/on; X's are closed/off.)

		Heating chamber vacuum valve	Quench chamber vacuum valve	Heating chamber nitrogen backfill valve	Nitrogen quench	Carburizing throughput	Partial pressure heat treat nitrogen or hydrogen backfill valve	Heat on	Cooling fan	Baffle servovalve	Pressure equalization servovalve	Carburizing fan
Reset (start)												
Differential pressure	Pump down	X	O	X	X	X	X	X	X	X	X	O
Chamber pressure 50 microns	Transfer load and pump down	O	O	X	X	X	X	X	X	X	X	O
Confirm	Heat to Setpoint 1	O	O	X	X	X	O	O	X	X	X	O
Time out	Soak Timer 1	O	O	X	X	X	O	O	X	X	X	O
Time out	Carburizing timer	X	X	X	O	O	X	O	X	X	X	O
Time out	Diffusion timer	O	O	X	X	X	O	O	X	X	X	O
Confirm	Transfer setpoint, Heat to Setpoint 2	O	O	X	X	X	O	O	X	X	X	O
Time out	Soak Timer 2	O	O	X	X	X	O	O	X	X	X	O
Transfer complete	Transfer load	O	O	X	X	X	X	O	X	X	X	O
Time out	Oil quench timer	O	X	X	O	X	X	O	X	X	X	O
Program complete	Oil drain timer	O	X	X	O	X	X	X	X	X	X	O
	Cycle over	X	X	X	X	X	X	X	X	X	X	O

Fig. 2. Modern programmable controller system, showing processor, I/O module, and program loader terminal. (Courtesy IPC division of Honeywell Inc.)

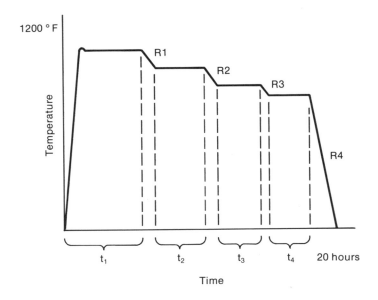

TED F. MEINHOLD, PE
Senior Editor

Programmable Controllers

... an overview of sizes and capabilities plus guidelines on selection, installation, maintenance, and use

Early programmable controllers (PC's) were basically used as sequential controllers, performing simple on-off functions associated with limit switches, pushbuttons, relays, timers, and solenoids. Today, PC's have moved far beyond their sequencing role, having evolved into rugged, versatile, powerful devices capable of performing sophisticated control tasks in virtually all industries, Fig. 1, 2.

What is a PC?—Definitions vary, but simply stated, PC's are general purpose microprocessor-based controls that accept inputs, evaluate them, and generate appropriate outputs to control machines or processes. The National Electrical Manufacturers Association (NEMA) prefers a more specific definition: a PC is a digitally operating electronic apparatus that uses a programmable memory for the internal storage of instructions for implementing specific functions such as logic, sequencing, timing, counting, and arithmetic, to control machines or processes through digital or analog input or output (I/O) modules.

But not everyone accepts NEMA's definition. Critics argue that many hybrid designs are excluded by NEMA's definition. They say that ladder logic controls, time and count-based sequencers, sequential step-advance controls, and industrial control computers are also classified as PC's.

Sequential control occurs when switches, motors, solenoids, lights, or instruments are turned on and off in preselected intervals or in combined sequence patterns controlling an operation.

Historically, sequential controllers could not perform event-based control functions. But things are changing. Sequential devices have recently been developed that can be programmed to do comparisons and combinational logic, as well as wait for specific events, before proceeding to or escaping from a point in the program if an event fails to occur.

Disagreement also exists concerning the classification of PC sizes. The entire industry is in a state of flux because of the rapid development that is occurring. Units that used to be classified as medium sized a few years ago are now considered to be small. Currently, a generally accepted categorization based on I/O points and memory size is: very small, small, medium, large, and very large. Each of these categories is discussed in greater detail later.

PC Components

Regardless of size, cost, or complexity, all PC's have the same basic elements and functional characteristics. The unit's major components are the central processing unit (CPU), memory, input/output (I/O) system, and power supply. Programming devices are usually also included except on certain very small devices, Fig. 3.

The CPU is the "brain" of the PC and organizes all controller activity. The CPU receives input data, performs logical decisions based upon a stored program, and drives outputs. Normally, each PC has one CPU that can service or control many I/O points.

A number (or a combination of letters and numbers) is used to code

Fig. 1. Pyramid depicts broad range of control tasks that PC's are capable of performing in industrial plants. The various levels of control react with each other and each level builds on the technology of the level below it. (Courtesy Allen-Bradley)

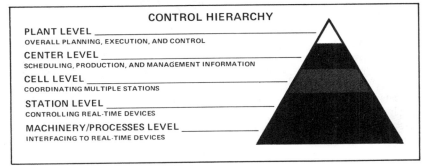

CONTROL HIERARCHY

PLANT LEVEL
OVERALL PLANNING, EXECUTION, AND CONTROL

CENTER LEVEL
SCHEDULING, PRODUCTION, AND MANAGEMENT INFORMATION

CELL LEVEL
COORDINATING MULTIPLE STATIONS

STATION LEVEL
CONTROLLING REAL-TIME DEVICES

MACHINERY/PROCESSES LEVEL
INTERFACING TO REAL-TIME DEVICES

Reprinted with permission from Plant Engineering, November 23, 1983, 16 pp, © 1983 Technical Publishing Co., 873 Third Avenue, New York, NY 10022

Fig. 2. Ability of PC's to withstand tough industrial environments is shown here. Throughout the welding period, the unit continued to control critical plant functions without error. (Courtesy Texas Instruments Inc.)

Fig. 3. Major components of a PC are the CPU, memory, I/O hardware, and power supply. Programming device is required to enter the application program into memory. (Courtesy Reliance Electric Co.)

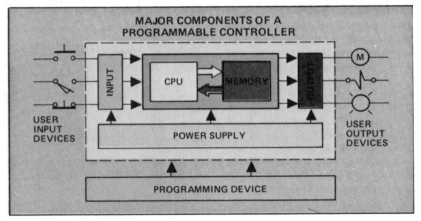

data locations known as addresses. The CPU uses these addresses to sort and fetch I/O and memory data. The user selects the I/O address by assigning an input or output circuit to a specific field device. In addition to straight logic processing, CPU's perform timing, counting, latching, comparing, and retentive storage functions. When operating, the CPU continually scans the status of all I/O circuits and the application memory, thereby determining what the output status should be.

The time in which a CPU completes one operating cycle is called the scan time. During a scan, all the inputs are read, logic is solved, and outputs are updated. Scan times typically run from 5 to 200 milliseconds. Scan time varies with particular models, memory size, amount of user logic, and the number of I/O's.

Based on each scan, the CPU may initiate one or more control actions, depending on the I/O conditions. Scans repeat so any changes in the I/O's can be acted upon. This action establishes a control loop consisting of input signals from sensors and switches, and output signals to solenoids, control valves, and motor starters.

The memory system provides a means of storing and retrieving data. The memory is divided into executive program and application program instructions. Executive program instructions that direct the CPU's activities are provided by the PC manufacturer. The user enters the application program instructions that control machines or processes. The complexity of the control plan determines the amount of application program memory needed.

The amount of memory required for an application is a function of program length and the number of I/O points involved. Required memory size can be estimated by writing out the program and counting the number of instructions used. Total memory needs are obtained by multiplying the number of instructions by the number of words used in each instruction. Memories are available in many forms and specific increments ranging from 64 to 192K words. But a word does not have an exact definition. Memory size is usually specified in bytes (see glossary of PC terms). Most memories are expandable in 1024 byte (1K) increments. Bits are also used to designate memory size. Eight bits of data are equivalent to one byte.

Currently, the most commonly used memory in PC's is a complementary metal oxide semiconductor (CMOS), referred to as CMOS-RAM (random access memory). This is an integrated-circuit-based memory that permits stored programs to be easily altered whenever necessary. Access is random, which means storage locations can be accessed in any order any number of times. No established sequence of accesses is required. The memory can be protected from power outages by lithium batteries having service lives as long as 4 yrs.

Programmable read only memory (PROM) is another popular semiconductor memory. PROM automatically retains its memory during power failures, so battery back-up power is not required. A shortcoming of PROM is that information cannot be easily erased once it is entered.

Erasable programmable read only memory (EPROM) permits data erasure. EPROM can be re-programmed and also provides long term program retention without battery backup power. Memory can be erased by exposure to an intense source of ultraviolet light. Electrically erasable read only memory (EEROM) is another popular non-volatile memory that can be erased electrically and reprogrammed. One of the newest memories is nonvolatile random access memory (NOVRAM). This memory acts like RAM but stores the data like EEROM. No battery backup power is required.

Other forms of memory include read only memory (ROM) and CORE (a magnetic core). ROM is usually used only to perform a specific function, such as providing executive program instructions. Data are permanently entered into it at the time it is manufactured.

CORE stores information in ferrite cores. It is not a semiconductor memory. Data are nonvolatile and can be changed. No battery backup power is required. However, its me-

chanical assembly is somewhat bulky. CORE is more expensive than RAM or PROM. Logic solving time is high in systems having over 16K bytes.

I/O systems have point capacities ranging from 32 to 4096. An I/O rack essentially consists of various groups of input and output circuit cards or modules mounted in a cabinet. Typically, I/O modules contain 4, 8, 16, or 32 circuits.

The input cards receive data from field input devices such as pushbuttons, limit switches, relay contacts, analog sensors, or selector switches. The cards accept the high incoming control voltages from these units and convert them to logic level voltages acceptable to the CPU.

The CPU controls the output card signals that drive field output devices such as motor starters, solenoid valves, indicator lights, light emitting diode (LED) displays, positioning valves, or relay coils. The output cards transform the logic output voltages into levels required to drive the field output units.

I/O racks may be positioned next to the CPU or up to 10,000 to 15,000 ft away from it, Fig. 4. Communication is provided by a twisted pair cable or fiber optic cable. Remote I/O rack installations exchange information with the CPU by a digital data link. This drastically reduces wiring when action is far removed from the CPU. Whatever the arrangement, typically a bank of I/O modules will be associated with a screw terminal barrier strip to which the user attaches the control wiring.

PC's may have digital modules, analog modules, or both. Digital input signals are typically supplied by pushbuttons, switches, contacts, etc. Analog input signals are provided by process instrumentations, transducers, etc. Digital cards do not accept or generate analog signals. They handle only high level on-off signals in the range of 5 v dc to 240 v ac. A wide variety of I/O

Fig. 4. I/O racks (right) may be positioned immediately next to a CPU or thousands of ft away from it. Electrician is shown making routine check of I/O cards. (Courtesy General Electric Co.)

Fig. 5. Programming units vary in size and capability, ranging from handheld devices to large, color CRT's. Portable devices such as this permit PC's to be programmed at remote sites. (Courtesy Eagle Signal Controls)

voltages, current capacities, and module types are available. PC's handle analog information in one of two ways: they compare the input signal against the preset level within an input module with the objective of developing an on/off limit signal; or they perform analog-to-digital (A/D) conversion, where the continuous input is transformed into digital code representing its numeric value.

Of special interest to process control users are proportional-integral-derivative (PID) boards. These units permit automatic closed-loop operation of multiple continuous process control loops. Important operating data can be monitored on a CRT terminal.

Power supplies provide all the voltage levels required for the PC's internal operations. The power supply can be packaged directly into the CPU or installed some distance away as a separate unit connected to the CPU by a cable. The power supply converts 120 or 220 v ac line power into the dc power required by the CPU and I/O modules. For example, power supplies may convert 120 v ac into 5 or 15 v dc, or 220 v ac into 24 v dc.

Programming devices are required to enter application programs into a PC's memory. Exceptions are very small, single-board ROM PC's programmed by the manufacturer to do specific jobs. Such programs are usually permanent and are never changed.

Programming units vary in size and capability. They range from small, handheld devices to color CRT's with monitoring and graphics capabilities, Fig. 5.

Programmers are usually only connected to the PC while the program is being entered. Once programming has been completed, the device is disconnected. However, some small PC's have permanently mounted membrane keypad programmers located in their panels.

The typical portable programmer is a handheld device about the size of a desk-top calculator, large enough to be used comfortably, but light enough to be carried in a briefcase. The programmer, which has its own power cord and nonvolatile memory, permits users to perform programming tasks at remote sites, in offices, or at home through telephone lines.

A relatively new concept is the use of microcomputers as program loaders. The microcomputer can be used for a variety of engineering and business purposes when it is not connected to the PC.

Programming devices can be used to verify previously entered programs, make changes to existing programs, or permit maintenance personnel to check the performance of various portions of a control system. Monitoring can be done in the control room or at remote locations. Some portable programmers feature two-color liquid crystal displays (LCD).

Very Small PC's—These units usually have a maximum of 32 I/O's and memories up to 1K bytes, Fig. 6. Typical cost is approximately $500. Very small PC's were initially used in fixed applications where little or no future expansion was planned. Such dedicated units

Fig. 6. Very small PC's house up to 32 I/O. This unit features plug-in modules of four I/O's each. Handheld programmer (not shown) can be powered from PC or, with optional power supply, plugged into 120/230 v ac outlet. (Courtesy Eaton Corp., Cutler-Hammer Products)

Fig. 7. Small PC's have 64 to 128 I/Os and memories to 2K bytes. This unit features a removable programming panel that permits programs to be entered directly into battery backed-up CMOS-RAM from existing ladder diagrams. Expander module (bottom) provides 10 I/O channel capability increase. (Courtesy Cramer Div., Conrac Corp.)

generally have read only memories (ROM). More recent versions have RAM or EPROM, proof of the increasing capabilities resulting from technological advancements. Scan times can be as short as 10 milliseconds. Diagnostic capabilities simplify and speed up program editing and debugging. An RS-232-C communications port (see glossary of PC terms) provides a data link to any standard printer.

Typical applications for very small PC's include controlling basic machine functions such as counting, sorting, cutting, indexing, sequencing, and stacking. Very small PC's are ideally suited for replacing hard-wired panels or other small control systems having as few as two timers and counters, or a few relays.

Small PC's—This category covers units with 64 to 128 I/O's and

memories to 2K bytes. Prices range from about $500 to $5000. Small PC's are capable of providing the versatility and sophistication needed for advanced levels of machine control. I/O functioning capability has improved significantly in recent years to a point where small PC's now have the capability of performing tasks done by medium PC's a few years ago.

Small PC's are capable of providing the versatility and sophistication required for advanced levels of machine control. Modular designs permit easy future expansion, Fig 7. Applications include molding, plating, heat treating, material handling, machining, and miscellaneous manufacturing processes.

Medium/PC's—These units generally have 256 to 512 I/O's and memories from 4K to 7K bytes, Fig. 8. Prices range from $5000 to $10,000 for basic systems. Medium PC's are commonly used in midsize applications, although there is some overlapping with small PC's. Modular designs permit expansions. Intelligent I/O cards make medium PC's adaptable to temperature, pressure, flow, weight, position, or any type of analog function commonly encountered in process control applications.

Large and Very Large PC's—These are the most sophisticated units of the PC family, Fig 9. In general, large units have 1024 to 2048 I/Os and memories from 8K to 32K bytes. Prices run from about $10,000 to $15,000 for basic systems. Very large PC's have over 4096 I/O's and memories from 128K to 192K. Costs range from $15,000 to $20,000. Considerable overlapping of capabilities exists beteen these two categories.

Both sizes boast virtually unlimited applications. They can control individual production processes or entire plants. Typical installations include metal, food, chemical, and oil processing facilities. Modular design permits systems to be expanded to control thousands of analog and digital points.

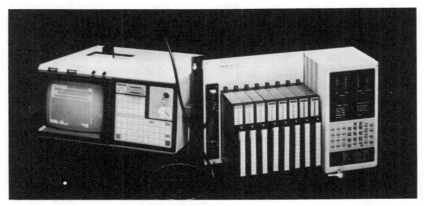

Fig. 8. Medium size PC has modular expandability up to 320 control I/O points or 512 data I/O points and 4K of user memory. Optional CRT graphic programmer is at left. (Courtesy Omron Electronics, Inc.)

Fig. 9. Large PC's such as this provide the capability for hundreds of fault tolerant analog I/O points, as well as virtually unlimited digital I/O capabilities. Applications include both continuous and batch operations. Multiple process CRT operator workstations shown in foreground may be locally or remotely located. (Courtesy August Systems, Inc.)

Programming a PC

PC programming languages use functions relating to machine and process control. Although various programming languages are being used today, relay ladder logic is by far the most dominant language. Unlike programming languages used in conventional computers, relay ladder logic is understood by most plant personnel. Relay ladder logic is easily learned even by those that have no programming training.

PC ladder diagrams are derived from series and parallel-connected electromechanical networks. The ladder diagrams, as well as the electromechanical networks they are derived from, can be a simple series, simple parallel, or a combination series parallel circuit. Electromechanical networks and PC logic diagrams are connected to drive various output devices based on specific input conditions, Fig 10. Normally, inputs are switch closures such as pushbuttons, limit switches, and circuit breakers. Output devices are those that initiate some kind of action, such as solenoids, pumps, or motor starters.

A ladder logic diagram is drawn in a series of rungs. Each rung contains one or more inputs and the output or outputs directly controlled by the inputs, Fig. 11. The rung instructs the PC to control a part of the electromechanical circuit. Basic input and output symbols are used to represent each specific input and output to be controlled (see accompanying section, "Ladder Logic Symbols for Inputs and Outputs"). However simple or complex the circuit, it can be enacted using the simple logic functions AND, OR, NOT. Common relay control functions and their logic equivalents are shown in the accompanying section, "Control Diagram Symbols."

The *series circuit* is one in which all the inputs in the electromechanical network (or the PC ladder diagram) must be activated (ON or

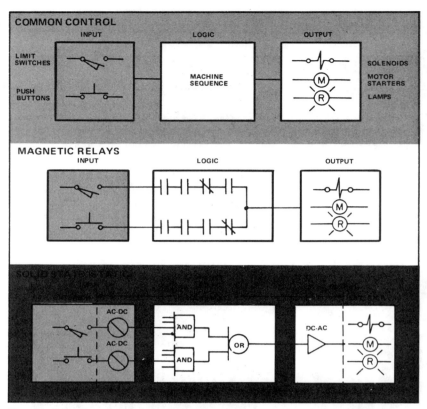

Fig. 10. Control systems share three common features: inputs, logic functions, and outputs. Common machine control block diagram is at top, conventional magnetic relay control at center, and solid state system is at bottom. (Courtesy Westinghouse Electric Corp.)

LADDER LOGIC SYMBOLS FOR INPUTS AND OUTPUTS

Ladder logic diagrams use symbols to represent inputs and outputs of the system to be controlled by the PC. All inputs to the system, whether hardwired to the PC or simulated by control relays within the PC, are represented by the basic symbol

$$\dashv\vdash$$

All outputs of the system, whether hardwired to the PC or a relay coil simulated by an internal PC control relay, are represented by the basic symbol

$$-\bigcirc-$$

Specific variations of the basic input and output symbols are used to represent each specific input and output of the system to be controlled. Each of these specific symbols is shown and explained below.

 ... Used to represent any input, hardwired to a PC input terminal, that must be closed (turned on) to activate an output.

... Used to represent any input, hardwired to a PC input terminal, that must be open (turned off) to activate an output.

 ... Used to represent a contact of a relay, to be simulated by an internal PC control relay, that must be closed to activate an output.

... Used to represent a contact of a relay, to be simulated by an internal PC control relay, that must be turned off to activate an output.

... Used to represent a simulated input of any hardwired output that must be turned on to activate another output.

... Used to represent a simulated input of any hardwired output that must be turned off to activate another output.

... Used to represent the coil of a relay that is to be simulated by an internal control relay.

... Used to represent any output device that is hardwired to a PC output terminal and controlled by the PC.

Source: Texas Instruments Inc.

CONTROL DIAGRAM SYMBOLS

COMMON RELAY CONTROL FUNCTIONS AND THEIR LOGIC EQUIVALENTS ARE SHOWN IN THIS CHART.

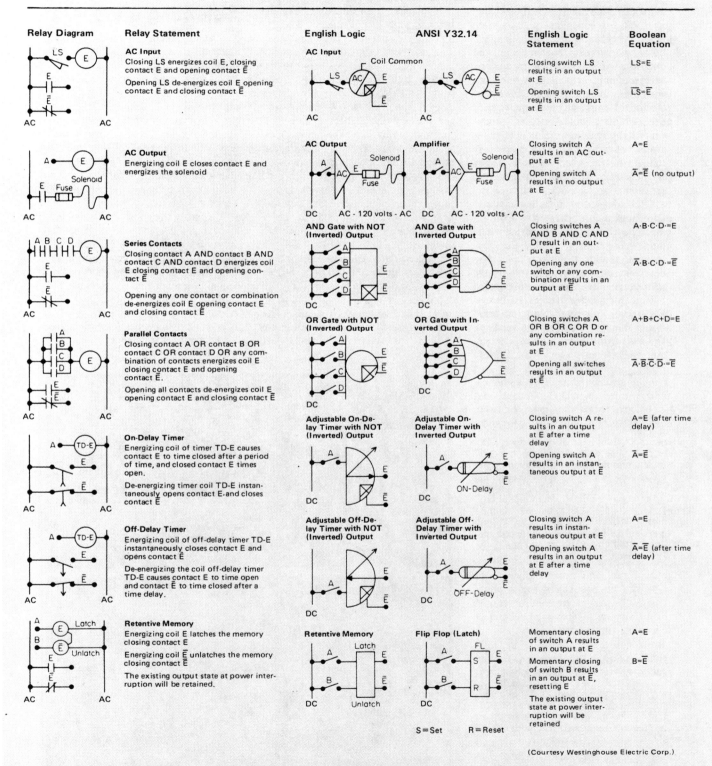

Relay Diagram	Relay Statement	English Logic	ANSI Y32.14	English Logic Statement	Boolean Equation
	AC Input	**AC Input**		Closing switch LS results in an output at E	LS=E
	Closing LS energizes coil E, closing contact E and opening contact Ē	Coil Common		Opening switch LS results in an output at Ē	$\overline{LS}=\overline{E}$
	Opening LS de-energizes coil E opening contact E and closing contact Ē				
	AC Output	**AC Output**	**Amplifier**	Closing switch A results in an AC output at E	A=E
	Energizing coil E closes contact E and energizes the solenoid	Solenoid / Fuse	Solenoid / Fuse	Opening switch A results in no output at E	$\overline{A}=\overline{E}$ (no output)
		DC AC - 120 volts - AC	DC AC - 120 volts - AC		
	Series Contacts	**AND Gate with NOT (Inverted) Output**	**AND Gate with Inverted Output**	Closing switches A AND B AND C AND D result in an output at E	A·B·C·D·=E
	Closing contact A AND contact B AND contact C AND contact D energizes coil E closing contact E and opening contact Ē			Opening any one switch or any combination results in an output at Ē	$\overline{A}·\overline{B}·\overline{C}·\overline{D}·=\overline{E}$
	Opening any one contact or combination de-energizes coil E opening contact E and closing contact Ē	DC	DC		
	Parallel Contacts	**OR Gate with NOT (Inverted) Output**	**OR Gate with Inverted Output**	Closing switches A OR B OR C OR D or any combination results in an output at E	A+B+C+D=E
	Closing contact A OR contact B OR contact C OR contact D OR any combination of contacts energizes coil E closing contact E and opening contact Ē.			Opening all switches results in an output at Ē	$\overline{A}·\overline{B}·\overline{C}·\overline{D}·=\overline{E}$
	Opening all contacts de-energizes coil E opening contact E and closing contact Ē	DC	DC		
	On-Delay Timer	**Adjustable On-Delay Timer with NOT (Inverted) Output**	**Adjustable On-Delay Timer with Inverted Output**	Closing switch A results in an output at E after a time delay	A=E (after time delay)
	Energizing coil of timer TD-E causes contact E to time closed after a period of time, and closed contact E times open.			Opening switch A results in an instantaneous output at Ē	$\overline{A}=\overline{E}$
	De-energizing timer coil TD-E instantaneously opens contact E and closes contact Ē	DC	ON-Delay DC		
	Off-Delay Timer	**Adjustable Off-Delay Timer with NOT (Inverted) Output**	**Adjustable Off-Delay Timer with Inverted Output**	Closing switch A results in instantaneous output at E	A=E
	Energizing coil of off-delay timer TD-E instantaneously closes contact E and opens contact Ē			Opening switch A results in an output at E after a time delay	$\overline{A}=\overline{E}$ (after time delay)
	De-energizing the coil off-delay timer TD-E causes contact E to time open and contact Ē to time closed after a time delay.	DC	OFF-Delay DC		
	Retentive Memory	**Retentive Memory**	**Flip Flop (Latch)**	Momentary closing of switch A results in an output at E	A=E
	Energizing coil E latches the memory closing contact E	Latch	FL / S / R	Momentary closing of switch B results in an output at Ē, resetting E	B=Ē
	Energizing coil Ē unlatches the memory closing contact Ē			The existing output state at power interruption will be retained	
	The existing output state at power interruption will be retained.	DC Unlatch	DC		
			S = Set R = Reset		

(Courtesy Westinghouse Electric Corp.)

Source: Plant Engineering, November 23, 1983, 16 pp

closed) to activate the output. The series circuit functions just as the logical AND.

The parallel circuit is one in which one or more of the inputs in the electromechanical network (or the ladder diagram) must be activated (ON or closed) to activate the output. The parallel circuit functions as the logical OR.

A series parallel circuit is a combination of the AND and OR logic functions. The circuit is a combination of two or more series circuits in parallel. A series parallel circuit can also be a combination of a number of parallel circuits in series with one or more inputs.

Relay ladder diagrams are expected to remain popular for many more years. However, increased use of PC's in process, integrated manufacturing, and total plant control applications is resulting in greater use of alternative languages.

Some PC manufacturers are offering add-on, plug-in, BASIC modules for use with their products. The user can continue to use relay ladder diagrams in his existing PC system, but if he prefers to program some tasks in BASIC, he simply inserts the module. PC operation continues without interruption.

Function blocks have been added to ladder diagrams to provide details of the control functions written into a program, Fig. 12. Prompted instructions simplify the programming of complex operations and less keystrokes are required to enter them. The enhanced programming instructions help ensure that all necessary data are included in the ladder diagram. Function blocks also expand the PC's flexibility and capability. The ability of PC's to do jobs other than replacing relays has become a key factor in achieving greater productivity in plants.

Selecting a PC

The key factor in selecting a PC is establishing exactly what the unit is supposed to do.

Current designs cover a broad range of sizes and capability. At the small end, PC's are primarily used as relay replacements to provide standard relay logic, timing, counting, and shift register functions. At the large end, analog I/O capability has made it possible for the units to become an integral part of process control systems.

Probably the most important step in sizing a PC system is determining what the I/O requirements are, including types, location, and quantity. If the application involves the replacement of relays, the user can usually quickly determine I/O needs. Establishing *analog* I/O needs is much more complicated and may require expert assistance.

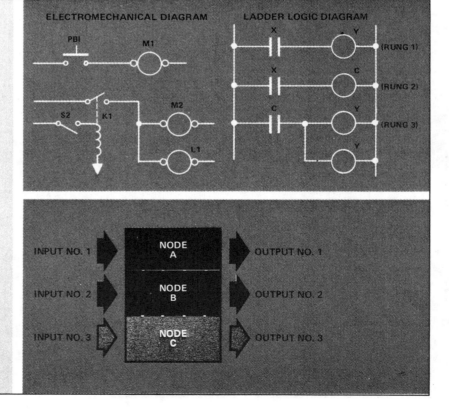

Fig. 11. To develop a PC program that will control a system, an electromechanical diagram of inputs and outputs (left) must be organized in a ladder logic diagram (right) of instructions for use by the PC. Rung 1 of the ladder logic diagram represents an instruction for the PC to turn on M1 (Y) when PBI (X) is closed. Rung 2 represents an instruction for the PC to energize the coil of an internal control relay (C) when S2 (X) is closed. Rung 3 represents an instruction to the PC to turn on M2 (Y) and L1 (Y) when the internal control relay is energized and its normally open contact (C) is closed.

Fig. 12. The number of stacked nodes for a particular type of function block (three in this case) varies according to the complexity of the function being performed. Each node will specify a numerical value or the location of such a value in the memory. The function block interfaces to standard relay ladder logic by means of its inputs and outputs.

Other requirements to be evaluated include memory type and capacity, programming procedures, and peripheral equipment needs. An I/O sizing example is presented in the accompanying section, "Determin-ing I/O Needs." Normally, a 10 to 20 percent spare or expansion capability should be allowed for each application.

Cost Considerations—Determining the cost of a PC is not easy. Many intangibles must be taken into consideration. System requirements dictate costs to a certain degree, but the true cost also depends on the value of increased production, improved quality, greater

---

DETERMINING I/O NEEDS

Sizing a PC for a specific application can often be done by the potential user without expert assistance. Probably the most important step in sizing a PC system is determining what the I/O requirements are. Consider the following process application.

Assume a total I/O count of 764, divided into 436 inputs and 328 outputs. By definition, this application falls into the medium size PC category. Since the majority of the field devices are located a considerable distance from the CPU, a PC with remote I/O is desirable. The I/O requirements by location are:

Process Area—390 I/O
254 inputs; 230 at 24 v dc, 24 at 4 to 20 milliampere analog
136 outputs; all 24 v dc
Tank Farm No. 1—98 I/O
62 inputs; all 120 v ac
36 outputs; all 120 v ac
Tank Farm No. 2—86 I/O
52 inputs; all 120 v ac
34 outputs; all 120 v ac
Loading Station—190 I/O
68 inputs; all 120 v ac
122 outputs; 24 at 120 v ac, 98 at 240 v ac

Assume that the PC system being considered has the following features: no constraints on input and output mixture; the I/O modules are available in two formats, 16 points per module and 8 points per module; and 10 percent spare I/O is required. For the sake of illustration, we will use the 16 point/module I/O structure in the process tank area, and the 8 point/module structure in the tank farms and loading station. The I/O distribution (including spares) per location would be as follows:

Process Area—390 I/O
24 v dc inputs—(230 + 10 percent) ÷
  16 points per module = 15.8 or   16
4 to 20 milliampere analog inputs—(24 + 10 percent) ÷
  16 points per module = 1.6 or   2
24 v dc outputs—(136 + 10 percent) ÷
  16 points per module = 9.4 or   <u>10</u>
         Total 16 point modules = 28

Tank Farm No. 1—98 I/O
120 v ac inputs—(62 + 10 percent) ÷
  8 points per module = 8.5 or   9
120 v ac outputs—(36 + 10 percent) ÷
  8 points per module = 4.9 or   <u>5</u>
         Total 8 point modules = 14

Tank Farm No. 2—86 I/O
120 v ac inputs—(52 + 10 percent) ÷
  8 points per module = 7.2 or   8
120 v ac outputs—(34 + 10 percent) ÷
  8 points per module = 4.7 or   <u>5</u>
         Total 8 point modules = 13

Loading Station
120 v ac inputs—(68 + 10 percent) ÷
  8 points per module = 9.4 or   10
120 v ac outputs—(24 + 10 percent) ÷
  8 points per module = 3.3 or   4
240 v ac outputs—(98 + 10 percent) ÷
  8 points per module = 13.5 or   <u>14</u>
         Total 8 point modules = 28

Assuming that one remote communication channel can service up to 128 I/O in groups of sixteen 8 point modules or eight 16 point modules, the system becomes:

Process Area—390 I/O; 28 16 point modules; 4 remote channels

Tank Farm No. 1—98 I/O; 14 8 point modules; 1 remote channel

Tank Farm No. 2—86 I/O; 12 8 point modules; 1 remote channel

Loading Station—190 I/O; 28 8 point modules; 2 remote channels

Source: Allen-Bradley Co.

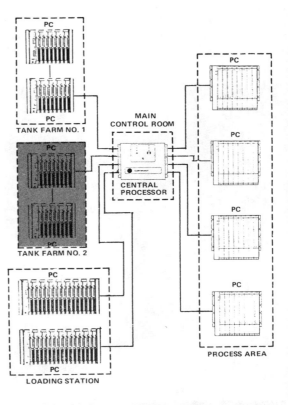

flexibility, and reduced downtime.

Installation, operation, and maintenance costs are important economic factors. Servicing can be expensive. However, if inhouse repair capability is available, servicing costs may be reduced.

In general, plant engineers are not advised to buy a PC system that is larger than what they need. But all phases of the project must be considered and future needs accurately anticipated to ensure that the system is the proper size and will fill the plant's needs. Pay out time for PC's depends on the application. In some cases it may be realized in less than 6 mo, in others it may take 2 yr or more.

Working with PC's

Ideally, inplant personnel—the individuals that will use the devices and maintain them—should install PC's, Fig. 13. This practice is not always possible, so PC manufacturers are ready to provide installation service if requested. Estimates are that approximately 70 percent of PC users perform their own installation on machine control applications. But only about 30 percent install PC's on process control applications, which are considerably more complex. Practically all retrofitting jobs are done by inplant personnel.

Most PC operating problems result from inadequate grounding or electrical noise shielding. PC's are also vulnerable to excessive voltages or temperatures. PC's can misinterpret transient voltages as legitimate signals, and electrical noise can alter the digital data upon which the PC operates. Transients in ac lines result from electrical storms, large solenoids switching on and off, etc. They can couple into the PC through its ac input transformer. Solutions to these problems include installing power line filters and isolation transformers.

Stray magnetic fields abound in industrial plants. Power lines, relays, solenoids, transformers, motors, and generators all contribute their share of interference. Any fluctuating magnetic field cutting through a control loop will induce noise, which will be superimposed upon the signal being transmitted.

Shielded twin-axial cable is recommended for use in troublesome areas. An outer conductive jacket serves as a shield surrounding the two inner signal carrying wires. The shield blocks electrostatic pickup. Magnetic pickup is prevented by twisting the inner signal carrying wires.

Signal wiring should be positioned away from motors, transformers, and ac power wiring whenever possible. If lines must cross, the intersection should be perpendicular to minimize the magnetic coupling effect. Magnetic shielding, such as steel conduit, should be used to encase shielded twin-axial cable if the line must pass near strong sources such as motors and generators.

Other suggestions for avoiding electrical problems include keeping ac and dc wiring in panels separate whenever possible, avoiding the placement of high voltage wiring

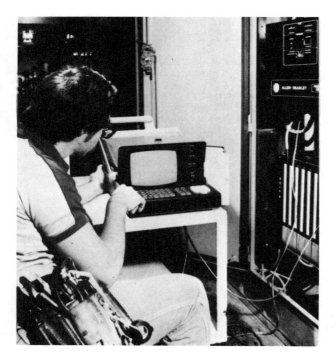

Fig. 13. Ideally, PC's should be installed by plant personnel—individuals that will use and maintain them. (Courtesy Allen-Bradley)

Fig. 14. Typical PC communication network connects various industrial devices together to exchange information between themselves or with computers or similar intelligent devices. Networks usually operate at moderate speeds of 100,000 to 10 million bits per sec and over distances of 100 yards to a few miles. (Courtesy Reliance Electric Co.)

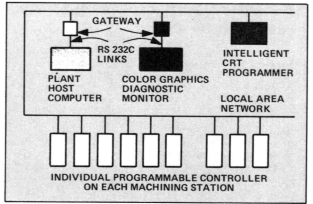

near control wiring, and grouping wire types by similar electrical signals. Low level signals should always be run separate from other field wiring. Direct current I/O field wiring should be kept away from ac field wiring. Splices in signal cables should be avoided whenever possible.

Networking — Communication networks for PC's, computers, printers, and CRT displays are currently being marketed under various tradenames. Computer interfacing permits distributed and supervisory control to be performed that greatly increases machine and process control capability, Fig. 14. PC's can talk to each other, to robots, or to other intelligent devices in the network.

Operations can be easily monitored from a centrally located computer. Troubles and trends can be spotted before they become serious problems. Communication links can transfer data through single shielded twisted pair cable at distances to 15,000 ft in some systems. RS-232-C communication ports and compatible modems permit monitoring over telephone lines.

Connecting a PC to a network requires inserting a special communications module in an I/O rack. Once connected, the PC becomes part of the data communications system. Units can be added or removed from the network without disturbing operations.

Peripheral Equipment—The most useful peripheral device is the programmer, mentioned earlier. CRT terminals can serve either as programmers or monitors. These units can provide visual indications of required operator actions and general system operating conditions. They can also be used for troubleshooting. CRT's can display the status of all field devices as well as the real-time power flow in any portion of a ladder diagram, Fig. 15. The units can be connected to any CPU or local I/O without disturbing the scanning. Since CRT's are usually only temporarily connected to PC's for short periods of time, they can be used to service several PC's.

Color graphics in CRT's are attractive, but may not be necessary for a given application. Price differential is significant too. Software for color CRT's may cost 5 to 10 times more than that for a black and white unit. Color graphics can be important in installations where extremely large volumes of information are required and graphic reports must be generated. Users must ask themselves how much information they really need to do the job. If the sole requirements are programming and monitoring, black and white CRT's are usually sufficient. But there is no question that color terminals will gain in popularity as prices drop.

For hard-copy printouts, PC's should be connected to printers or plotters. These units are available in many shapes and sizes. Printers are especially useful if printouts of ladder diagrams or programs are needed. A full 80 to 132 column printer is recommended if complex production and test reports are required with text statements. A relatively inexpensive strip or short-column printer may be sufficient if data logging with simple descriptions is only required.

OUTLOOK

The future for PC's appears to be bright. Succinctly stated, future designs will provide users with more "bang for the buck." Memory sizes will increase and prices are expected to drop. Larger memories will permit PC's to monitor dozens of variables to further extend process control capabilities. Scan times will be faster too. The ability to perform multiple scans will be common.

Large memories available in today's big PC's will be standard in the smaller PC's of tomorrow. The net result will be more computing power. The progress will be similar to what has occurred in the personal computer field. But the basic PC design will remain the same. The units already are small and rugged and are expected to stay that way.

Ease of use will increase. The use of multiple programming languages in PC's will be common. Distributed control will gain in popularity. Individual, stand-alone PC's will be replaced by numbers of small PC's, talking to each other and doing things, bringing control much closer to the machines or processes being controlled.

Speech modules will be common. PC's will operate paging systems. For example, if a problem is detected on a production machine, the PC will activate an alarm. If no acknowledgement or remedial action is received or initiated within 3 min, the PC will telephone the area supervisor. And instead of only reporting that something is wrong, the PC may actually solve the problem, and tell the supervisor what action it took.

Networking capabilities and applications will continue to grow as standards are developed. PC networks will be linked into shipping departments, production lines, and management information systems. PC's will perform bookkeeping tasks such as comparing incoming parts with those scheduled to arrive, and then directing them to the proper location for immediate use.

The trend towards greater inplant PC repair capability will continue. Some of this effort is already being encouraged by manufacturers whose servicing costs are rapidly increasing. Greater use of inhouse repair capability will minimize expenditures for all parties concerned.

The use of fiber optic cables is also expected to increase significantly as prices come down. Fiber optic cables have significant advantages over traditional coaxial transmission cable. For example, in addition to being small in size and light in weight, the cable has excellent noise immunity, is not affected by high voltage, presents no spark hazard, and is capable of high rates of data transmission.

Source: Plant Engineering, November 23, 1983, 16 pp

Fig. 15. CRT terminals can serve either as programmers or monitors. When used for troubleshooting, CRT's display the status of all field devices as well as the real-time power flow in any portion of a ladder diagram. (Courtesy Modicon Programmable Control Div, Gould Inc.)

Fig. 16. PC's require little, if any, maintenance. Efforts should be made to keep PC panels free of dirt and dust. Peripheral equipment must be checked occasionally too. (Courtesy Allen-Bradley)

Fault Tolerance—Some confusion exists between redundancy and fault tolerance, although the terms are often used interchangeably. Redundancy involves an *active* operating system in a PC that is continuously duplicating the work of a primary system. If a failure occurs in the primary system, the backup system takes over.

Fault tolerance means that if a failure takes place, an *inactive* backup system is automatically activated and it performs the task of the failed system. In essence, fault tolerance is a bumpless, automatic switching process.

Various degrees of redundancy and fault tolerance are available in PC systems. Either one is important in continuous processes or critical applications where downtime is intolerable or safety is mandatory. Initial PC costs can easily double, depending upon the extent of fault tolerance or redundancy incorporated in the system. Because of the additional expense, these features

are usually used only in large and very large PC's.

Fault tolerance must not be confused with error detection and fault isolation (EDFI) systems available in most PC's. These systems provide internal self-diagnostic capabilities that protect the equipment against accidental damage and help maintenance personnel troubleshoot the equipment. Important error indicators are associated with the PC's power supply, CPU, and I/O modules.

When a PC shuts itself down as a result of an internal or external fault, diagnostic indicators in the hardware identify the problem area. Moreover, PC's can assist maintenance personnel to pinpoint the cause of the failure. In some cases, the unit will even identify the failed component. Modular plug-in designs permit quick replacement of the defective part.

Maintenance—PC's require little, if any, maintenance. Efforts should be made to keep PC panels

free of dirt and dust. For example, excess accumulations of conductive metal filings can cause problems, Fig. 16. Specially designed cabinets may be necessary for PC's located in harsh environments.

Batteries used for backup power must be checked regularly and replaced when necessary. Many batteries have visual or audible alarms that become activated when replacement is necessary. Peripheral equipment must also be checked. Tape loader heads should be cleaned regularly. Memory disk drives require periodic lubrication. Manufacturers normally supply customers with manuals that discuss operating and maintenance procedures. Some users have sufficient inplant capability to repair PC boards, but the majority simply return the failed ones to the supplier and ask for replacements.

Training—Practically all PC suppliers offer some degree of training to their customers. But the amount of training needed depends on the complexity of the PC. Very small PC's, for example, can usually be mastered within a few hours. The manual or booklet that accompanies such units normally contains all the information that the user requires to operate and maintain the equipment.

Training programs for larger PC's may take from 3 days to 3 wk. Many vendors offer formal training courses in well-equipped training centers, or will conduct courses on-site in the user's plant. These courses cover engineering basics, PC installation and use, and troubleshooting. Usually, there is a nominal charge for such training, which may be included in the sales contract or service agreement. PC purchasers are urged to take advantage of manufacturer's training programs. There is a big difference between learning how to turn a PC on, program it, and make it do something, and knowing how to get the maximum benefit from it.

Security and Safety—Most PC's have built-in security features to prevent unauthorized persons from tampering or gaining access to them. A lock and key arrangement is the minimum security level that should be accepted. However, electronic access codes have become very popular. These electronic locks require that specific data be entered into the programmer to gain access to it. Various code levels are used depending upon the degree of program control penetration desired. The access codes are assigned to authorized personnel on a need-to-know basis.

Standardization Efforts—Some progress has been reported in developing national and international standards for PC's. But progress has been slow. Most of the work has been limited to local network standards. The lack of overall standards presents a significant obstacle in achieving necessary compatibility between PC's made by different manufacturers.

Pressure from users for manufacturers and associations to continue their efforts to develop standards is increasing. However, experts predict that the earliest date for the introduction of practical PC standards is the mid 1990s. Scan times will be faster too. The ability to perform multiple scans will be common.

Special thanks and recognition is given to the following individuals for contributing valuable information for this article: Jodie Glore, Systems Div., Allen-Bradley, Highland Heights, OH; Rod Khanna, Modicon Div., Gould Inc., Andover, MA; M. McDonnell, Industrial Systems Div., Texas Instruments, Inc., Johnson City, TN; and Angelo Vinch, Numa-Logic Dept., Westinghouse Electric Corp., Madison Heights, MI.

PLANT ENGINEERING magazine acknowledges with appreciation the programmable controller suppliers listed here who provided information for this article.

ASEA Industrial Systems, Inc. (A,B,C,D)
P.O. Box 732
Milwaukee, WI 53201

Action Instruments Co., Inc. (B,C)
8601 Aero Dr.
San Diego, CA 92123

Adatek, Inc. (B,C)
Box 1339
Sandpoint, ID 83864

Allen Bradley (A,B,C,D,E)
Systems Div.
747 Alpha Dr.
Highland Heights, OH 44143

The Anderson Cornelius Co. (A,B,C)
6750 Shady Oak Rd.
Eden Prairie, MN 55344

August Systems, Inc. (D,E)
18277 S.W. Boones Ferry Rd.
Tigard, OR 97223

Automation Systems Inc. (A,B)
208 N. 12th Ave.
Eldridge, IA 52748

Barber-Colman Co. (A,B,C)
1354 Clifford Ave.
Loves Park, IL 61132

Comptrol, Inc. (B,C)
9505 Midwest Ave.
Cleveland, OH 44125

Compucon Corp. (A,B)
9307 Science Center Dr.
New Hope, MN 55428

Control Technology Corp. (A)
82 Turnpike Rd.
Westboro, MA 01581

Conrac Corp. (A,B,C,D)
Cramer Div.
Mill Rock Rd.
Old Saybrook, CT 06475

Datricon Corp. (A,B)
155 B Ave.
Lake Oswego, OR 97034

Divelbiss Corp. (A,B)
97776 Mt. Gilead Rd.
Fredericktown, OH 43019

EMICC (B,C)
2871 Avondale Mill Rd.
Macon, GA 31206

Eagle Signal Controls (A,B,C,D)
736 Federal St.
Davenport, IA 52803

Eaton Corp. (A,B,C,D,E)
Cutler-Hammer Products
4201 N. 27th St.
Milwaukee, WI 53216

Encoder Products Co. (C,D)
1601B Dover Hwy.
P.O. Box 1548
Sandpoint, ID 83864-0868

Entertron Industries Inc. (A,B,C)
3857 Orangeport Rd.
Gasport, NY 14067

The Foxboro Co. (C,D)
Bristol Park
Foxboro, MA 02035

General Electric Co. (A,B,C,D,E)
Programmable-Control Dept.
Box 8106
Charlottesville, VA 22906

Giddings & Lewis Electronics (B,C)
666 S. Military Rd.
Fond du Lac, WI 54935

Gould Inc. (A,B,C,D,E)
Modicon PC Div.
P.O. Box 3083
Andover, MA 01810

GTE Products Corp. (C,D,E)
Electronic Control Operation
100 First Ave.
Waltham, MA 02154

Holmco Inc. (A,B,C,D)
169 Route 206
Flanders, NJ 07836

IDEC Systems & Controls Corp. (A)
3050 Tasman Dr.
Santa Clara, CA 95050

Inconix Corp. (A,B,C,D)
10 Tech Circle
Natick, MA 01760

Industrial Solid State Controls, Inc. (C,D)
435 W. Philadelphia St.
Box 934
York, PA 17405

International Products & Technologies, Inc. (A)
541 Davisville Rd.
Willow Grove, PA 19090

Kaye Instruments Inc. (A,B)
15 De Angelo Dr.
Bedford, MA 01730

Klockner-Moeller Corp. (C,D)
4 Strathmore Rd.
Natick, MA 01760

Kuhnke Controls (B,C)
26 W. Highland Ave.
Atlantic Highlands, NJ 07716

MTS Systems Corp. (B,C)
Machine Controls Div.
Box 24012
Minneapolis, MN 55424

McGill Manufacturing Co., Inc. (A,B,C)
1002 N. Campbell St.
Valparaiso, IN 46383

Minarik Electric Co. (A,B)
Box 54210
Terminal Annex
Los Angeles, CA 90054

Mitsubishi Electric Sales America (A,B)
3030 E. Victoria St.
Rancho Dominguez, CA 90221

Omron Electronics, Inc. (A,B,C)
Control Components Div.
650 Woodfield Dr.
Schaumburg, IL 60195

Phoenix Digital Equipment Corp. (B,C)
2315 N. 35th Ave.
Phoenix, AZ 85009

Reliance Electric (A,B,C,D)
Marketing Info Center
P.O. Box 99457
Cleveland, OH 44199

Siemens-Allis, Inc. (C,D)
P.O. Box 89
Wichita Falls, TX 76307

Solid Controls (A,B)
6925 Washington Ave., S.
Minneapolis, MN 55435

Square D Co. (A,B,C,D)
4041 N. Richards
Box 472
Milwaukee, WI 53201

Struthers-Dunn, Inc. (B,C)
Box 1327
Bettendorf, IA 52722

Tenor Co., Inc. (A,B,C)
17020 W. Rogers Dr.
New Berlin, WI 53151

Texas Instruments Inc. (A,B,C)
P.O. Box 402370
Dept. ILAE43P2
Dallas, TX 75240

Trius Corp. (A,B)
2904 Corvin Dr.
Santa Clara, CA 95051

Unico, Inc. (A,B)
3725 Nicholson Rd.
Franksville, WI 53126

Veeder-Root Co. (A)
70 Sargent St.
Hartford, CT 06102

Westinghouse Electric Corp. (A,B,C,D)
Industry Electronics Div.
Numa-Logic Dept.
1521 Avis
Madison Heights, MI 48071

Reference code:
A. Very small PCs
B. Small PCs
C. Medium PCs
D. Large PCs
E. Very large PCs

Glossary of PC Terms

A compilation of some of the more common hardware and software terms that apply to PC applications follows:

Acquisition—a function that obtains information from PC memory locations or data files for use in data manipulation or data handling.

Adapter module—a printed circuit card that provides communications between an I/O rack and the processor. It transmits I/O rack input terminal status to, and receives output data from, the processor.

Address—a numeric value used to identify a specific I/O channel and/or module.

Address index pin—a technique used to establish proper identification of I/O modules.

Analog I/O module—a module (input, for example) that receives an analog signal from a user device. An analog signal is one that is continuously varying, such as a voltage or frequency. The module performs an analog to digital conversion and provides the digital result to the programmable controller. An analog output module converts the digital output from the PC and an analog signal for the user device.

AND (Logical)—a mathematical operation between two bits. The result of the logical AND will be a one (ON) bit only if both bits (one from each group) are one bits; otherwise, the result will be a zero (OFF) bit. (This operation can be performed between groups of bits with each pair of bits—one from each group—examined by their relative location within each group.)

Arithmetic function—a type of logic used to add, subtract, multiply, or divide two numeric values. The status of the output is governed by the result of the arithmetic operation (additional overflow, comparisons, and illegal division.)

ASCII—a seven-bit digital coding of standard alphanumeric characters as established by the American National Standards Institute. ASCII stands for the American Standard Code for Information Interchange.

Baud—a unit of data transmission speed equal to the number of code elements (bits) per second.

Binary-coded decimal (BCD)—a system of numbers representing decimal digits (0 to 9) using four binary digits (1 or 0). BCD is a recognized industrial standard; BCD input (for example, thumbwheels) and output (for example, numerical displays) are readily available.

Binary—a numeric system wherein values are represented only by numerals 1 and 0 (ON/OFF). Also called "base 2". This system is commonly employed in electronic hardware since circuits can be economically designed for ON/OFF status.

Bit—contraction of binary digit. A single number whose value can be either a one or a zero. The smallest division of a PC word.

Boolean algebra—shorthand notation for expressing logic functions.

Boolean equation—expression of relations between logic functions.

Bus—an electrical channel used to send or receive data.

Byte—a string of binary digits operated upon as a unit. Unless otherwise specified, a byte normally contains eight bits.

Channel—a group of I/O modules that are separately connected to the mainframe. For example, a channel of I/O can contain up to 128 input points and 128 output points.

Chassis—a housing that contains subassemblies. As a PC unit, it becomes a rack whenever modules are installed.

Checksum—One of several types of error detection techniques for ensuring the security of data during storage in memory or during serial transmission. The data is generally summed up as bytes or words (with or without wraparound and carries) and the result is then stored for comparison at a later time.

Circuit card—a printed circuit board containing electronic components.

Clock—pulse generator that synchronizes the timing of various logic circuits and memory in the processor.

Complementary metal oxide semiconductor (CMOS)—an integrated circuit family that has high threshold logic and low power consumption, thus making it especially useful in remote applications where supplying power becomes expensive.

Code—a system of symbols (bits) for representing data (characters).

Coil—a discrete logical conclusion to a series of logical operations performed by the PC. The results can be output to the real world via an output module to activate motor starters, solenoids, relays or pilot lamps. Coils are turned OFF when power is removed from the mainframe. (See *Latch*.)

Communication network—a serial link that provides communication among multiple stations which may be separate PC's, computers, or data terminals. It eliminates the need for separate, independently wired data links. Whether communicating or not, all stations can function independently.

Counter—a type of logic that is used to simulate the operation of external counters. In relay panel hardware, an electromechanical device that can be wired and preset to control other devices according to the total cycle of one ON or OFF function. In a PC, a counter is internal to the processor, which is to say it is an electronic function controlled by a user programmed instruction.

CPU (Central Processing Unit)—the brain of the controller system, wherein the program logic and the system executive is stored. All logic solving and decision making is performed by the processor.

Crosstalk—the electrical noise that occurs as the result of signals in one circuit causing interference with the signals in an adjacent circuit.

Cursor—visual movable pointer used on a CRT or programming panel by the programmer to indicate where an instruction is to be added to the ladder diagram. The cursor is also used for editing functions.

Cycle—(1) A sequence of operations that is repeated regularly. (2) The time it takes for one such sequence to occur.

Decay time—time required for the trailing edge of a pulse to reach a stated fraction of its initial value.

Delimiter—a special ASCII character that terminates or ends an ASCII communication: normally a carriage return.

Digital—having discrete states. Digital logic can have up to 16 states, but most digital logic is binary logic, with two states (ON or OFF).

Disruptive discharge—the sudden and large increase in current through insulation due to the complete failure of insulation exposed to intense electrostatic stress.

Distributed system—any combination of PC's, computers, and data terminals intercommunicating by means of a communication network.

Double precision function—the technique of storing a single numerical value in two consecutive registers. Since each register can store up to four digits (maximum value 9,999), double-precision allows magnitudes of up to 99,999,999 to be stored.

Documentation—an orderly collection of recorded hardware and software data such as tables, listing, diagrams, etc., to provide reference information for PC application operation and maintenance.

Dump—recording the entire contents of user memory onto a storage medium (e.g., magnetic tape, floppy disc, etc.—.

Duplex—a means of two-way data communication.

Edit—to deliberately modify the user program.

Encoder—(1) a rotary feedback device that transmits a specific code for each position. (2) A device that transmits a fixed amount of pulses for each revolution.

Glossary continued on following page

Exclusive OR (XOR)—a mathematical operation between two bits. The result of the exclusive OR will be a one (ON) bit only if either bit (one from each group) is a one bit. Only if they are both zeros (OFF) or both ones (ON) will the result be a zero. (The exclusive OR is similar to the inclusive OR except for the case where both bits are ones.)

Fault current—the current in a circuit that results from loss of insulation between conductors or between a conductor and ground.

Feedback—the signal or data sent to the PC from a controlled machine or process to denote its response to the command signal.

Force—the function that can be used to change the state of a disabled reference. The reference can be changed from OFF to ON. This allows the user to energize any input or output by means of the program panel independent of the PC program. The reference can also be changed from ON to OFF thereby allowing the user to deenergize any input or output.

Frequency—the number of times a given event occurs within a specified period. It most commonly refers to the number of pulses per second occurring in various electronic devices. The standard unit of measure is Hz (hertz), for cycles per second.

Hexadecimal—the numbering system that represents all possible ON/OFF combinations of four bits with sixteen unique digits (0 to 9 then A to F).

Host computer—a computer that monitors and controls other computers and peripheral devices.

Image table—a table in PC memory that contains the status of all inputs, coils, and registers.

Inclusive OR—a mathematical operation between two bits. The result of the inclusive OR will be a one (ON) bit if either bit (one from each group) is a one bit or both bits are ones; only if both bits are zeros (OFF) will the result be a zero. (This operation can be performed between groups of bits with each pair of bits—one from each group—examined by their relative location within each group.)

Input—a signal that provides information to the controller; can be either discrete input (pushbutton, relay contacts, limit switches, etc.) or numeric input (thumbwheel, external solid-state device, etc.).

Input module—a device that is used to connect the PC with the input devices. The input module contains the circuiting required to convert the incoming voltages to signal levels compatible with processor.

Instruction—a command or order that will cause a PC to perform one certain prescribed operation. The user enters a combination of instructions into PC memory to form a unique application program.

I/O scan time—the time required for the PC processor to monitor all inputs and control all outputs. The I/O scan repeats continuously.

Language—a set of symbols and rules for representing and communicating information (data) among people, or between people and machines.

Latch—the type of coil that is retentive upon power failure. Can be used similar to a latching relay. Normally, coils are reset to OFF conditions upon powerup; those coils selected by the user as latched (L) will not be altered and thus retain their previous condition (ON/OFF).

Logic—a means of solving complex problems through the repeated use of simple functions that define basic concepts. Three basic logic functions are AND, OR, and NOT.

Logic diagram—a drawing that represents the logic functions AND, OR, NOT, etc.

Logic element—any one of the elements that can be used in a ladder logic diagram. The elements include relays, coils, shunts, timers, counters, and arithmetic functions.

Logic line—a line of user logic used to construct the unique logic for the application.

Matric function—matrices are defined as sequential registers, each as 16 bits, up to a maximum of 99 registers (1,584 bits). A group of consecutive registers referred to by logic, such that individual bits can be utilized in lieu of numerical values. Bit operations that can be performed include: AND, OR (inclusive), COMPARE, CLEAR, SET, SENSE, COMPLEMENT, OR (exclusive), ROTATE LEFT, and ROTATE RIGHT.

Memory protect—the hardware capability to prevent a portion of the memory from being altered by an external device. This hardware feature is under keylock control.

Modem—the term modem is a contraction from MOdulator/DEModulator. It is so called because it converts the digital signals coming from this process to analog signals suitable for transmission over a telephone line and vice-versa.

Multiplexing—the time-shared scanning of a number of data lines into a single channel. Only one data line is enabled at any instant.

Node—the smallest possible programming increment in a ladder logic diagram. (Most logic elements require only one node, others require two or more nodes.)

Noise—extraneous signals or any disturbance that causes interference with the desired signal or operation.

One-shot—a discrete reference, typically a logic coil, that is energized (valid) for exactly one scan of the controller's logic.

Optical coupler—a device that couples input and output using a light source and detector in the same package. It is used to provide electrical isolation between input circuitry and output circuitry.

Output—a signal provided from the controller to the "real world"; can be either discrete output (solenoid, relay, motor starter, indicator lamp, etc.) or numerical output (for example, display of values stored within the controller).

Output module—a device that is used to connect the PC with the user's devices. The output module contains the circuitry required to convert the PC output signals to voltage levels compatible with the user's device.

Parity—method of verifying the accuracy of recorded data.

Parity bit—an additional bit added to a memory word to make the sum of the number of 1's in a word always even parity or odd parity.

Parity check—a check that tests whether the number of 1's in an array of binary digits is in a word odd or even.

Proportional, integral, derivative (PID) module—an optional module that provides automatic closed-loop operation of multiple continuous process control loops. For each loop, this module can perform any or all of the following control actions: proportional control causes the output signal to change as a direct ratio of input signal variation; integral control causes the output signal to change according to the summation of input signal values sampled up to the present time; and derivative control causes the output signal to change according to the rate at which input signal variations occur during a certain time interval.

Port—an I/O connection on a processor or peripheral device.

Preset—the upper limit specified for a counter or timer function. When the specified (preset) value is reached, an output will be energized indicating the status of a counter or timer.

Protocol—a defined means of establishing criteria for receiving and transmitting data through communication channels.

Read—to sense the presence of information in some type of storage, that includes RAM memory, magnetic tape, punched tape, etc.

Reference numbers—numbers that identify the elements of the relay ladder logic. References can be either discrete (logic coils, inputs, or sequencer steps) or register (input or holding).

Register—a location within the controller allocated to the storage of numerical values. All holding registers are retentive on power failure. There are three types of registers: input whose contents are controlled by the "real world" outside the controller; holding registers whose contents are controlled from within the controller; and output registers, which are special holding registers because their contents can also be provided to the "real world."

Register module—a device used to select, convert, and condition binary coded decimal (BCD) signals that pass between a user's device being controlled and the PC.

Relay—an electromagnetic device operated by a variation in conditions of an electric circuit. When so operated, it operates other devices such as switches.

Relay element—a logic symbol used to simulate the effect of a relay. Contacts can be normally open, normally closed, or transitional.

RS-232-C—Electronic Institute of America (EIA) standard for data communications, RS-232 type C. Data is provided at various rates, eight data bits per character.

Scan—the technique of examining or solving logic networks one at a time in their numeric order. After the last logic network is solved, the next scan begins at network one; logic is always solved in this fixed cyclic process.

Self-diagnostic—the hardware and firmware within a controller that allows it to continuously monitor its own status and indicate any fault that may occur within it.

Shield—any barrier to the passage of interference-causing electrostatic or electromagnetic fields. An electrostatic shield is formed by a conductive layer surrounding a cable core. An electromagnetic shield is a ferrous metal cabinet or wireway.

Skip function—this function allows a group of consecutive networks to be skipped or omitted in the scanned logic solution. The status (ON/OFF) of all coils and register content controlled by these networks are not altered when they are skipped.

Strip printer—a peripheral device used in conjunction with a PC to provide a hardcopy of process numbers, status and functions.

Synchronous—data is transmitted continuously against a time base that is shared by transmitting and receiving terminals. If no legitimate data are available to be sent at a given time, synch or idle characters are sent to keep the transmitter and receiver in time synchronization.

Table—a group of consecutive registers used to store numerical values; maximum table size is 99 registers.

Traffic cop—a portion of the PC executive that controls how input and output data is interpreted relative to its channel number and address index position.

Transistor/transistor logic (TTL)—a family of integrated circuit logic. (Usually 5v is high or "1" and 0v is low or "0"; 5 v = 1, 0 v = 0.)

Transmitted data (TD)—The data line over which data is transmitted. (Pin 2 of an RS-232-C connector.)

Twisted pair—two insulated wires (signal and return) that are twisted around each other. Since both wires have nearly equal exposure to any electrostatic or electromagnetic interference, the differential noise is slight.

Unit of load—the internal dc current required to drive an I/O module.

Word storage—an unused data table word that may be used to contain numerical information without directly controlling any outputs. Any storage word may be monitored as often as necessary by the user program.

Write—process of loading information into memory.

(This list of terms was adapted from information supplied by Allen-Bradley Co. and Modicon Programmable Control Div., Gould, Inc.)

Programmable controller applications roundup

A look at some of the many ways in which PCs are being used today.

John Hall
Senior Technical Editor

Since the middle 1960's, programmable controllers (PCs) have grown from being simple relay replacement equipment to modern control systems that rival a minicomputer in power and sophistication. At the same time, the number of ways in which PCs are being put to work in manufacturing and process operations is also growing.

This article highlights some of the latest applications we have come across. Hopefully, it will suggest ways in which you can use PCs in your factory or plant to improve control and increase productivity.

CASTING STEEL

A new six strand continuous caster at U. S. Steel Corporation's Lorain, OH plant produces semi-finished, high quality steel rounds (between 6 and 10 inches in diameter) for a seamless pipe mill. The caster has a capacity of more than 550,000 tons/year and uses approximately 50% less energy than the ingot pour method.

The casting machine is controlled by an Allen-Bradley PLC-3 PC (Fig. 1) with a redundant hot back-up processor. Each processor has an expander chassis to accommodate 32K memory modules, S4A scanning modules, S4B report generation modules, and KA Data Highway modules. Automatic transfer to the back-up processor to retain continuous control occurs in the event of a self-diagnosed system malfunction. The program to control 8000 I/O circuits contains 4175 rungs of ladder diagrams, and uses XL2 and XL3

software documentation packages.

The caster can handle up to 10 heats/day. Ladles containing 220 tons of steel are shuttled from the melt shop and placed into a two-arm ladle rotator. The ladles are rotated into position over the upper caster area and the tundish car is filled with molten steel. After filling, the six strands of steel are started and the PLC-3 sequences the proper machinery and coordinates with an L&N MAC I system to activate cooling/spray valves.

Maintaining the proper level in the mold is critical for quality control. Using what is perhaps the world's most sophisticated mold level control system, radioactive source and sensors provide mold level inputs to PID intelligent I/O modules. Each two loop PID module monitors two strand levels and provides signals to three Mini PLC-2/15 PCs. These control servo valves which hydraulically activate and regulate pouring rates from the tundish to the six molds to maintain the level at one of two set points.

The motor drives for each of the two tundish cars, the six mold oscillator motors, the six straightening roll motors, and starter bar extractor motors are all controlled by variable speed dc drives digitally coupled to the PLC-3 control system using analog I/O cards.

The six rounds drop down through the cooling molds and into a cooling spray chamber, and emerge through the straightening rollers in a red hot semi-solid state.

Optical encoders attached to the final straightening roll-

Fig. 1 (left): Red hot, continuously cast steel rounds emerge from the caster and move through three rows of straightening rolls. The variable dc drives moving the rolls are under PC supervision.

Fig. 2 (right): More than 100,000 bottles per week from a blow molder travel along this conveyor under PC control at Calgon's commerical division in St. Louis.

Fig. 3 (left): Wooden picture frame moldings move along a conveyor at Ivy Industries, Charlottesville, VA. PC I/O racks are at the left.

Fig. 4 (right): One of Enterprise Products' pumping stations in Texas. Unattended PC is in instrument house at left.

Fig. 5 (far right): Filament winding follows a complex shaped mandrel under PC control at Fiber Science, Salt Lake City.

er motor shaft provide precise speed and accumulated length input to the PLC-3 through encoder counter cards. This data is used to tell the flame cutters when to cut off the piece. Length data is manually entered via thumbwheel switch settings. Sample lengths, when needed, are automatically added to the roll.

After the torch cut, the finished piece is sequenced to automatic ID stampers, which are controlled by PLC-2/30 PCs. A unique ID number is stamped on the end of the piece and read back for verification.

OLD INTO NEW

A more than 20-year-old plastic extrusion blow molder was saved from the scrap heap at Calgon's Commercial Division plant in St. Louis, MO by the use of PCs to handle the control function. Saving the machine eliminated a large capital expense and, at the same time, increased productivity.

The machine produces upwards of 100,000 bottles a week, which are filled with commercial cleaning products. Within weeks after the retrofit, downtime due to maintenance was cut from 20% to virtually nothing.

The blow molder was originally controlled with electromechanical timers, relays, and temperature controls. Because of the machine's age, control parts were impossible to obtain. Before scrapping the machine, however, Calgon decided to talk to a local industrial controls company, the Richard Green Co., who recommended an Eagle Signal PC control system (Fig.2).

The PCs handle several processes. The heater temperature is controlled by an EPTAK 240 PC with 14 analog loops; a second 240 PC controls cycling as well as water and oil temperature. Both PCs are expandable, and additional functions can be added as needed.

Safety interlocks are also very important in the system to eliminate all possibility of danger to the operators. And, damage to extraneous parts and functions of the blow molder was not acceptable. Both requirements were met in the PC software programs.

The retrofit was a success. Bottles produced were of predictable quality and passed inspection. The main benefit of the new system, according to Calgon, is that the machine works right all the time. In addition, the PCs troubleshoot themselves and maintenance mechanics aren't tied up for hours. Finally, they like the visible I/O modules that indicate which circuits to troubleshoot.

SPEEDING PACKAGE PRODUCTION

The R. A. Jones Co., Cincinnati, OH, manufactures custom packaging machinery that makes cartons for the food packaging industry. The cartons are used to package a wide variety of products, ranging from facial tissues to spaghetti.

Prior to adopting a new control system easily adaptable to any machine, Jones engineers designed a control system for each machine from the ground up—an arduous and time consuming task.

The control system chosen was a Giddings & Lewis Dialog central control system, which uses two PiC 409s with four memory banks of 30K each and a 150 I/O points. The 409s provide a wide range of motion control including eight axes of full proportional servos and full networking capability.

The control system stores and displays the necessary data for production changeovers. In addition, it can monitor machine speed, production, accumulated time, and other functions requested by the operator.

The cartoner production system is a constant motion, hand-loaded one using an existing mechanical drive system controlled by the 409s. This allows Jones engineers to create a control system for each custom cartoner by simply reprogramming the 409. The machines can handle up to 300 cartons/min and accommodate a wide variety of sizes and styles of cartons including four flap, sealed end, window, and internal dividers. With the 409s, Jones engineers estimate they save from 30 to 50% in engineering time in the creation of a custom control system for each customer

FRAMED

Ivy Industries, Inc., of Charlottesville, VA, manufactures high quality wood molding frames in what is considered to be the most automated factory of its kind in the world. Miles of wood picture frame moldings come off the production line each week for nationwide distribution.

A GE Series Six Model 60 PC performs a variety of functions in the plant and had an installed cost of less than a conventional relay system. The PC allows one operator control of a sophisticated conveying system with 450 linear ft of conveyors moving 100 racks of picture frame moldings around the plant (Fig. 3). The Model 60 has two I/O racks and approximately 125 inputs and 125 outputs. Only 2.5K of the 4K memory is being used.

Photoelectric cells along the conveyors send inputs to

the PC. This allows the operator to keep track of the racks and move them to various work areas where the frames are finished, rubbed, and polished manually. The photoelectric cells and the PC also prevent the racks from bumping each other and damaging the moldings.

Other PC controlled operations tied in with the conveyors include painting, storing, and curing. The PC regulates the temperature in the ovens where the frames are cured.

The Series Six PC is also being integrated into the plant energy management system. It will monitor temperatures outside and inside the plant, and control HVAC systems.

TRACKING A PIPELINE

Just outside of Houston, Enterprise Products runs a pipeline beneath almost 170 miles of rural area. Unattended pumping stations are located 25 and 100 miles from the control room. Operators supervise operations using a large Gould Electronics 584 PC at each of the stations (Fig. 4), and a Modvue programmable touch sensitive color monitor in the control room. This provides a window to the process so that an operator can monitor the pipeline in a control room near the plant. The monitor communicates with the PCs and remote communications are done through a modem connected to a Modbus communications port.

At the control center, a continuously updated trend generated by the PCs is displayed on the monitor screen that shows suction and discharge pressures at the pumping stations. Operators can turn a pump on or off by pressing its graphic on the screen. A confirm or cancel button prevents turning a pump on or off by accidentally hitting the screen.

The monitor also generates documentation of the process that is required by the Department of Transportation. Each hour, it polls the PCs at the pumping stations to get pump pressures. The PCs send this information to a printer in the control room. The control system's automatic documentation reduces tedious paperwork and reduces human errors. The monitor also shows status changes and alarm conditions by sounding an alarm and displaying the alarmed station. A flashing message on the screen alerts the operator to the alarm and the pumping station PC sends detailed information to the control room printer.

FILAMENT WINDING

The Fiber Science Div., EDO Corp., Salt Lake City, UT, uses an MTS INCOL/470 PC system to control filament winding for use in composite structures. PC controlled electrohydraulic stepping motors on the spindle and carriage axes, and an electric servo drive controlling the winding eye position replaced a complicated mechanical control system.

Unlike many spindle and carriage processes, where the workpiece is cylindrical, filament winding requires the winding eye to follow a curved pattern (Fig. 5) as a function of spindle position as it is traversing from one end of the vessel to the other. If the vessel is asymmetrical in shape, it requires a different winding motion at each end. In addition, each traverse of the winding eye must be offset a precise amount so that succeeding bands layup properly.

This was an extremely complex task with mechanical controls. With the PC system, however, it is relatively simple. An operator enters a precalculated sequence of carriage and eye positions through a keyboard as a function of the spindle's radial position. Or, he may set the program up directly on the production machinery. In addition, only one end of symmetrical parts need be programmed because the processor automatically generates the correct motion profile for the other end.

Not only is the product quality consistently high, but there has been a substantial reduction in the labor hours required. Although the final numbers are still to come, a 15% gain in speed, along with a 25-30% gain in output, is expected. In addition, maintenance and downtime expenses have dropped from more than $5000/month to nearly zero.

CUTTING POWER COSTS

Manan Manufacturing Co., Skokie, IL, builds Omron SYSMAC-S6 PCs into their line of plastic molding machines. The PC's 64 I/O points take input from photoelectric and proximity sensors and operate different motors on the machine. The standard power source for the PC is used to operate 9 to 15 solenoids that, in turn, drive air valves in the injection process. The use of low power consumption solenoids combined with the 2 A capacity of the power supply cut power consumption from close to $2.50./hr to 25¢/hr.

AUTOMATIC MOVIE MAKING

At the Federal Aviation Administration's (FAA) Civil Aeromedical Institute in Oklahoma City, OK, motion pictures are being made to show the results of testing aircraft seats and restraints. Lighting for the test and camera operation is controlled by a SY/MAX Model 300 PC from Square

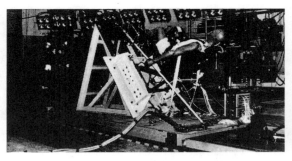

Fig. 6 (right): Typical test test configuration at FAA Civil Aeromedical Institute, Oklahoma City, OK. The PC controls batteries of lights and a high speed movie camera.

Fig. 7 (far right): PC controls aeration of cement powder in these 2300 ton capacity silos at Riverside Cement at Oral Grande, CA.

D Co. The PC, which uses four inputs and 36 outputs, replaced a drum sequencer whose operation was described as a nightmare by FAA engineers.

To simulate the effects of a plane crash, the seat and a number of instrumented dummies are mounted on a sled (Fig. 6), which is accelerated down a track and brought to a sudden stop by wires stretched across the track. The sled moves at speeds up to 65 ft/s and applies up to 60 G's of force on the seats. High speed movies provide researchers with a slow motion view of the dynamic impact test.

Filming requires automatic control for several reasons. First, very brilliant lighting is needed for the rapidly exposed film. More than 20 banks of lights are used, each of which draw 6000 W. To prevent electrical overloads, the PC activates the lighting contactors for a 5 s period immediately before the release of the sled.

The lights generate an enormous amount of heat and if left on would affect the sensors and melt the rubber flesh of the $15,000 dummies. The PC prevents this by shutting off the lights a few seconds after the sled crashes.

The high speed cameras devour film at the rate 500 or 1000 frames/s and run empty very quickly. The PC runs the cameras just long enough to record the impact. The exact camera start and stop times can be entered through a SY/MAX Loader/Monitor mounted in the PC enclosure. Accumulated camera run time is stored by the PC, allowing operators to gauge the amount of remaining film.

STERILIZING SUTURES

Ethicon Inc., Div. Johnson and Johnson, Sommerville, NJ, believes its PC controlled ethylene oxide (EO) sterilization systems are the most efficient anywhere. The sterilization process consists of several steps. First, the suture products are put into humidity chambers and held at a specified temperature for many hours to encourage micro-organism growth. The sutures, full of micro-organisms, are put into a sterilization chamber where EO is introduced. The EO atmosphere is held at a prescribed temperature until all of the micro-organisms are killed.

Next, the sutures are moved through a sterile room into a degassing chamber. This process removes the poisonous EO from the product. Packaging, resterilization, and another packaging step follow. Finally, the product winds up double packaged and sterile from the outer package in.

For control purposes, the company uses one Texas Instrument PM550 PC per EO sterilizer, one PM550 per humidifier, and one PM550 per degassing unit. The PCs replace batch process controls—chart recorders, loop controllers, relays, timers, and so on.

75% reduction in spoiled product. In addition, record keeping is vastly improved. Records must be kept on every product, both to satisfy the FDA as well as on the company's Quality Assurance review. The use of PCs eliminated record keeping by hand and the human error that can result from it.

An additional benefit was the elimination of the need to design a new sterilization process for each new product. Instead, the PCs are simply reprogrammed. And finally, maintenance time was reduced because the PCs have self-diagnostics.

MAKING CEMENT

The Riverside Cement Co., Oral Grande, CA, has automated their cement processing operation using Westinghouse Numa-Logic PCs with remote I/O. The system replaced one using electromechanical relays at silo sites. Even though these were hermetically sealed, they required weekly replacement due to dust build-up. The PC control system paid for itself in four months with fuel savings alone.

Cement making starts with crushed limestone or shale being ground into a powder that has the consistency of talc. The powder is stored in silos for blending. After blending, an elevator and screw conveying system air-pushes the powder along a conveyor belt to a kiln. The product is baked at 2700°F and turns into a clinker resembling a small, black ball bearing. The clinker is ground and the finished product is cement (Fig. 7).

The PCs control the process from the raw ball mill grind, to the silos, to the kiln. While in the silos, the product is continually aerated or fluffed, which keeps the fine powder at an even consistency. This helps it to burn uniformly in the kiln and saves fuel. An important part of the aeration process is sequencing, which maintains air pressure and silo volume within safe limits to prevent explosion. The math and compare functions of the PC perform this task. ∎

Programmable controllers in the factory

Today's PCs are ideal solutions to a host of factory control problems.

Lee Farrar
Technical Support Manager
Eagle Signal Controls
Davenport, IA 52803

Last month's installment discussed how Local Area Networks can link the various "Islands of Automation" in a factory. Many of these "Islands" are based on programmable controllers, because these powerful devices are well suited to the kinds of control applications found on the factory floor. This month's installment of our Automated Factory Series concentrates on programmable controllers, and shows how they can be used for factory control tasks.

• • • • •

Fig. 1: This hydraulic press machine bends and forms sheets of metal into plates for timers, switches, and other components. A programmable controller handles the machine's hydraulic and pneumatic functions. The PC was installed to replace a hard-wired relay panel.

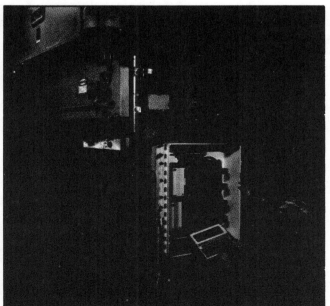

There is no doubt that programmable controllers (PC's) can perform all the necessary control functions for most, if not all, factory applications. And they can do so cost effectively. The only major question facing a user is how to best put together a PC-based control system. The answer to this question depends mostly on the needs of the specific application. In this article, we'll look at typical factory applications and PC capabilities, and show you how to match them up.

Typical applications

Examples of factory automation tasks and the types of control functions performed by the PC include:

• *Machine sequencing*—Almost any machine sequencing operation—for a single or multiple devices—can be handled by a PC using standard PC control functions, such as relay logic, analog controls, counting, totalizing, drum sequencing, math, and data manipulation (Fig. 1).

The specific functions used in a given application depend upon the exact machine requirements. For example, injection molding machines use analog and temperature controls, while drilling machines are more logic intensive.

In machine control, the PC sends outputs to operate the machine, provides status to operators via displays, and prints parts totals and status information. It may control a single machine or an entire group of machines (Fig. 2) in an area or department. Often, a PC is a big improvement over existing discrete controls because of its increased speed, lower power consumption, increased operator and management information reporting, and reduced maintenance requirements.

• *Energy-management*—PC's can now assist in optimizing energy consumption for a department, floor, or entire

235

Fig. 2 (above): Two EPTAK 700 PC's control 84 rubber molding presses as part of an injection molding process. The PC's control temperature, time curing cycles, and monitor production data.

Fig. 3 (right): Simulator modules for digital and analog inputs help debug hardware and software during development.

plant complex. The logic is based upon the load-shedding method, which prevents premium electric billing from the power company. The PC monitors power consumption and selectively sheds and restores loads. This reduces billing while maintaining critical loads needed for plant operations.

• *Plant event control*—A PC can also control the work events within a plant. Using a basic time-of-day clock function, the PC can control bell signals for work shifts, breaks, and meal times. In addition, it can be used to handle overall plant activities, such as security systems, HVAC, (heating, ventilating and air conditioning) and lighting controls. In most cases, it's best to assign the control of plant events to the PC doing energy management control.

• *Motor start/sequencing*—This application generally involves cycling the startup sequences of motors, solenoids, valves, etc. to equalize wear on all components over a period of time. The PC logic is usually written around a simple shift register that starts from a different interval for each start-up sequence.

• *Conveyors and material handling*—Conveyors and material handling applications may involve machine sequencing tasks, motion control, and sorting functions. Sorting may involve categorizing by color, size, model, customer order, product code, or other criteria to differentiate between parts on the conveyor. A single conveyor or material handling line may be responsible for parts being sent to twenty or more locations.

Control inputs here include sensors to help identify individual parts or material, logic to detect backup or blockage, and alarms or status indicators. Data shift registers can control blockage or backup annunciation. In addition, the PC needs some way to input parts orders—so it knows what part or material is to go to which location—and needs a large memory to store part locations. In some cases, these functions are performed by an external computer, and the locations are provided to the PC via an interface.

• *Robotics*—Industrial robots can perform functions that previously were done by human operators. These include loading and unloading operations, moving parts from one location to another, welding, painting, assembling, and many other tasks. PC's are being widely used for robot control, in both new robots and in retrofits (see "Industrial robots: Getting smarter all the time, "*I&CS*, July 1982). A standard PC can easily handle the sequencing requirements of point-to-point and pick-and-place robots. For complex velocity and acceleration control, PC's with special motion controller modules are now available.

• *Waste treatment and material reclamation*—Many plant operations generate by-products that must be eliminated or reclaimed. A PC can provide the logic and computing power needed to sample, test, evaluate, and start up the necessary processing equipment. For example, the PC can control the periodic shaking of dust filters; order a chemical analyzer to take a sample and analyze it; and, finally, print out the results. Also, the PC can compare the results against limits to determine if other actions are needed, such as calling in maintenance people.

Know your application

Once the decision has been made to use a PC for your factory task, you must select the right one for the job. The universal key to this is *knowledge*—of the application, and the available controllers and their manufacturers.

First, overview your application in terms of functions. Most similar applications have the same general characteristics, but details may cause variations in the final solution. If the application is presently being controlled by discrete components, the PC may be able to duplicate component functions.

For example, a parts washing application using electromechanical step switches will probably use a PC shift register as its basic control scheme. The shift register can be considered for any application which is basically sequential in nature.

Fig. 4 (above): Industrial PC's can be mounted without enclosures, as shown here. In most cases, if human operators can survive the plant's environment, so can a PC.

Fig. 5 (right): To prevent unauthorized tampering, the PC can be hidden away in a cabinet. In this console for an automated test facility at a refrigerator/freezer plant, the PC is mounted in the cabinet at lower right (see extreme right).

Your application overview will give general system requirements and allow you to decide if a certain PC is suited to your job. In the overview, you must determine if any of the following PC functions are needed:

- Timers (single or repeat cycle?),
- Counters,
- Time or count totalizers,
- Analog or temperature controls,
- Data handling,
- Math functions,
- Data logging,
- Computer interface.

The next step is to determine the "numbers" for I/O and memory. If your system requires 16 inputs and 20 outputs, you don't need a PC that handles 4028 I/O. Conversely, your PC should have enough I/O capacity plus a reasonable allowance for possible expansion or future modifications. This concept should be applied to all functions. Try to keep the numbers relatively close to your requirements, and don't pay for a machine drastically larger than you can reasonably expect to use.

Next, make sure that the PC's digital and analog I/O modules are built and rated for the current or voltage in your application. Inputs usually present few problems, but machine control outputs deserve special consideration.

For load output requirements, compute the current draw for each of the loads and compare it to the output ratings of the PC's output drivers. Remember to consult the driver's derating curve if the output modules are to operate above room temperature. This is necessary because solid state triacs, as well as all power solid state devices, must carry less current at high temperatures to avoid excessive internal heating.

Also, be aware that triacs have a minimum current requirement. If low current devices, such as neon bulbs, are being used, you may have to put an additional resistor in parallel with the load to cause the minimum current flow.

If you need mechanical, isolated contact outputs, con-sider using relay modules specifically designed to be used with the PC's I/O system. Relay contacts can be used for external loads requiring up to 10 A current capacity, for isolating outputs with true open contacts, or for interfacing to inputs requiring extremely high, "off" input impedance.

Use internal control relays (in software) for logic—save your I/O for driving external devices. Consider if control relays, timers, counters, shift registers, etc. are to reset on power failure or remain in their present state.

Memory requirements

The selected PC must have enough program memory capacity for your user software. This can be estimated in advance either from information supplied in vendor manuals, or though consultation with the manufacturer's application engineers. Advertised memory size figures may be somewhat misleading. Make absolutely sure you understand the capacities and their exact meaning.

Memories are available in various types such as E²PROM, RAM, and UVPROM. Each has certain advantages and disadvantages:

- *Electrically Erasable Programmable Read-Only Memory (E²PROM)*—The latest memory technology, it is slower to program and edit, but does not require batteries. E²PROM will store programs without power for years.
- *Random Access Memory (RAM)*—Fast but volatile, a RAM requires battery backup to hold memory during power failure. These batteries must periodically be replaced. For this replacement to be achieved, the machine may need to be shut down during a maintenance period. Some PC's overcome this disadvantage by providing two battery terminals so that a fresh battery can be installed before the old one is removed.
- *Ultra-Violet Programmable Read-Only-Memory (UVPROM)*—This type of memory is nonvolatile, and requires no backup battery. Selective editing or deletion of UVPROM locations is difficult or impossible, and requires additional software. A separate ultra-violet lamp is re-

quired for program erasure. Total erasure under the lamp takes a few minutes to achieve.

In all cases, make sure you know the memory technology being used and consider the impact of each type on your particular application.

Communication needs

Your PC probably will have to communicate with one or more external devices. These can include standard peripheral devices, such as printers, CRT's, chart recorders, or hand-held programmers. They can also be devices such as data loggers, panel displays, annunciators, computers, "smart front ends," remote I/O multiplexers, personal computers, or analog controllers. Make sure the PC you have in mind has the proper interface and communications protocol to handle the devices your application requires.

Various data links and communications protocols are being used in factories these days. The RS-232C serial link with ASCII data is the most popular, and can be used for distances up to about 50 ft. A similar method, RS-422, can go up to 1000 ft.

Because the RS-232C/RS-422 system is the most popular, many vendors advertise that they support the interface. However, make sure that all devices *really are* compatible. There are many stories in the industry about engineers who spent hours trying to make two RS-232C-compatible devices talk to each other. The vendor should demonstrate or guarantee that the interface between the devices actually works.

Several PC manufacturers offer data highways that connect their own PC's with each other, and with computers or other intelligent devices. Be warned that almost none of the PC data highways are compatible with each other. It is possible, however, to purchase interface devices that connect one PC highway to another.

Software development

After selecting a PC, one of the first steps in its installation is software development. If you are not familiar with your PC's programming methods, write short programs to perform various functions such as timing, counting, shift register manipulators, ladder logic, mathematics, and so on. Make sure you understand what your PC is capable of, and how to program it, before you try to approach the entire programming task.

Write your program in sections, if possible, and check out each program section on the PC as you go. Using a duplicate system or a development PC that can be set up in your office or lab is a good way to check out software. In some cases, the development system can be used to simulate the expected machine conditions during on-line debugging.

Setting up some kind of simulation is a good way to approach checkout. If you can't use the development system, build a simple switch panel for on-off inputs or a set of pots to simulate analog inputs. Some PC vendors supply simulator panels or modules as accessories or as part of the PC I/O (Fig. 3).

If you are replacing or duplicating existing discrete re-lay logic, make sure you have the *latest* documentation and schematics available. Otherwise, you may wind up duplicating a relay control system that existed ten years ago, but has been changed several times since. Also, watch out for relay logic that contains "race circuits." These circuits work only because of the relatively slow response of certain relays, and will not work in a PC.

Don't be afraid to ask questions of your PC vendor. If you can't figure out the proper approach to take regarding any aspect of the programming or installation process, stop and resolve the problem before proceeding.

Secure installations

Give careful consideration to how you will mount the PC at your applications site. If mounted in an enclosure, check the types of magnetic, power, and control components that may also be mounted in the box. Make sure the close mounting of such devices will not interfere with the PC.

If the enclosure is to be mounted near the machine it will be controlling, you may be able to put machine controls and indicators—such as selector switches, alarms, annunciators, display readouts, etc.—on the enclosure door. This simplifies much of the wiring.

In all cases, consult the user's manual or vendor documentation concerning mounting, grounding, and wiring practices. Although PC's can withstand most industrial noise conditions, you may have unanticipated or unusual conditions at your site. If all wiring precautions are followed and problems still occur, consult your vendor's field service or application engineer for assistance.

In some factory applications, the PC can be left exposed (Fig. 4). This is because factories rarely have environmental or corrosive problems, and the rugged housing supplied with most industrial PC's often provides adequate protection. Unfortunately, PC's left exposed may be subject to tampering. For various reasons, ranging from idle mischief to malicious vandalism, some people may be tempted to alter the PC's control settings. And sometimes, tampering goes further than just altering a few switch settings.

Consider the risk factors that may exist in your plant. Then develop a plan for securing your system by using passwords, key locks, sealed enclosures, concrete posts, etc. A good guideline for security is: Out of sight, out of mind (Fig. 5).

Future trends

As a product class, PC's are about 15 years old. During that time, they have greatly impacted the direction and breadth of automated industrial control. As we look to the future I believe we'll see the vast majority of factory control applications using PC-type products. The discrete products, as we know them today, will be reduced to very minor roles in these automated systems.

However, for this scenario to develop, the PC must continue to evolve and prove its effectiveness in terms of function, cost, size, and capability. And, as this is accomplished, plant personnel, design engineers, and the industrial community in general will become more complacent with this technology. Consequently, PC's will be increasingly taken for granted. And rightly so. ■

Computerized Heat Treating

SECTION VII
Computerized Heat Treating

The development of the microprocessor technology made it inevitable that conventional heat processing control systems take advantage of digital instrumentation to obtain more accurate and functional furnace control.

The rising cost of energy, coupled with tighter requirements for quality control in the heat treatment of metals, has paralleled the development of the microprocessor. It has thus been economically feasible to choose digital instrumentation over the lower initial cost of conventional instrumentation for the monitoring and control of furnace process variables.

A digital, or "computerized," control system provides:

1. *Multiple function capability.* This means that even single-loop controllers can incorporate self-checking circuitry, automatic tuning, and other functions.
2. *Higher speed of measurement and response* to the change in process variables during furnace operation. Long an advantage of electronic controls, speeds are enhanced even more by the microprocessor.
3. *Higher precision of measurement.* For example, accuracies of ±1 degree over ranges of 0 to 2400 °F are common.
4. *Greater control flexibility.* Computer memories can hold hundreds of different control "algorithms."
5. *More options* for record-keeping, or data-handling, and readout for reporting and monitoring purposes.
6. *The ability to communicate* electronically between controllers and computers in hierarchical information management systems.

Various applications of computer technology as a means of increasing furnace operating efficiency are illustrated by the articles in this section. On a time basis, these range from pioneering efforts in the mid-1970's, as documented in "Computerized Systems for Heat Treating," which records the concept of "total furnace management," to some of the latest approaches

New interfaces, born of the computer age, are appearing in the familiar confines of the conventional heat treating shop. Mounted in the panel at above left is a new microprocessor-based device controlling the temperature of the furnace at the right in accordance with a time-vs.-temperature program that was entered into the instrument's digital memory by the operator, using the pushbuttons. (Honeywell).

as reported in "Microprocessor Control in Metallurgical Furnaces."

In addition to increased production and quality standards, many scheduling problems become more manageable with computerized methods. Greater flexibility in the use of equipment to meet production requirements is realized, as well as the ability to schedule furnace operations effectively so as to capitalize on plant services and utilities. Improvements in furnace loading and unloading are easily matched to manpower availability.

Computerized monitoring also leads to more accurate and efficient reporting of problems and irregular conditions that may occur in normal heat treating practice. Timely and effective reporting of alarm conditions increases energy savings and decreases downtime.

Fig. 1 — Copy of shop job card which travels with each batch of work as it moves from work station to work station. Each operation line which is a multiple of 10 will be statused by the heat treater performing that operation.

The heat treater will enter his or her initials, job number, batch number, furnace number, and any pertinent comments into the CRT, then complete the appropriate information on the job card. The card then becomes a permanent history backed up by magnetic tape history.

```
*** M A S T E R   J O B   C A R D ***
                                                          01/17/80
JOB NUMBER            021875        JOB WEIGHT           2400
NUMBER OF PARTS          144        PROCESS CODE         001
CUSTOMER NUMBER       018500        P O NUMBER           423-908
TYPE OF MATERIAL 4140
BILL TO: SPACE TEK MFG         SHIP TO: SPACE TEK MFG
         4001 N. LEWIS                  8740 E 46TH

         TULSA            OK             TULSA            OK

DESCRIPTION:      24-405-3878 ENGINE MOUNT BRACKET
                  STRAIGHTEN TO .005 TIR

                  CERTIFICATION REQUIRED

PROCESS INFO:     HEAT TREAT 160-180 KSI PER MIL-H-6875

BATCH NUMBER  1  BATCH WEIGHT  400   FINAL HARDNESS  375-385 Rc

*********** THIS IS THE STANDARD ROUTER FOR THIS JOB *******************
LINE   OPCD OPERATION          QUALIFICATION      MM/DD/YY HH:MM ST NO INITIALS
00004 0008 MIL SPEC JOB        MIL SPEC JOB       --/--/-- --:-- ----- --------
00006 0200 PLACE LOAD          IN FIXTURE         --/--/-- --:-- ----- --------
00008 0433 RUN 4140 TEST SLUG  .                  --/--/-- --:-- ----- --------
00012 0404 RUN SHIM EACH HOUR  .40 C              --/--/-- --:-- ----- --------
00014 0009 RUN SHIM            BEFORE LOADING     --/--/-- --:-- ----- --------
00016 0102 NORMALIZE           1650 F  1 HR       --/--/-- --:-- ----- --------
00018 0329 CHAMBER COOL        1 HR               --/--/-- --:-- ----- --------
00020 0001 TIME CHARGED                           1/17/80 21:10 104  DEG
00022 0002 TIME AT HEAT                                   21:12
00024 0003 TIME TO QUENCH                                 22:12
00026 0301 AUSTENITIZE, HOLD 155UF 1 HR           --/--/-- --:-- ----- --------
00028 0215 OIL QUENCH          150F  15 LOW        --/--/-- --:-- ----- --------
00030 0001 TIME CHARGED                           1/18/80 1:05 104  JEM
00032 0002 TIME AT HEAT                                    1:45
00034 0003 TIME TO QUENCH                                 4:45
00036 0416 TEMPER AT ONCE      RUSH TO TEMP FURN  --/--/-- --:-- ----- --------
00038 0311 TEMPER             950F  4 HRS          --/--/-- --:-- ----- --------
00040 0001 TIME CHARGED                           1/18/80 4:35 08  OSD
00042 0002 TIME AT HEAT                                   4:55
00044 0003 TIME TO QUENCH                                 8:55
00050 0005 PARTS IN AND READY FOR INSPECTION      --/--/-- --:-- ----- --------
00052 0121 INSPECT HARDNESS                       --/--/-- --:-- ----- --------
00054 0425 LAB TEST           LAB RELEASE REQD     --/--/-- --:-- ----- --------
00056 0449 STRAIGHTEN                              --/--/-- --:-- ----- --------
00058 0106 STRESS RELIEVE     900F 2 HRS#76 FURN   --/--/-- --:-- ----- --------
00060 0001 TIME CHARGED                           1/14/80 5:15 ----- USO
00062 0002 TIME AT HEAT                                   5:25
00064 0003 TIME TO QUENCH                                 8:15
00066 0123 SHIP                                   --/--/-- --:-- ----- --------
```

Computer System at Hinderliter Heat Treating

By John D. Hubbard

THIS IS THE STORY of how and why a commercial heat treater is using a computer now and where he plans to expand its application.

First, some background: we are bombarded daily with different materials, configurations, customer specifications, and unique requests. Such variables are common to the commercial heat treating industry, as are customer inquiries concerning the status of parts. The customer expects delivery on the date and time promised. He also expects all parts to meet specification.

Giving the customer what he wants has been our basic philosophy for over 30 years, but living up to this objective became difficult as we grew in size and new generations of people joined us. We found we had to spend more money and exercise more managerial effort to maintain control of performance.

Several alternatives to controlling costs on the one hand and meeting customer service requirements on the other were evaluated. One of them, computerizing our system, at first appeared to be either impossible or too expensive. But after an in-depth investigation, we found that the alternative is both a reasonable investment and a long range solution to our problem.

What It is, How It Operates

Our system was patterned after the airline reservation setups we see when we purchase a ticket and get a boarding pass. The problems faced by both industries are similar — finite capacity, generally unpredictable demand, and the need for fast delivery.

We had one advantage at the outset. We had been using a batch type computer for over eight years to

Fig. 2 — Example of Hinderliter on-line inquiry system as it appears when the customer calls seeking the status of a job as shown in Fig. 1. Operator enters the customer's account number, then the computer lists all of the customer's jobs in house at the time. After the job in question is identified, its number is entered, and the complete router will be displayed; each batch number is displayed beside the operation where the batch is currently under process. Weight of the batch and furnace number are displayed beside the operation.

Reprinted from Metal Progress, July 1980, 46-48, © 1980 American Society for Metals

```
                              ***HINDERLITER HEAT TREAT***
                              ***ONLINE INQUIRY SYSTEM ***
```

```
SELECT
1=inquire by customer number
2=inquire by job number
E=end program
```

1 Operator enters 1 to inquire by customer number

```
ENTER CUSTOMER NUMBER.
```

18500 Operator enters customer account number

```
SPACE TEK MFG
4001 N. LEWIS

TULSA          OK

IS THIS THE CUSTOMER YOU WISH? (Y/N)
```

Y Operator verifies the correct account number has been entered.

```
DO YOU WISH TO SEE:
1--ALL JOBS FOR CUSTOMER
2--UNINVOICED JOBS FOR CUSTOMER
```

2 Operator selects 2 to view all uninvoiced jobs for this account.

```
JOB-NO P.O.#        #PARTS   DESCRIPTION OF PARTS
021675 423-9087         144 24-405-3878 ENGINE MOUNT BRACKET
              01/17/80
NO MORE JOBS FOR THIS CUSTOMER

DO YOU WANT A FULL BREAKDOWN OF A JOB? (Y/N)
```

Y Operator wishes more detail on a particular job.

```
ENTER JOB NUMBER.
```

21675 Operator enters job number in question.

```
24-405-3878 ENGINE MOUNT BRACKET
STRAIGHTEN TO .005 TIR

     CERTIFICATION REQUIRED

/ - - - - - - - R O U T E R - - - - - - - - // - - - - - B A T C H - - - -
R-LINE RW OPERATION-NAME       OPERATION-DESCRIPT B# WT-LBS WORKSTATION-NAME
00004  0  MIL SPEC JOB         MIL SPEC JOB
00006  0  PLACE LOAD           IN FIXTURE
00008  0  RUN 4140 TEST SLUG
00012  0  RUN SHIM EACH HOUR  .40 C
00014  0  RUN SHIM             BEFORE LOADING
00016  0  NORMALIZE            1650 F  1 HR
00018  0  CHAMBER COOL         1 HR
00020  0  TIME CHARGED
                                            06     400 IPSEN 94

00022  0  TIME AT HEAT
00024  0  TIME TO QUENCH
00026  0  AUSTENITIZE,  HOLD 1550F  1 HR
00028  0  OIL QUENCH           150F   15 LOW
00030  0  TIME CHARGED
                                            05     400 IPSEN 104

00032  0  TIME AT HEAT
00034  0  TIME TO QUENCH
00036  0  TEMPER AT ONCE       RUSH TO TEMP FURN
00038  0  TEMPER               950F   4 HRS
00040  0  TIME CHARGED
                                            04     400 ELECT PIT #1
```

provide accounting, financial, and managerial reports and analyses. The computer was an accepted, proven tool.

The step from batch operations to a continuous on-line system required more modern computer hardware and a system analyst familiar with on-line systems. Once we had both on board we converted the old accounting, finance, and managerial programs to the new computer; then we ran parallel systems for about two months.

With the conversion completed, there was no noticeable difference between old and new except the new computer took up less floor space and was able to print faster. Although the old reports looked the same, we were no longer using keypunched cards to enter data. They were replaced by a CRT (cathode ray tube) which looks like a television screen tied to a typewriter keyboard.

At this point we were ready to create our on-line system. The system had to improve our internal control of jobs flowing through our plant, simplify our paperwork, and provide customers with fast, accurate information concerning status order. The result was: HOTS (Hinderliter On-Time Services).

Status of jobs is maintained by heat treaters out on the floor. A master job card (Fig. 1) travels with each batch of work as it moves from work station to work station. When the heat treaters move the job from receiving to the next step, they enter that information on the master job card and directly into the computer — 24 hours a day, seven days a week. It is no more difficult to enter data into our computer than to use the 24-hour tellers that many banks now have. With about 15 minutes of training, you can become a full-fledged computer operator.

When a customer calls in to inquire about a job we simply key in the account number, and the computer displays every job the customer has in house, providing a job number, purchase order number, number of pieces, weight, description, and date received (Fig. 2).

Once the specific job in question is identified, the computer will display the shop router for that job in line-by-line detail of all operations the parts must go through (Fig. 2). Each individual furnace load will be shown in the actual furnace where it is being run or the exact location of that load, such as furnace number, receiving, inspection, and shipping.

Armed with this information, we can give the customer fast (2 min average) delivery information without being put on "hold" while our people run around the plant in an Easter egg hunt for the parts in question.

Side benefits are: shop routers have now been standardized so that, automatically, successful routers can be duplicated or custom routers built. Invoices,

acknowledgements, and shipper forms are also automatically generated. Backlog reports are generated as needed with no effort; special tailored reports can be available in a matter of a few hours where before such reports would take days to manually compile.

How Computer System Is Being Expanded

The system has been so successful at our plant in Tulsa that we are installing the same thing in our Oklahoma City plant. With each additional facility we will not have to add a new computer because our initial planning included the purchase of a main frame capable of handling many facilities in widely separated locations. The minimum we need is a CRT, a printer, and a phone line to computerize each new plant.

The systems we have installed have been a big success with our customers and with our own people. We are now starting to focus our attention on the next problem we need to solve: process control. Again, the key consideration is cost versus benefit. This time, the variable we are working with is delivery reliability. It includes consistency of quality.

We will be working to tie the computer to the control of key metallurgical variables while parts are being heat treated. As we see it now, they will include temperature, carbon control, time, heating-cooling rates, quench temperatures, and velocity of agitation.

Because the computer already specifies and stores those key metallurgical variables each time it develops a router, it "knows" what should happen so our problem is to develop the hardware and software logic which will allow the computer to take control of the equipment and satisfy our metallurgical requirements.

We feel the basic technology exists for the computer to completely control an atmospheric integral quench furnace. The primary challenges are the integration and retrofitting of existing equipment. We will use a present furnace as a prototype. Once computer control is achieved, we will expand that capability to all of our other furnaces.

Computerized furnace control will remove the cause of approximately 80% of our rework and redraw — human error. However, computerization does not downgrade the human aspect of heat treating. In fact, the computer enhances the capability of our heat treaters to concentrate on the other critical areas that cause the other 20% of our rework and redraw. ⊕

For More Information: You are invited to contact the author directly by letter or telephone. Mr. Hubbard is president of Hinderliter Heat Treating Inc., 1240 N. Harvard, P. O. Box 4699, Tulsa, Okla. 74104; tel: 918/939-0855.

Microprocessors and Instrumentation in Furnace Control

By J. L. ROBERTSON and
P. R. CLARKE
BNF Metals
Technology Centre
Wantage, Oxon.

Editor's Note: Parameters that influence energy conservation in thermal processing are considered in order to take them into account in the development of a microprocessor-based automatic control system for furnaces. These parameters and the control system are described in this article which is from a paper presented at Furnace Symposium 1980 (Thermal Processing—A Fresh Look at Energy Conservation) Birmingham, England. The symposium was organized by International Symposia & Exhibitions Ltd., and sponsored by "Metallurgia", The Society of Industrial Furnace Engineers, and The British Contract Heat Treatment Association.

In the production and semi-fabrication of non-ferrous metals, approximately 80% of the energy consumed is directed towards the melting operations. These are carried out mainly in open-flame reverberatory furnaces, fired with natural gas or petroleum oil. The reverberatory furnace has several factors to recommend it, namely: (1) it is simple in construction; (2) it handles a large variety of feedstock, from bulk material to finely compacted scrap; (3) charging, treatment and the pouring of liquid metal are accomplished easily; (4) it is easy to operate.

Previously, the combination of low fuel prices with the above benefits has masked its generally low thermal efficiency. Continually rising prices and limited fuel supplies have now caused companies to re-examine the role of the reverberatory furnace.

Improvements in furnace control and design can lead to significant reductions in energy consumption and in turn improved economy.

BNF Metals Technology Centre has been active in the improvement of furnace design and in the development of fully automatic furnace controllers for open flame furnaces. This article illustrates the factors which affect furnace performance and shows how these parameters have been included in a micro-processor control system which is undergoing test in a commercial application.

Heat Transfer

Heat transfer in the high temperature furnace is predominantly by radiation: (1) directly from the flame and the hot gases; (2) by re-radiation from the high temperature refractory surfaces.

Radiation heat transfer contributes approximately 95% of the energy required, the remaining 5% being derived from convection directly from the hot combustion gases.

The rate of radiant heat transfer, q_g, is given by the Stefan-Bolzmann equation:

$$q_g = A\,T_g4$$

where A = constant, which is a function of emissivity, absorptivity and includes the Stefan constant.

From this equation note that the rate of radiant heat transfer is proportional to the fourth power of the gas temperature. Hence, variations in flame temperature can affect furnace performance seriously. For example, a 1000°C drop in flame temperature from, say, 1600°C, results in a 23% reduction in radiation heat transfer (it will be shown later that an increase by 10% of the excess air in the furnace chamber can cause a lowering of the flame temperature by at least 100°C).

Factors Affecting Furnace Efficiency

To improve the melting operation it is necessary to exercise control over several factors: (1) combustion/fuel-air ratio control; (2) furnace pressure; (3) insulation; (4) waste heat recovery.

For a heating process to operate, the combination gases must leave the hot zone at a temperature in excess of that to which the stock is raised. This premise makes it possible to calculate the maximum thermal efficiency, without recuperation, for open-flame furnaces according to the maximum stock temperatures required, as shown in Fig. 1. The graph indicates that a theoretical thermal efficiency of 60% is possible when melting aluminum; for melting zinc an efficiency of 80% is predicted. For comparison typical operating efficiencies for several non-ferrous processes are shown; it is evident that there is much room for improvement, although the predicted high efficiencies for melting the low temperature metals will not be possible without a radical change in furnace design.

Fuels and Their Combustion

In the simplest terms, the furnace is a refractory chamber built to retain the charge and to contain the energy generated by the burning in air of either gas or fuel oil. In the U.K., natural gas from the North Sea, with a chemical composition almost identical to methane, CH_4, is used; more common on the continent is Groningen gas (82% CH_4, 15% N_2). The oil may be one of several grades; the most common are given in Table I.

Gaseous fuels are readily burnt by ignition of a pre-mixed gas/air jet, or by injection of separate gas and air jets into the furnace chamber. In the latter case, mixing of the gas and air takes place along the furnace length, producing a diffusion flame. Few problems are encountered in producing gas flames, however, it is necessary to select the burner to suit the refractory chamber, in order to provide the correct gas flow distribution.

Combustion of fuel oils is more complex. Initially, the oil must be atomized (typically droplets less than 100 microns in diameter) and injected into

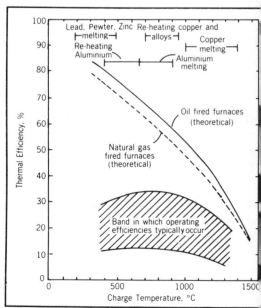

Fig. 1 Theoretical thermal efficiencies and practical efficiencies.

the main turbulent combustion air stream. The droplets are heated by radiation from within the chamber via the hot combustion gases, evaporation follows and ignition of the vapor phase occurs. The flame front moves through the droplets so that the fuel burns progressively.

In order to achieve efficient combustion of fuel oil, it is essential to pre-heat the oil to the correct atomizing temperature. The range of recommended temperatures for the various grades of oils is given in Table I. Insufficient atomization creates large oil droplets which tend not to burn within the furnace chamber. In the furnace chamber, inadequate atomization of the oil may be recognized by the characteristic sparks formed at the extremities of the flame front.

Table I Common Petroleum Fuels (Shell Manual)

	Gas/Diesel Oil	Light Fuel Oil	Medium Fuel Oil	Heavy Fuel Oil
Class	D	E	F	G
Viscosity centi Stokes (100°F)	1.6-6.0 (100°F)	12.5 (180°F)	30.0 (180°F)	70.0 (180°F)
Redwood No. 1. secs. (100°F)	34	250 max	1000 max	3500 max
Gross Calorific Value kJ/kg	45590	43496	43030	42800
Therms/gal	1.64	1.74	1.74	1.77
Oil Preheat °C, Temp. required for atomizing	Ambient	70/80	90/100	120/130

Correct mixing and proportioning of the fuel and air is essential; firing the furnace with too much or too little air (i.e. excess fuel) leads to excessive waste of fuel.

When furnaces are fired with too much air, the melting performance declines because the maximum flame temperature obtainable is much reduced, as shown in Fig. 2; this in turn reduces the rate of heat transfer to the charge from flame radiation.

Secondly, energy developed in the combustion process is employed unnecessarily in heating the considerable volumes of excess air before this is expelled in the stack at high temperature. Fig. 3 shows the amount of heat carried out in the flue gas as a percentage of the "heat in" versus the excess air for different waste-gas temperatures. Note that for a stack gas temperature of 1000°C, reducing the excess air from 35% to 15% reduces the heat loss as percentage of the "heat in" by approximately 5%.

Furnace Pressure Control

A major cause of unwanted cold air in a furnace arises from operating the furnace with a negative pressure at, and above, hearth level. The effect of the negative pressure is to induce cold air through all the available stacks in the superstructure. Unfortunately, construction of the majority of furnaces tends to create negative pressure at hearth level; because of the buoyancy of hot air there is always a pressure close to the roof. In the majority of furnaces where the stack exit is sited in, and close to, the roof the natural draft created by the thermal head in the furnace and by the gases leaving at high velocity through the stack, tends to produce a negative pressure at melt level.

Maintaining adequate positive pressure at all firing rates is, therefore, another objective. Adequate furnace pressure control is maintained by the use of a mechanical automatic damper situated in the stack. Because of the high temperature environment encountered, normal damper systems tend to fail and for this reason an air curtain damper control system has been developed at the BNF. Simply, an air curtain is fired at a suitable position higher in the stack and

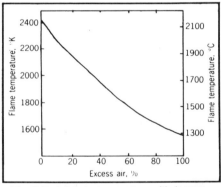

Fig. 2 Variation in temperature with increasing excess air.

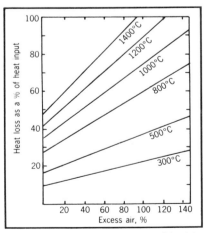

Fig. 3 Heat loss as a percentage of heat input, vs. excess air for different stack gas temperatures.

this acts as a down-draft and restricts the flow of hot gases up the stack. This form of damper control is ideal for use in an automatic system, as there are no moving parts.

Furnace Insulation

In bulk melting furnaces, because of the demands placed on the furnace linings it is still necessary to employ refractory firebricks of one form or other. Although it is possible to use moldable or castable refractories, the tendency in melting furnace design is to use a hot-face firebrick, backed by insulating brick, with some form of insulation, all contained in an outer metal casing for rigidity. The heat losses through the walls and roof generally amount to only 1% to 5% of the total heat input and, therefore, the extra expense of providing more effective insulation does not seem warranted. It is possible that with further increase in fuel prices increased insulation might become viable.

It is only in the case of reheating and heat treatment furnaces, where wear is considerably less arduous, that the lighter weight ceramic fiber linings may be used. Because of the lightness and rigidity of the fiber material it is possible to construct furnace shapes without the extensive steel work involved in the more traditional designs. The reduced overall weight along with the improved insulation properties create a much more thermally efficient furnace. In particular where intermittent operation is involved, the reduced mass decreases the thermal inertia of the furnace (ceramic fiber linings have a heat storage approximately one-quarter that of insulating firebrick) hence, the heating cycle is shortened. For these cases, the replacement of traditional refractory bricks with ceramic fiber has resulted in fuel savings of between 10-15%.

Waste Heat Recovery

By far the largest contribution to losses in furnaces arises from the high temperature of the exhaust gases. Typically, for aluminum, 60-70% of the heat input would be carried out by the waste gases, and even for the optimum theoretical case, 40-50% would still be lost. Even so, to reduce the stack gas temperature to this minimum would provide a valuable improvement in operating efficiency.

The temperatures at which gases leave the hot zone depend on the efficiency of heat transfer from the gases to the charge. Allowing that, the burner and furnace aerodynamics are designed for maximum heat transfer, the furnace stack losses still contribute upwards of 50% of the overall loss in furnace performance. To further improve furnace performance, it is necessary to provide waste-heat recovery from the stack gases.

Recuperation of heat from the waste gas may be accomplished either directly, by pre-heating the combustion air, or indirectly by raising steam. If the combustion air is pre-heated a direct saving in fuel results as shown in Fig. 4. For example, preheating the combustion air to 400°C prior to entry to the furnace will reduce fuel consumption by approximately 20%. Against such fuel reductions, must be balanced the increased costs for the maintenance of the heat exchanger.

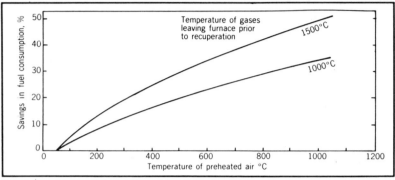

Fig. 4 The saving of fuel obtained by preheating of the air.

If steam is raised by recuperation, there is no direct fuel saving and the overall furnace performance has not been improved; it becomes necessary to "sell" the steam raised, or use it elsewhere, for it to contribute to fuel cost reduction. Maintenance costs increase considerably over the simple heat exchanger used for air pre-heat, and this method of recuperation must be evaluated to determine its attractiveness.

Automatic Control of a Fuel-Fired Furnace

Taking into account the factors affecting furnace performance, BNF has developed a microprocessor-based automatic control system for furnaces. The controller has been designed to handle the following.

(1) Monitor the fuel and air flows supplied to the burners.

(2) Calculate the fuel/air ratio and the excess air from (1).

(3) Analyze the stack gas for oxygen content.

(4) From (3), determine the excess air in the stack gas; it is possible by comparing (2) and (4) to obtain an indication of both burner mixing and performance as well as whether air ingress through furnace openings occurs.

(5) Furnace pressure is measured at hearth level.

(6) Control of the fuel flow rate in

different parts of the melting/re-heating cycle.

These measurements, (1) − (6), may be stored on file; selected values as required may be printed out (i.e. fuel rate; total fuel) for data collection; the digital displays provide immediate visual checks.

There is a multiple directing safety system included in the microprocessor control program, as follows: (1) flame failure automatically overrides all program statements; (2) a check is always made to ensure that air flows to the burners on a priority interrupt schedule; (3) the correct fuel/air ratio is maintained between pre-set limits;

(4) a maximum refractory temperature control reduces or switches off the burners, to protect the furnace.

Operation of Automatic Controller

The controller is programmed with a variety of fuel-firing rates, according to the typical furnace operation involved; the operator may then select any suitable firing operation. After selection, and the passing of control to the automatic system, the furnace is fired accordingly. For example, in the case of a melting furnace, during charging the burners would be held on low fire until completion, after which the burners would be raised to an intermediate firing rate. A flow rate lower than the maximum is suggested in order to allow for the reduced combustion space caused by bulk charging. As melting takes place, the combustion space increases and the firing rate may be raised. In programming, the change in fuel flow rate would be on a time basis, until melting had been completed and it was required only to hold the liquid metal.

Throughout the firing cycles, the fuel/air ratio is monitored and optimized according to the signals received by the controller. If a negative pressure is determined within the furnace, the controller attempts to maximize the furnace

pressure by introducing the furnace damper control system. In the BNF furnace, a forced air curtain (damper control) is directed across the stack exit; this permits an easy mode of control through opening of an air valve. Furthermore, since a lowered furnace pressure is only encountered when the burners are switched to low fire, the damper may be operated from the same fans that supply the combustion air; that is, part of the fan power is diverted to air damper service.

To provide for communication between the microprocessor and the melting shop manager, a display unit is included in the system. It may be situated either next to the furnace or some distance away, say in the manager's office.

The display unit can receive information from up to eight satellite furnace control units and transmit instructions back to the satellites. In conjunction with a hard-copy printer, the system forms an ideal tool for production control. Continuous display of fuel consumption, combustion conditions, furnace pressure and temperatures, and process states are available on demand. The display unit has the facility to analyze the data over periods of time, allowing regular reports to be printed directly from the computer. In addition, the continuous record of operations will indicate any deterioration in furnace performance caused by refractory wear; highlight break-downs in production; and produce comparisons of performance between different furnaces.

Conclusions

Fuel savings can be considerable, particularly when one remembers that since August 1973, the price of light oil in the U.K. has risen from 5p/gallon to 60p/gallon.

In the short term, increased thermal efficiency of existing furnaces may be achieved by: (1) better control over combustion parameters; (2) using the waste gas effectively; (3) the application of modern burners in existing furnaces; (4) employing automatic furnace control.

An automatic furnace control system can be programmed for melt at optimum efficiency for a given melt-rate. A modification of the system has been installed in one company melting copper and is in the process of being installed in other companies refining lead and copper, in the U.K. and Italy. During the trials and commissioning period of one industrial system, energy savings in the region of 15% have been obtained over an extended period. In other circumstances considerably larger savings of fuel have been achieved. ■

Computerized Systems for Heat Treating

By Theodore K. Thomas and Richard I. Gruber

This article is based on the chapter of the same name edited by Mr. Thomas for Vol 4 on "Heat Treating" of the Ninth Edition of the ASM *Metals Handbook*.

FROM THE development and successful introduction of ENIAC (Electronic Numerical Integrator and Computer) in 1946, up to the latest "generation" of product offerings from computer manufacturers, this technology has exhibited several clear and distinctive trends:

1. The steady and continuing decrease in the size of computer components and equipment.

2. The continuing decrease in the price of computer equipment.

3. The steadily increasing cost of the programming, or software, necessary to make the computer perform its desired functions.

The steadily rising cost of energy, coupled with more exacting requirements for quality controls, are parallel trends in the heat treating industry. Today, the rising energy cost curve is nearing a crossover of the dropping computer cost curve. It is now economically feasible to consider the digital computer's advantages over conventional instrumentation in the control of furnace process variables. These advantages are:

1. Multiple-function capability.

2. High speed measurement and response to process change.

3. High accuracy because of digital measurement and readout as compared with analog operation.

4. High reliability of solid state electronic components and circuitry.

5. The ability to communicate electronically between computers and other elements of computerized systems.

The application of computer technology to make furnace operation more efficient and cost effective foreshadows a more productive era for the heat treating industry.

The balance of this article includes a discussion of the growing need for computerization in heat treating; a primer on computers; a look at a typical microprocessor based control system; and a description of how one forge shop has computerized the control of its heat treating operations.

Evolution of the Need for Computerized Systems

The main reason computerized control systems have not been applied to heat treat furnaces until recently is economic. Conventional instruments, such as recorder-controllers with thermocouple inputs and indicating controllers with special reset options for high limit control, have been much improved by solid state circuitry, digital setpoint indexing, digital display, and other innovations designed to make their operation more accurate, faster responding, and more reliable. Anything connected with the word "computer" has been considered expensive and complicated, especially by furnace builders who have to sell the control package to their customers. Therefore, the updated versions of traditional and more familiar measurement and control instruments have been favored by both OEM's and users, particularly for single-chamber furnaces performing one or two operations.

However, consider the example of the multichamber vacuum furnace in today's era of tightening quality requirements, pressure for increased production, and ever-increasing energy costs. Proper operation of the furnace requires complex temperature ramps and soaks, mass flow measurement and control of atmosphere gas admission, varying vacuum levels between chambers, sequences of door interlocks, limit alarms, load and transfer adjustment sequences, and other functions within specific time limits. Under these conditions, the advantages of a digital based electronic control system — in other words, computerization — not only appear desirable, but are becoming mandatory.

The new microprocessor based controllers can perform the functions with electronic speed, high reliability, and pushbutton ease. The digital display capability presents setpoint and process variable readings in a format which can hardly be misinterpreted and can be presented with readabilities of one and two decimal places. Another advantage of the new

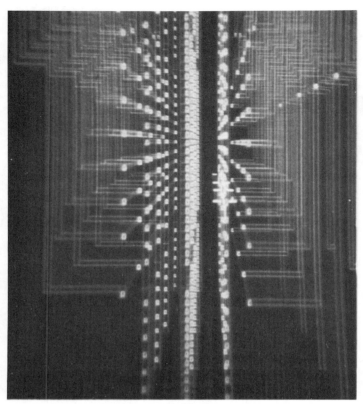

digital based instruments is their relatively small size, which saves panel space and cuts installation cost. The microprocessor based controllers and programmers can also interface and communicate with a mini-computer for performance monitoring or setpoint supervisory control. All of the functions of relay logic, for interlock sequences and digital, or on-off, inputs, can be provided by digital circuitry contained in the PLC (programmable logic controller) provided as part of the modular hardware in a microprocessor based furnace controller package.

In some of the more advanced control philosophies, a multiple-furnace facility's operation can be monitored from one supervisory computer, while each furnace is simultaneously controlled by its own microcomputer. This is the essential concept behind the term "distributed control" which is now widely used in discussions of computerized manufacturing applications.

Justification — One writer, commenting on forging operations, sums up the need for computerized systems this way: [1]"More and more, the forging operation is being taken out of the hands of the operator, however skilled he may be, and the control turned over to a computer, a microprocessor, or other type of control. Forging operations have become too sophisticated, too integrated, too costly, and too complex for an individual operator to decide what the forging techniques should be.

"Computer control, for example, allows a forging plant to deliver forgings that are far more accurate and uniform from workpiece to workpiece, and especially so with forgings of complex shapes.

"Computer control also lends consistency to the forging technique, eliminating variations that are bound to exist among individual operators.

"Product quality can be maintained more readily by the fact that the computer issues precise instructions at each stage of the forging operation.

"Also, the fast response of the computer sharply reduces manipulator time between passes, giving the forging press more time to work while the workpiece is within range. This reduces the number of reheats, conserves energy, and thereby helps reduce over-all cost.

"The storage capability of the computer speeds the assessment of new work and provides reliable data on the essential parameters used for previous forgings. The compilation of forging records over a period of time could serve as the basis to generate new forging schedules."

Many, if not all, of these reasons could, but with slight modification, also be used to justify the acquisition of a computerized heat treating control system.

Computer Basics: Micro's, Mini's, Mainframes[2]

Computers operate by digital methods, rather than the analog techniques traditionally associated with industrial instrumentation. An analog instrument handles data in terms of electronic or pneumatic signals which are proportional in size — analogs — to the quantity of the variable being measured, such as temperature or flow. Digital logic expresses the same quantities electronically in terms of binary digits, or bits, each with a value of 0 or 1. With digital logic, immense amounts of information can be accumulated and processed in a small space with a very high degree of accuracy and reliability.

A computer has three main parts: a CPU, or central processing unit; a data storage unit, or memory; and I/O (input/output) equipment. The CPU contains an arithmetic unit, where the actual computing takes place, a control unit which tells the arithmetic unit what to do, and storage registers which accumulate the numbers involved in the calculations as they step through the predetermined (programmed) sequence following the directions of the control unit.

Three Classes — Every computer has a processor, memory, and some type of I/O. From there, computers can be classified as microcomputers, minicomputers, or mainframes. Each class overlaps the other. Generalized definitions of the three follow:

A microcomputer contains a microprocessor, usually located on one LSI (large-scale integrated) chip or a small LSI chip set. The processor's word length is between 4 and 8 bits. Most microcomputers have at least 256 words of memory.

A minicomputer's word length is between 12 and 16 bits. And its minimum memory size is 4096 (4K) words.

A mainframe, or so-called large scale computer, usually has a word length of 32 bits or more and a minimum memory capacity of 16 384 words.

A microprocessor (or microcomputer) at the sensor level can increase system accuracy and the ease of transmitting signals. Whereas analog signals may suffer from variations in amplification or other losses when transmitted over long distances, digital data is transmitted in binary, which is less subject to error. A simple microprocessor or even simpler LSI circuit in a digital converter can gather data with minimum system loss and can also preprocess some of the data before sending it on in binary form. Nevertheless, the microprocessor or microcomputer is dedicated to a specific task or group of tasks within an application.

In an industrial control system such as one for a fuel-fired furnace, a microcomputer can sample and evaluate temperature at several different points in the furnace, compare the results with acceptable limits, and make necessary adjustments. It can also continuously measure and relate fuel and combustion air flows at a pre-set ratio, which will yield more energy efficient control.

A microcomputer or minicomputer can also tie together a number of instruments. In this case, it can act as a data logger, preprocessor or "a smart front end" to a larger computer.

Large-Scale Computers — Whereas a large mainframe may have been used a few years ago for more complex control or measurement applications, today's minicomputer usually fills the bill.

More commonly, a large-scale computer is used for after-the-fact analysis, and can be connected to a smaller computer. This is, in turn, connected to a process through its instrumentation. In this network fashion, a large computer either monitors the small computer's activities or it can back up the smaller devices.

Memories — Using any computer in an application involves other devices along with the central processor. A computing/control system also includes interfaces to attach the computer to the application, programs to make the system run (software), and devices to store or communicate data.

In any computer, the region which stores program instructions and data for instant use is the memory. Memories can be either read-only or read/write. If the data or instructions in a computer's memory do not need to be altered, read-only memories (ROM's) are used since the computer can't alter their contents. Read/write memories, on the other hand, can be

altered and accessed by the computer. Read/write memories are also known as random access memories (RAM's) since any data location within memory can be accessed as easily as any other.

Memories can also be classified according to the way they are constructed — either with ferrite cores or semiconductors. Technically, core memory is a matrix having elements composed of tiny toroids of ferrite materials. These elements are magnetized to store information. Semiconductor memory is a memory matrix composed of tiny semiconductor circuits.

All core and semiconductor ROM memories retain data whether power is off or on. Such is not the case with semiconductor RAM's. If they lose power, they lose memory. RAM can be protected by connecting batteries to the memory package for use in the event of a power failure.

Semiconductor memory is usually faster, more compact, and less expensive than core memory. So in applications where volatility is not a key factor, or where battery backup can be used, semiconductor memory is more often selected. Some computers do, however, use a combination of semiconductor and core memory.

Which type of memory is used with a computer depends in part on the application. For instance, a microcomputer based controller with a limited number of functions would probably incorporate a ROM as the main part of the memory and a small RAM to temporarily store data.

Signal Conversion — In control applications, computers must connect either to a binary-type signal (open or closed relay), an analog signal, or in some cases a digital signal. Sometimes the connection between the digital computer and a binary circuit involves only matching voltage levels. If an instrument or valve has a digital output, connecting it to the computer again involves only voltage or current matching.

However, if the control has an analog output, which is usually the case, the analog signal from the sensor to the computer must be converted to digital form. The interface, in this case, is an analog-to-digital (A/D) converter placed between the computer and the analog control device.

If the computer must send an analog signal to a control loop, then a digital-to-analog (D/A) converter acts as the interface. The D/A converter constructs an analog signal from the digital information supplied by the computer.

Both A/D and D/A interfaces can be found in even very small processes, since they are usually the only communication links possible between processes and analog control devices.

Programming — Without software, the computer, no matter how complex, will not work. Software gives the computer its flexibility and also many of its problems. These problems are often caused by human error which results because software is usually not written in English, but in a special language that the computer can translate into the binary language it recognizes. The software language written by the user can range from "machine-like" to "English-like." Machine-like languages are called assembly or low-level languages, and English-like languages are called high-level or compiled languages.

An assembly language is a collection of mnemonics that refers to the exact functions a computer must perform. In PDP-8 assembly language, for example, CLA stands for "clear the accumulator," which is far easier to work with than its binary equivalent: 111110000000.

Assembly languages are used mostly with microcomputers because they conserve memory — typically, the largest hardware expense of a microcomputer system.

While assembly languages are organized to match the way computers operate, high-level languages are organized more along the lines of a spoken language. High-level programs do not need to specify every single operation the computer performs; instead, they must only indicate what function to perform (e.g., add two fields).

Both assembly and high-level languages require a special program to convert user codes into binary codes. This special program, called either a compiler or interpreter, occupies a significant portion of memory (more for a high-level language), thus increasing the size of memory required. More memory is also needed for high-level languages because they're not optimized for one specific machine. Hence, they are not as efficient in their translation as assembly languages.

Although high-level languages are not designed for specific machines, they are designed for specific functions. Of the popular languages, BASIC and FORTRAN were developed for math and scientific problem solving. FORTRAN, particularly FORTRAN IV, tends to have higher arithmetic precision than an equivalent BASIC, but BASIC is easier to learn.

High-level languages are usually associated with minicomputers and larger computers because they require more memory. (They are now, however, becoming available for microcomputers.) Minis and larger computers also have standard, packaged "operating systems" written in assembly or high-level language. By using an operating system, users need to be concerned only with the application.

Selection — Computers can be used nearly everywhere. But for a computer to provide the advantages of accuracy, flexibility, and reliability, users must also look at available software, memory requirements, storage, peripherals, and of course,

interfaces. It's important to keep in mind that the diversity of processor types, peripherals, and software makes it possible to apply computer operations effectively to almost any control or instrumentation application.

A Typical Microprocessor Based System

This section describes a microprocessor based instrument designed to control a time vs temperature program for a heat treating furnace. It's a product of Process Control Div., Honeywell Inc.

The Old — A typical process profile for a heat treating furnace is shown in Fig. 1. This same type of program could also apply to environmental chambers, weld stress relieving, and other applications where the process variable (PV) must be controlled as a function of time.

To achieve this kind of programmed control in the past, it was necessary to assemble and interwire a separate setpoint programming source, a controller, and interrupters and timers into a system.

The types of programming sources available were limited in their ability to provide the degree of resolution and setpoint accuracies needed to optimize process control. These devices included cam programmers, photoelectric sensors, capacitive sensors, and other instrumentation for generating setpoint vs time profiles. All had to be interfaced with some sort of controller that could accept a remote setpoint input, and with several timers and interrupters. The result in terms of process control was (and still is) often marginal. Lack of characterization of the generated setpoint signal is one of the largest sources of error in programming controllers using nonlinear process sensors such as thermocouples.

The New — Today, however, a totally digital approach can be incorporated into a single instrument that combines all the traditional elements of a programmed control system.

One such instrument contains the setpoint vs time program signal source, a digital three-mode controller with automatic/manual operating modes, and 12 programmable event switches.

All operating data (setpoint, process variable, ramp rates, and percent controller output) are displayed continuously by seven-segment digital readouts. Light emitting diode (LED) indicators show operating mode and event switch status.

Programs are entered through the use of a front panel keyboard, and are then permanently stored in nonvolatile memory which requires no battery backup (stored programs will not be lost in the event of a power failure).

System Memories — Figure 2 shows a process control loop in which the microprocessor based instrument is used to measure, indicate, and control the temperature according to a preset schedule (a

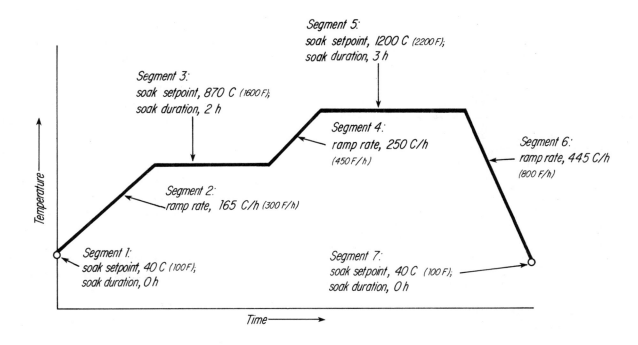

Segment 5:
soak setpoint, 1200 C (2200 F);
soak duration, 3 h

Segment 3:
soak setpoint, 870 C (1600 F);
soak duration, 2 h

Segment 4:
ramp rate, 250 C/h
(450 F/h)

Segment 6:
ramp rate, 445 C/h
(800 F/h)

Segment 2:
ramp rate, 165 C/h (300 F/h)

Segment 1:
soak setpoint, 40 C (100 F);
soak duration, 0 h

Segment 7:
soak setpoint, 40 C (100 F);
soak duration, 0 h

Temperature

Time

Fig. 1 — Typical process profile for a heat treating furnace. Cycles like this can today be handled by a single, microprocessor based instrument that combines all the traditional elements of a programmed control system.

program). The microprocessor is the instrument's heartbeat and brain.

Three types of memory devices are depicted in Fig. 2. The ROM (read-only memory) is the permanent memory. All of its information is "burned in" by the device manufacturer and cannot be changed.

The other two types — RAM (random access memory) and EAROM (electrically alterable read-only memory) are not permanent. Their information is changed, as required, when the system is operating. The main difference between the two is that RAM is volatile (memory is lost whenever the power is turned off or fails) whereas EAROM is nonvolatile.

The less expensive RAM is used as a "scratch pad" to make temporary calculations before storing data in EAROM. These data include: temperature vs time program, tuning constants for the algorithm, and "machine state" data such as current setpoint, controller output, and elapsed times.

Inputs/Outputs — All of the information in the microprocessor and memories is in digital form, which is unintelligible to the process and operator. Therefore, input/output (I/O) devices are required to communicate among them. These are mentioned in the following simplified description of how the

controller diagrammed in Fig. 2 operates.

1. The millivolt output from the thermocouple is conditioned and amplified in the "range card" section to produce an output of 1 to 2 V dc.

2. The analog 0 to 2 V signal is then changed to a 12-bit digital signal by the A/D (analog-to-digital) converter.

3. Every 300 ms, the microprocessor samples the output of the A/D and stores it in RAM.

4. Data in RAM are compared with the thermocouple vs emf data in ROM to determine the actual temperature being sensed.

5. The actual temperature is then sent to the operator's display and also compared with the setpoint being generated by the program stored in EAROM. (The comparison and calculation of the control algorithm actually takes place in RAM which gets data from ROM and EAROM, and then stores the results in EAROM.)

6. Every 300 ms, the microprocessor samples the resultant of the control algorithm and sends it to a ten-bit D/A (digital-to-analog) converter having an output of 4 to 20 mA.

7. The output of the D/A converter can be used to directly control the process or may, if desired, be converted to either a time-proportioning or

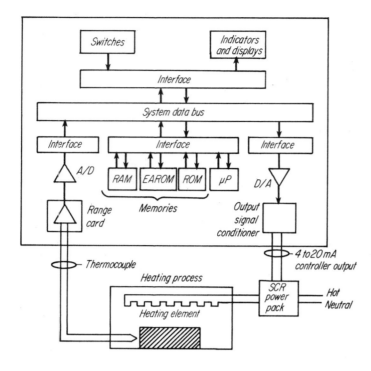

Fig. 2 — The microprocessor (μP) is the heartbeat and brain of modern process control instrumentation. This schematic of a process control loop shows three types of memories — random access (RAM), electrically alterable read-only (EAROM), and read-only (ROM) — and both analog-to-digital (A/D) and digital-to-analog (D/A) input/output (I/O) devices.

position-proportioning control form.

Computerization In a Multifurnace Forge Shop

This case history describes the application of a general purpose minicomputer to a multifurnace forging facility for job scheduling, temperature monitoring and control, and furnace pressure monitoring. The forger produces critical jet engine and other aerospace industry parts, many of which are heat treated before shipment. Assurance of proper heat treatment and accurate record keeping are extemely important.

Furnaces — The heat treating plant consists of 21 box furnaces (see Fig. 3). The single-zone furnaces are fired with excess air to meet temperature uniformity requirements. There are five thermocouples — one for control and four to indicate temperature uniformity — positioned in each furnace. In addition, critical loads are thermocoupled and monitored. Furnaces are surveyed on a scheduled basis and certified for temperature uniformity.

The former control system consisted of a controller, circular chart recorder, and a four-point strip chart recorder for each furnace. An additional multipoint recorder was connected to the thermocouples monitoring critical loads. This array of recorders and controllers led to problems with maintenance, calibration, and pens, and mixups of charts having different temperature ranges.

Increasing requirements for closer temperature control and more and better records plus the problems cited above were the main reasons for switching to computer control.

What It Does — The main function of the computer is to provide temperature control and temperature records for the 21 furnaces. It does this by monitoring up to 14 thermocouples in each furnace. The computer automatically compensates for the deviation of the thermocouple wire so that all readings are actual temperature.

The advantage of computer control is that the computer can be programmed to perform numerous functions and to have several levels of response for each function. In a real-time system such as this one, the computer responds to events as they occur, and, via programming, it responds according to a set of priorities. This means that if an event occurs that has been assigned a high priority — such as furnace over-temperature — it will respond to that event first.

Components — The system consists of two Modcomp computers, a Modacs III process input/output interface system, a tape drive, two printers, a Textronics graphic terminal and plotter, a pair of ten-megabyte discs, and 12 CRT terminals. Only one computer is needed to run the system. The second is a backup. A spare control board and power supply are also installed in the Modacs III interface unit. The redundancy is justifiable in light of the high cost of a shutdown and/or loss of process control.

Communication between operator and computer can take place via any of the terminals or printers. Most data are entered from terminals in the control room. The seven terminals located by the furnaces are typically used by operators on the shop floor to quickly check the status of a furnace or to rapidly respond to alarms.

How It Works — In setting up a load for heat treating, the operator first enters data about the load (such as part number and serial numbers), and then information about the load heat treat cycle (such as type of cycle, control temperatures, heatup rate, and length of cycle). Once a load has been assigned to a furnace and the computer has been told that the furnace has been loaded, the computer takes over. The system is flexible enough, however, that the operator can make changes to the heating cycle while it's underway.

The computer is programmed with several control

schemes for the heatup and hold portions of the cycle. The specific control scheme is automatically selected by the computer, based on information entered by the operator. If an alarm condition occurs, the computer will automatically change the control scheme until the problem has been resolved.

Capabilities — The computer also periodically monitors and stores the maximum and minimum readings of each thermocouple during the selected time interval. All alarm signals are also stored. At the end of each shift, all the information entered by the operator, plus all temperature and alarm data, are dumped from disc onto magnetic tape for storage as a permanent record. A shift report is also generated.

In addition, operators may request and receive temperature vs time curves via the graphic terminal and plotter. Either the high and low reading for each thermocouple, or the average of these two readings can be plotted.

Other functions performed by the computer include tracking furnace operating hours and the time until the next scheduled temperature uniformity survey. It's also programmed to read gas meters and print out a daily report of natural gas consumption. Finally, it controls furnace pressure and limits the amount of excess combustion air to just that needed to maintain temperature uniformity.

Benefits of computerization in this application: better temperature control, better records, increased production, and energy savings.

For More Information: You are invited to contact the authors directly by letter or telephone. Messrs. Thomas and Gruber are with Process Control Div., Honeywell Inc., 1100 Virginia Dr., Fort Washington, Pa. 19034; tel: 215/641-3112.

References
1. "How Forging Has Put New Punch Into Its Act," *Iron Age*, 5 May 1980.
2. Materials for this section taken in part from "Micro, Mini, and Mainframe Basics," by Peter Masucci, *Instruments & Control Systems*, June 1977.

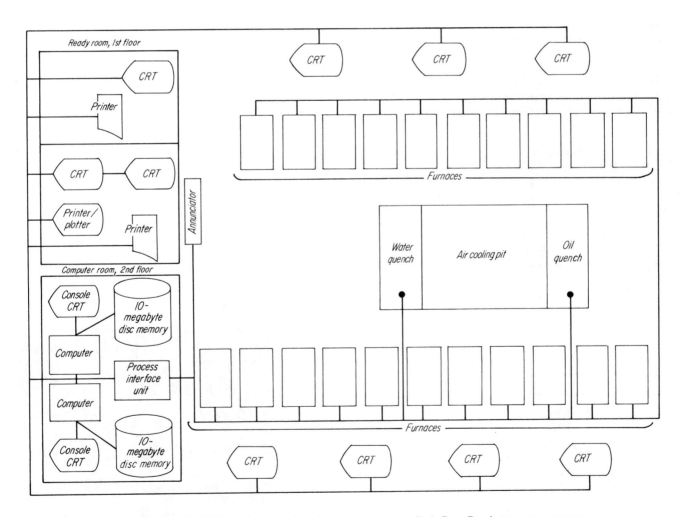

Fig. 3 — This 21-furnace installation in a forging plant is computer controlled. Benefits: better temperature control, better records, increased production, and energy savings. Only one of the two computers shown is needed to control the system. The other's a backup.

Controllers help billet furnace run at 54% efficiency

James M. McQueen
Senior Electrical Engineer
Michael J. DellaRocco
Electrical Project Engineer
Arco Metals Co., American Brass Div.
Ansonia, CT

American Brass recently installed a totally automatic billet heating furnace in its computer-controlled extrusion mill in Ansonia, CT. The mill's indirect extrusion press handles 59-in. billets and produces coiled round rod, round, and hexagonal bar products.

The furnace is a five zone, gas-fired unit with a pre-heat zone that functions entirely off recuperated flue heat. Its computer-based control system operates all functions of the furnace and monitors temperatures throughout the system. Major savings in energy are produced via an auxiliary control system that monitors and adjusts temperatures in the furnace's middle zone according to work flow. To date, the furnace has been operating at an efficiency of 54%.

Controlling temperatures

American Brass uses seven Honeywell UDC 500 Universal Digital Controllers and an Eptak 700 computer in the furnace control system. One UDC is used in each heat zone and two controllers are used to minimize energy consumption.

The microprocessor-based single-loop controller accepts two inputs: a primary and a secondary. Typically, the primary input accepts analog signals from thermocouples, current, and millivolt sources. The secondary input can ac-cept various types of signals such as a remote setpoint, data acquisition signal, alarm parameter, or a process variable. The UDC samples these signals and updates its output three times a second. Outputs include on-off, on-off duplex, time proportional, time proportional duplex, current proportional and position proportional.

The furnace heats billets via direct flame. Two infrared cameras—one on either side of the furnace—scan and measure the surface temperature of a billet in each zone. Camera 1 (Fig. 1) sends its 0-20mA signal directly to input 1 of the UDC. Camera 1's signal is then sent to the computer, via the UDC's auxiliary output. The signal from camera 2 goes to the central computer for processing, then to input 2 of the UDC. The controller uses the two signals to determine average temperature and either opens or closes valves for more or less heat.

Because of direct flame heating, overheating and billet melt down can occur very quickly. Using two IR cameras in each zone helps minimize this hazard. Each camera's signal is monitored by both the computer and the UDC for overtemperature conditions that would shut down the zone.

The computer compares the two signals to see if a large difference exists, which would indicate a possible IR cam-

Fig. 1: Infrared cameras scan billet skin temperatures and transmit readings to UDC 500 contollers and computer.

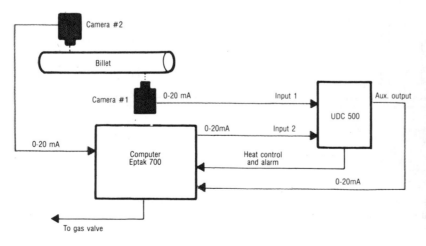

Fig. 2: Two infrared cameras are mounted in each zone of the furnace.

Reprinted with permission from I&CS, March 1984, 57-58, © 1984 Chilton Co.

era malfunction. If there is a large difference, the computer then determines the correct signal and alerts a technician via a message on the diagnostic CRT. The technician then configures the UDC to work off the one good signal until the problem is solved.

Auxiliary controllers

The auxiliary control system consists of two UDC 500s that scan billet temperatures in zone three, and adjust maximum heat input according to furnace load conditions. One UDC is set to provide 80% of maximum heat input; the other is set to 60%. Both obtain temperature information from camera 1 in zone 3.

Both auxiliary UDCs have setpoints slightly below the main UDC's setpoint. For example, if the zone 3 setpoint is 800°C, the 80% UDC would be set at 760°C, and the 60% controller would be set at 780°C. Alarms occur automatically at 10°C higher than the setpoint. When temperature hits 770°C, the first controller would shut down the gas valve to 80% of capacity. Then, if the temperature continues to rise and reaches 790°C, the other controller would signal the valve to shut down to 60%. With furnace load conditions varying widely, the resulting savings in energy consumption could be as much as 20-25%.

One major advantage of the furnace's computer-supervised control system is that it enables the company to run various alloys back-to-back throughout the production shift without interruption. The company extrudes coiled rod and bar stock from a variety of different copper and

Fig. 3: Five UDC 500 controllers monitor and adjust furnace temperature. The bottom readouts are setpoints; the top readouts are actual billet skin temperatures. Two other controllers handle furnace load control.

brass alloys at temperatures of 580-1000°C. Once an order of specific alloy is entered, the computer will automatically signal the controllers to process the alloy at the correct temperature. In the future, the signal from the IR cameras to the UDC will be conditioned by the computer to compensate for variations in the emissivity of the different alloys.—Honeywell Inc., Process Control Div., Ft. Washington, PA 19034.

Applying Digital Equipment for the Control of Industrial Heating Equipment

by James Sullivan

Furnace control systems no longer need to be connected with a morass of wires running hither and yon between the proper mix of available instrumentation. At least not since the microprocessor (or chip as it has come to be known) hit the factory floor.

The ubiquitous chip offers designers and users of industrial heating equipment an awesome increase in functionality over traditional analog equipment. What's more, the chip permits the use of standard digital communication techniques which in turn also offers an overwhelming increase in reliability and security.

Yet most industrial heating equipment users are still leery of the digital world, and for good reason. What with all the jargon and the latest buzzwords, the digital domain can be tremendously confusing. Yet an awareness of the fundamentals of digital communication, as well as the strengths and weaknesses of each method along with its optimum application, is necessary for a reliable, safe and secure system design.

As new as it may seem, digital communication has been around for quite some time. Paul Revere's ride started with digital signals: "one if by land and two if by sea." Samuel F. B. Morse's telegraph was also digital.

In other words, digital or data communication is nothing more than information coded against time. In the case of the telegraph, it's dots and dashes. In today's digital world, it is the timed transition from one level of signal strength to another.

The smallest unit of information in digital communication is the bit, which stands for "binary digit." A bit is a 1 or a 0, a high or a low voltage (see Figure 1). Several bits are usually lumped together to form higher level units of information, just as letters of the alphabet are lumped together to form higher level units called words.

Bit to Byte

In the digital world, the most common grouping is eight bits, which is usually referred to as a byte or character (just for the record, a half-byte is called a nibble). The letter A in "digital," for example, is 11000001. The number 8 is 00111000.

There are basically two ways of moving bits from one place or device in a system to another. The first method is to move them serially, or one little bit at a time. That's the way the telegraph worked. The second method is to move information in bigger bites, or bytes, called parallel transmission (see Figure 2).

Obviously, moving bytes is much faster than moving bits. Since speed is what it's all about, parallel transmission is the method used within digital devices such as computers.

Moving data from one digital device to another, however, is another story. Parallel transmission is easily disrupted by outside interference and noise. Unwanted signals jump into the data stream and change the information. Protecting the data as it snakes over the many required wires, especially in a "hostile" environment, can be an exercise in futility.

So serial communication is the preferred method of linking dispersed digital devices. There are essentially two formats for serial transmission. Asynchronous transmission uses a varying idle time between the transmission of successive characters.

James M. Sullivan is Industrial Heating Market Manager for Honeywell's Manufacturing Control Systems Business.
An 18-year veteran with the company, Mr. Sullivan began his career as a Field Engineer in Boston.

A native New Englander, he holds a degree in Electrical Engineering from Northeastern University. He is a member of the American Society for Metals and is the official Honeywell representative to the Industrial Heating Equipment Association.

Reprinted with permission from Heat Management, June 1985, 6 pp, © 1985 Industrial Heating Equipment Association

DIGITAL DATA

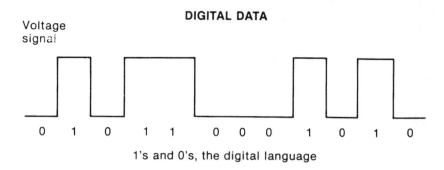

Voltage signal

0 1 0 1 1 0 0 0 1 0 1 0

1's and 0's, the digital language

Figure 1

Also, both the beginning and end of each character are indicated by a start bit and one or more stop bits.

Synchronous transmission sends an uninterrupted block of data. Each block is preceded by one or more sync characters in order to synchronize the receiver with the incoming data stream. Each have their advantages and disadvantages (see Figure 3).

Async Transmission

Asynchronous is probably the most familiar as it is commonly used in teletype and telegraph communications. There are a couple of standards for asynchronous transmission. One is called ASCII (which stands for American Standard Code for Information Interchange) and was developed by the American National Standards Institute. The other is called RS-232C. Actually, it is a recommended standard developed by the Electronics Industries Association.

ASCII (pronounced "askee") is both a data code and a standard. The ASCII code uses seven bits for information and one parity bit. The parity bit is included to check whether or not an error has occurred during transmission.

Figure 4 shows the ASCII code set. Since there are seven information bits, there are 2^7 or 128 characters. These include both upper and lower case letters of the alphabet, numbers 0 to 9, punctuation marks, various symbols, and control characters.

The control characters do not appear in the text of a message, but perform such functions identifying the beginning of a message or acknowledging receipt of a message. The functional definitions of the control characters are spelled out for the ASCII

SERIAL VS PARALLEL

Serial

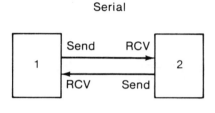

- Data is sent bit by bit
- Only two wires needed
- Slower speeds
- Longer distances

Parallel

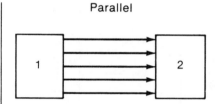

- Entire 8-bit or 16-bit word is sent at once
- Computer to computer
- Short distances (<10 ft)
- Fast speeds (up to 1 MBAUD)

Figure 2

SYNCHRONOUS/ASYNCHRONOUS

SYNCHRONOUS

- Serial transmission
- Uses internal clock to synchronize characters
- Most efficient for high volume, constant transmission
- High-speed

ASYNCHRONOUS

- Serial transmission
- Uses start and stop bits to separate characters
- PC's CRT's, printers and other factory devices
- Lower speed

Figure 3

Source: Heat Management, June 1985, 6 pp

259

MULTIDROP CONFIGURATION USING RS422

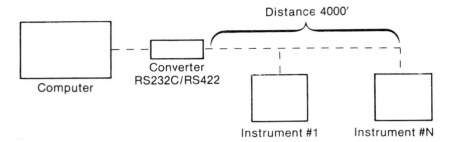

Figure 6

DMCS HEAT TREAT MANAGEMENT SYSTEM

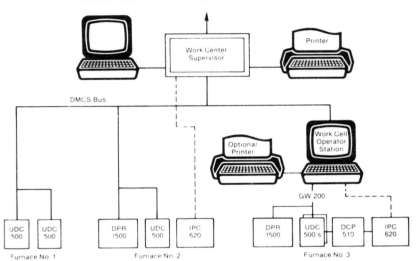

Figure 7

also adopted by the Electronic Industries Association, is a balanced system (in that it includes separate ground wires) which offers greater transmission distances and higher speeds. For example, the maximum RS-422 rate is 6,000,000 bps (6M bps) at 50 feet. At 4000 feet, the maximum distance permitted, the baud rate is 100K bps. Furthermore, RS-422 permits several devices, called multidrop connections, to be interconnected. It also provides superior protection against noise (see Figure 6).

That protection against noise turns out to be terribly significant, especially in the operational control of industrial heating equipment. If a temperature controller receives a command to reset the atmosphere cutoff to 145.0°F when the message sent said 1450°F, for example, a disaster could result.

Clearly one could protect against that sort of occurrence by proper monitoring of the system in operation. But reliable noise protection makes that job a little easier.

There are, then, essentially three basic requirements for a good industrial heat treatment digital communications network:

- Reliable communications security (or, simply, noise protection)
- Multidrop interconnections
- Highest transfer rate possible

Unfortunately, there really is no existing standard which satisfies the requirements for reliable information exchange in a hostile environment. Consequently, each vendor ends up developing a proprietary communications system.

In the recently introduced DMCS 3000 heat treat management system, for example, the communication format and protocol used is a synchronous, modified RS-422 version with a High Level Data Link Control (HDLC) protocol. HDLC is an International Standards Organization protocol that uses a defined frame format, zero-bit insertion and deletion and cyclical redundancy check error checking. Without going into a long explanation, HDLC is defined so as to detect data transmission errors in an extremely reliable fashion.

DMCS (which stands for Distributed Manufacturing Control System) requires a high degree of data integrity because its architecture allows up to 124 controllers, recorders, programmers and programmable controllers to connect to the work center supervisor (WCS), a management level system.

The system allows the heat treating operation to collect sensor-based heat treating and production information in real-time. That means it requires very fast data transfer rates. And, since the system is designed to control all sizes of industrial heating operations, including those with multiple furnaces, it requires long distances between devices (see Figure 7).

Personal Computer to Programmable Controller

One other thing that the DMCS offers is an interface capability to most personal computers. But most personal computers use RS-232C for data communications. That means that somehow the RS-232C ASCII format and protocol need to be translated to this proprietary system. The device that

	Bits			7	0	0	0	0	1	1	1	1
				6	0	0	1	1	0	0	1	1
4	3	2	1	5	0	1	0	1	0	1	0	1
0	0	0	0		NUL	DLE	SP	0	@	P	\	p
0	0	0	1		SOH	DC1	!	1	A	Q	a	q
0	0	1	0		STX	DC2	"	2	B	R	b	r
0	0	1	1		ETX	DC3	#	3	C	S	c	s
0	1	0	0		EOT	DC4	$	4	D	T	d	t
0	1	0	1		ENQ	NAK	%	5	E	U	e	u
0	1	1	0		ACK	SYN	&	6	F	V	f	v
0	1	1	1		BEL	ETB	'	7	G	W	g	w
1	0	0	0		BS	CAN	(	8	H	X	h	x
1	0	0	1		HT	EM	)	9	I	Y	i	y
1	0	1	0		LF	SUB	*	:	J	Z	j	z
1	0	1	1		VT	ESC	+	;	K	[	k	{
1	1	0	0		FF	FS	'	<	L	\	l	:
1	1	0	1		CR	GS	–	=	M	]	m	}
1	1	1	0		SO	RS	.	>	N	∧	n	~
1	1	1	1		SI	US	/	?	O	—	o	DEL

code set, although some equipment manufacturers redefine the functions of some of these control characters for their particular piece of equipment.

Since each character in asynchronous transmission uses a specific start and end bit sequence to identify the beginning and end of a character, the overhead required per character is significant. Asynchronous is therefore most suited where short, slow messages are predominant.

RS-232C

The RS-232C standard covers the mechanical, electrical and functional characteristics of asynchronous transmission. The mechanical aspect specifies the type of plug (male or female), the pin assignments of the plug and the number of wires for data, timing and control interchange. The electrical portion defines acceptable signal levels and grounding requirements. Finally, the functional section specifies 25 interchange circuits although the number actually employed depends on the type of system used.

RS-232C electrical characteristics are such that data transmission can occur only between a single point and another (called point-to-point) that are no more than 50 feet apart. What's more, the maximum transmission rate is 20,000 bits per second (or 20K bps or 20K baud). That minimal distance can be extended if signal boosters (called line drivers) or modems are plugged in. A modem is a device which converts (MOdulates) transmitted digital signals into analog form and reconverts (DEModulates) received analog signals back into digital form (see Figure 5).

There is an alternative to the limitations of RS-232C, however. RS-422,

Example

```
Bits:  P*  7 6 5 4 3 2 1
       1   1 0 0 0 0 0 1  = letter "A" (Odd Parity)
       0   0 1 1 1 0 0 0  = number "8" (Odd Parity)
```

P* = Parity Bit

Figure 4

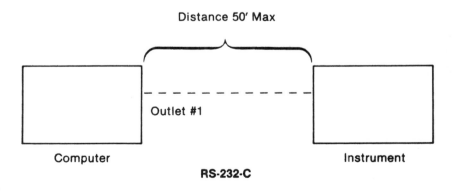

Distance 50' Max

Outlet #1

Computer   Instrument

RS-232-C

Figure 5

does that is called a Gateway (see Figure 8).

Gateway devices translate at various levels, ranging from the relatively simple RS-232C to RS-422 conversion up to transmission checks for errors and self-diagnostics.

The Gateway allows a selectable transmission rate of 110 bps to 19.2K bps on the RS-232 side. Data security procedures are such that the RS-232 security approaches that of the DMCS transmission line (or bus). Since the gateway can sit very close to the personal computer it translates for, the potential for a hostile environment is less of a problem.

To sum it all up, then, there are essentially three considerations of utmost importance when using digital communications in industrial heating applications:

• Reliable communications security (or, simply, noise protection)

• Multidrop interconnections

• Highest transfer rate possible

The system designer or user who allows for a tremendously hostile environment while ensuring the highest data transfer rate possible between the broadest number of devices will find that the improved efficiency and reliability of the digital heat treat management system will more than make up for the time put into designing the system. ∎

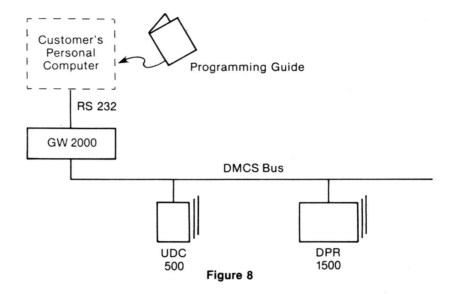

Figure 8

Microprocessor Control in Metallurgical Furnaces

Stavros A. Argyropoulos, Osama T. Albaharna and Bernard M. Closset

INTRODUCTION

Microprocessors are invading almost every area of our lives and frequently turn up in unexpected applications. While new equipment often features built-in microprocessor systems, the upgrading of existing processes has developed slowly. There are a variety of reasons for this—many economic. One of the principal reasons, however, is the general reluctance of many organizations to strongly commit to put this new technology into existing processes.

Microprocessor systems developed "in lab" not only measure the thermophysical events taking place in metallurgical furnaces, but simultaneously control the furnaces.

Hardware

This work incorporates a design philosophy involving the "in-situ" digitization of analog signals in an intelligent satellite microperipheral. The μMAC-5000 control system was chosen in this application which required distributed intelligence of a local front end.[1] It has the Intel 8088 microprocessor, 80 K bytes of ROM and 64 K bytes of RAM. Analog-to-digital conversion is performed on the μMAC-5000 via an integrating converter providing 14-bit resolution. The analog input handling capability offers reliable operation in the harsh, electrically noisy, metallurgical environment. The system scales, linearizes the converts the input data to engineering units. Both analog and digital control capability are provided by the μMAX-5000 control system. In supervisory control applications, the μMAC-5000 can be used either as a local front end or can be located up to 10,000 feet from a host. It also operates at speeds up to 19.2 K baud in an asynchronous mode. While the μMAC-5000 provided the intelligence for the decision-making required in controlling metallurgical furnaces, different hardware was interfaced with each μMAC-5000 controlled furnace.

Software

A structured programming philosophy was followed in developing the software which systematically evaluates the correctness of the smaller independent modules to minimize the overall program testing phase. Self-documenting, structured programs enable a reduction in the often overlooked, and sometimes significant, maintenance costs. In addition, errors are easily isolated since they are contained within a module and not the entire program. Structured programming also aids productivity since several programmers can work on the various modules independently. The μMACBASIC is stored in 80 K bytes of ROM in the μMAC-5000 and provides these structured programming benefits. It is especially designed to perform real-time measurement and control functions using simple BASIC commands. The μMACBASIC permits: communications interrupts (generated by user defined character); periodic interrupts (for periods of 100 ms to 1 year); and event interrupts (occurring when the μMAC-5000, event counters count down to zero).

RESISTANCE FURNACE/ Al-Sr PHASE DIAGRAM

There is disagreement in the literature concerning the Al-Sr phase diagram. Vakhobov et al[2] found one intermediary compound ($Al_4 Sr$) which melts congruently at 1000°C. Bruzzone et al[3] identified three compounds, one melting congruently ($Al_4 Sr$), and two melting incongruently ($Al_2 Sr$ and $Al_2 Sr_3$). A Kanthal wound resistance furnace was used to further study the Al-Sr phase diagram by a thermal analysis technique. The reactivity of Strontium at high temperatures with elements such as oxygen, nitrogen and water vapor has necessitated that the thermal analysis of the different Al-Sr alloys should be done under Argon atmosphere. Alloy samples of 100 gr were prepared, and the different transformation points of

The two microprocessor based controllers featured in this paper were developed with the philosophy of the distributed computing concept. The analog signals were digitized very close to the source, as well as the real-time functions.

The following data demonstrate the continued trend towards taking the computational capability to the process instead of bringing the process to the computer. As the digital hardware continues to become more economical, it is expected that greater utilization of these types of systems will be seen in controlling and optimizing existing processes.

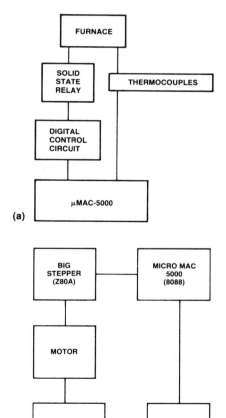

▲
Figure 1. Overview of the control systems. (a) resistance furnace. (b) Induction furnace.

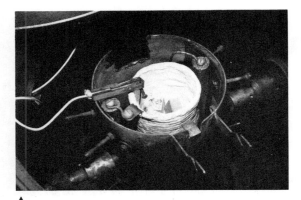

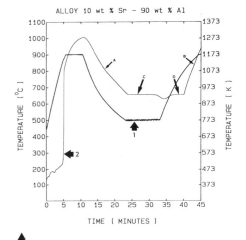

Figure 2. Detailed description of the digital control circuit for the Kanthal wound resistance furnace. The photograph shows the inside of the vacuum chamber with the control thermocouple.

Figure 3. Experimental results from thermal analysis of a 10% Sr-90% Al alloy. Curve 1 shows the measured temperature by the control Thermocouple. Curve 2 presents the measured temperature of the alloy melt. Arrows A and B illustrate the liquidus transformation points. Arrows C and D depict the eutectic temperature.

each alloy were determined during controlled cooling and heating cycles. Several compositions from 0 to 100% Strontium were investigated. At low Strontium levels, an alumina crucible was used. With high Strontium content alloys, an iron crucible was utilized. Two K-type thermocouples were used—one to control the electrical furnace and consequently the cooling and heating cycle, and another immersed into the molten alloy to measure the transformations taking place during heating and cooling. The second thermocouple was attached to an assembly and immersed into the alloy when it was liquid. The relative positions of these two thermocouples were not always the same.

As seen in Figure 1, a solid state relay was used to rapidly switch the furnace on and off. The control signal of the relay is derived from the digital control circuit. Figure 2 presents the digital control circuit in detail. The circuit controls the average power supplied to the furnace. Three bits of the digital output port of μMAC-5000 are fed into this circuit. Depending on the type of bits the μMAC-5000 sends, the control circuit generates square waves of different duty cycles. These square waves are used to control the relay. A high in the wave causes the relay to close and vice-versa. By varying the duty cycle of the wave, the relay's switching time is changed. Since the square waves are periodic signals with a relatively high frequency, the average power supplied to the furnace can be changed by varying the duty cycle. Since three bits are supplied to the digital control circuit, eight different sequences of different duty cycles can be generated. The binary numbers appearing on the μMAC-5000's digital output port can be changed by the appropriate program command. As a result of the computations done in the program, the specific sequence decision is made and sent to the relay.

The resistance furnace follows a specified temperature profile, shown in Figures 3 and 4 (Curve 1). Table I presents all the parameters input by the user at the program's beginning. The experiment is divided into six states. The algorithm differs from each of these states since they occur at different temperatures—where the heat transfer characteristics of the furnace are different. Thus, the program should be tuned accordingly. The software consists of a main program and an interrupt procedure. The procedure is called by the main program every 250 ms. It reads the temperature, identifies the state number (which, for programming purposes, ranges from 0 to 5), compares the actual temperature with the expected one and then decides for how much time the relay will be open or closed. At the end of the experiment, the results are transferred to a host microcomputer system for further processing.

Table I. List of the Initial Parameters Used During Program Initialization

MS:	Duration of experiment in seconds.
HR:	Specified first heating rate.
HS:	Specified second heating rate.
TT:	First temperature limit.
TS:	Duration in seconds at first limit.
CR:	Specified cooling rate.
TL:	Lower limit.
TM:	Duration a lower limit.
UT:	Second temperature limit.

Table II. Partial List of ASCII Strings Which Are Accepted from the BIG STEPPER

S,1,200,F,20	Step motor 1 forward 200 steps/sec for 20 steps.
S,1,200,R,10	Step motor 1 reverse 200 steps/sec for 10 steps.
A,1,600	Alter rate on motor 1 to 600 steps/sec on the fly.
F,1	Halt motor 1.

Table III. Chemical Analysis of Ferrotungsten and Silicon Powder Used.

Ferrotungsten powder	W	79.40
(30 mesh × down)	Si	0.55
	P	0.03
	C	0.01
	S	0.05
	Fe	Bal.
Silicon powder	Si	98.64
(50 mesh × down)	Fe	0.55
	Ca	0.21
	Al	Bal.

Figure 3 presents typical experimental results for an aluminum-10% weight Strontium alloy. The liquidus transformation temperatures at 810°C and 820°C for the corresponding cooling and heating of the alloy. The eutectic temperature is 655°C. The two phases identified by X-ray diffraction and metallography are Al and Al_4 Sr. Figure 4 shows similar experimental results with an aluminum Strontium alloy containing 90% Strontium in weight. The eutectic temperatures determined during cooling and heating are respectively 575°C and 580°C. At this eutectic composition, the two phases are Sr and AlSr.

INDUCTION FURNACE/ MICROEXOTHERMIC FERROALLOYS

Microexothermic (Autoexothermic) ladle additives make up a new family of ferroalloys, which exhibit superior recovery and uniformity characteristics when conventional alloying or microalloying schemes are implemented.[4,5,6,7] In this family, the heat, released from the intermetallic formation, assists their assimilation into liquid steel. This results since the released heat alters the assimilation process of ferroalloys from dissolution to melting.[4] During this family's development, an experimental precedure was devised to measure their "microexothermicity." This was done by heating samples from these additions in an air induction coil and keeping the power input to the coil manually constant. Figure 5 shows a typical experimental result obtained from the controller. The term "microexothermicity" was coined during an earlier work to indicate the type of exothermicity released when the intermetallic formation in powder alloy compacts takes place.[4]

The microprocessor-based system presented schematically in Figure 1 replaced the manual control with an automatic one. As shown in Figure 6, the power knob of the induction furnace was replaced with a stepping motor. The design leaves the induction furnace's electrical systems intact and facilitates furnace operation when manual control is required. As Figure 1 illustrates, the μMAC-5000 is the host microcomputer and is connected with an intelligent motion controller, the BIG STEPPER, via a standard RS-232C interface. The BIG STEPPER is controlled via a series of ASCII command strings (Table III) passed through the connecting RS-232C cable from the μMAC-5000.[8] the stepping motor is a multiple pole DC permanent magnet motor with multiple windings. When a correctly sequenced voltage pattern is applied to the windings, the motor can rotate a precise number of steps in either direction or stop in an exact position. Stepping motors do not just rotate, but instead move in discrete repeatable steps (for this particular motor, each step is 1.8 degree).

The μMAC-5000 measured the temperature of the specimen inside the coil. The control thermocouple was inserted in a small groove at the specimen's edge. The other thermocouple was located in the specimen's centre. This controller applied a constant linear heating rate at the specimen's edge with simultaneous measurement of the temperature at the specimen's centre. When the specimen edge reached a given temperature, the controller kept the temperature constant. The user specifies the heating rate as well as the upper temperature limit. A proportional-integral-derivative (PID) algorithm was implemented in the software. The algorithm was designed to achieve true anticipation of heating changes resulting in tighter control and minimized setpoint overshoot. Figure 6 shows the block diagram of the induction furnace control. This feedback servo-control system performs three control actions. First, it measures the difference between the commanded and actual temperatures and uses a function of

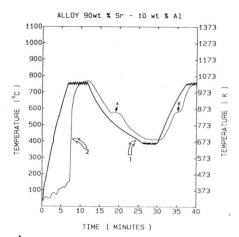

Figure 4. Experimental results from thermal analysis of a 90% Sr-10% Al alloy. Curve 1 shows the measured temperature by the control thermocouple. Curve 2 presents the measure temperature of the alloy melt. Arrows A and B indicate the eutectic temperature.

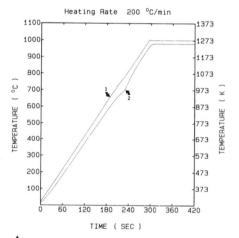

Figure 5. Experimental results obtained by heating a solid steel cylinder inside the induction coil. Curve 1 shows the measured temperature at a location of 0.4 cm from the cylinder's edge. Curve 2 presents the temperature at the cylinder's vertical axis.

Figure 6. Block diagram of the induction furnace control system. The μMAC-5000, BIG STEPPER and Stepping Motor are attached to the control panel.

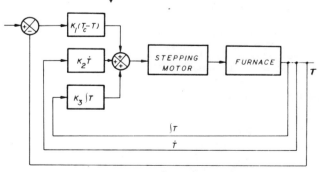

T = temperature
Ṫ = rate of temperature change
∫T = integral of temperature
T_c = desired value of temperature

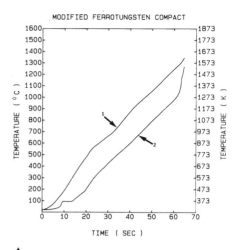

MODIFIED FERROTUNGSTEN COMPACT

Figure 7. Experimental results from the heating of modified ferrotungsten compact in air induction coil. Curve 1 is the controlled temperature at the edge of the compact. Curve 2 is the temperature at the centerline of the compact.

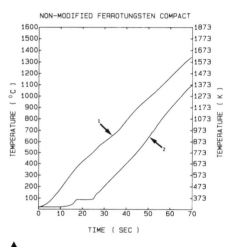

NON-MODIFIED FERROTUNGSTEN COMPACT

Figure 8. Heating of a non-modified ferrotungsten compact in an induction coil with the microprocessor based controller. Curve 1 is the controlled temperature at the edge of the compact. Curve 2 is the temperature at the centerline of the compact.

ABOUT THE AUTHORS ▉▉▉▉▉▉

Stavros A. Argyropoulos received his Ph.D. in metallurgy from McGill University. He is currently NSERC Assistant Professor in the Department of Mining and Metallurgical Engineering at McGill University in Montreal, Quebec, Canada.

Osama T. Albaharna received his B. Eng. in electrical engineering from McGill University in 1984. He is currently pursuing his M.Eng. in Electrical Engineering at McGill University.

Bernard M. Closset received his Ph.D in metallurgy from the School of Mining in Nancy, France. He is currently Manager of Technical Services of CHROMASCO Ltd., Toronto, Ontario, Canada.

that difference to drive the error-reducing action of the furnace. Second, it introduces a "rate damping" to the system. This is represented by the symbol T. Third, the addition of a quantity proportional to the integral of the controlled quantity reduces error when the system reaches equilibrium. This is represented by the symbol $\int T$. The different values of K_1, K_2 and K_3 were determined with a series of tests done by heating a solid steel cylinder 3 cm in diameter. In these tests, the control thermocouple was located 0.4 cm from the cylinder's edge. Different heating rates, ranging from 200°C/min to 1000°C/min, were used with a constant upper temperature of 1000°C. As shown in Figure 5, the controller imposed a perfect linear heating rate on the steel cylinder. When it reached the maximum, the overshoot of the temperature was quickly damped out.

The microprocessor based induction furnace controller was used to measure the "microexothermicity" of different types of ferrotungsten cylindrical compacts. Table II gives the chemical analysis of the materials used. Two types of compacts were made: first, the modified compacts (the ferrotungsten powder was mixed with silicon powder); and second, the non-modified compacts (only ferrotungsten powder was used). More details on the compact preparation is featured in reference 4.

Figures 7 and 8 present typical experimental results for modified and non-modified ferrotungsten compacts, respectively. In both tests, the microprocessor-based controller imposed practically the same heating rate 1200°C/min. In these figures, Curve 1 shows the temperature at the edge of the compact, and Curve 2 shows the temperature in the center line of the compact. In both cases, there is a delay in the temperature increase at the compact centerline (Curve 2). This is due to the fact that the moisture existing in the compact takes time to evaporate. Subsequently, both compacts follow similar temperature increases up to the 60th second. After this time, however, there is a dramatic increase in the slope of Curve 2 for the case of the modified compact. This abrupt change in the temperature slope can only be explained on the basis of the heat generated in the body of the compact. Clearly, the formation of different tungsten-silicon intermetallics is exothermic. Consequently, heat is generated from the interior of the compact. In the case of the non-modified ferrotungsten (Figure 8, Curve #2), where the compact was made only from ferrotungsten without any silicon, the aforementioned phenomenon is neither observed nor expected. Curve 2 in Figure 7 possesses an important feature—the exothermic reaction is initiated when the centerline temperature is about 1000°C. After a few seconds, however, there is a corrosive attack on the thermocouple, and when the temperature rises to 1270°C, the thermocouple is destroyed. The heat released from the intermetallic formation is enough to melt the tungsten-silicides. Consequently, this kind of addition can be dispersed into liquid steel following a melting pattern rather than a dissolution one. The whole mass transfer kinetics can thus be tremendously enhanced. In addition, the relatively low temperature at which the microexothermic reaction initiated, means this family of additions can be used for both alloying liquid steel and alloying cast iron.

ACKNOWLEDGEMENTS

The authors would like to express their sincere appreciation to the National Sciences and Engineering Research Council of Canada and the Faculty of Graduate Studies and Research of McGill University of providing support for this work. This work was also supported in part by the Shieldalloy Corporation by a grant to S.A. Argyropoulos.

References

1. *μMAC-5000, Operation, Concepts, Command Reference Programming,* User's Manuals, Analog Devices Inc., Norwood, Massachusetts, 1984.
2. A.V. Vakhobov, T.D. Dzhurayev, V.A. Bardin, G.A. Zademiko, "Constitution Diagrams of the Aluminum-Strontium and Lead-Strontium Systems," *Russian Metallurgy (Metally),* No 1, pp. 163-166.
3. G. Bruzzone and F. Merlo, "The Strontium-Aluminum and Barium-Aluminum Systems," *Journal of Less-Common Metals,* 39 (1975), pp. 1-6.
4. S.A. Argyropoulos, "The Effect of Microexothermicity and Macroexothermicity on the Dissolution of Ferroalloys in Liquid Steel," *42nd ISS-AIME Electric Furnace Proceedings,* Toronto, 1984, pp. 133-148.
5. P.D. Deeley and S.A. Argyropoulos, "Autoexothermic Ferroalloys a New Ladle Metallurgy Tool for Improved Continuous Casting Success," Proceedings of Continuous Casting '85, 22-24 May 1985, The Institute of Metals, London, England, in press.
6. S.A. Argyropoulos and P.D. Deeley, "Exothermic Alloy for Addition of Alloying Ingredients to Steel," United States Patent, Patent Number 4472196, September 18, 1984.
7. European Patent Application No. 84303942.1 (Publication No. 0129390), Claiming Priority from USSN 504217: "Exothermic Alloy for Addition of Alloying Ingredients to Steel," filed December 6, 1984.
8. Stepping Motor Cookbook, Centre Computer Consultants, P.O. Box 739, State College, PA 16894, U.S.A.

SECTION VIII
The Computer in Materials Engineering

SECTION VIII
The Computer in Materials Engineering

The computer's great power and speed in acquiring, storing, coding, classifying, sorting, and accessing many thousands of items of information have made it useful in the important field of materials testing in metalworking.

Mechanical testing machines are important in metalworking for both the establishment and the maintenance of agreed-upon quality standards, as well as for meeting metallurgical requirements in critical customer applications. The market for mechanical testing instruments is long-established, with traditional methods of testing for hardness, elasticity, and other tensile parameters.

In recent years, the run-up of energy costs, together with shortages in some materials, has resulted in a search for alternative materials in order to save weight and costs while still meeting metallurgical requirements and standards. In addition, the tightening of product liability rules and regulations has required more testing, with better repeatability at higher data speeds. Older methods of manually recording test data have become less cost-effective than they once were.

Not surprisingly, the way to more efficient, accurate, and recordable mechanical testing lay in the development and application of the digital computer.

Before this could take place, however, a means had to be found and applied to enable the testing machines to convert their mechanically derived information into analog electronic information, and then into digital data for inputting to microprocessor devices. The invention of the bonded strain gage, with its numerous solid-state descendants, opened the way for the computerization of materials testing, as the following articles will show.

Before the availability of the digital computer, the design and selection of alloy and carbon steels for heat treated applications was a science mixed with not a little art. One approach is described in the series of four articles, "Computer-Based System Selects Optimum Cost Steels."

A similar, but more recent application is described in "Computer Program Predicts Jominy End Quench Hardenability."

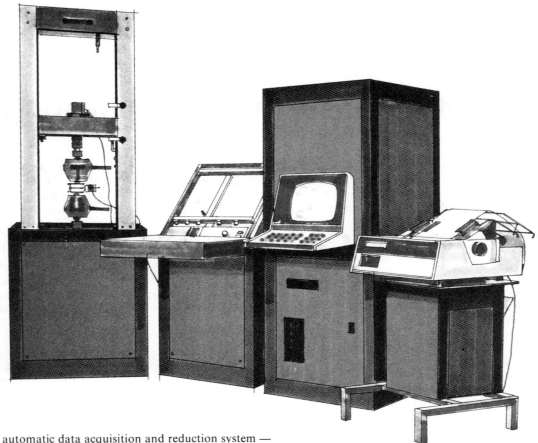

An automatic data acquisition and reduction system —
the Data-Matic III. From left to right: an
electromechanical testing machine; an X-Y
plotter-recorder; a console containing a microcomputer,
automatic machine control, computer display terminal
screen, tape deck, punched card reader, and floppy
disc system; and the last unit, a line printer.

Application of Microcomputers
to Mechanical Testing of Materials

By Paul M. Mumford

IN 1938, TWO MEN came up independently — and
almost simultaneously — with the idea of sticking a
filament wire to a test specimen with plastic cement,
and then reading strain by the change of resistance of
the wire. This "bonded wire strain gage" was the first
means of measuring actual strain in parts during
usage. Development of the strain gage has been hailed

by most people in the mechanical testing of materials
arena as the "start of the engineering revolution."

A similar revolution has been changing the
electronics world over the past several years. Invention of the integrated circuit and its development into
the microprocessor has certainly revolutionized the
electronics industry.

Today, two major factors create the need for more
and better tests in the materials field.

Reprinted from Metal Progress, August 1980, 46-50, © 1980 American Society for Metals

First, energy and materials shortages have created tremendous interest in alternate materials to save weight and to avoid use of rare or expensive materials.

Second, strict product liability decisions have made additional testing an economic necessity for many companies. Application of computer technology in these tests has been growing over several years, but cost has been a severe limitation.

Development of the microprocessor has made it possible to provide a dedicated small computer at an economically attractive price. The machines can provide several technical improvements to yield better test results, and to save enough time to reduce costs associated with testing programs.

An Overview of This Technology

In summary, advantages and limitations of the microcomputer as applied to a universal testing machine follow:

Advantages Relating to Better Test Data

1. A standard test procedure is preprogrammed. No operator skill or judgment is required.

2. Data are printed directly in the test report, or sent to a large computer system data base without human intervention — no transposed figures or unreadable data sheets.

3. Analysis of data to obtain modulus of elasticity, yield strength, etc. is preprogrammed. Operator skill or judgment does not affect results.

Advantages Relating to Economics

1. For a high volume test requiring determination of modulus of elasticity, 0.2% offset yield strength, and ultimate strength, a single testing machine and operator can conduct about three times as many tests per shift as the same operator with the same testing machine could without the computer.

2. The operator need not be skilled in analysis of data.

Limitations of the System

1. The economic advantage is reduced as the volume of similar tests is reduced. Time saved by the computer is essentially in data reduction time. The more data, the more savings.

2. The cost and time needed for preparation of programs may be too large to be practical for a small number of tests. The problem is combated two ways. First, all user programs are in BASIC, making program development relatively quick and easy when compared with the assembly language programs used by earlier machines. Second, by offering these programs to multiple customers, we share the cost among several users.

3. Our present microcomputer can process only a finite amount of data. Its capacity is adequate for control and processing on a single machine at normal test speeds. It is not suitable for fatigue testing or other tests needing data sampling rates faster than the range of 5 to 15 samples per second.

The Evolution of a New System

Our Data-Matic family of data acquisition and control systems started with discussions between myself and Jere Watson, president of United Calibration, back in 1972. He saw the need for a data acquisition system, and I saw a way to build it.

Our first system was built as a programmable calculator, using a four function calculator integrated circuit combined with a large board full of less complex integrated circuits, a calculator style keyboard, digital display, and a tape printer. This system, called Data-Matic I, was shown at the 1973 Metal Show at Cobo Hall in Detroit. For the first time, a 0.2% offset yield strength test could be run without manual data reduction from the stress-strain curve.

Data-Matic I was moderately successful, but it had several important limitations, specifically:

1. The tape printer could print only numeric results, so data still were transcribed to the test report.

2. The machine was programmed in its own language, which made program development slow and required special training.

3. Programs were stored in read-only memory. Revisions required that new memory devices be programmed — a task requiring special equipment.

4. The processor was limited in speed. Some important analysis techniques could not be used simply because the calculator device could not do the required math in a reasonable time. This was by far the most important limitation. We could process only about one data point per second, using very simple techniques which required no more than four to six multiply operations plus eight to twelve add or subtract operations per point.

Data-Matic II was developed in response to a

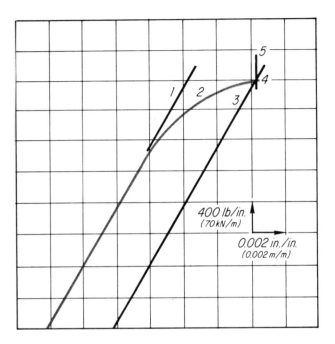

Fig. 1 — Typical X-Y recorder plot for tensile test, 0.2% offset yield. Code: 1, best fit modulus line; 2, stress-strain curve; 3, percent offset line (0.2% of 2 in. [51 mm] = 0.004 in. [0.1 mm]); 4, offset yield point (3200 lb [14.2 kN]); 5, ultimate stress (3525 lb [15.7 kN]).

requirement to record data from up to five channels at rates of 100 points per second. No data processing was done internally, but a communication link to a large computer (IBM 370) allowed highly sophisticated processing by the large machine. The requirement for this system coincided with the availability of a much more powerful microcomputer, the Motorola M6800. This new device was utilized to control the data acquisition, recording, and communications system.

Next Step — Our experience with applying the M6800 in Data-Matic II led to the conclusion that it made possible a much more powerful self-continued data processor, so the Data-Matic III was conceived. This System provided:

1. A multiply time of about 0.003 s vs 0.120 s for Data-Matic I.

2. BASIC language, to speed development of the required programs.

3. Cassette tape storage for all programs, to make the system flexible, and to ease the problem of sharing programs among several users.

4. A full page printer so that a complete test report can be printed in final form.

5. A CRT terminal to allow prompting of the operator for necessary input information and other actions.

6. Complete communications capability to allow direct links to any large computer system.

Refinement — Once these new tools were available, we set out to develop a better program for 0.2% offset yield determination. We wanted to eliminate manual entry of an estimated proportional limit and also improve the precision.

Harold Stroebel of Aluminum Co. of America suggested and programmed a best straight line, least squares fit of the initial data to establish modulus of elasticity. This program has been refined over the past two years and gives excellent results (Fig. 1).

Two areas proved critical:

1. Where to start looking at the curve. Due to imperfections in samples and extensometry, the start of the stress-strain curve is frequently distorted. To avoid this possibly distorted area, we do not start fitting data to the straight line until some strain and load thresholds are reached (typically 200 to 500 microstrain and 0.1 to 1% of load range).

2. Where to terminate the straight line fit of the data. After the strain and load thresholds are reached, a fixed number of data points (typically ten) is put into straight line fit to establish an approximate modulus slope. Subsequently, each new data point is added to the curve fit and a new slope computed for the whole. This new slope is then compared with the preceding value to determine when the data deviates from the straight line. Of course, this is not quite as simple as it sounds. At the beginning, with only a ten-point fit, a substantial uncertainty in the slope value exists. If the curve continued straight, and we add more points to the least squares fit, then the uncertainty is reduced, so we need to make the comparison limits closer as more points are added to fit (Fig. 2).

Typically, the comparison equation is:

If $B < ((1 + 0.02/N) \cdot B (N - 1))$, then count one point off the straight line. Where B is the slope, and N is the number of points fit to the curve.

When some number of points (typically three) have fallen off the straight line, we back up to the last point which fell within the linearity limits and use the slope and intercept values existing at that point in the curve fit.

The intercept value from the least squares fit is also

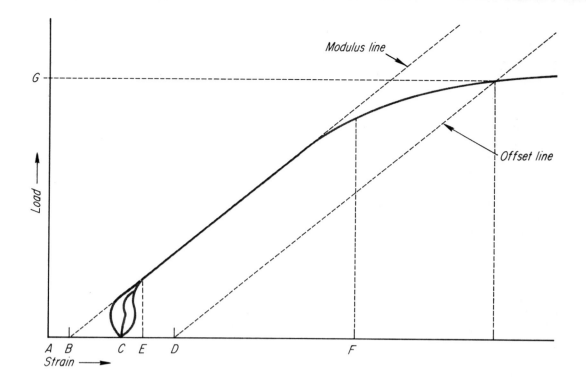

Fig. 2 — Load-strain plot shows zero error sources and compensation. Point C is initial reading (tare) for strain. Value is subtracted from all subsequent readings to correct measurement zero error. Point E is strain threshold for start of least squares fit. Fit operation continues between Points E and F. Point B is intercept point, where zero strain would have been if sample and extensometer were perfect. Point G is load for offset yield.

very important. Because of the imperfections in the stress-strain curve noted earlier, the actual zero value for strain may differ significantly from the "zero" noted at the start of the test. The intercept value does, in fact, establish a zero strain value at the point where the extended straight line portion of the curve crosses the zero load line.

When we subsequently test for a yield or extension under load value, the intercept value is included in the computation to locate the offset line value. This computation is detailed below:

Stress=Load/Area.

Theoretical Strain=Stress/Modulus.

Offset Strain=Theoretical Strain+Offset.

Measured Strain=Strain Reading−Initial Tare Value−Intercept Value.

If measured strain is less than offset strain, then we have not yet reached yield.

This process is, of course, limited in accuracy by the change in load and strain from one data point to the next. The Data-Matic III is capable of adequate sampling rates to make the error insignificant for most situations. An interpolation routine has been used to further reduce this error in cases where the curve normally crosses the offset line at an acute angle, making yield determination difficult, or where the test is to be pulled at high speeds, causing larger than usual changes in load and strain within the data sampling time.

Ultimate load determination is a straightforward process of comparing each successive reading with a value which is updated if the new reading is larger, and retained if not.

Total energy absorbed has also been determined by piecewise integration of the area added by each data sample. This technique is easy to implement and quite accurate.

Several other types of tests have been implemented; among them, flexural modulus, short beam shear, and ultimate compressive strength, to name a few. To date, the curve fitting technique for modulus determination is the most complex problem encountered.

Computer Levels — In addition to the data processing capabilities of the Data-Matic III, two levels of computer control of the load frame are in service.

The first level of control allows the computer to start, to stop the loading process, and to return the crosshead to an initial position determined by place-

Source: Metal Progress, August 1980, 46-50

ment of a limit switch. The control scheme simply allows the computer, via solid-state relays, to "push" the control buttons on the load control console.

A second level of control has been applied to both mechanical and hydraulic load frames and allows complete control of loading rate by the computer. In this setup, the computer is able, via a digital to analog converter, to command a rate for the load frame drive system.

The computer may be programmed to provide control based on measured strain, load, or position data. The system is able to process the complete servo plus data program at a rate of 5 to 15 points per second. This update rate has proved adequate to provide very good control of test rates both in hydraulic and mechanical drive load frames. This data rate would, of course, be insufficient for fatigue-type machines having fast response capability.

Digital computation of the servocontrol equations provides several major advantages over analog servoamplifiers, including:

1. Constants are easily changed at the press of a few keys.

2. The program can change its own constants easily to conform to any definable relationship. For example, the stiffness of a setup could be measured by using crosshead position and load data when taking up the slack in the system. The required load servoloop gain constant could be accurately determined from the stiffness data, and automatically put into the program.

3. The program can simultaneously keep track of load, strain, and position data. Transfers from one parameter to another are made smoothly with no bumps.

The digital servocontrol adds much to the utility of the system:

1. It allows complete test programming to drastically reduce human operator variables in the test.

2. It speeds testing by consistently running the test at specified speeds without operator intervention.

3. Automatic overload protection is easily included in any program. Load limit, strain limit, position limit, or any combination of these limits, may be programmed.

4. The machine may be stopped after yield is reached, if desired, to make certain that the extensometer is removed prior to sample break. (An audible alarm also may be sounded to remind the operator to remove the extensometer.)

5. On screw driven machines, crosshead position is obtained by installing an optical encoder to measure screw rotation. This yields high resolution crosshead position information with an accuracy as good as the screw. For example, 0.0005 in. (0.01 mm) for a 0.25 in. (6.3 mm) pitch screw.

A Look Into the Future

We are presently active in exploring the Motorola MC6809 microprocessor and the Advanced Micro Devices AM 9511 floating point math processor. These two new devices will allow us to offer two additional increments of computing power sometime in 1981. These faster, more powerful processors will be made available in our standard board configuration and could be easily retrofit to existing systems by a simple board change.

The MC6809 and AM 9511 combination will reduce multiply time from the present 0.003 s to about 0.0001 s, about 1000 times faster than the Data-Matic I achieved in 1973. This reduction in computation time will allow more sophisticated processing and also increase the rate at which data points may be processed.

To accompany the increase in processing capability, an analog to digital converter board with faster conversion capability will probably be needed. Again, we would make it essentially interchangeable with the present 15 conversions/s unit.

Many users wish to have a data retention capability for statistical analysis, or correlation of data from various types of tests on a lot of material. This capability is currently provided by an optional floppy disc system. The future will certainly see a change to the new fixed and sealed media, hard disc, which provides much greater storage capacity and faster data access.

On the programming side, we expect to improve and expand our library of programs to cover most all test requirements. We also plan to continue to use BASIC language for all test programs. We may, however, be able to expand the BASIC interpreter to include some features not now available. ☸

For More Information: You are invited to contact the author directly by letter or telephone. Mr. Mumford is director, data systems engineering, United Calibration Corp., 12761 Monarch St., Garden Grove, Calif. 92641; tel: 714/638-2322 or 893-1821.

Computer-Based System Selects Optimum Cost Steels

By DALE H. BREEN and GORDON H. WALTER

This is the first in a series of articles describing a new approach for the design and selection of alloy and carbon steels for heat treated applications. It is a systems approach incorporating the use of a digital computer. Within International Harvester, it is called the CHAT system, an acronym standing for Computer Harmonized-Application Tailored. These articles will describe, in some detail, the techniques used in the CHAT system. The first article deals mostly with background material, while the remaining articles provide a detailed description of application tailoring (the defining of metallurgical requirements), and computer harmonizing (designing optimum cost steels using a digital computer).

Since the turn of the century, the general nature of steel has continually changed through increased understanding of the metallurgical factors which control reactions in heat treatment and subsequent engineering performance. With increased understanding of the primary influences of microstructure came also the gradual development of quantitative knowledge concerning alloying elements and their influence over microstructures. The importance of the classic works of Shepherd[1] in grain size, Bain and Davenport[2] in transformation rationale, Grossmann[3] in calculated hardenability, and Jominy[4] in hardenability measurement is indisputable. Results of early work, mostly prior to 1959, are essential to current efforts of economical alloy utilization and alloy conservation.

If one would plot the various metallurgical milestones along a time line, he would note that the general trend of alloy usage and selection became more rational with increasing metallurgical insights. Change is healthy, and should be motivated by the desire for efficiency — and, of course, it should be done rationally. The steel to use is that with an alloy combination which will most effectively provide the engineering performance at least cost. In essence,

Mr. Breen is chief, Materials Div., and Mr. Walter is chief engineer, Metallurgical Section, Materials Div., Engineering Research, International Harvester Co., Hinsdale, Ill.

this is the heart of the CHAT procedure.

Hardenability Concepts

Since hardenability is essential to this concept of steel design, it is desirable to review the fundamental principles. Hardenability is defined as "a measure of the ease with which one can obtain hardness." Figure 1 illustrates the cross sectional view of steel bars of different alloy content. Their increasing hardenability is shown by the increasing depth of hardness, left to right.

The interrelationship between hardenability, hardness depth, and cooling rate can also be illustrated by the same bars. Assuming that all bars are of the same hardenability, the same pattern of hardness depths could be obtained by increasing the quench severity. That is, if the bars (of equal hardenability) were quenched, left to right, in hot oil, cold oil, water, and brine, the same variation of hardened depth could be obtained. With hardenability and quenching medium held constant, the same variation of hardened depth could also be obtained by decreasing bar diameter, left to right — which, in effect, increases the cooling rate.

Fortunately, the Jominy end-quench test unites the three important parameters into one test, allowing the development of quantitative information. The Jominy bar provides a series of cooling rates, from extremely fast at the quench end to very slow at the air-cooled end. Recording the hardness values obtained along the length of the bar provides a complete hardenability "fingerprint" of a given steel composition.

Figure 2 illustrates Jominy hardenability curves for six different compositions. The curves represent two carbon levels, 0.12 and 0.48%, and three hardenability levels for each of the carbon contents. Three critical features are illustrated. First, the hardness obtained at any given cooling rate, J position, is a function of hardenability and carbon content. Second, the initial hardness (IH, measured at $J = \frac{1}{16}$ in.) is a function of carbon content and independent of hardenability. Third, the Jominy hardenability curve can be characterized, within limits, by a single number known as the "ideal diameter", or D_I[3].

The larger the D_I value, the greater the hardenability. Through standard tables[5], one can determine the hardenability (D_I) required to obtain any specified distance hardness (DH) at any specified J position.

Calculating Hardenability

The next issue concerns the use of alloy multiplying factors to calculate D_I from chemical composition and grain size. This calculation procedure relies on a series of hardenability factors for each alloying element in the composition: multiplied together, they give a D_I value. The D_I value can then be translated into a Jominy curve by using the IH/DH tables[5]. Over the years, values of alloy multiplying factors have been improved. The factors proposed by Doane[6] are utilized here, but sometimes either the Grossmann[3] or Kramer[7] factors can be used advantageously.

Another important concept to be considered in devising new steels is the distinction between case hardenability and base (core) hardenability, designated herein as D_{Ic} and D_{Ib} respectively. The effect of a given alloy addition on hardenability is significantly different for low carbon than for high carbon steel. For carburizing steels, this difference becomes important because the steel will harden as a hypereutectoid steel in the carburized region and as a hypoeutectoid steel in the remaining area. Thus, a steel must possess adequate base and case hardenability.

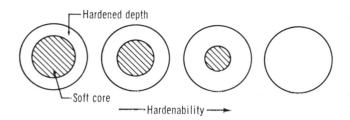

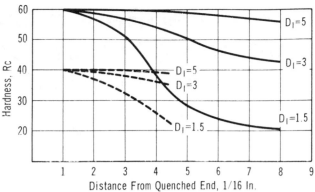

Fig. 1 — Bars of different steels with the same diameter harden to varying depths—hardenability increases from left to right. The same effect is created in bars of the same steel if they are quenched in increasingly effective cooling mediums.

Fig. 2 — Determined by end-quench tests, these curves show hardenabilities of six different steels, three with 0.12% C (dotted lines) and three with 0.48% C. Increasing D_I indicates greater quantities of alloying elements other than carbon.

The case hardenability is calculated in exactly the same manner as D_{Ib} except that the multiplying factors derived by Jatczak and Girardi[8] are used. A major difference, however, is that the Jominy curve for case hardenability cannot be calculated from the D_{Ic} value by use of the IH/DH tables. The D_{Ic} value only locates the Jominy position at which 10% transformation to nonmartensitic products has occurred, approximately Rc 60. Because transformation products are usually not tolerated in quality carburized components, another manipulation is required to assure 100% martensite plus austenite structures, as is discussed in a forthcoming article of the series.

The boron effect on hardenability is dependent upon carbon content. Although a boron multiplying factor[9] has been developed for use with Grossmann factors, a new boron factor is needed for use with the Doane multiplying factors. Based on a limited amount of data, the following equation represents the effect of boron in steel containing 0.20 to 0.40 C:

Boron Multiplying Factor
$$= 1 + 1.76 (0.74 - \%C)$$

The multiplier approach is not entirely satisfactorily applied to boron. Additional work utilizing regression analysis seems to hold promise for improvements.

Development of Replacement Steels

Now it is appropriate, as an informative exercise, to consider the development of the nickelfree carburizing steels EX 15, 16, 17, and 18, known as Equalloy steels. During the nickel shortage of 1969, the concept and techniques which embody the CHAT system were not fully developed. This example serves to illustrate some of the basic factors and computational techniques to be considered in designing a steel, a replacement steel in this instance. The problem during the nickel crisis involved devising a series of steels

containing only residual nickel to replace the International Harvester grades EX 5, 6, 7, and 8, a series of 8600-type carburizing steels with low molybdenum and 0.40 to 0.60 Ni. Other elements in the replacement compositions, however, would have minimum changes so that hardenability, mechanical properties, distortion characteristics, and machinability would be essentially the same as those of the steels replaced. The criterion of "least-change" suggested that carbon, phosphorus, sulfur, and silicon contents employed in the previously used EX series remain unchanged, compositional changes in the respective replacements being limited to manganese, chromium, and molybdenum.

Heat treatable steels have been defined in terms of H-bands. A base hardenability band and a case hardenability band define, within limits, a carburizing steel. Figure 3 shows the standard base hardenability band for 8622H and EX 6. Superimposed are minimum, midrange, and maximum hardenability curves calculated

from the minimum, midrange, and maximum composition of the EX 6 grade. Note that the midrange composition gives a hardenability curve which lies in the middle of the base hardenability band. However, computations using the minimum and maximum compositions provide hardenability lines which fall outside of the standard band. The reason for this discrepancy relates to the procedure used to develop H-bands. An empirical procedure, it provides limits in low and high hardenability which assumes remote probabilities for all elements reaching their respective extremes simultaneously. Hence, to compute bands from composition, other means were adopted. This entailed obtaining the statistical distribution of each element in normal practice (for the IH Wisconsin Steel Div.), then utilizing ±2 standard deviations (±2σ), as the equivalent high or low compositions.

Use of the +2σ and −2σ compositions as the limits provided high and low hardenability curves which

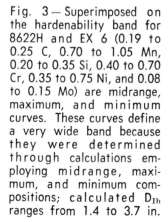

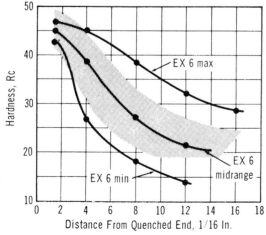

Fig. 3 — Superimposed on the hardenability band for 8622H and EX 6 (0.19 to 0.25 C, 0.70 to 1.05 Mn, 0.20 to 0.35 Si, 0.40 to 0.70 Cr, 0.35 to 0.75 Ni, and 0.08 to 0.15 Mo) are midrange, maximum, and minimum curves. These curves define a very wide band because they were determined through calculations employing midrange, maximum, and minimum compositions; calculated D_{Ib} ranges from 1.4 to 3.7 in.

matched the band reasonably well. This is shown in Figure 4, which portrays the calculated hardenability based on the ±2σ compositions for the EX 6 steel, and the comparison of this calculated hardenability with the published bands for 8622H and EX 6. Though somewhat empirical, the ±2σ technique has proved to be satisfactory in many instances of steel design. Some compositions may require the use of different multiples of sigma for adequate accuracy.

Matching Case and Base Hardenabilities

Before alloy ranges can be manipulated to produce a steel with a calculated D_{Ib} equivalent to that of the original steel, the case hardenability, D_{Ic}, must be considered. This is a more difficult problem because a replacement designed with equivalent base hardenability may possess insufficient case hardenability, or vice versa. In general, it is difficult to match case and base hardenability simultaneously because the calculated hardenabilities of a given composition are usually unique. However, some latitude exists in attempting to devise a replacement steel with essentially the same case and base hardenabilities as those of the original material.

Base hardenability of the replacement steel should be equal, as closely as possible, to the base hardenability of the steel replaced so that the same hardness occurs at approximately the same J distance. Greater or lesser base hardenability in a replacement steel is undesirable because a part made from the new steel could have significantly different strength properties and possibly different heat treat distortion characteristics.

It is not detrimental for the replacement composition to exhibit greater case hardenability than the original material; however, insufficient case hardenability is unacceptable. In the hypereutectoid region, quenching will yield martensite-retained austenite structures at maximum hardnesses of Rc 60 to 65. Excess case hardenability does not result in higher hardness. However, a lower case hardenability results in the maximum-hardness structure forming to a lesser J-distance than encountered with the original steel. Thus, there is a risk of not hardening the surface of a part or of materially changing the hardness gradient. The

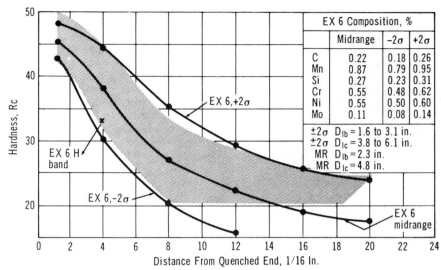

EX 6 Composition, %			
	Midrange	−2σ	+2σ
C	0.22	0.18	0.26
Mn	0.87	0.79	0.95
Si	0.27	0.23	0.31
Cr	0.55	0.48	0.62
Ni	0.55	0.50	0.60
Mo	0.11	0.08	0.14

±2σ D_{Ib} = 1.6 to 3.1 in.
±2σ D_{Ic} = 3.8 to 6.1 in.
MR D_{Ib} = 2.3 in.
MR D_{Ic} = 4.8 in.

Fig. 4 — When hardenability limits for EX 6 are calculated on the basis of the listed ±2σ limits for the maximum and minimum compositions (rather than with the standard maximum and minimum compositions listed for EX 6 in Fig. 3), maximum and minimum curves coincide closely with the standard band limits for 8622H and EX 6.

replacement composition should possess a calculated D_{Ib} virtually equal to that of the original material, and a calculated D_{Ic} equal to or greater than that of the original material.

When changing an alloy composition, it is important to consider both the specific values of the alloy hardenability multiplying factors and the change in the range of the multiplying factors. This is demonstrated schematically in Fig. 5, which illustrates how changing the mean concentration of an element from a lower to a higher value acted to widen the multiplying factor range, ΔMF. Even though the chemical analysis range remains constant, the nonlinear nature of the multiplying factor function causes a wider spread of the multiplying factor for the new composition. As a result, the hardenability band would be wider for the new steel.

Regarding the nickelfree EX grades described here, analysis of the individual ΔMF ranges indicated a slight reduction in the cumulative ΔMF range for the replacement alloy. Mill experience to date has shown these steels to have hardenability bands which are the same as those of standard 8600 bands.

In designing nickelfree grades to

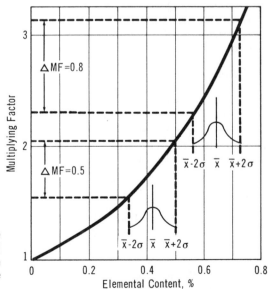

Fig. 5 — Changing the amount of an individual alloying element in a steel can increase the multiplying factor range.

Source: Metal Progress, Reprint 1974, 30 pp

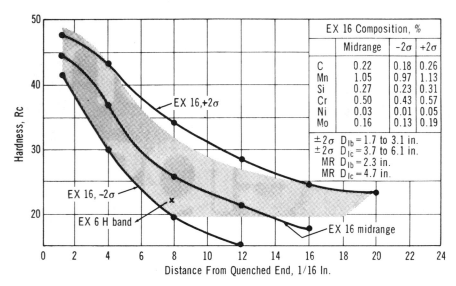

EX 16 Composition, %			
	Midrange	−2σ	+2σ
C	0.22	0.18	0.26
Mn	1.05	0.97	1.13
Si	0.27	0.23	0.31
Cr	0.50	0.43	0.57
Ni	0.03	0.01	0.05
Mo	0.16	0.13	0.19

$\pm 2\sigma$ D_{Ib} = 1.7 to 3.1 in.
$\pm 2\sigma$ D_{Ic} = 3.7 to 6.1 in.
MR D_{Ib} = 2.3 in.
MR D_{Ic} = 4.7 in.

Fig. 6 — Calculated from ±2σ limits for specific elements of the nickelfree EX 16 composition, maximum and minimum hardenability curves agree well with the hardenability band for EX 6. Respective values for ideal diameters of case and core, also calculated, are similar to those of EX 6 and 8622H (Fig. 4).

replace 8600 types, nickel was removed, and chromium, manganese, and molybdenum contents were adjusted to provide calculated base and case hardenabilities matching those of the previously used EX steels. Figure 6 shows the resultant base hardenability calculated for a ±2σ analysis for the nickelfree grade, EX 16. Minimum and maximum hardenability curves, calculated, fit the standard hardenability band well, and there is an equivalence between case hardenability values.

Martensite start temperatures (M_s) were not used as criteria in the search for replacement compositions. Once a replacement based on D_{Ib} and D_{Ic} was found, however, the M_s temperatures of this steel at both core carbon and 1% C levels were computed (755 and 297 F, respectively) and compared to those of the steel replaced (752 and 296 F). The M_s values influence the sequence of transformation, which affects residual stress and distortion in heat treatment.

When designing this replacement steel, a basic assumption was applied — metallurgical factors governing the engineering performance of a heat treated part are primarily the carbon content, the microstructure, and the residual stress. This is especially true when dealing with alloy steels of relatively low alloy content. Specific alloy effects, such as the molybdenum effect on temper embrittlement, are recognized. They will be further discussed in the Computer Harmonized section.

It should be noted that, due to the "least change" limitation and the complex requirement of simultaneous matching of D_{Ib} and D_{Ic}, the composition of the replacement is not optimum from a cost viewpoint. The utility of a computer in solving this problem will be shown later.

Evaluation of Performance

Therefore, with equivalent carbon content, microstructure, and residual

stress, it was anticipated that these nickelfree steels would give the same engineering performance as previously obtained. This was confirmed by tests conducted on standard 6-pitch test pinions. They showed that the nickelfree steels developed resistance to both bending and pitting fatigue which was equivalent to that of the previously used 8600 steels.

In an additional evaluation, large hypoid rear axle ring gears were run on a chassis test designed to simulate the heavy shock loading endured by rear axles of trucks in severe off-highway service. In all instances, the nickelfree grades performed as well as the previously used EX or 8600 types. Of more significance, however, is three years of successful use with these nickelfree grades of steel, during which no changes have been made in either machining or heat treating practice. Even though this is not an optimum-cost composition, a more efficient utilization of alloys has resulted in substantial savings because of the approximate $20.00 per ton price differential in addition to the conservation of nickel.

Summary

To provide an environment for further discussion, this first article has briefly reviewed some of the fundamental and historical aspects of the hardenability of steel. The discussion concerning the replacement steel, EX 16, demonstrated the utility of computational techniques for designing steels, and showed that engineering performance can be maintained if the design of the alloy is approached judiciously. This discussion also clearly defined a steel in terms of multiple hardenability relationships (range of D_{Ib} and D_{Ic}), and noted the shortcomings of present techniques in arriving at optimum-cost alloys. The next article will present the method used to arrive at and define engineering requirements. ✪

References

1. "The P-F Characteristic of Steel", by B. F. Shepherd, Transactions, American Society for Metals, Vol. 22, 1934.
2. "Transformation of Austenite at Constant Subcritical Temperature", by E. C. Bain and E. S. Davenport, Transactions, AIME, Vol. 90, 1930.
3. "Hardenability Calculated from Chemical Composition," by M. A. Grossmann, Transactions, AIME, Vol. 150, 1942.
4. "A Hardenability Test for Shallow Hardening Steels", by W. E. Jominy, Transactions, ASM, Vol. 27, 1939.
5. "Calculation of End-Quench Hardenability Curve", by Boyd and Field, booklet issued by American Iron & Steel Institute, February 1946.
6. "Predicting Hardenability of Carburizing Steels", by A. F. deRetana and D. V. Doane, Metal Progress, September 1971.
7. "Factors for the Calculation of Hardenability", by I. R. Kramer, S. Siegel and J. G. Brooks, Transactions, AIME, Vol. 167, 1946.
8. "Multiplying Factors for Calculation of Hardenability of Hypereutectoid Steels Hardened from 1700 F", Transactions, ASM, Vol. 51, 1959.
9. "The Effect of Carbon Content on the Hardenability of Boron Steels", Transactions, ASM, Vol. 40, 1948.

Computer-Based System Selects Optimum Cost Steels - II

By DALE H. BREEN, GORDON H. WALTER, and CARL J. KEITH JR.

Determining hardenability requirements for heat-treated components is the subject of this installment (Part I appeared in the December 1972 issue). To define minimum hardenability (D_I) for a steel which will allow a given part to develop a specified strength under production quenching conditions, the Jominy equivalent cooling (Jec) Rate must be determined. Once the D_I is established, standard or special-analysis steels may be selected for the part, depending on such factors as tonnage and inventory requirements.

The first article in this series, which appeared in the December 1972 issue, defined the CHAT (Computer Harmonized Application Tailored) approach to steel design. Briefly, the CHAT system consists of two basic parts (Computer Harmonizing and Application Tailoring), either of which can be executed and applied independently, although they are most effective when used in combination.

In this article, Application Tailoring (AT) will be applied to components requiring the use of quenched and tempered steels — that is, through-hardened steels. Application Tailoring for carburized components will be considered in the next article of this series.

Definition and Basis

Application Tailoring, AT, is the quantitative determination of hardenability requirements, in terms of D_I, for a given application. It is a formalized process in which cooling rates are used to define the hardenability required to meet engineering design criteria such as hardness and microstructure.

A key factor in AT is the determination of the Jominy equivalent cooling (Jec) Rate for a specific part given a production heating and quenching cycle. The Jec Rate at a given location in the material of a part is the cooling rate, expressed

Mr. Breen is chief, Materials Div., Mr. Walter is chief engineer, and Mr. Keith is research metallurgist, Metallurgical Section, Materials Div., Engineering Research, International Harvester Co., Hinsdale, Ill.

as J distance, on a Jominy bar of the same material that produces the same hardening response.

The fundamentals of the AT concept can be illustrated by a simple example of a stepped shaft. It is assumed that design analysis of the shaft shown in Fig. 1 has determined that it should be of a 0.35% C steel. Also, it should have 90% martensite at the center and be tempered to Rc 30 to perform its anticipated load carrying function for an acceptable time.

These requirements (0.35% C and 90% martensite) translate, as shown in Fig. 2, into an as-quenched hardness of Rc 46. AT consists of answering the question, "What minimum D_{Ib} is required to produce Rc 46, at the center of the shaft, with a 0.35% C steel?" For expediency assume that the Jec Rate for the shaft center has been found, by experi-

Fig. 1 — In the text, AT analysis is used to determine the correct hardenability for the 0.35% C steel specified for the diagrammed shaft. Also specified for the cross section indicated by the shaded area is at least 90% martensite at the center. For this section, as-quenched hardness is Rc 46 min, and tempered hardness is Rc 30 min at the center.

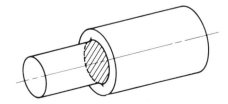

mental procedures, to equal J4. Shown in Fig. 3 is a series of hardenability curves for 0.35% C steel. This graph reveals that the minimum D_{Ib} required to yield Rc 46 at J4, the shaft center, is 2.7 in. Though any steel with a D_{Ib} greater than 2.7 in. also can be used, such steels have excessive hardenability, and are thus a waste of costly alloying elements.

As Fig. 4 illustrates, the Jec Rate is determined experimentally by comparing hardness values, obtained on the cross section of a part subjected to the production heat and quench cycle, with hardness values obtained on Jominy bars of the same steel. Three critical items must be considered when determining the experimental curves:

1. The Jominy bar should be given the same thermal cycle as the part because hardenability is influenced by thermal history (Fig. 4b).

2. Enough parts should be examined to evaluate effects of load size, position in the load, and changes in quench severity with usage.

3. The steel used for Jec Rate evaluation must have an end-quench curve which drops steeply through the critical cooling rate range so that hardness changes give a sensitive indication of cooling rate.

In some instances, part size, costs, or time prohibit the experimental determination of Jec Rates. Though estimations can be made by using published cooling rate curves[1,2,3] caution should be exercised because accurate quench severity values (H values)[1] are not usually known.

The use of Jec Rates to predict hardness in a quenched part is not a perfect technique. Carney[4] and others[5,6] have clearly indicated a potential shortcoming, presumably related to the dependence of austenite transformation on the state of stress developed during quenching. In most instances, however, the procedure gives accurate, useful results.

Applying AT Analysis

Application Tailoring varies from a very simple procedure which utilizes established design requirements

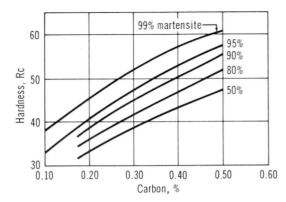

Fig. 2 — As-quenched hardnesses of steel vary with carbon content and the amount of martensite developed by quenching.

in conjunction with Jec Rates, to a more comprehensive procedure which includes a determination of required strength level and microstructure based on an engineering analysis of expected stress spectrum and desired life. The efficiency and reliability of an over-all CHAT analysis is, of course, a direct function of the accuracy of the initial engineering analysis. The AT procedure for quenched and tempered parts with varying degrees of intricacy will be demonstrated by three examples.

Simple Application Tailoring

Shown in Fig. 5 is a wheel spindle, an example of a part for which hardness and heat treatment have previously been established. The procedure for simple Application Tailoring, in this example only, involves determination of Jec Rate and definition of the D_{Ib} required to develop the mechanical properties. The spindle is forged, austenitized, oil quenched, tempered and machined to final dimensions.

Assume the engineering specification requires a 0.40% C steel quenched to 90% martensite at the finish machined surface of the critical section followed by tempering to a hardness range of Rc 30-37. As Fig. 2 shows, the requirement of 90% martensite for a 0.40% C steel necessitates an as-quenched hardness of Rc 50 minimum at the finish machined surface. The Jec Rate is evaluated by determining the cross-sectional hardnesses of a number of production quenched parts, and comparing these values to equal hardnesses on Jominy bars from the same pieces. The resulting minimum Jec Rate curve for the critical cross section is shown in Fig. 6, where the specific Jec Rate for the machined surface is J9.5.

The hardenability may now be de-

termined. One method to do this is by using IH/DH rations [1,8,9], as shown in Table I. The method is as follows:

1. Determine the initial hardness (IH) for a 0.40% C steel. As shown in the second column, the IH, which is equivalent to 99% martensite at the J1 position, is Rc 56.

2. Using distance hardness (DH), which is Rc 50 by definition, calculate the IH/DH Ratio — it is 56/50, or 1.12.

3. Using the defined distance, J9.5, extrapolate between the vertical columns labeled ½ and ¾ in. (J8 and J12). As will be noted, a D_{Ib} of 4.90 in. corresponds to a IH/DH of 1.12 in.

The AT Procedure has thus determined (1) the critical location cooling rate, and (2) that a minimum hardenability of $D_{Ib} = 4.90$ in. is required to develop 90% martensite at that location when the part is processed in the production heat treatment facilities. (A graphical means to determine hardenability, D_{Ib}, will be demonstrated in a subsequent example.)

The values found, carbon level and D_{Ib}, may be used as the input to design a CHAT steel or as the basis for selecting a standard steel. Where the tonnage of steel is sufficient to warrant the purchase of heat lots, a CHAT steel will provide the most efficient and economical steel selection.

When dealing with established heat treat practices, it is expected that the practice itself should be questioned. In this instance, the combination of oil quenching and medium-carbon alloy was acceptable at one time. The advent of new technology has made it appropriate to consider the water quenching of lower-carbon steels as a way to gain additional economies without loss of performance. Because the spindle is machined after heat treatment, distortion does not appear to be a signif-

Fig. 3 — This series of hardenability curves for 0.35% C steels indicates that Rc 46 min at J4 (center of shaft of Fig. 1) corresponds to a D_{Ib} requirement of approximately 2.7 in.

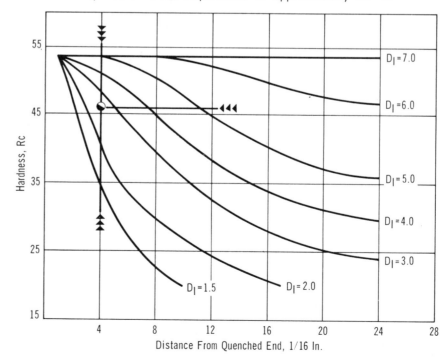

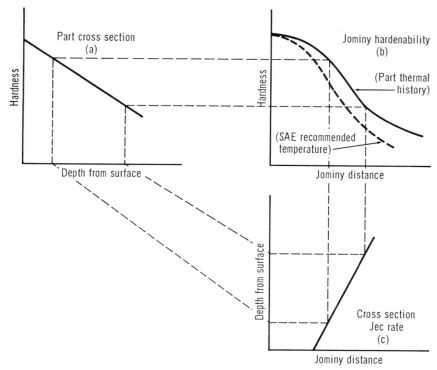

Fig. 4 — Jominy equivalent cooling (Jec) Rates are determined by comparing hardnesses of cross sections of parts receiving the established production heat treatment to hardnesses obtained on end-quenched bars of the same steel.

Fig. 5 — Application Tailoring is used to select the steel for this wheel spindle. Design requirements call for the critical section to harden to Rc 50 min and 90% martensite at the surface when the spindle is quenched. Tempered hardness on the surface should be Rc 30 min.

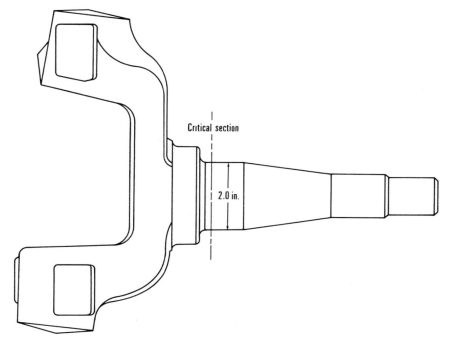

icant factor. Of course, it is generally not desirable to water quench a medium-carbon steel because quench cracks can occur.

However, the wheel spindle is tempered to a Rc 30 minimum, a hardness which is readily attainable in a steel with less carbon. Thus, AT analysis can be reapplied to the wheel spindle with a view toward the use of a water-quenching steel with less carbon and hardenability. The essential design requirements of the wheel spindle are (1) a tempered surface hardness equal to Rc 30 min and (2) a surface microstructure of at least 90% martensite.

As a rule, the as-quenched hardness must be at least 7 Rc points higher than the specified tempered hardness to allow for tempering. (A more exact approach to determine the as-quenched hardness is available from the work of Hollomon and Jaffee[10].) However, in this example, the required as-quenched hardness is taken to be Rc 37.

The first step in the AT procedure is to define the carbon range. Figure 2 indicates that a 0.17% C level is sufficient to develop a hardness of Rc 37 with a microstructure of 90% martensite. However, carbon is the most efficient element for increasing a steel's hardenability. Increasing the carbon content, therefore, reduces the amount of alloy required. Balancing this consideration against the problem of quench cracking, a minimum of 0.25% C is established. A structure of 90% martensite with 0.25% C content will develop an as-quenched hardness of Rc 42 (Fig. 2). The carbon range is established by reference to the applicable AISI Steel Products Manuals,[11, 12] which indicate a ladle composition range of 0.06% C.

The next step is to evaluate the production Jec Rate. If the severity of the production quench (Grossmann H Value)[1, 13] is known or can be estimated, the Jec Rate can be evaluated from a series of curves such as shown in Fig. 7 and 8. For a well-agitated water quench (H= 1.5), the Jec Rate would be J8.2 at the center of a 2.4 in. round, and the Jec Rate would be J5.3 at a depth equal to 70% of the center-to-surface distance. By this method, the Jec Rate of the cross section is evaluated, as shown in Fig. 6. The machined surface of the section has a Jec Rate of J3.8 (assumed to be

Source: Metal Progress, Reprint 1974, 30 pp

J4 for ease of calculation). The minimum hardenability required may be determined as before.

1. From Table I, the initial hardness, IH, of a 0.25% C steel, is Rc 47.

2. Because the distance hardness is Rc 42 (necessary for 90% martensite), the IH/DH ratio is 47/42, or 1.12, at the J4 position.

3. By extrapolation in the ¼ in. (J4) column, the required D_{Ib} is 2.95 in.

The minimum hardenability may also be determined directly from Fig. 9, which has been prepared from ratios of IH/DH and plots of carbon content against hardness for different percentages of martensite. Figure 9 indicates that 90% martensite at the J4 position can be obtained with a hardenability of D_{Ib} = 3.0 in. The following metallurgical requirements have been determined:

1. Carbon content—0.25 to 0.31%.

2. Hardenability — D_{Ib} = 2.95 in. min, which will develop a hardness of Rc 42 (90% martensite) at a depth equivalent to the J4 position.

Maximum hardenability allowed is not so significant insofar as the wheel spindle is concerned because a relatively high tempering temperature

is used to achieve the final hardness. In some instances where the temperature is very low, however, maximum hardenability can be a factor. In these situations, the maximum D_{Ib} to be permitted would be evaluated in the same manner as the minimum D_{Ib}, based upon maximum carbon content and maximum hardness allowed.

The above requirements are used as the basis for designing a CHAT steel or for selecting a standard steel. If production is low or inventory

problems exist, a standard steel should be used. A standard steel should also be evaluated for comparison purposes even when a CHAT steel is being considered.

In the event that a CHAT analysis steel is indicated, the parameters, carbon level, and minimum D_{Ib} are used as input data for Computer Harmonizing (CH) to develop a steel with optimum alloy content. To select an appropriate standard steel, the SAE hardenability bands are consulted. Selection of one steel from

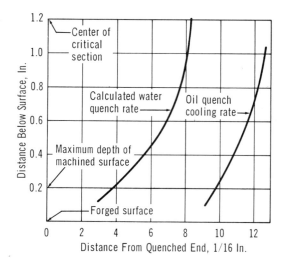

Fig. 6 — Minimum Jec Rates for the wheel spindle shown in Fig. 5 are employed to determine the minimum D_{Ib} required.

Fig. 7 — This chart shows locations on end-quenched hardenability test bars corresponding to centers of round bars.

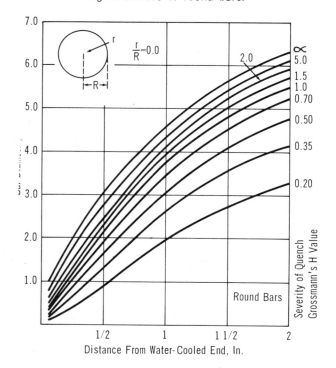

Fig. 8 — Similar to Fig. 7, this chart relates locations on end-quenched hardenability test bars corresponding to 70% from centers of round bars.

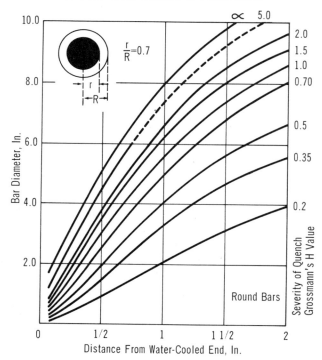

Table I — Values for End-Quench Hardenability Calculation

Carbon, %	Initial (Max.) Hardness (IH), Rc	Ideal Critical Diameter (D_I) In.	Mm	¼	½	¾	1	1¼	1½	1¾	2
				Distance From Quenched End, In. — RATIO: Initial Hardness / Distance Hardness							
0.10	38	0.50	12.7	4.90							
0.11	39	0.55	14.0	4.42							
0.12	40	0.60	15.2	4.03							
0.13	40	0.65	16.5	3.70	6.00						
0.14	41	0.70	17.8	3.47	5.15						
0.15	41	0.75	19.1	3.25	4.50						
0.16	42	0.80	20.3	3.07	4.18						
0.17	42	0.85	21.6	2.90	3.88	6.00					
0.18	43	0.90	22.9	2.75	3.68	5.13					
0.19	44	0.95	24.1	2.61	3.50	4.70					
0.20	44	1.00	25.4	2.48	3.33	4.40					
0.21	45	1.05	26.7	2.33	3.20	4.13	5.28				
0.22	45	1.10	27.9	2.17	3.08	3.93	4.75	5.70			
0.23	46	1.15	29.2	2.05	2.96	3.76	4.40	4.95	5.75		
0.24	46	1.20	30.5	1.96	2.86	3.60	4.15	4.58	5.00	6.00	
0.25	47	1.25	31.8	1.88	2.76	3.45	3.95	4.32	4.65	5.15	6.00
0.26	48	1.30	33.0	1.80	2.66	3.32	3.78	4.13	4.40	4.72	5.25
0.27	49	1.35	34.3	1.73	2.57	3.21	3.65	3.95	4.18	4.45	4.83
0.28	49	1.40	35.6	1.67	2.49	3.10	3.53	3.77	4.02	4.26	4.53
0.29	50	1.45	36.8	1.62	2.42	3.01	3.41	3.65	3.87	4.08	4.29
0.30	50	1.50	38.1	1.57	2.34	2.93	3.30	3.53	3.73	3.91	4.10
0.31	51	1.55	39.4	1.53	2.27	2.84	3.20	3.43	3.61	3.78	3.96
0.32	51	1.60	40.6	1.49	2.21	2.75	3.10	3.33	3.51	3.67	3.83
0.33	52	1.65	41.9	1.46	2.16	2.66	3.01	3.24	3.42	3.57	3.71
0.34	53	1.70	43.2	1.43	2.11	2.59	2.93	3.16	3.33	3.47	3.59
0.35	53	1.75	44.4	1.40	2.06	2.52	2.86	3.08	3.25	3.38	3.49
0.36	54	1.80	45.7	1.38	2.01	2.45	2.80	3.00	3.17	3.29	3.40
0.37	55	1.85	47.0	1.36	1.96	2.38	2.74	2.94	3.10	3.21	3.32
0.38	55	1.90	48.3	1.34	1.91	2.33	2.68	2.88	3.04	3.14	3.25
0.39	56	1.95	49.8	1.32	1.87	2.27	2.63	2.83	2.97	3.08	3.18
0.40	56	2.00	50.8	1.30	1.83	2.23	2.58	2.78	2.92	3.02	3.11
0.41	57	2.10	53.3	1.26	1.75	2.13	2.50	2.69	2.82	2.91	3.00
0.42	57	2.20	55.9	1.24	1.69	2.06	2.42	2.61	2.73	2.83	2.91
0.43	58	2.30	58.4	1.22	1.64	1.99	2.35	2.53	2.65	2.75	2.83
0.44	58	2.40	61.0	1.20	1.60	1.93	2.27	2.47	2.58	2.67	2.75
0.45	59	2.50	64.0	1.18	1.55	1.88	2.22	2.40	2.51	2.60	2.68
0.46	59	2.60	66.0	1.17	1.52	1.84	2.16	2.34	2.44	2.53	2.61
0.47	60	2.70	68.6	1.15	1.48	1.80	2.10	2.28	2.38	2.47	2.54
0.48	60	2.80	71.1	1.14	1.45	1.76	2.05	2.23	2.33	2.41	2.48
0.49	60	2.90	73.7	1.13	1.42	1.72	2.00	2.18	2.28	2.35	2.42
0.50	61	3.00	76.2	1.11	1.39	1.68	1.94	2.12	2.22	2.28	2.36
0.51	61	3.10	78.7	1.10	1.37	1.65	1.90	2.08	2.18	2.24	2.32
0.52	61	3.20	81.3	1.09	1.35	1.61	1.86	2.04	2.13	2.20	2.27
0.53	62	3.30	83.8	1.08	1.33	1.58	1.83	2.00	2.08	2.15	2.22
0.54	62	3.40	86.4	1.07	1.31	1.55	1.80	1.95	2.04	2.11	2.17
0.55	63	3.50	88.9	1.07	1.29	1.51	1.76	1.91	2.00	2.07	2.13
0.56	63	3.60	91.4	1.06	1.27	1.48	1.72	1.87	1.96	2.03	2.08
0.57	63	3.70	94.0	1.06	1.25	1.46	1.68	1.83	1.92	1.98	2.04
0.58	64	3.80	96.5	1.05	1.23	1.43	1.65	1.80	1.88	1.94	2.00
0.59	64	3.90	99.1	1.05	1.22	1.41	1.62	1.76	1.84	1.90	1.96
0.60	64	4.00	101.6	1.04	1.20	1.38	1.59	1.72	1.80	1.86	1.92
		4.10	104.1	1.04	1.18	1.36	1.56	1.68	1.77	1.82	1.88
		4.20	106.7	1.03	1.17	1.34	1.53	1.65	1.73	1.78	1.84
		4.30	109.2	1.03	1.16	1.32	1.50	1.62	1.70	1.75	1.80
		4.40	111.8	1.02	1.15	1.30	1.47	1.58	1.66	1.72	1.76
		4.50	114.3	1.02	1.14	1.28	1.44	1.55	1.63	1.68	1.73
		4.60	116.8	1.02	1.12	1.26	1.41	1.52	1.59	1.64	1.69
		4.70	119.4	1.01	1.11	1.24	1.38	1.49	1.56	1.61	1.65
		4.80	121.9	1.01	1.10	1.22	1.36	1.46	1.53	1.57	1.62
		4.90	124.5	1.00	1.08	1.20	1.33	1.43	1.49	1.53	1.58
		5.00	127.0	1.00	1.07	1.18	1.31	1.40	1.46	1.50	1.54
		5.10	129.5	1.00	1.06	1.17	1.28	1.37	1.43	1.47	1.51
		5.20	132.1	1.00	1.05	1.15	1.25	1.34	1.39	1.43	1.47
		5.30	134.6	1.00	1.04	1.13	1.23	1.31	1.36	1.39	1.43
		5.40	137.2	1.00	1.03	1.12	1.21	1.28	1.33	1.36	1.40
		5.50	139.7	1.00	1.03	1.10	1.18	1.25	1.29	1.33	1.37
		5.60	142.2	1.00	1.02	1.09	1.16	1.22	1.26	1.28	1.33
		5.70	144.8	1.00	1.02	1.07	1.13	1.19	1.23	1.25	1.29
		5.80	147.3	1.00	1.01	1.06	1.11	1.17	1.19	1.22	1.25
		5.90	149.9	1.00	1.01	1.04	1.09	1.13	1.16	1.18	1.21
		6.00	152.4	1.00	1.00	1.03	1.07	1.10	1.13	1.15	1.18

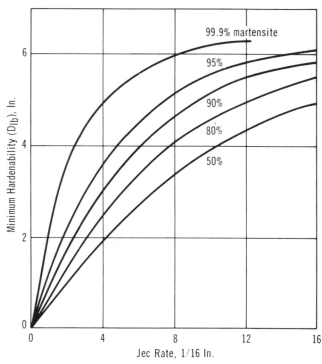

Fig. 9 — Knowing the Jec Rate, it is possible to determine the minimum hardenability (D_{Ib}) needed by a steel to develop the indicated martensitic structures. The graph is suitable for steels containing 0.15 to 0.50% C.

Fig. 10 — This part, an equalizer bar for a crawler tractor, is being tested at International Harvester's laboratories.

the many available with proper carbon and D_{Ib} values is made on the basis of cost, machinability, availability, and other considerations.

Comprehensive Application Tailoring

This example considers the part in the design stage prior to production. The part (shown in Fig. 10) is an equalizer bar for a crawler tractor. Engineering analysis indicates the tapered beam will be subjected to bending (stress ratio of R=0), with a maximum surface stress of 75,000 psi and zero stress at the neutral axis. A fatigue life of 10 million cycles is desired.

The required surface hardness is determined by reference to a modified Goodman diagram (Fig. 11). As is shown, a tempered surface hardness of Rc 40 will provide a fatigue safety factor of 1.3 for N = 1x10⁷ cycles. To provide for tempering to Rc 40, a minimum as-quench hardness of Rc 47 is required. In addition, it is desirable to provide 99% martensitic surface structure for optimum fatigue resistance.

The next step, therefore, is to determine the carbon content. Fig-

ure 2 indicates that steel with a minimum carbon content of 0.30% will develop Rc 50 when quenched to 99% martensite; this carbon level is thus considered satisfactory for the application. At 0.30% C, a conservative estimate[14] of the tempered center hardness is Rc 15. By comparing the applied stress gradient with the expected strength gradient, Rc 40 to Rc 15, on a fatigue basis, it is determined that a conservative strength gradient will be obtained by requiring Rc 45 as quenched to 1 in. below the surface (Fig. 12).

The metallurgical requirements are:

1. As-quenched surface hardness — Rc 50 min, 99% martensite.
2. Carbon content — 0.30 to 0.36%.
3. As-quenched hardness — Rc 45 min (90% martensite) at a depth equal to 1.0 in. below the surface.
4. Tempered surface hardness — Rc 40 min.

In this preproduction example, prototype parts for direct evaluation of the Jec Rate are not available. It is necessary, therefore, to estimate this rate by using published curves. This is done by assuming that the

tapered beam may be approximated by a 7 in. round. Figure 8 indicates that in a well-agitated water quench (H Value = 1.5) the round will experience a cooling rate equivalent to J18 to J19 at the 1.0 in. depth.

The hardenability, D_{Ib}, required will be calculated by using the ratio of initial hardness to distance hardness in the manner applied before (Table I).

1. For a 0.30% C steel, the initial hardness, IH, is Rc 50.
2. At the distance, J18, distance hardness (DH) is Rc 45, making the IH/DH ratio = 50/45, or 1.10.
3. By extrapolation between the Table I columns labeled 1 and 1¼ in. (J16 and J20), the value of 1.10 at J18 corresponds to a D_{Ib} of 6.2 in. The requirements of C = 0.30 min and D_{Ib} = 6.2 in. min are used to select a standard alloy steel. If sufficient tonnage is involved, these values may be used for "Computer Harmonizing" to design a more efficient steel.

The AT System in Summary

As has been pointed out, excess hardenability often represents excess cost. Application Tailoring, through use of engineering type analysis and Jec Rates, can be used to determine hardenability requirements accurately. When large tonnages are involved, considerable experimental work can be justified because of

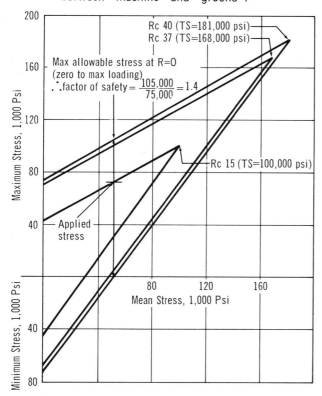

Fig. 11 — Allowable diagram is used for determining the strength requirement for the equalizer bar in Fig. 10. This diagram is based on a surface finish between "machine" and "ground".[15]

Rc 40 (TS=181,000 psi)
Rc 37 (TS=168,000 psi)

Max allowable stress at R=0 (zero to max loading)
∴ factor of safety = $\frac{105,000}{75,000}$ = 1.4

Rc 15 (TS=100,000 psi)

Applied stress

Maximum Stress, 1,000 Psi

Minimum Stress, 1,000 Psi

Mean Stress, 1,000 Psi

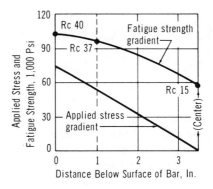

Fig. 12 — Comparisons of gradients for fatigue strength (for R = 0 and N = 10^7) and applied bending stresses are needed for evaluating the durability of a steel part in service.

the potential of sizable savings. Though Application Tailoring can be quite simple and straightforward, a more comprehensive approach encompassing engineering design and applied mechanics techniques is applicable in some instances.

In the examples described herein, the base hardenability required for particular applications has been the principal concern. Part III of this series will consider the AT procedure as applied to applications utilizing carburized components where core and case hardenability requirements (D_{Ib} and D_{Ic}) are involved.

References

1. "Carilloy Steels," United States Steel Corp., Carnegie-Illinois Steel Corp., 1948.
2. "How to Estimate Hardening Depth in Bars," by J. L. Lamont, Iron Age, 14 October 1943.
3. "The Flow of Heat in Metals," by J. B. Austin, ASM, 1942.
4. "Another Look at Quenchants, Cooling Rate, and Hardenability," D. J. Carney, ASM Transactions, Vol. 46, 1954.
5. "An Estimation of the Quenching Constant, H," by D. J. Carney and A. D. Janulionis, ASM Transactions, Vol. 43, 1951.
6. "Effect of Applied Stress on the Martensitic Transformation," by S. A. Kulin, Morris Cohen, and B. L. Averbach, Journal of Metals, June 1952.
7. "Relationship Between Hardenability and Percentage of Martensite in Some Low Alloy Steels," by J. M. Hodge and M. A. Orehoski, AIME, TP 1800, 1945.
8. "Calculation of the Standard End-Quench Hardenability Curve from Chemical Composition and Grain Size," by L. C. Boyd and J. Field, AISI Contribution to the Metallurgy of Steel, No. 12, 1945.
9. "Republic Alloy Steels," Republic Steel Corp., Handbook Reprint, 1961.
10. "Ferrous Metallurgical Design," by J. H. Hollomon and L. D. Jaffee, John Wiley & Son, New York, 1948.
11. "Alloy Steel: Semi-Finished; Hot Rolled and Cold Finished Bars," AISI Steel Products Manual, AISI, August 1970.
12. "Carbon Steel: Semifinished for Forging; Hot Rolled and Cold Finished Bars; Hot Rolled Reformed Concrete Reinforcing Bars," AISI Steel Products Manual, AISI, May 1964.
13. "Elements of Hardenability," by M. A. Grossmann, ASM, 1952.
14. "Modern Steels and Their Properties," Bethlehem Steel Co., Sixth Edition, Bethlehem Steel Co., Bethlehem, Pa., 1961.
15. "Application of Stress Analysis," by C. Lipson, University of Michigan, Ann Arbor, June 1959.

Computer-Based System Selects Optimum Cost Steels-III

By DALE H. BREEN, GORDON H. WALTER, and CARL J. KEITH JR.

Determination of hardenability requirements for carburized components
is the subject of this installment. The discussion complements
Part II of this series (Metal Progress, February 1973),
which concerned Application Tailoring (AT) of steel components that are
normally through hardened. Application Tailoring, a system for
determining the exact hardenability requirements necessary
for producing specified strength (hardness and microstructure)
properties using production heat treat facilities, employs the
Jominy equivalent cooling (Jec) rate as an important factor in the analysis.
In this article, carburized gears are used as examples.
Both the base hardenability (D_{Ib}) and the case hardenability (D_{Ic})
must be considered as well as the carbon gradient. Subsequent articles
will tie this information to the Computer Harmonizing aspect
of International Harvester's CHAT system, and demonstrate
how it can optimize steel costs.

Carburized gears represent a masterpiece in metallurgical and engineering design. Because of the complexities involved, however, they are often made of inappropriate materials from a cost viewpoint.

Many rules of thumb have been substituted for rational engineering approaches. They have had some degree of success, of course. In addition, attempts have been made to set down blanket rules for material selection, but by and large, they too have limitations.

For cost-conscious industries, the best rule is "to use the lowest cost combination of material and heat treatment that will perform the functions for the prescribed time." This entails some detailed analysis, and can result in the use of special compositions.

The analysis should start on a rather elemental level. To be sure, such factors as percentage of retained austenite, intergranular oxide penetration, microcracks, carbides, and inclusion content must be considered in the final analysis. When selecting steels for gear applications, however, the basic problem is to

Mr. Breen is chief, Materials Div., Mr. Walter is chief engineer, and Mr. Keith is research metallurgist, Metallurgical Section, Materials Div., Engineering Research, International Harvester Co., Hinsdale, Ill.

provide a composition which, when carburized and quenched, will produce strength gradients in critical locations (pitch line and root fillet, for example) which will accommodate imposed stresses. This involves, of course, considering carbon level of the base material, the carbon gradient produced in carburizing, the hardenability in the case zone (D_{Ic}) and the core region (D_{Ib}), and the quenching rate.

The procedures used to define the steel characteristics for a given gear application are explained in some detail in the following example. Though some assumptions are made and shortcuts taken in the interest of brevity, the main features related to hardenability are handled.

In this example, the Application Tailoring (AT) procedure will be applied to a six-pitch spur gear (Fig. 1). It is assumed that the design engineer has completed the design analysis, and has found that the specified strength gradient can be obtained with the following metallurgical requirements: surface hardness, Rc 60 to 65; effective depth at pitch line, 0.030 to 0.050 in. (distance to Rc 50); core hardness, Rc 30 to 40. Illustrated graphically in Fig. 2, these three requirements serve as guidelines for the AT procedure.

The intent in this example is to

develop a surface microstructure containing only martensite and austenite. It is recognized, however, that intergranular oxidation[1, 2] may cause decreased hardenability at the extreme surface due to localized alloy depletion. The development of nonmartensitic transformation products due to this effect will be discussed again in Part V of this series.

Determination of Jec Rate

The first, and key, step in the AT procedure is to develop information on the Jominy equivalent cooling (Jec) rate for critical locations on parts subjected to the production heat and quench cycle. This involves machining Jominy bars and gear samples of the required geometry out of a suitable steel — that is, a steel with an end quench curve which drops steeply through the estimated critical cooling rate range so

Fig. 1 — A six-pitch spur gear, such as shown here in two views, is used to demonstrate AT analysis of carburized parts, as described in this article.

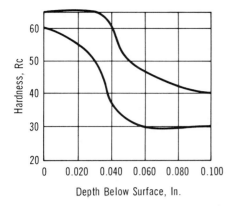

Fig. 2 — The design for the spur gear to be analyzed by AT methods calls for the illustrated hardness-depth limits in the case. Specification of effective case depth applies to the active profile of the gear teeth as measured at the pitch line. However, the root fillets of the gear teeth are required to have a minimum effective case depth of 50% of the minimum pitch line case.

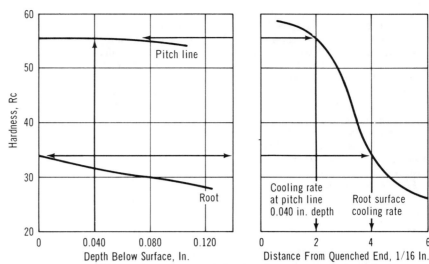

Fig. 4 — Given test gears and Jominy bars representing the same bar of 1040, correlations of hardness gradients in gears with end quench data show that gear hardnesses for the pitch line at 0.040 in. and the root surface correlate with J2 and J4 positions, respectively. These positions give the Jec rates.

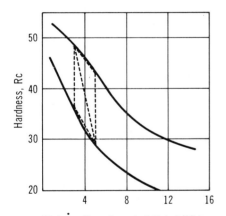

Fig. 3 — Variations in quenching rates contribute to the spread in core hardness. For example, hardness varies from Rc 36 to 48.5 at the J3 position. However, a variation in quenching rate from J3 to J5 will result in a hardness variation of Rc 29 to 48.5 (indicated by diagonal dotted line).

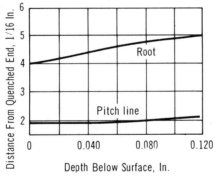

Fig. 5 — The Jec rates, which vary with distances below the surfaces (particularly for the teeth roots), are used to determine minimum and maximum D_{Ib}'s for the gear.

that hardness changes provide a sensitive indication of cooling rate. At this point, the interest is in the quenching rate. Therefore, it is desirable to obtain hardness gradients from parts that have been subjected to the complete heat treatment cycle except for carburization.

To determine the true gradient hardness due to cooling rate variation, then, it is necessary to prevent carbon penetration into the sample part. The simplest method of doing this is to "mock" carburize the part by using copper plating or other stopoff medium which will not interfere with quenching.

Gear samples are copper plated and heat treated at production carburizing and quenching facilities. The location in the furnace tray is

judiciously made to determine the range of operational quenching rates. (Figure 3 demonstrates the effect of a typical range.) The Jec rate of the part is then evaluated by comparing hardnesses on the critical section of the part to hardnesses on the Jominy bar. This bar is prepared from the same steel (1040 in this example) from which the gears were machined, and from the same bar if possible. It is subjected to the same thermal cycle (see Part II of this series, February 1973, p. 76) as the test gears, and quenched in a standard Jominy fixture. The test gears are sectioned, and the Jec rates of the gears are determined. For example, Fig. 4 indicates that the Jec rate at the gear tooth root surface is J4, while the Jec rate at the pitch line core is J2.

D_{Ib} Requirements

Having established the Jec rate, the next step is to determine the hardenability levels required. First, the carbon level must be ascertained. Then, using the Jec rate shown in Fig. 5, minimum and maximum D_{Ib}'s are determined.

Selecting the optimum base car-

Source: Metal Progress, Reprint 1974, 30 pp

bon content involves an engineering compromise. Generally speaking, the lower the base (core) carbon content, the higher will be the residual compressive stress developed in the surface of the carburized case.[3] However, the lower the base carbon content, the higher the D_{Ib} necessary; thus, greater amounts of alloying elements will be needed to achieve the required core hardness. Also, longer carburizing times are required.

For this particular example, a car-

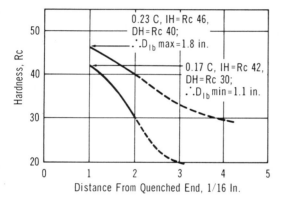

Fig. 6 — The minimum and maximum D_{Ib} values, 1.1 and 1.8 in., are shown as described in the text. At a Jec rate of J2, the minimum D_{Ib} of 1.1 in. will harden to Rc 30, and the maximum D_{Ib} of 1.8 in. will harden to Rc 40 at the pitch line core.

bon content typical for a carburizing grade of steel has been selected — 0.17 to 0.23% C. It must be noted that the carbon level selected for this example may not be the optimum carbon level from a cost standpoint; in practice, higher carbon contents should be considered. If the use of higher carbon contents represents a significant departure from previous experience, engineering tests may be required.

In combination with the carbon range selected, the core hardness specified controls the needed base hardenability. Carburized parts are generally martempered, or conventionally oil quenched and tempered at a relatively low temperature which does not significantly change the hardness level. For this reason,

it is assumed that the as-quenched and as-tempered hardnesses are approximately equal. In computations, then, use of the final tempered hardnesses (Rc 30 to 40 in this instance) is justified.

The method for computing base hardenability values, D_{Ib} min and D_{Ib} max, will be demonstrated with the aid of Fig. 6. These hardenability values are based upon the core hardness requirements of the gear. At a given cooling rate, the maximum and minimum core hardnesses relate to the extremes of the composition range.

For example, the minimum core hardness will occur in conjunction with the minimum chemical composition, which will contain 0.17% C. A steel with 0.17% C develops an initial hardness, IH, of Rc 42 (from Table I of Part II, columns 1 and 2). Because the distance hardness (DH) required at J2 is taken as the minimum core hardness allowed, Rc 30, the ratio of initial hardness to distance hardness (IH/DH) is 42/30, or 1.42. By extrapolation on Table I (linear extrapolation between J1 and J4), the minimum D_{Ib} value having an IH/DH ratio = 1.42 at the J2 position is 1.10 in.

In a similar manner, the maximum D_{Ib} value is based upon the maximum core hardness (Rc 40) with a composition containing 0.23% C (IH = Rc 46). (IH/DH = 46/40 or 1.15.) The D_{Ib} value having an IH/DH ratio of 1.15 at the J2 position is found by extrapolation, again in Table I of Part II, to be 1.80 in.

Minimum and maximum base hardenability limits (D_{Ib} = 1.1 to 1.8 in.) and the carbon content have now been established. Note: Though these limits (Fig. 6) describe a steel in terms of its hardenability (in other words, they indicate a hardenability band), an actual chemical composition is not defined as yet.

Using these limits, and the Jec rates at a number of locations below the surface at both the pitch line and the root, it is possible to plot the as-quenched hardness of an uncarburized gear made from a steel having the carbon content and hardenability limits described. This is demonstrated in Fig. 7, which shows the minimum hardness gradient expected on the pitch line cross section. In this manner, the minimum and maximum hardness gradi-

ents at both the pitch line and root may be ascertained.

The D_{Ic} Requirements

The next major step is the determination of the necessary case hardenability. The other specified hardnesses (Rc 60 to 65 at the surface, and Rc 50 at 0.030 to 0.050 in. below the pitch line surface) will be obtained by carburizing. Recall that a martensite plus austenite structure of Rc 60 to 65 is the desired surface condition.

To determine the D_{Ic} required to

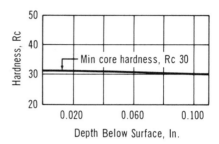

Fig. 7 — Through calculation, the minimum hardness gradient expected at the pitch line of an uncarburized gear can be determined. The minimum core hardness is Rc 30.

yield such a surface structure, the curve shown in Fig. 8 is utilized. This graph is a plot of the thermodynamic relationship[4] between the diameter of an "ideally" quenched round bar having the same cooling rate at its center as a Jominy bar at a given distance along its length. For example, a 2 in. round bar given an "ideal" quench will cool at the same rate at its center as does a Jominy bar $\frac{9}{16}$ in. from the quenched end. Jatczak[4,5] used this curve, taken from Carney's work[6], to define D_{Ic}, and to determine the effect of alloying elements on case hardenability. Multiplying factors were developed in a manner similar to the base factors previously discussed. A 10% nonmartensitic transformation criterion was used in this work, making its direct application inadvisable since, as in this instance, many applications require 100% martensite plus austenite structures. The additional step necessary to account for this difference is incorporated into the analysis below.

To determine the minimum case hardenability (D_{Ic}), refer again to Fig. 5. This figure indicates that

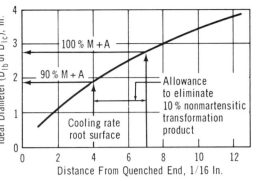

Fig. 8 — This curve relates the ideal diameter to the distances along a Jominy bar. Though J4 indicates the cooling rate at the tooth surface, 3/16 in. is added to increase the D_{Ic} to 2.7 in. This increase acts to eliminate the 10% nonmartensitic transformation products, which would develop in the case if the original D_{Ic}, 1.8 in, were used.

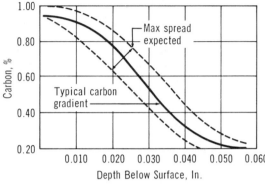

Fig. 9 — In a typical light case, carbon content will vary with the depth below the surface as shown. Base carbon is 0.20%. The carbon gradient spread is approximated from data given in the ASM Metals Handbook, 8th Edition, Vol. 2, p. 101 to 103.

the gear root cools more slowly than the pitch line. Hence, the minimum D_{Ic} value will be selected so that the case microstructure will be only martensite and austenite in the root area. By establishing a case microstructure of 100% martensite plus austenite at the slower-cooling root area, a microstructure of 100% martensite plus austenite will also be developed at the pitch line surface, which cools more rapidly.

In Fig. 8, the minimum D_{Ic} value required is determined by a vertical line drawn from the root surface Jec rate, J4, to the curve. A horizontal line from this point on the curve to the ordinate indicates that a steel with a minimum D_{Ic} of 1.8 in. would develop 90% martensite plus austenite at the root surface.

Because this example does not allow the presence of nonmartensitic transformation products — only a structure of martensite plus austenite is permitted — an additional step is necessary in the sequence. Limited experience gained by the study of carburized hardenability bars has indicated that adding 3/16 in. to the experimentally determined Jec rate and using the same thermodynamic relationship (Fig. 8) gives a new D_{Ic} value generally sufficient to produce a surface case structure free from transformation products. (This concept is in need of additional refinement.) When 3/16 in. is added to the root Jec rate, J4 + 3, and a vertical line is drawn at the J7 position, the new D_{Ic} value is 2.7 in.

In other words, a steel with a minimum D_{Ic} value of 2.7 in. will develop the desired microstructure

of 100% martensite plus austenite. It should be noted that, due to the retained austenite, some gears may develop surface hardness values below Rc 60.

Carbon Gradient Selection

Under production conditions, only a limited number of carburizing cycles are usually available due to practical considerations. Carbon gradients produced by these cycles are potentially fairly controllable. Most cycles are set up to give surface carbon contents in the 0.90 to 1.00% range. The problem then becomes one of selecting a time cycle which will result in a carbon level at the critical depth (in this instance, 0.030 to 0.050 in.) which, in combination with the previously determined D_{Ib}, will give Rc 50 under the expected quenching conditions.

As a first trial, and to demonstrate the computational principles, the typical "light case" carbon gradient shown in Fig. 9 has been selected. Showing the relationship of carbon content to distance below the pitch line surface, this graph enables the calculation of a new D_{Ib} value at each point below the surface. At each point in the cross section, it is possible to consider the increment of hardenability increase (ΔD_{Ib}) due to the increased carbon content. Because the Jec rate, the D_{Ib}, and the carbon content are known at each point below the gear surface, the appropriate hardness values can be calculated. When considering minimum and maximum hardness values, it is necessary to consider the range of carbon, range of quenching rate, and range of hardenability at that location.

For the sake of brevity, a single cooling rate is assumed. For example, consider now the minimum case depth requirement of Rc 50 to 0.030 in. below the pitch line surface.

To say that a steel has a D_{Ib} of 1.1 in. means that the product of the multiplying factors of the alloying elements which make up its composition is also equal to 1.1 — that is, D_{Ib} = MF (carbon) × MF (manganese) × . . . Dividing the D_{Ib} by the multiplying factor (MF) for minimum carbon gives the contribution of the alloys.

Figure 9 shows that 0.41% C is the minimum expected at 0.030 in. The D_{Ib} for the composition at 0.030 in. is found by multiplying the MF for this higher carbon content by the alloy D_{Ib} product, giving (in this instance) a D_{Ib} of 2.05 in. To calculate the minimum expected hardness

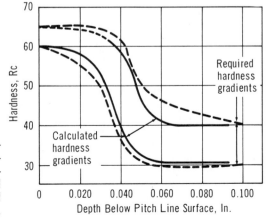

Fig. 10 — Pitch line hardness gradients calculated according to AT precepts correlate with those required for the gear. Minimum and maximum lines are shown.

Source: Metal Progress, Reprint 1974, 30 pp

at the 0.030 in. depth, reference is again made to Table I, of Part II. From Column 1 and 2, the initial hardness for 0.41% C is Rc 57. Though the Jec rate for the gear in the pitch line varies from J1.9 to J2.2, it is assumed constant at J2 for ease of calculation. By extrapolation, the IH/DH ratio is 1.075 for a steel with $D_{Ib} = 2.05$ in. at the J2 position. Dividing the IH value (Rc 57) by the IH/DH ratio (1.075) gives a DH value of Rc 53.

This value, Rc 53, represents the minimum hardness expected at the 0.030 in. depth with a minimum composition and the minimum carbon penetration of this cycle. In a similar manner, the maximum hardness level to be expected at the 0.050 in. depth is calculated to be Rc 46. It must be pointed out that, in calculations of this type, the use of D_{Ib} values and the IH/DH tables is restricted to hypoeutectoid compositions.

Figure 10, which shows results of a series of these calculations, compares the required hardness gradients at the pitch line with the calculated hardness gradients, based on the carbon concentration gradient for a typical light case and the base hardenabilities. In this instance, the trial gradient proved satisfactory. Had it not, another trial would have been necessary. It should be noted that, if one had maximum freedom in selecting carbon gradients, an optimum gradient could be selected by manipulating the above principles.

Insofar as the root fillet is concerned, experience has indicated that, when the hardenability of the steel is adequate to satisfy the pitch line strength gradient, it will also harden enough to provide an acceptable strength gradient at the root fillet.

Alternatives for Determining Composition

The Application Tailoring procedure has been used to find the minimum and maximum D_{Ib}'s (1.1 and 1.8 in.) and the minimum D_{Ic} (2.7 in.) required for the steel needed for a particular six-pitch spur gear (Fig. 1). The minimum hardness gradient to be expected in production has also been described. (Note: a max D_{Ic} is not needed; refer to Part I of this series.) Again, attention is called to the fact that the characteristics of the steel have been completely described but the composition has not yet been mentioned.

There are two routes open in this regard. One is to find a standard steel composition that fulfills the requirements. Since standard steels are not designed to have uniform changes in hardenability from grade to grade and are not systematized in terms of D_{Ic}, a compromise of some sort is likely to be necessary. The second route is to design a special steel. Procedures to develop the exact composition fitting multiple requirements comprise part of the CHAT system, and will be developed in future articles.

Selection of a standard steel for the six-pitch gear is, in part, similar to selection of a standard steel for a through-hardened part in that it is based upon the carbon content and base hardenability. Of course, case hardenability must also be considered. A standard steel is selected on the basis of carbon content (0.17 to 0.23% C) and its Jominy curve (required Rc 30 to 40 at J2 for core hardness). Finally, the expected D_{Ic} value calculated for the standard steel must be 2.7 in. minimum.

Summary

Basically, AT is that part of the generalized CHAT procedure which is used to determine the minimum and maximum hardenability levels required to develop the desired hardness and strength levels in a part for a given application. Application Tailoring could include other items such as the M_S temperature or other metallurgical or processing variables. One objective of AT analysis is improved efficiency and accuracy throughout the entire design procedure. Accuracy is emphasized in the design analysis stage because the more accurately that loads and stress levels are known, the more efficient and reliable will be the steel designed for the part.

Experience has shown that AT analysis of a part tends to promote refinements in the design analysis. In addition, AT analysis causes one to look critically at the manufacturing process, particularly the heat treatment. Alterations in quench rates, use of faster quenchants, and selection of optimum heat treating or carburizing cycles frequently result.

Finally, determination of the minimum hardenability requirements leads to a more efficient usage of alloying elements (or combinations of alloying elements) in the selection of a standard steel or the design of a steel by the CHAT approach. In ensuing articles, the logic and some of the algorithmic aspects of Computer Harmonizing (CH) will be discussed.

References

1. Mitsuo Hattori, "Heat Treatment Practice in Japanese Automotive Industry." Private correspondence.
2. "Effect of Surface Condition on the Fatigue Resistance of Hardened Steel," by G. H. Robinson, Fatigue Durability of Carburized Steel, Special ASM Publication, 1957, p. 11-47.
3. "The Distribution of Residual Stresses in Carburized Cases and Their Origin," by D. P. Koistinen, ASM Transactions, Vol. 50, 1958, p. 227-238.
4. "Multiplying Factors for the Calculation of Hardenability of Hypereutectoid Steels Hardened from 1700 F," by C. F. Jatczak and D. J. Girardi, ASM Transactions, Vol. 51, 1959, p. 335-349.
5. "Hardenability in High Carbon Steels," by C. F. Jatczak, preprint of paper presented at Materials Engineering Congress, 17 October 1972, Cleveland. To be printed in Metallurgical Transactions of ASM-AIME, 1973.
6. "Another Look at Quenchants, Cooling Rates, and Hardenability," by D. J. Carney, ASM Transactions, Vol. 46, 1954, p. 882-925.

Computer-Based System Selects Optimum Cost Steels-IV

By DALE H. BREEN, GORDON H. WALTER, and JOHN T. SPONZILLI

This is the fourth in a series of five articles covering a steel design and selection procedure for heat treated components known as CHAT (an acronym for Computer Harmonized Application Tailored). Part IV concerns the metallurgical and mathematical foundation for Computer Harmonizing (CH). Used in designing special optimum cost steels, Computer Harmonizing is the computerized determination of the least costly combination of alloying elements which will meet the property requirements called for by Application Tailoring (AT).

Computer Harmonized (CH) steels are designed with one of three objectives. These are (1) to provide a special steel that meets a component's Application Tailored (AT) requirements; (2) to aid in the selection of the most economical standard steel that meets a component's AT requirements; or (3) to develop "replacement" steels for standard AISI H steels. The distinction between using Computer Harmonizing for the first or second objective lies in whether or not the tonnage or other requirements of the component justify a special steel. For example, if the component under consideration is a low tonnage item, the engineer may lean toward a standard steel that will be readily available from a steel mill or warehouse. Using Computer Harmonizing for the third objective provides a systematic method for devising chemical compositions for "replacement" steels, which (as defined in the first article of this series) are steels which match the base and case hardenability and other characteristics of the original steel.

In each instance, the resulting CH steel has a chemical composition that is optimized with respect to the cost of alloying elements. This usually results in a steel that will also have an optimum price in terms of its "chemistry grade extra", as covered in the steel product price book. The differ-

ence between "cost" and "price" in this context must be understood; there is not always a one-to-one correspondence between the two, although a close correlation does exist.

The need for using a computer approach to develop a least-cost steel becomes apparent when the various aspects of the problem are considered. A least-cost steel which only needs to meet a specified D_{Ib} value could be designed fairly easily by a manual method with tables (or nomographs) containing alloy costs and hardenability multiplying factors. When a carburizing grade is designed, however, at least one additional restriction, a minimum D_{Ic} value, is added to the problem. Because hardenability factors for the individual alloying elements are different for the case and the base composition, a steel designed to have a least-cost base composition most likely would not satisfy the D_{Ic} requirement. If further restrictions, such as the martensite start temperatures (M_S) of the case and base analyses are added, it becomes clearly impossible to find, manually, the least-cost combination of alloying elements that satisfies the multiple requirements. Because of the multiple restrictions, as will be shown later, this problem of cost optimization requires the use of separable programming techniques.

Before embarking upon the complexities of linear and separable programming, it is worth while to take a rather simplistic look at the problem of cost optimization. Although the computer system does not use hardenability efficiency directly, the concept of hardenability efficiency is a useful strategem to aid the understanding of

the cost optimization procedure.

Hardenability Efficiency

Although alloying elements' costs and their effects on hardenability have been qualitatively linked to one another almost since steelmaking began, one of the first published quantitative attempts to use this information to devise an optimum steel composition is credited to H. E. Hostetter in the 1940's[1]. At that time Hostetter indicated that the relative costs of the alloying elements chromium, nickel, and molybdenum were about 1:2:5, respectively. Today, the figures indicate that this ratio is about 1:5:9.

A careful and continuing survey of the cost and availability of the various alloying elements used in steelmaking is essential for their efficient utilization in steel design. Quantitative knowledge of the influence of individual alloying elements on such properties as case and base hardenability enables the mathematical coupling of an alloying element's cost with its effect on properties. This is the basis of the procedure for optimizing the chemical composition of a steel.

Ideally, hardenability efficiency should reflect the hardenability contribution of an alloying element with respect to the cost of the element. Because the contributions to base and

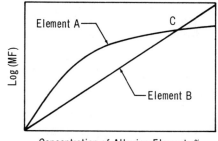

Fig. 1 — These hypothetical curves show the logarithm of the hardenability factor as related to the concentrations of two equal-cost alloying elements, A and B. Point C indicates the point at which the hardenability contribution and cost (relative efficiency) of each is identical.

Mr. Breen is chief, Materials Div., Mr. Walter is chief engineer, and Mr. Sponzilli is materials research stress engineer, Metallurgical Section, Materials Div., Engineering Research, International Harvester Co., Hinsdale, Ill. Parts I, II, and III of this series appeared in Metal Progress in December 1972, February 1973, and April 1973.

Source: Metal Progress, Reprint 1974, 30 pp

case hardenability (D_{Ib} and D_{Ic}) are both important in designing a case hardening steel, two efficiencies must be considered for each element. For a through-hardening steel, only base hardenability efficiency is important.

Relative Hardenability Efficiency

Relative hardenability efficiency is defined as the logarithm of the hardenability multiplying factor (MF) divided by the cost (C) of the element:

Relative hardenability efficiency =
$$\frac{\text{Log (MF)}}{C}$$

Here, the multiplying factor is the factor at any per cent, X, of the element, and C is the cost of X per cent of the element. Logarithms of multiplying factors are used to facilitate computational techniques. The cost is derived from the cost and composition of the addition made to the melt and its recovery level. Cost factors vary somewhat from steel mill to steel mill.

To illustrate the concept of relative hardenability efficiency, suppose that the logarithms of the multiplying factors for two elements, A and B, are as plotted in Fig. 1. Furthermore, assume that the cost per per cent of element A equals that of element B. The relative hardenability efficiency of the two elements would be identical to one another at point C because, at that concentration level, they provide the same contribution toward hardenability at an equal cost. At any percentage less than point C, the relative efficiency of element B is less than that of element A, while at percent-ages greater than C, the relative efficiency of element B exceeds that of element A.

This concept was used in constructing curves (Fig. 2 and 3) of relative base and case hardenability efficiency for manganese, silicon, nickel, chromium, and molybdenum. (For multiplying factors used, see references 2 and 3.) Note that the relative hardenability efficiencies of manganese and chromium are generally high in the base and case compositions, while those of nickel and molybdenum are generally low. Silicon in the 0 to 0.6% range however, has zero efficiency in the base composition, but has an appreciable efficiency over the same range in the case composition. Thus, silicon can be important when it is necessary to increase case hardenability without affecting base hardenability.

Although carbon and boron are much more effective in low-carbon steels than other alloying elements, curves are not shown for them in Fig. 2 and 3 for several reasons. In "blowing down" a heat of steel, carbon is removed from the melt, so that a cost-vs-hardenability effect is difficult to ascertain. As for boron, its hardenability effect is considered to be largely independent of the quantity of boron present, providing it is maintained within certain limits. Also, the boron effect varies with the carbon level, so that it too does not lend itself to the simple representation of Fig. 2 and 3.

Incremental Hardenability Efficiency

Since the hardenability multiplying factors are generally nonlinear, the relative hardenability efficiencies are not linearly related to the percentage of alloying element in a composition. Use of the efficiency concept in qualitatively understanding the determination of an optimum steel composition is furthered, therefore, by defining an incremental hardenability efficiency. This factor is defined as:

Incremental hardenability efficiency = $\triangle$log (MF)/C

Here, $\triangle$log (MF) represents the incremental change in the logarithm of the multiplying factor, and C is the alloy cost to obtain the $\triangle$log (MF).

This concept is illustrated graphically in Fig. 4 by the slopes of the two multiplying factor curves described previously (Fig. 1). As before, the cost per per cent of element A equals that of element B. The important point to note here is that, beyond X%, the incremental efficiency of element A is significantly lower than that of element B. That is, it requires 2 $\triangle$X of element A to obtain the same hardenability contribution ($\triangle$Y) provided by $\triangle$X of element B.

Although both A and B have the same relative hardenability at point C, and steels with a hardenability requirement of Log (MF) = Y_1 could be made with either element at the same over-all cost, it is significant that neither steel represents the least-cost analysis. The least-cost analysis would be obtained by adding element A until its incremental efficiency was just less than that which characterized element B. At this point, element B would be added to the analysis until the desired effect was obtained.

One can now visualize the optimization procedure — briefly, the hardenability efficiency factors, relative and incremental, for base and case composition are continuously scanned, and compared with the input requirements of D_{Ib} and D_{Ic}. The system will add elements in the proper amounts according to their hardenability efficiencies in such a way as to meet the D_I requirements and still minimize the total cost of alloying elements.

This analysis of the problem is an oversimplification of the facts concerning cost optimization. To provide a least-cost analysis that satisfies all the requirements involves more than a simple expression of hardenability efficiencies. Instead, linear and separable programming techniques, discussed next, provide the computer means for least-cost alloy design.

Fig. 2 — Given equal amounts of alloy, manganese has the greatest relative base hardenability followed by chromium, molybdenum, silicon, and nickel.

Fig. 3 — In carburized cases, the relative case hardenability efficiency is greatest for silicon, followed by manganese, chromium, molybdenum, and nickel. Compare with Fig. 2.

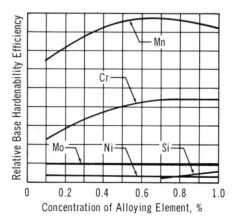

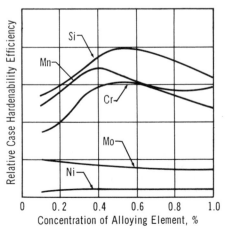

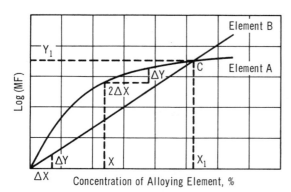

Log (MF) ... Y₁ ... Element B ... C ... Element A ... ΔY ... 2ΔX ... ΔY ... X ... X₁ ... ΔX ... Concentration of Alloying Element, %

Source: Metal Progress, Reprint 1974, 30 pp

Linear and Separable Programming

The concept of cost minimization while simultaneously satisfying a number of other requirements is a familiar one that has been dealt with through the use of linear programming. Basically, the process entails setting up a set of equations that will model the problem accurately, and then solving the system by linear programming.

Two restrictions must be placed on a set of equations for them to be solvable by this method. First, an objective, such as cost, must exist to be optimized, and it must be expressable as a linear function. Second, there must be restrictions on the attainment of the objective, and these restrictions must be expressed as a system of linear equalities[4].

To use a linear programming approach to determine the chemical composition of a steel that will meet AT requirements, it is necessary to quantify the properties under consideration. For discussion purposes, the groundwork for Computer Harmonizing will be developed with only two limited objectives. These are (1) minimizing the cost of alloy additions (objective function), and (2) satisfying the hardenability requirements (restriction functions). The latter are given in the form of ideal critical diameters of the base and the carburized case compositions, D_{Ib} and D_{Ic}.

When accurate equations become available for expressing other properties — nil ductility transition temperature, Charpy V Notch impact energy, fracture toughness, weldability, and others — in terms of chemical composition, they too may be included in the system of equations as restriction equations. For example, M_S temperatures for base and case compositions can be expressed as functions of the chemical composition.

The basic system of equations needed to develop least-cost carburizing steels will enable a resulting composition to fulfill cost, D_{Ib}, D_{Ic}, and other requirements simultaneously.

Fig. 4 — Concept of incremental efficiency is illustrated with the curves in Fig. 1. Note how equal increments in log (MF) are obtained through different incremental additions of the equal-cost elements A and B.

This accomplishment would be virtually impossible without the aid of a computer. The generalized form of some of these equations follows:

1. $Cost = K_1 \cdot x_I + K_2 \cdot x_2 + K_3 \cdot x_3 + \ldots$ (Objective function)
2. $D_{Ib} = f_1(x_1) \cdot f_2(x_2) \cdot f_3(x_3) \ldots$ (Restriction function)
3. $D_{Ic} = g_1(x_1) \cdot g_2(x_2) \cdot g_3(x_3) \ldots$ (Restriction function)

Here, cost = total cost of alloy addition; K = cost per per cent of alloying element; x_1, x_2, ... = per cent carbon, manganese, etc.; and f, g = multiplying factor functions for calculating base and case hardenability respectively — these are generally nonlinear. The optimum solution would be one that minimizes the alloy cost equation while meeting the D_{Ib} and D_{Ic} requirements.

As part of the second requirement, to enable the problem to be solved using linear programming, the terms must be additive. Though Equations 2 and 3 do not meet that requirement, they can be rewritten as:

4. $Log(D_{Ib}) = \log f_1(x_1) + \log f_2(x_2) + \log f_3(x_3) + \ldots$
5. $Log(D_{Ic}) = \log g_1(x_1) + \log g_2(x_2) + \log g_3(x_3) + \ldots$

The system of equations now becomes solvable by a modified linear programming technique known as separable programming. Separable programming requires that some additional conditions must be met. For example, each nonlinear function (multiplying factor function) must be a function of only one variable or a linear combination of such functions. Furthermore, each function must be polygonal, or replaceable by a polygonal representation of it. In other words, it must be capable of representation by a piecewise linear function[5].

The first requirement is met by all of the equations. Dividing each multiplying factor function into a number of linear segments will meet the second requirement. (The utility of using logarithms to define hardenability efficiencies should now become evident.) Of the computer programs available for solving this general type of problem, the one chosen was the IBM Mathematical Programming System/360[5], which can handle several types of linear and separable programming problems.

Computer Program Input

The input data section for the program contains the cost factors, hardenability multiplying factor functions, and M_S temperature equations. The D_{Ib} values are computed with multiplying factors developed by

Fig. 5 — Employing a graphics console, author John T. Sponzilli is shown defining D_{Ib} requirements from Jominy hardness requirements.

Doane and DeRetana[2]; D_{Ic} values, with factors developed by Jatczak[3]; M_S temperatures of base compositions, with an equation developed by K. W. Andrews[6]; and M_S temperatures of case compositions, with an equation developed by Payson and Savage[7].

To solve the system of equations and obtain a least-cost combination of alloys now only requires that specific requirements be given to D_{Ib}, D_{Ic}, and the M_S temperature equations, and that allowable limits be placed on the chemical elements in the composition when necessary. Changes in the cost of alloy additions can be made simply by revising the necessary terms in the cost equation. If only one requirement, such as D_{Ib}, needs to be satisfied, requirements for D_{Ic} and M_S temperatures can be set equal to zero.

When more than one requirement is placed on a composition, this programming approach provides the only sure way of designing a steel composition that can simultaneously meet all of the requirements with the least-cost combination of alloying elements. As indicated previously, small changes in alloying elements can represent surprisingly sizable savings.

Although they are not yet developed, other computer techniques are being studied for further automating the entire CHAT system. In one such approach, the graphics console shown in Fig. 5 is used. It provides a direct interface with the computer that could enable the operator to carry out both Application Tailoring and Computer Harmonizing functions.

The System in Summary

The CH procedure is designed to optimize one property while simultaneously satisfying multiple restrictions. In the work described here, the CH system determines the least costly alloy combination which will satisfy specified values for D_{Ib}, D_{Ic}, and M_S. Other properties can be optimized and other restrictions added, provided all features can be expressed as quantitative functions of the chemical elements. Working with hardenability and M_S temperatures as a basis, the CH process solves the least-cost alloys problem by using a modified linear programming technique known as separable programming.

In the final article of this series, the selection of standard grades, the designing of special steels for particular applications, and the procedure for developing replacement steels by Computer Harmonizing will be covered in some detail. Specific examples of CHAT components, including test results on the experimental steels, will be presented.

References

1. "Determination of Most Efficient Alloy Combinations for Hardenability," by H. E. Hostetter, AIME, Vol. 167, 1946, p. 643-652.
2. "Predicting Hardenability of Carburizing Steels," by A. F. deRetana and D. V. Doane, Metal Progress, September 1971, p. 65-69.
3. "Determining Hardenability from Composition," by C. F. Jatczak, Metal Progress, September 1971, p. 60-65.
4. "Introduction to Operations Research," by C. West, R. L. Ackoff, and E. L. Arnoff, New York, John Wiley & Sons Inc., 1957, p. 281.
5. "Mathematical Programming System/360, Version 2, Linear and Separable Programming — User's Manual," IBM Application Program, p. 165.
6. "Empirical Formulae for the Calculation of Some Transformation Temperatures," by K. W. Andrews, JISI, July 1965, p. 721-727.
7. Discussion by A. E. Nehrenberg of "The Temperature Range for Martensite Formation," by R. A. Grange and H. M. Stewart, Transactions of AIME, Vol. 167, 1947, p. 467.

Computer-Based System Selects Optimum Cost Steels - V

By DALE H. BREEN, GORDON H. WALTER, and JOHN T. SPONZILLI

This is the final article in a five-part series on procedure for designing and selecting steels for heat treated parts. Part V details the development and application of specific steels.

The first four articles in this series detailed the concepts and applications of the Computer Harmonized Application Tailored (CHAT) steel design and selection system. Included were examples of the Application Tailoring (AT) analyses of through and case-hardened components, as well as a discussion of the metallurgical and mathematical foundation for Computer Harmonizing (CH).

In this article, AT requirements are now used to either select a standard SAE-AISI steel, or serve as input for the computer program that develops an optimum cost steel to meet the requirements of the application. In addition, methods and examples of the development of economical steels for use as replacements to standard H-band steels will be presented.

Alloy Restrictions

Before developing Computer Harmonized steels, it is appropriate to cover the subject of restrictions on the final total quantities of individual alloying elements in a CH composition. Important metallurgical factors that need to be considered when designing a CH steel for through hardening applications include machinability, temperability, forgeability, weldability, and brittle transition temperature. In designing CH steels for case hardening, carbides, retained austenite, intergranular oxidation, case M_S temperature, distortion, machinability, forgeability, and microcracks must be considered. The significance

Mr. Breen is chief, Materials Div., Mr. Walter is chief engineer, and Mr. Sponzilli is materials research stress engineer, Metallurgical Section, Materials Div., Engineering Research, International Harvester Co., Hinsdale, Ill. Parts I through IV appeared in Metal Progress in December 1972, February 1973, April 1973, and June 1973.

of these factors in governing the final properties of the product will result in certain alloy restrictions, and perhaps the imposition of additional restriction functions, such as martensite start temperatures (M_S).

For example, problems with temper embrittlement have generally been alleviated by adding molybdenum. Accordingly, a minimum molybdenum content might be placed as an input restriction in the computer program to insure against temper embrittlement.

When designing a case-hardening steel, the final alloy balance must be chosen to limit amounts of carbide and retained austenite in the case. The amount of retained austenite can be partially controlled by imposing M_S temperature restrictions, and the tendency to form alloy carbides can be limited by placing maximum values on amounts of strong carbide formers, such as molybdenum and chromium.

Concerning intergranular oxidation which occurs at the surface of carburized steel, it has been recently shown[1] that silicon, manganese, and chromium oxidize, in that order, preferentially at the austenite grain boundaries. For this reason, maximum concentration restrictions may be placed on silicon, manganese, and chromium. These, in turn, could add nickel and molybdenum to the composition, even though the relative base and case hardenability efficiencies of nickel and molybdenum are generally lower than those of silicon, manganese, and chromium.[2]

With respect to distortion, the metallurgical factors most applicable are the relative case and base M_S temperatures, and the relative hardenabilities between the case and base. Since the values of M_S temperatures, D_{Ib}, and D_{Ic} may be specified as computer input requirements, the distortion factor can be substantially controlled.

Another important factor that can influence restrictions on allowable alloying elements, either minimum or maximum, is centered on the problem of price breaks in "chemistry grade extra" sections of steel price books; one such section is shown in Table I.

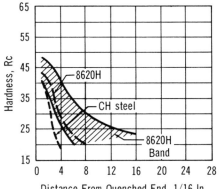

Fig. 1 — The base hardenability of the experimental CH steel designed specifically for a spur gear application is lower than that of a typical 8620H steel, which could also be used. The band for 8620H, a conventional gear steel, is also shown for comparison. The Jec rate at the pitch line core of the gear is J2.

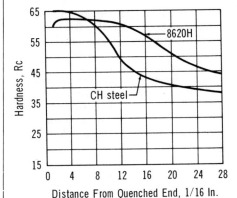

Fig. 2 — Though the case (1% C) hardenability of the experimental CH steel is lower than that of 8620H, it is adequate for the application, a spur gear. The Jec rate at the root fillet surface is J4.

Source: Metal Progress, Reprint 1974, 30 pp

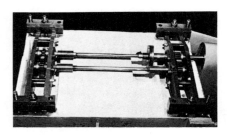

Fig. 3 — Gears made of the experimental CH steel (Fig. 1 and 2) were tested on this dynamometer. See Fig. 4.

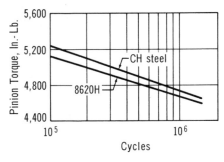

Fig. 4 — Weibull analysis of bending fatigue data from gear tests indicate that gears made from either the experimental CH steel or 8620H have equivalent durability.

Note how chemistry grade extras depend on manganese, chromium, and carbon contents.

The importance of price breaks is better understood through the use of an example. Suppose a nonboron steel with 0.18% C min and a minimum D_{Ib} requirement of 1.33 in. is needed, and the only special composition restriction is for the resulting composition to contain 0.20% Si min. For explanatory purposes, suppose that the computer program was given these requirements, and produced the following optimum minimum composition: 0.18 C, 1.13 Mn, 0.20 Si, and 0.37 Cr, with nickel and molybdenum being 0.01% minimum residuals.

Residual contents of alloying elements must be considered because they contribute toward hardenability. They are therefore specified as minimum alloy input restrictions for each computer problem. Although present in all heats of steel, residual alloys are not specified in the resulting composition range to which the steel is purchased. If a steel is to be ordered from a particular mill of known high or low residuals, this fact can be taken into account when developing its composition.

Applying standard chemistry ranges to the minimum analysis developed above provides a composition spread to which this steel could be ordered: 0.18 to 0.23 C, 1.10 to 1.40 Mn, 0.20 to 0.35 Si, and 0.35 to 0.55 Cr.

From Table I, the manganese-chromium grade extra would be $1.15 per 100 lb. The CH analysis that resulted from the first computer run could be advantageously revised by placing a maximum restriction of 1.05% on the minimum manganese allowed, and rerunning the problem, leaving other requirements as before.

Restricting manganese results in the following analysis; 0.18 C, 1.05 Mn, 0.20 Si, and 0.44 Cr. This steel could be ordered to the following range: 0.18 to 0.23 C, 1.00 to 1.30 Mn, 0.20 to 0.35 Si, and 0.45 to 0.65 Cr. Again from Table I, the manganese-chromium grade extra is $0.90, which represents a $5.00 per ton savings over the original steel.

In some instances, it is possible to adjust alloy composition to move the final steel from the alloy steel price base to the carbon steel special-quality price base. This results in substantial savings, amounting to a minimum of $19.50 per ton.

In general, restrictions on alloying elements in CH compositions are made and revised carefully, always keeping in view the ultimate effect on the resulting steel or production part. (Many of the alloy restrictions necessary for equivalent "replacement" steels were discussed in the first article of this series.) Whenever possible, adjustment of alloy additions should be considered to insure the design of an economical steel, not only from the alloy additions standpoint but also from the standpoint of consumer grade extra price.

Table I — Typical Grade Extras for Manganese and Chromium in Alloy Steels

Carbon Range	Chromium Range					
	to 0.40% incl.	0.41 to 0.65% incl.	0.66 to 0.90% incl.	0.91 to 1.20% incl.	1.21 to 1.50% incl.	1.51 to 1.80% incl.
Manganese, 0.20% max						
to 0.10% incl.	$0.65	$0.85	$1.05	$1.25	$1.55	$1.80
0.11 to 0.20% incl.	0.60	0.75	0.95	1.15	1.40	1.65
0.21 to 0.24% incl.	0.40	0.55	0.70	0.90	1.15	1.40
0.25 to 0.28% incl.	0.40	0.55	0.65	0.80	1.05	1.25
over 0.28%	0.40	0.55	0.65	0.80	1.00	1.20
Manganese Range, 0.21 to 0.40%						
to 0.10% incl.	0.65	0.85	1.05	1.25	1.55	1.85
0.11 to 0.20% incl.	0.55	0.70	0.90	1.10	1.35	1.65
0.21 to 0.24% incl.	0.40	0.55	0.70	0.90	1.15	1.45
0.25 to 0.28% incl.	0.40	0.50	0.65	0.80	1.05	1.30
over 0.28%	0.40	0.50	0.65	0.80	1.00	1.20
Manganese Range, 0.41 to 0.70%						
to 0.10% incl.	0.70	0.90	1.10	1.35	1.65	1.95
0.11 to 0.20% incl.	0.55	0.70	0.90	1.15	1.40	1.65
0.21 to 0.24% incl.	0.35	0.50	0.70	0.95	1.25	1.50
0.25 to 0.28% incl.	0.35	0.50	0.65	0.85	1.10	1.30
over 0.28%	0.35	0.50	0.60	0.80	1.00	1.20
Manganese Range, 0.71 to 1.00%						
to 0.10% incl.	1.00	1.20	1.45	1.70	1.95	2.25
0.11 to 0.20% incl.	0.75	0.95	1.20	1.45	1.65	1.95
0.21 to 0.24% incl.	0.55	0.70	0.95	1.20	1.40	1.70
0.25 to 0.28% incl.	0.50	0.60	0.80	1.05	1.25	1.55
over 0.28%	0.50	0.60	0.80	1.00	1.15	1.45
Manganese Range, 1.01 to 1.30%						
to 0.10 % incl.	1.25	1.45	1.70	1.95	2.20	2.45
0.11 to 0.20% incl.	1.00	1.20	1.45	1.70	1.90	2.15
0.21 to 0.24% incl.	0.70	0.90	1.15	1.40	1.60	1.85
0.25 to 0.28% incl.	0.65	0.80	1.00	1.25	1.45	1.70
over 0.28%	0.65	0.75	0.95	1.15	1.35	1.60
Manganese Range, 1.31 to 1.60%						
to 0.10% incl.	1.55	1.80	2.00	2.25	2.50	2.75
0.11 to 0.20% incl.	1.20	1.45	1.65	1.90	2.10	2.35
0.21 to 0.24% incl.	0.90	1.15	1.35	1.60	1.80	2.05
0.25 to 0.28% incl.	0.75	1.00	1.20	1.45	1.65	1.90
over 0.28%	0.75	0.95	1.10	1.35	1.55	1.75

Note: Extras, listed in dollars per 100 lb, apply if manganese and chromium are specified, but only when: (a) max carbon is 0.75% or under and max Cr is 1.80% or under, or (b) max carbon is over 0.75% and max Cr is 0.45% or under, or (c) max Mn does not exceed 2.20%.

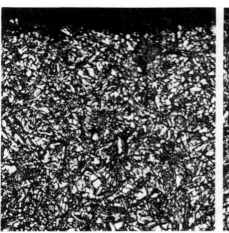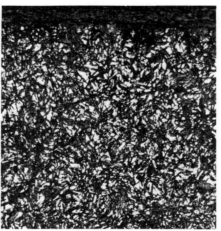

Fig. 5 — Microstructures at roots of 8620H (left) and the experimental CH steel (right) gears indicate both quench to 100% martensite plus austenite at that zone even though the latter has lower hardenability. Etchant, 2% Nital; 500×.

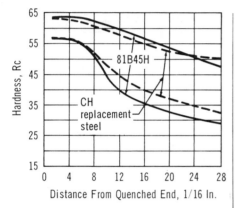

Fig. 6 — The calculated hardenability band for the CH equivalent of 81B45H agrees with that for standard grade.

Selecting and Designing Optimum Steels

Given engineering design requirements, several components were Application Tailored, as described in the second and third articles of this series. Specifying the appropriate steel for two of the components — a wheel spindle and a carburized spur gear — will now be considered in detail.

● **Oil-Quenched Wheel Spindle:** One method of specifying a steel using the hardenability requirements established through AT analysis would be to examine published hardenability bands for the most economical steel that meets the requirements. For comparative purposes, this approach, as well as the Computer Harmonizing approach, will be applied to the development of a steel specification for the wheel spindle.

The engineering requirements for the spindle were a 0.40 to 0.47 C steel, oil-quenched to 90% martensite at the finished machined surface, and tempered at Rc 30 to 37. Through the AT analysis, these requirements translated to a 0.40 C min steel with Jominy hardenability requirements of Rc 56 min at J1 and Rc 50 minimum at J9.5. The equivalent minimum D_{Ib} requirement was found to be 4.90 in.

Listed in SAE Standard J407c, "Hardenability Bands for Alloy H Steels," are three suitable steels: 4145H, 4147H, and 86B45H (Table II). Of the three, 4145 would be the preferred steel because of its price and carbon content, which is closest to the required 0.40 to 0.47%.

If tonnage requirements are sufficient, a special CH steel could be developed for this component. Because a boron-containing steel is generally the most economical steel for a through-hardening application, a boron steel was Computer Harmonized for the wheel spindle. Its composition follows: 0.40 to 0.46 C, 1.20 to 1.50 Mn, 0.20 to 0.35 Si, 0.55 to 0.75 Cr, and 0.0005 B min. Minimum calculated Jominy hardness values are Rc 56 at J1 and Rc 50 at J10; the grade extra is $1.60 per 100 lb.

This steel has the advantage over 4145H in that it has a carbon content identical to that required, along with a grade extra that represents a $7 per ton savings. Before adopting the new steel for the wheel spindle, laboratory evaluations of its Jominy hardenability and other metallurgical characteristics would be made.

● **Carburized Spur Gear:** In the third article of the series, a spur gear (6 diametral pitch, 3 in. pitch diameter) was Application Tailored. Engineering requirements for this carburized, oil-quenched part were a surface hardness of Rc 60 to 65, an effective case depth of 0.030 to 0.050 in. at the pitch line, and a core hardness range of Rc 30 to 40. The AT procedure indicated that a 0.17 to 0.23 C steel with the following hardenability requirements would be adequate: Rc 30 to 40 at J2 (1.10 to 1.80 in. D_{Ib}). On the 1% C hardenability bar, 100% martensite plus austenite at J4 min was required (2.70 in. D_{Ic} min).

The objective of Computer Harmonizing is to develop an optimum cost steel that would meet the above requirements. Input requirements for the computer program were 0.17 C content, and minimum D_{Ib} and D_{Ic} of 1.1 and 2.7 in. respectively. No other special composition restrictions, except for minimum residuals, were specified. The resulting steel contained 0.17 C, 0.70 Mn, 0.30 Si, 0.50 Cr, plus minimum residuals of 0.01 Ni and Mo; D_{Ib} and D_{Ic} values were 1.1 and 2.8 in., respectively.

A second steel was developed according to the same requirements plus a restriction of 0.02 Cr max — composition was as follows: 0.17 C, 1.10 Mn, 0.55 Si, 0.01 Ni, 0.02 Cr, and 0.01 Mo (the last three are residuals). Though the total cost of alloying additions was increased slightly, the second steel could be priced as a carbon steel rather than as an alloy steel.

Both compositions represent minimum analyses that just meet acceptable hardenability levels called for by the AT procedure. A quantity of the second steel was procured to evaluate for the spur gear. The composition of the experimental steel was 0.16 C, 1.24 Mn, 0.58 Si, 0.07 Ni, 0.05 Cr, and Mo; D_{Ib} and D_{Ic} values were 1.1 and 2.8 in., respectively.

(This experimental steel is referred to as "CH steel" in the rest of the article, and in Fig. 1, 2, 3, 4, and 5.)

Figure 1 shows base hardenability curves for the experimental steel and a heat of 8620H; the 8620H hardenability band is also given. The base hardenability curves indicate J2 hardnesses of Rc 35 for the experimental CH steel and Rc 41 for the 8620H steel. Gears made from these steels would be expected to have pitch line core hardnesses of Rc 35 and 41. The hardenability comparison is made between 8620H and the experimental steel because 8620H (or one of its EX steel replacements) is commonly used in this type of application.

Figure 2 illustrates case hardenability curves (1% C) for the experimental CH steel and for the heat of 8620H. These curves indicate that gears made from either steel would meet the 100% martensite plus austenite microstructure requirement at the root fillet surface, which has a Jec rate of J4. Note that the experimental CH steel met the AT Jominy hardness requirements for the pitch line core and the root fillet surface, as expected. Although the 8620H steel is acceptable for this part, it actually has excess base and case hardenability. For this gear, therefore, it represents metallurgical overdesign.

Normal machining and heat treating procedures were used for making about 30, 6-pitch test pinions, which were then tested in the 4-square dynamometer test (Fig. 3). As Fig. 4 reveals, the experimental CH steel had fatigue resistance equivalent to that obtained with gears made of 8620H.

Metallurgical data gathered on these gears (including information on microstructure, microhardness, residual stress, surface finish, and fractography) helped establish the adequacy of the new steel. Figure 5 shows microstructures of the root fillet surface of gears made of 8620H and the experimental CH steel. Although the experimental steel had a significantly lower case hardenability, it quenched out to a 100% martensite plus austenite structure at the root fillet surface. Obviously, it had adequate, although not excessive, case hardenability.

Designing Replacement Steels

Generally speaking, two steels with equivalent carbon contents, and equivalent case and base hardenability bands as well, will produce end products with equivalent microstructures and hardness gradients. This concept lays the foundation for Computer Harmonizing a standard SAE steel to develop a lower-cost, lower-price replacement with equivalent carbon content and hardenability to the original.

Any standard SAE-AISI steel can

HISTORY OF ALLOY STEEL USAGE

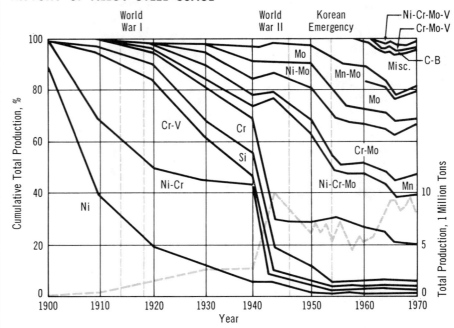

Fig. 7 — Through the years since 1900, the pattern of alloy steel usage has changed greatly. The dotted line indicates alloy steel production.

be computer-harmonized. That is, it is possible to calculate, for each grade, the average base and case hardenability which is characteristic of that steel, and place these values into the computer program, coupled with whatever restrictions are necessary on minimum and maximum values of certain elements. The result will be a replacement steel of the same carbon content which will develop the same average hardenability, and thus the same hardness gradients and microstructure, in the end product.

Once the average composition is developed, the minimum and maximum expected hardenability and composition can be determined from a statistical analysis of past mill practice in producing many heats of the numerous standard grades. (This technique for calculating hardenability bands — the "plus and minus 2σ method — was discussed in the first article of this series.)

As an example, chemical analyses and midrange base and case hardenabilities for 4817H and its CH equivalent are shown below:

	4817H	CH Equivalent
C	0.14-0.20%	0.14-0.20%
Mn	0.30-0.70	1.25-1.55
Si	0.20-0.35	0.40-0.60
Ni	3.20-3.80	—
Cr	—	0.80-1.00
Mo	0.20-0.30	—
D_{IH}	2.6 in.	2.6 in.
D_{IL}	6.4	6.4

Comparative chemistry grade extras are $8.55 per 100 lb for 4817H and $2.45 per 100 lb for the CH equivalent, representing a saving of approximately $122 per ton.

Because this particular example is for illustration purposes, only midrange base and case hardenabilities are shown. The final analysis would be developed by using the $\pm 2\sigma$ analysis technique to assure that the calculated base hardenability band matched the base hardenability band for 4817H. Experimental parts of steel similar to the 4817H replacement are currently undergoing feasibility studies to determine whether they have similar processing and performance characteristics.

This approach was recently applied to design a replacement composition for 81B45H. The chemical analysis range for the original steel and the CH version follow:

Table II — Suitable Wheel Spindle Steels

Steel	Carbon Range	Min. Jominy Hardness, Rc		Grade Extras per 100 Lb
		J1	J10	
4145H	0.42-0.49	56	50	$1.95
4147H	0.44-0.51	57	53	$1.95
86B45H	0.42-0.49	56	51	$2.95

	81B45H	CH Equivalent
C	0.43-0.48%	0.43-0.48%
Mn	0.70-1.05	1.00-1.30
Si	0.20-0.35	0.20-0.35
Ni	0.15-0.45	—
Cr	0.30-0.60	0.45-0.65
Mo	0.08-0.15	—
B	0.0005 min	0.0005 min

Figure 6 shows that hardenability bands for 81B45H and its CH equivalent conform closely. The bands are slightly different because one is calculated while the other is based on Jominy tests of many heats. In practice, the replacement steel could be consistently produced to fall within the published band for 81B45H.

The replacement represents a consumer price saving of $20 per ton. Careful evaluations of these steels similar to those discussed for the wheel spindle example may be necessary, depending on the end use of the replacement steel.

Concluding Summary

This last article culminates a series which has presented a comprehensive look at a system for arriving at optimum alloy usage for heat-treated components, especially those utilizing large tonnages of steel. Even so, many important issues were only tagged and not fully handled. The contribution to engineering properties by individual alloys per se, for example, was not argued. Neither were residual stress effects, a factor known to be important in metal fatigue. The significant effect of grain size on hardenability was not discussed, all computations being done assuming an ASTM No. 7 grain size to ease calculation. Also, no attempt was made to answer the questions brought forth by the use of a multiplier for boron. Other approaches to this problem, including the development and usage of regression analysis equations such as those of Table III, are being investigated.

We do not believe that the shortcomings noted detract from the utility of the CHAT system, but simply indicate it is not a panacea. As is the situation for many useful tools, the system should be manipulated and applied with discretion, utilizing all available technology as a base.

Two basic concepts have been presented in the CHAT system. One was termed CH — the computer was used to develop an optimum cost steel meeting specific hardenability requirements. The other was termed

Table III — Equations for Calculating Boron Steel Hardenability Curves

$$J1 = 37.5 \times (\% \text{ C}) + 39.5$$
$$J2 = 37.9 \times (\% \text{ C}) + 38.6$$
$$J3 = 37.8 \times (\% \text{ C}) + 38.1$$
$$J4 = 41.1 \times (\% \text{ C}) + 36.3$$
$$J5 = 44.6 \times (\% \text{ C}) + 8.0 \times (\% \text{ Mn}) + 10.2 \times (\% \text{ Cr}) + 23.6$$
$$J6 = 58.0 \times (\% \text{ C}) + 16.2 \times (\% \text{ Mn}) + 30.4 \times (\% \text{ Cr}) + 5.3$$
$$J7 = 65.5 \times (\% \text{ C}) + 35.1 \times (\% \text{ Mn}) + 66.0 \times (\% \text{ Cr}) - 27.6$$
$$J8 = 54.4 \times (\% \text{ C}) + 42.0 \times (\% \text{ Mn}) + 93.6 \times (\% \text{ Cr}) - 40.7$$
$$J10 = 39.2 \times (\% \text{ C}) + 37.3 \times (\% \text{ Mn}) + 73.9 \times (\% \text{ Cr}) - 37.5$$
$$J12 = 37.5 \times (\% \text{ C}) + 29.9 \times (\% \text{ Mn}) + 54.8 \times (\% \text{ Cr}) - 31.2$$

Note: Calculated values indicate Rc hardnesses. To use the equations, all chemical elements in a composition must fall within the following effective ranges: 0.25 to 0.35 C, 1.20 to 1.38 Mn, 0.23 to 0.31 Si, 0.02 to 0.04 Ni, 0.03 to 0.12 Cr, 0.01 to 0.02 Mo, and 0.0005 B min.

AT — engineering requirements were translated into quantitive metallurgical requirements.

In addition to these prime engineering and cost aspects, the definition of the CHAT system creates an awareness of the explicit definition of a steel. A steel is not a simple single analysis, as might be implied by the standard numbering system and as it is considered to be by many design engineers. Instead, it is a large, finite number of analyses having quantifiable statistical characteristics in terms of such items as base and case hardenability. This comprehension by designers as to what comprises a steel is an important link in a viable, flexible materials engineering activity.

The principal of "optimum steel usage" has never been completely ignored by the maker of machine parts and assemblies. Many times, however, it has been relegated to a subordinate role, especially when times are good and the climate for change unfavorable. This climate has complex aspects, even to the point of being influenced by the personality of the final decision maker.

As the historical view of alloy steel usage shows (Fig. 7), many metallurgical developments stem from national or international crises.[3] The connotation this carries is not necessarily favorable to the metallurgical community. With the sure knowledge that history repeats itself, an acceptable system to make changes from one alloy system to another will certainly be put to good use in the future.

Maybe the day will come when heat treatable steels will be defined almost solely on the basis of response to heat treating rather than composition. Then, periodic adjustments in alloy content would be made without hesitation, taking immediate advantage of

either or both changing alloy costs and availability. The CHAT system should help bring us closer to the ultimate goal — perpetual efficient utilization of alloys in heat treatable steels.

The strategic alloy problem has been the subject of much discussion. In a recent article[4] A. C. Sutton gave an interesting discussion which pointed out that our dependency on other countries for ores could be an important factor in future international relationships. He described a "weak link" principle — an advantage could be gained by a potential enemy who subtly promoted dependency on a strategic alloy import. Then, alternate approaches would be necessary during a war. Having principles established and disseminated can have considerable influence on the reaction time when the need for change arises. This series of articles, we believe, partially serves this purpose.

Finally, the rationale and tools discussed in this series of articles are effective in lowering production costs without sacrificing quality. The concept, though, is important as an ongoing philosophy in terms of efficiency in original designs, preparedness for change, and conservation of strategic materials. ⊕

REFERENCES —
1. Mitsuo Hattori, "Heat Treatment Practice in Japanese Automotive Industry," private correspondence.
2. "Computer-Based System Selects Optimum Cost Steels — IV," by D. H. Breen, et al, Metal Progress, June, 1973.
3. "The Sorby Centennial Symposium on the History of Metallurgy," Cyril Stanley Smith, ed., Gordon & Breach Science Publishers, New York, 1963, p. 475.
4. "Soviet Strategy," by A. C. Sutton, Ordnance Magazine, Nov.-Dec., 1969.

John Tartaglia
AMAX Materials Research Co.

Computer program predicts Jominy end quench hardenability

Predictive program allows researchers to design new carburized steels without melting any metal.

HARDENABILITY, a measure of the propensity for a steel to harden during quenching, is one of the most important properties to consider when designing low-alloy steel products. The principal factors that determine steel hardenability are its composition and its grain size.

When designing a new steel alloy for an application, the metallurgist or materials engineer needs to know if a steel will have the hardening response required to meet other property requirements such as strength and fatigue resistance.

To help meet this need, we have developed a computer program that allows the materials designer to make preliminary hardenability calculations using a proposed chemical composition without actually performing expensive and time-consuming experiments.

Hardenability is the factor that controls the amount and kind of microstructural constituents that will form when a piece of steel is quenched. For a specific alloy content and austenite grain size, the volume fractions of martensite (the hardest steel phase) and bainite, pearlite, and ferrite (the soft constituents) dictate the hardness level that will exist at a specific location in a quenched part.

Our computer program does not predict the kinds and amounts of phases in a hardened steel but, rather, predicts hardness at the Jominy positions that have the same cooling rates as the locations of interest in the steel sample.

Jominy curves are a commonly used tool for quantifying steel hardenability. In a Jominy test, a test bar of the steel is heated for about 30 min. at 925 C (a typical austenitizing temperature for 0.2% carbon steels).

The Jominy specimen then is placed in a support and one end is subjected to a water spray of a specific pressure and flow rate. Since the bar cools mainly by conduction from the end, slower cooling rates are obtained at greater distances from the quenched end (the end sprayed with water).

After end-quenching, longitudinal flats are ground on the bar and the Rockwell C distance hardness (DH) is measured as a function of the distance from the quenched end (Jominy distance).

A commonly used hardenability index is the ideal critical diameter (D_I)—the diameter of a cylindrical steel bar that will form 50% martensite at its center when subjected to an "ideal" quench. The D_I value of a steel can be determined from its Jominy curve.

Not only can D_I values be measured, they can be calculated from the steel composition using multiplying factors. The ideal critical diameter is related to alloy content using equations I, II, and III.

The alloy multiplying factors (MF_x) used in the calculations are based on Jominy curves obtained from hundreds of steel mixture analyses done in our laboratory. These factors have been shown to be very accurate for alloy steels containing 0.2% C.

The initial, or fully martensitic, hardness (IH) can be related to carbon content using equation IV. The hardness at other Jominy positions can be related to initial hardness and to the ideal critical diameter using equation V.

Reprinted with permission from Research & Development, April 1984, 108-110, © 1984 Technical Publishing Co.

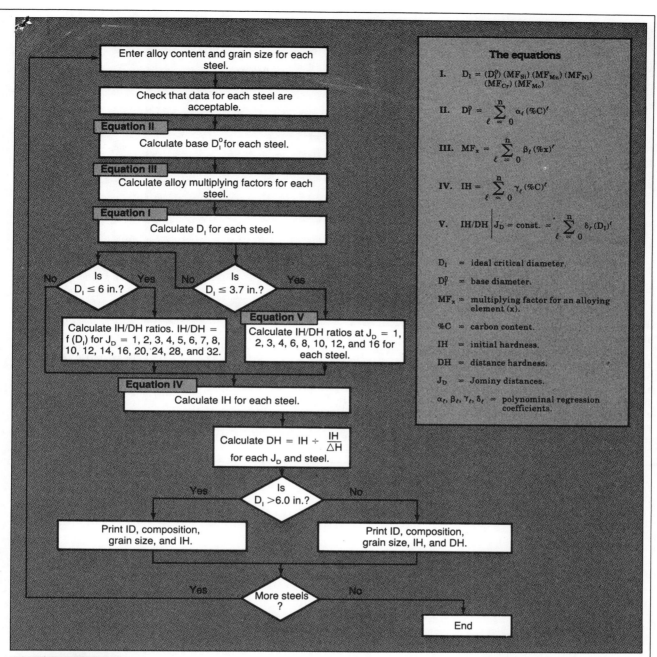

Flow chart shows how program predicts Jominy end quench hardenability and how the program relates to the five equations normally used in "manual" hardenability calculations.

To calculate the expected hardness at a specific Jominy distance (J_D) for a particular steel, the IH value is divided by the IH/DH ratio for the specified Jominy distance to obtain the distance hardness at that Jominy position. By applying this technique to several Jominy distances for a particular steel, an entire Jominy curve can be calculated.

The five equations provide a convenient algorithm for computerization of Jominy curve calculations. The only data needed beyond the five equations are the steel type, alloy content, and grain size. Data entry can be interactive, *i.e.*, the computer prompts the user for each entry, or a sequential disk file containing all of the necessary input data can be created.

After the data are entered, the computer program checks that all data are in the validity range for calculating D_I values. Specifically, the program allows maximum entries of 2% Si, 1.4% Mn, 1.5% Cr, 3.75% Ni, and 1% Mo. Grain sizes must be between ASTM Nos. 4 and 10.

Although the alloy multiplying factors were developed for steels containing 0.15 to 0.25% carbon (the typical range for core carbon concentrations in carburized steels), the program will allow entry of carbon contents up to 0.5%.

After the ideal critical diameter for the steel is calculated, the IH/DH ratios at selected Jominy positions are determined. Since the polynomial regression equations for IH/DH *vs* D_I are valid only up to $D_I = 97$ mm, the computer program uses a table of IH/DH ratios for steels with higher harden-

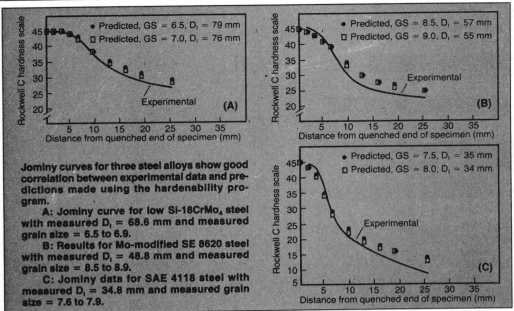

Jominy curves for three steel alloys show good correlation between experimental data and predictions made using the hardenability program.

A: Jominy curve for low Si-18CrMo₄ steel with measured D_I = 68.6 mm and measured grain size = 6.5 to 6.9.

B: Results for Mo-modified SE 8620 steel with measured D_I = 48.8 mm and measured grain size = 8.5 to 8.9.

C: Jominy data for SAE 4118 steel with measured D_I = 34.8 mm and measured grain size = 7.6 to 7.9.

THE AUTHOR
John Tartaglia is a senior research associate at AMAX Materials Research Co., Ann Arbor, MI, where he specializes in the physical and mechanical metallurgy of wrought low-alloy steels. He earned his PhD at Rensselaer Polytechnic Institute.

abilities (D_I up to 150 mm).

Finally, the program calculates IH and DH and prints a table showing the composition, grain size, ideal critical diameter, and initial hardness in addition to Jominy curves for each steel. If a graphics device is available, the results can be plotted and the predicted Jominy curves for several steels instantly compared.

The computer program uses the results of past research in our laboratory on the hardenability of 0.2% carbon steels and presents them in an easily used format. The Fortran program can be run on mainframes and minicomputers. A Basic program also is available for the "Apple" and IBM "PC" personal computers. Listings of these programs can be obtained, without charge, from the author (Box 1568, Ann Arbor, MI 48106).

R&D

System for engineering materials selection

by N Swindells and R J Swindells*

A new analysis of the problem of the selection of engineering materials has shown that there are four alternative situations which have to be resolved, arising from the interaction between the variables: Materials properties, manufacturing processes, and the shape and failure mode of a component for a given functional requirement. This analysis leads to a methodology for a knowledge-based computer system, called PERITUS, which has been developed for use at the innovation stage of design. This article presents examples of the operation of the system and a discussion of its relationship with large data bases of materials properties.

The problem of making a reliable and rational choice of materials for creating new or improved products is becoming increasingly important as the number of new materials with unfamiliar properties increases. In 1973, Hanley and Hobson[1] estimated the number of materials available to designers to be in excess of 15,000; that number is certainly greater now, when the cost of making an inadequate or wrong choice has also become very large.

At the present time, in all the world's industrialised countries, attention is being given to the use of computer systems to store and process data about the properties of materials so that the large storage capacity and rapid retrieval achievable with computer data base systems can provide easy access to the materials data which are available. Three recent conferences have concentrated on the problems confronting the development of large, computerised, materials data systems: A national review of the situation in the USA was held in 1982 at Fairfield Glade[2]; an international CODATA conference was held in June 1984 in Jerusalem[3]; and a European Community Workshop, held at Petten, the Netherlands in November last year, reviewed the European situation[4]. The report 'Materials properties data management – approaches to a critical need'[5] from the National Materials Advisory Board in the USA concentrates on drawing attention to the value of materials data as a national resource and advocates a major programme in the USA to achieve effective managment of, and access to, this resource. A 'Directory of Databases for Materials Properties' has been compiled by Hampel et al[6] and there is also a review of international information networks for materials properties from the same source[7].

However, a system which provides access to materials data is not necessarily a materials selection system, although access to the data is obviously essential to enable a selection to be made. In order to use a source of data effectively the user should have some knowledge base which can be used to formulate an intelligent approach to the search and to provide a framework within which the data can be used. 'Selection' implies 'decision', and decisions can only be made against a background of knowledge which would enable the data to be used in an intelligent way.

It is this aspect of the problem which has led some to suggest that only an expert can deal with materials selection problems because only an expert can have acquired the knowledge base which is necessary to make use of the available data. Users of this argument apparently fail to recognise that large numbers of engineers carry out materials selection functions from a limited base of knowledge of the nature of materials. Furthermore, it must be acknowledged that the number of such decision-making occasions is very large and the number of so-called experts is relatively small. The experts would be better and more effectively employed dealing with long term, difficult or strategic problems, delegating relatively routine questions to a system which engineers and designers could use directly. Such a system would form an 'intelligent knowledge-based system' (IKBS) for the selection of materials. Modern developments in computer power and the availability of specialised languages have eased the problem of providing such a system.

Before such a system can be devised it is necessary to adopt a methodology on which the system can be based. However, there does not appear to be a consensus on a generalised method of approaching the problem of materials selection. Plevy[8] attempted to establish a methodology but the scope of his concept is too broad to achieve a practical working method. Kusy[9] devised a rigorous procedure for guiding the selection of plastics which was a major step forward in technique because it sets out in detail the factors and characteristics for which data are required in order to arrive at a decision. The steels selection procedure devised by Breen et al[10] could

be considered to be a knowledge-based system since it contains a great deal of understanding about the relationships between composition, hardenability and cost-effectiveness, and includes an optimisation method for reaching a final decision.

The problem of materials selection

In order to derive a methodology for dealing with the problem of selection it is necessary to carry out an analysis of the problem to identify the steps which the methodology has to follow. The analysis which follows is based on the observation that the problem of materials selection for the design of the components involves a consideration of:

- The duty or function required of the component;
- the materials properties;
- the manufacturing route; and
- the shape, dimensions and failure mode of the component.

The three factors of properties, manufacturing and shape (P, M and S) are interactive variables. From these variables, there are four situations which arise from having some or none of the variables pre-determined and hence fixed, or else not important. The four situations are as shown in Table 1.

All materials selection problems, at least for mechanical design and some electrical problems, appear to be covered by these four situations if one uses the 'shape' variable to include the failure mode; 'cost' can be a parameter in P and M and could be calculated for S. This analysis shows that a conventional data base of materials properties would only be capable of aiding a solution to situation 1 in Table 1.

The requirement for a materials selection system to be used by materials engineers and design engineers is therefore that it should be a knowledge-based system which recognises these four situations and which provides procedures for dealing with them. The system should also possess the following desirable characteristics, which are not necessarily in the order of their importance:

a It should be rapid in use;
b it should require little learning;
c it should be accessible at different levels to suit different levels of user;
d its structure should match the structure of available knowledge for ease of creation and development;
e it should have text and graphical output;
f it should record transactions for reference; and
g the sources of data which it uses should be known.

New knowledge-based system

A knowledge-based system for the selection of engineering materials has been devised

Table I: *Situations which can arise from the existence of three variables in a selection problem*

SITUATION	VARIABLE	FIXED (OR UNIMPORTANT)
1	P	M,S
2	P,M	S
3	P,S	M
4	P,S,M	—

*N Swindells BSc, MSc, PhD, CEng, FIM is Lecturer in Metallurgy and Materials Science at the University of Liverpool and R J Swindells is Technical Director of Matsel Systems Ltd, Birkenhead, Merseyside.

Reprinted with permission from Metals and Materials, May 1985, 301-304, © 1985 Institute of Metals

which can deal with the four situations outlined, and which satisfies the desired characteristics listed at the end of the previous section. The system has three main stages.

The director stage: At this level the user can start from knowing in general outline only what has to be designed and what it has to do. The purpose of this stage is to direct the non-specialist user to data and knowledge modules for particular classes of materials and processes.

The pre-sort stage: The aim of this section is to achieve a short list of candidates from within a restricted class of materials indicated as a possibility by the director stage. Selection can be by previous use, duty requirements, properties limits or manufactured form.

Evaluation and optimisation stage: This section evaluates the candidates in the short list in more detail and produces an ordered list based on the optimum evaluation of important properties or of desired shapes and design limits or failure modes.

The system is modular and so extensions and developments can be added as required and each stage can be used independently of the others if desired. At its present stage of development it represents a 'skilled' system for the selection of materials because it contains knowledge and experience and a method of applying that knowledge and experience to a particular problem. As the system grows, 'intelligence' is being added to provide the ability to interpret the requirements of the user and to anticipate factors which the user may not otherwise have considered. The system is called PERITUS[11] and it is intended for use at the innovation stage in design when the general feasibility of a new design concept needs to be established.

Director stage

The structure of the director stage of the PERITUS system is shown in fig 1. It is a development of the method proposed by Kusy[9], which has been generalised to deal with metals and plastics. The user of this stage sets out the requirements at each step in the form of a list. The program matches this list of requirements to its store of knowledge and it outputs possible materials and processes for more detailed consideration. As a simple example, the list of requirements for a spring contact for making an electrical connection at the component category step could include the features: Spring, electrostructural, conductor and wear surface. This step has an influence mainly on the broad choice of material.

Taking the requirements for this particular component into the next step, the shape category, would also use the alternative list of either 'spirals, bent' or 'flat, bent', depending on whether a flat or a coil spring was envisaged. This step has an influence mainly on the type of process used to form the material. Subsequent steps in the director stage examine the materials and process

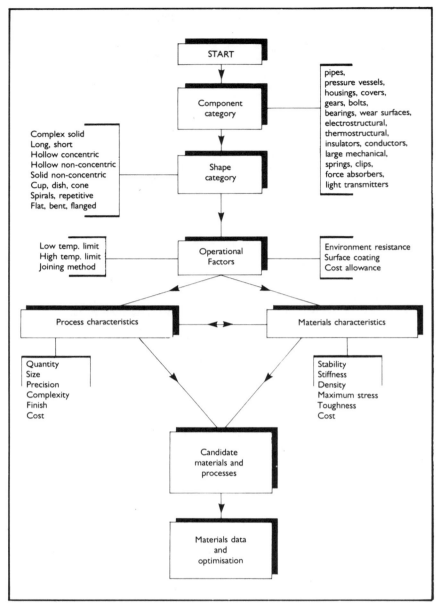

Fig I *The structure and features of the director stage of the PERITUS knowledge-based system for the selection of engineering materials*

characteristics established in the first two steps in more detail, checking that the earlier lists of requirements are not violated. The results from this stage would therefore draw the designer's attention to materials classes and fabrication processes which should suit his requirements in general terms and the pre-sort stage enables these suggestions to be examined in more detail.

Pre-sort stage

The pre-sort stage of the PERITUS system can be entered independently of the director stage. Within a chosen class of materials it provides a detailed match between the duty requirements and the material characteristics for a preferred manufacturing route in order to generate a short list of candidate materials. The user specifies his requirements for each general characteristic, such as weldability, corrosion resistance, fluidity, pressure tightness, etc, by setting a level of

importance for each characteristic on a scale from 0 (not important) to 3 (very important). The same procedure is used to specify the required manufacturing route, such as extrusion, sand casting, bright bar, etc, depending on which processes are appropriate for the particular class of materials. An efficient branching algorithm then identifies which materials in the class meet or exceed these requirements. This process is a generalisation and a development of the grid or matrix method described by Gillam[12] and by Crane and Charles[13].

The general application or common use of a material encapsulates all the duty requirements and possible processes for the material in one description. Therefore, an alternative way of generating a short list is to match the desired application of the component with previous uses and manufacturing routes.

Data and application modules have been developed, or are being developed, for: Cast

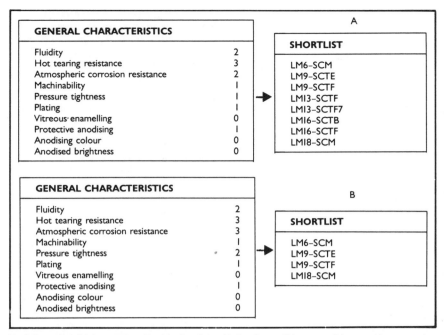

GENERAL CHARACTERISTICS

Fluidity	2
Hot tearing resistance	3
Atmospheric corrosion resistance	2
Machinability	1
Pressure tightness	1
Plating	1
Vitreous enamelling	0
Protective anodising	1
Anodising colour	0
Anodised brightness	0

A

SHORTLIST

LM6–SCM
LM9–SCTE
LM9–SCTF
LMI3–SCTF
LMI3–SCTF7
LMI6–SCTB
LMI6–SCTF
LMI8–SCM

GENERAL CHARACTERISTICS

Fluidity	2
Hot tearing resistance	3
Atmospheric corrosion resistance	3
Machinability	1
Pressure tightness	2
Plating	1
Vitreous enamelling	0
Protective anodising	1
Anodising colour	0
Anodised brightness	0

B

SHORTLIST

LM6–SCM
LM9–SCTE
LM9–SCTF
LMI8–SCM

*Fig 2 Examples of the generation of a short list of cast Al-alloys by specifying a set of duty requirements at the pre-sort stage. **B** shows the consequences of tightening the requirements for corrosion resistance and fluidity which were specified in **A**. In each case the diagrams show the appearance of the screen of the terminal.*

and wrought aluminium alloys; cast and wrought copper alloys; wrought copper; bar steels; sheet steels; cast irons; cast steels; stainless steels; titanium alloys; general thermoplastics; engineering thermoplastics; and long and short-fibre composites. These units of the system will also have to function as a data base if desired and provide the usual data base operations of searching for materials within given property limits and outputting the properties of a given material. This would solve the problem posed by situation 1.

Since the system is intended for use at the innovation stage of design when the feasibility of a proposed approach is being explored, the highest standards of data may not be required. Therefore, the PERITUS system uses standard data or typical data, whichever are most appropriate.

Examples of the operation of the presort stage are shown in figs 2 and 3. Figure 2 shows the effect on the short list of tightening the requirements for atmospheric corrosion resistance and pressure tightness of cast Al alloys. The alloy designations in the short list are those used in BS 1490 and the characters after the hyphen represent the casting process (SC is sand cast) and the heat treated condition (M, TE, TF, etc.).

Figure 3 shows the effect on the short list of relaxing the formability requirement for wrought Al alloys. The list of characteristics is specific to the particular type of alloy being investigated.

Evaluation and optimisation stage
The evaluation and optimisation stage of the short list produced by the pre-sort stage is achieved by two alternative approaches. The simplest evaluates the envelope of a set of properties, chosen by the user, for each item in the short list by reference to a set of ideal or standard values. The short list can then be ordered with respect to this standard and, if

required, the deviation of the candidates from the ideal can be displayed as an image for ease of assessment. This approach is a development of the method used by Hanley and Hobson[1]. The pre-sort stage and this unit therefore solve the problem identified as situation 2.

The solution to situation 3 requires a more complex optimisation process involving failure modes, component dimensions and the properties of the materials in the short list. Each

engineering situation requires its own treatment, but modules have been developed and tested for common sections forming simply supported beams in bend and for tubes in torsion. Plates made from long-fibre composites and failing by buckling due to end loading can also be accommodated. In this case the program module optimises the fibre orientation in each lamella making up the plate.

If the short list were generated by the pre-sort stage dealing with duty requirements and manufacturing methods, then the optimisation of shape completes the solution for situation 4. A diagrammatical representation of the pre-sorting and evaluation stages is shown in fig 4.

Discussion
The methodology proposed here is only one of several methodologies which could be derived from the analysis of the materials selection problem which was established earlier. The approach used in the PERITUS system has the advantages that it represents a series of developments and extensions of well-established part-methods and that it leads to a practicable computerised system. This system is modular and so more units of knowledge can be added as they are developed.

At the director stage, and in the section on previous uses at the pre-sort stage, the system takes the input from the user and matches it to its stored knowledge in the form of words. In these cases there is an obvious requirement for the system to be aware of synonyms and related descriptions for the terms stored in its knowledge base, otherwise the user would need to know what was in the system before he could formulate a question which could be answered. An extension to this aspect would be a foreign language inter-

*Fig 3 Examples of the generation of a short list of wrought Al-alloys by specifying a set of duty requirements at the pre-sort stage. **B** shows the consequences of relaxing the requirements for formability specified in **A**.*

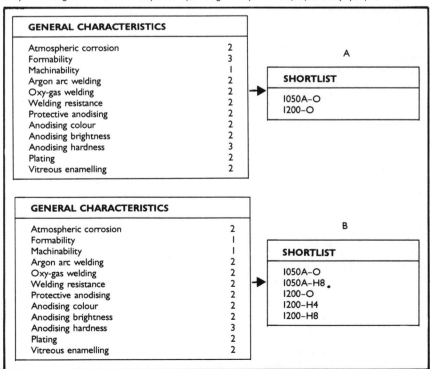

GENERAL CHARACTERISTICS

Atmospheric corrosion	2
Formability	3
Machinability	1
Argon arc welding	2
Oxy-gas welding	2
Welding resistance	2
Protective anodising	2
Anodising colour	2
Anodising brightness	2
Anodising hardness	3
Plating	2
Vitreous enamelling	2

A

SHORTLIST

1050A–O
1200–O

GENERAL CHARACTERISTICS

Atmospheric corrosion	2
Formability	1
Machinability	1
Argon arc welding	2
Oxy-gas welding	2
Welding resistance	2
Protective anodising	2
Anodising colour	2
Anodising brightness	2
Anodising hardness	3
Plating	2
Vitreous enamelling	2

B

SHORTLIST

1050A–O
1050A–H8
1200–O
1200–H4
1200–H8

preter. Dealing with this problem does not, fortunately, require the full scope of a natural language interpreter because the sub-set of terms used in a given country to describe materials components and engineering situations is restricted. Grammatical and syntactical complexities can be reduced by the design of the prompts to which the user has to respond. Nevertheless, a thesaurus of terms is obviously a necessary feature of an intelligent knowledge-based system for the selection of materials, and the computer language used for parts of the knowledge systems enables this feature to be incorporated.

The importance of the evaluation stage is in its assessment of the principal causes of failure, or the design limits. When considering possible causes of failure some important modes such as creep, fatigue, corrosion or wear are complex problems in their own right. Even if a reasonably careful choice of material has been made bearing these possibilities in mind at an earlier stage of the selection process, these failure modes involve the concept of a time-to-failure influenced by the failure mechanism, the operating conditions and the environment. The proper way to deal with these problems would be by dedicated evaluation modules which could be intelligent knowledge-based systems in their own right. The organisation of the PERITUS system into modular units and the capabilities of modern computer operating systems enable such units to be incorporated into the system computer or housed in another computer linked by a data network.

At the evaluation stage the user will require some guidance in making an optimised choice, and optimised decision-making is a necessary part of a materials selection system. Breen et al[10] used linear programming methods[14] in their system, and the numerical algorithms proposed by Hanley and Hobson[1] and Farag[15] can be used to produce single numerical values which can be ordered to indicate an optimum choice. Experience with the PERITUS system has shown that optimisation procedures have to be chosen to suit the particular aspect of the problem which is being evaluated; one method cannot be used universally.

It is an important feature of the design of this system that no attempt is made to produce only a single answer at the end. Instead, the aim is to present a short list of possible solutions with enough supporting information for the user to make a final decision or to carry out more detailed investigation if required. This approach has been adopted because it is strongly believed that the engineer using the system should not feel that he or she is being replaced by a machine, and also that the engineer should play an important role in the decision-making process, and hence have the opportunity to learn from the system. Also of importance is that the output is dependent, additional to the capabilities of the system, on the input provided by the user.

Finally, it is necessary to consider the relationship between knowledge-based systems, such as PERITUS, and the large data

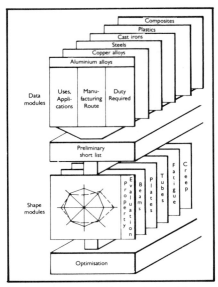

Fig 4 A representation of the organisation of the presort, evaluation and optimisation stages of the PERITUS knowledge-based system. (Copyright ©Matsel Systems Ltd, 1985).

bases currently under development. The importance of a data base springs from the origins of its data and its relationship with the system which has been established to generate those data. Gutteridge and Turner[16] have shown how a materials data base can be linked to the research effort which generated the information. The High Temperature Materials Data Base of the Commission of the European Communities[17] has demonstrated how a modern system can provide a means of collecting, collating and disseminating data about advanced materials which are the subject of international research efforts.

These systems and all the others, listed by Hampel et al[6], have been designed to respond to the generation of information in the form of numerical data and to provide an environment in which the data can be stroed, evaluated in some cases, and retrieved when required. Given the explosive growth rate of engineering information it is obviously essential that high-grade information which is up-to-date and traceable to its origins should be easily available. However, viewed in the context of the design process as a whole, and using the analysis of materials selection problems given earlier, it appears that this kind of information is of most value at the end of the design process, when most other decisions have been made. It would be at this final stage when the detailed performance of a particular material from a particular supplier, its price and availability would be most needed. The structure and organisation of the large data bases make them well suited to providing information of this kind.

It should be possible, therefore, to envisage that knowledge-based systems for materials selection can fulfil a complementary role to the large data bases of materials properties. The knowledge-based systems would be of most use at the innovation stage of the design process and would act as intelligent interfaces to the data bases, enabling the latter to

be used more often and more effectively. Modern telecommunications systems and developments to enable data bases to be linked together[18] make this concept feasible within the next few years.

Conclusions

An analysis of the process of materials selection has been used to establish a methodology for approaching this problem in the context of the design process. The approach which has been devised is based on using a knowledge-based system at the beginning or at the innovation stage of the design process, and data bases of specific materials and other detailed information at the later stages.

References

1 Hanley, D P and Hobson, E. *J Eng Mater and Technol.* 1973: 197–201.

2 Westbrook, J H and Rumble, J H, Jnr. Computerised Materials Data Systems, Proceedings of a Workshop held at Fairfield Glade, Tennessee, November 1982, Steering Committee of the Computerised Materials Data Workshop, 1983.

3 Proceedings of 9th International CODATA Conference, Jerusalem, June 1984. North-Holland Physics Publishing, Amsterdam, to be published.

4 Proceedings of CEC Workshop of Factual Materials Data Banks, Petten, The Netherlands, November 1984, to be published.

5 Materials Property Data Management, National Materials Advisory Board Report NMAB – 405, September 1983.

6 Hampel, V E, Bollinger, W A, Gayner, C A and Oldani, J J. UCRL Report No 90276, May 1984.

7 Hampel, V E, Gayner, C A, Hegemann, B E, Sanner, R D and Wolman, Y. UCRL Report No 90942, June 1984.

8 Plevy, T A H. *Metallurgist and Mater Technol.* 1981: **13**, 469–474.

9 Kusy, P F. *J Soc Automotive Eng.* 1976.

10 Breen, D H, Walter, G H and Sponzilli, J T. *Metal Progress.* 1973: **103**, 83–88.

11 Matsel Systems Ltd, 14 Mere Farm Road, Birkenhead, Merseyside L43 9TT, England.

12 E Gillam. *Metallurgist and Mater Technol.* 1979: **9**, 521–525.

13 Crane, F A A and Charles, J A. The Selection and Use of Engineering Materials, Butterworth, London, 1984.

14 Gillett, B E. Introduction to Operations Research, McGraw-Hill, New York, 1976.

15 Farag, M M. Materials and Process Selection in Engineering, Applied Science Publishers, London, 1979.

16 Gutteridge, P A and Turner, J. *Mater and Design.* 1982: **3**, 504–510.

17 First Demonstration Report on the High Temperature Materials Data Bank of JRC, Commission of the European Communities Report EUR 8817 ENfs.

18 Atkinson, M P, Bocca, J B, Elsey, T J, Fiddian, N J, Flower, M, Gray, P M D, Gray, W A, Hepp, P, Johnson, R G, Milne, W, Norrie, M, Omololu, A O, Oxbarrow, E A, Shave, M J R, Smith, A M, Stocker, P M and Walker, J. Proceedings of the Third National Conference on Data Bases: BNCOD 3 (ed J Longstaffe), Cambridge University Press, to be published. ■

Towards Failure Analysis Expert Systems

An example of the power of logic programming

Virtually all expert systems work on the same principle. They try to boil down knowledge into a highly structured outline, which then is converted into a number of "if-then" rules.

Logic programming is a powerful means for automated reasoning, that is, using the computer to aid in problem solving that requires reasoning.[1] The principal languages available for this purpose are PROLOG[2,3] and LOGLISP.[4] Both languages are based on a solid foundation of mathematics and logic. LOGLISP, which has been implemented on DEC-10, VAX-780, IBM 4341, and SYMBOLICS computers, is a combination of logic, that is, predicate calculus and resolution,[5,6] and LISP, the language of choice of the artificial intelligence research community. LISP is a flexible, sophisticated, programming language that can treat its own programs and control mechanisms as data objects. Weiss and Aha[7] have demonstrated the use of LOGLISP for materials selection, alloy substitution, fatigue data evaluation, and a research notebook. Weiss[8] has also shown how logic programming might be used in connection with materials characterization. A further use of LOGLISP is for the development of failure analysis expert systems. Because of the availability of the structural materials knowledge base MATISS,[7] this small demonstration system has been designed to deal with structural materials failures. However, LOGLISP expert systems are applicable to a far greater range of problems.

A LOGLISP knowledge base consists of sets of assertions, either statements of facts or relations, that is, rules. The facts have the form:

(¦ −(Is ST4340 400 TS 270)),

which states that the tensile strength of 4340 steel in condition 400, signifying tempering temperature, is 270 ksi. The leading term "Is" is called the predicate. To obtain information from the knowledge base, one of the search commands, THE, ANY n, ALL, is used.

(THE x (Is ST4340 400 TS x))

produces 270.0 as the response.

Rules can be formulae or other relationships. For example, the formula for the initiation of yielding in a thin walled pressure vessel, in psi, under internal pressure p (psi),

$$p = (2300 \cdot t \cdot ys)/ d,$$

where d is the diameter and t, the wall thickness, in inches, and ys, the yield strength of the material, in ksi, has the form:

(¦ — (PressureVessel-Ppsi mat cond dia th
　　(% (* 2300 th ys) dia))
< — (Is mat cond YS ys)).

It should be noted that the yield strength of the material is not entered in the query statement, but is obtained from the knowledge base through the (Is mat cond YS ys) statement.

Ad hoc queries of the knowledge base can readily be formulated. For example, to obtain the list of materials that can be used to build a thin walled pressure vessel, wall thickness 0.0625 in., diameter 18.0 in., operating pressure of 650 psi, safety factor 3.0, write in LOGLISP:

(ALL (mat cond ys sf)
　(PressureVessel-Ppsi
　　mat cond 18 .0625 p)
　(Is mat cond YS ys)
　(= sf (% p 650)
　(> = sf 3)),

where All is the deduction command, followed by the answer template specifying the information sought (material, condition, yield strength, actual safety factor). The key rule is the formula for the pressure to cause yielding in a thin walled vessel, shown as the "PressureVessel-Ppsi..." relation above. The formatted results are:

MATL	COND	YS	SFPCT
ST9255	QT400	297.	365.
CSIC-AL	VF50	262.	322.
ST5160	QT400	260.	319.
ST9255	QT600	260.	319.
ST50B60	QT600	257.	316.
ST5160	QT600	257.	316.
ST5150	QT400	251.	308.
ST81B45	QT400	250.	307.
ST4150	QT400	250.	307.
ST6150	QT400	245.	301.

LOGLISP for Failure Analysis

The structure of logic programming languages like PROLOG and LOGLISP is ideally suited for implementing expert systems.[9] The principal ingredients for failure analysis, as for many medical applications, are symptom-cause relationships and rules, and facts and rules about the system under consideration. Probabilities of such symptom-cause relationships can also be included. For example, one of the symptoms of a tensile failure is the existence of a neck:

(SYMPTOM VIS-neck) → (MODE
　Tension-overload)

is LOGLISP shorthand for the assertion that from the visual observation of a neck, the failure mode was tension overload. Another statement,

(SYMPTOM FSO-normal) → (MODE
　Tension-overload)

in the knowledge base indicates that the same conclusion could have been reached from another observation, namely that the orientation of the fracture surface was normal to the bar axis or the presumed direction of the principal stress. Although the observation of two different symptoms leads to the same conclusion, the probabilities of the correctness of these conclusions are quite different. While there is little doubt that tension overload failure from the observation of a neck can be concluded, the observation of a "normal" fracture surface orientation may suggest quite a number of different failure modes, for example, bending or torsion. Thus, it may be desirable to include information about the probability of the assertions, such as

(SYMPTOM sym) → (MODE mode
　probability) or
(SYMPTOM VIS-neck) → (MODE
　Tension-overload 1.0) and
(SYMPTOM FSO-normal) → (MODE
　Tension-overload .25),

if it can be concluded that a normal fracture surface orientation caused by tension overload is .25.

For this demonstration, symptoms are related to failure modes, such as tension, torsion, bending, and fatigue; and failure attributes, such as brittle, ductile, impact, and high temperature. Probability statements have not been included. A more compact form has been chosen for the assertions, for the failure modes and attributes:

(¦ — (Mode Tension-overload
　(SEM-cleavage SEM-dimples
　　VIS-neck VIS-shearlip
　　VIS-ludersbands FSO-normal
　　FSO-shear))) or
(¦ — (Attribute Ductile
　(VIS-neck VIS-plasticdeformation
　　VIS-shearlip SEM-dimples
　　SEM-sheardimples))).

In the absence of a thesaurus as part of the knowledge base, dictionaries of the al-

by Volker Weiss

Volker Weiss is vice-president for research and graduate affairs and professor of materials science at Syracuse Univ. He is the author or coauthor of over 90 publications on the mechanical properties of materials, fracture and fatigue, physical metallurgy, nondestructive testing, and advanced forming techniques. He is the editor of the Sagamore Materials Research Conference Series published annually since 1961, by Syracuse University Press and, since 1977, by Plenum Press.

lowable symptoms must be provided. A short sample of such a dictionary is:

Failure Mode Symptoms	Failure Attribute Symptoms
FSO-normal	FSO-normal
FSO-shear	FSO-spiral
FSO-spiral	SEM-cleavage
SEM-cleavage	SEM-dimples
SEM-dimples	SEM-intergrannular
SEM-sheardimples	SEM-sheardimples
SEM-striations	VIS-Chevrons
VIS-beachmarkings	VIS-HToxide
VIS-ludersbands	VIS-crackbranching
VIS-neck	VIS-grooves-near-weld
VIS-plastic-deformation	VIS-neck
VIS-plastichinge	VIS-pits
VIS-secondarycracks	VIS-plastic-deformation
VIS-shearlip	VIS-shearlip
VIS-twistlines	

With this system the failure analyses of a pressure vessel fracture may now be attempted. The vessel in question had fractured explosively along the circumference, near a circumferential weld. The data supplied are:

Material: Type 301 Stainless Steel, Annealed

Dimensions: Diameter 18 in., wall thickness 0.0625 in.

Maximum internal pressure at failure: 100 psi.

Upon receiving the failed vessel, the investigator may want to check the material and any applicable specifications and drawings. The knowledge base may, for example, be queried about the material. From the small demonstration knowledge base, MATISS, all information available on chemical composition and mechanical properties on SS301, annealed, is obtained,

(¦ — (Iss SS301 ANN ((TS110)
    (YS 40))))
(¦ — (Chem SS301 ((Cr 16 18)
    (NI 6 8) (C .134 .16)
    (Mn 1.8 2.2)))

Applicable ASTM or American Society of Mechanical Engineers (ASME) specifications and standards may also be called at this stage of the study.

Next, a visual inspection is conducted, measurements are made and samples containing the fracture surface are prepared, and studied under a scanning electron microscope (SEM). It is observed that the fracture surface orientation is normal and that the SEM studies reveal an intergrannular structure (rock candy appearance) near the fracture surface. The query of the expert system

(ALL (m a) (Modes m FSO-normal)

APOLLO COMPUTER INC.

(Attributes a SEM-intergrannular)))
results in 20 possibilities:

((Tension-overload High-Temperature)
(Tension-overload Weld-sensitized)
(Tension-overload SCC)
(Tension-overload Corrosion)
(Bend-overload High-Temperature)
(Bend-overload Weld-sensitized)
(Bend-overload SCC)
(Bend-overload Corrosion)
(Torsion High-Temperature)
(Torsion Weld-sensitized)
(Torsion SCC)
(Torsion Corrosion)
(Fatigue High-Temperature)
(Fatigue Weld-sensitized)
(Fatigue SCC)
(Fatigue Corrosion)
(LCFatigue High-Temperature)
(LCFatigue Weld-sensitized)
(LCFatigue SCC)
(LCFatigue Corrosion))

On closer examination, it appears that the fracture is at the center of a necked down region. That information, added to the previous observations,

(ALL (m a) (Modes m FSO-normal)
(Modes m VIS-neck)
(Attributes a SEM-intergrannular)

reduces the choice for the failure mode—attribute combinations to four:

((Tension-overload High-temperature
Weld-sensitized
Corrosion
SCC)).

All modes other than tension-overload are eliminated. Finally, the observation of grooves on either side of the circumferential weld, added to the other observations

(ALL (m a) (.) (.) (.) (.) (Attributes Grooves-near-weld)) leads to the conclusion that failure was due to tension overload of a corroded region, the intergrannular corrosion being due to weld sensitization,

(Tension-overload Weld-sensitized
Corrosion).

With the help of the (NOT) statement, the system also allows negative input. For example, if visual inspection reveals a normal fracture surface orientation, but no signs of a plastic hinge, that is

(ALL m (Modes m FSO-normal)
(NOT (Modes m VIS-plastic-hinge)), the possible failure modes can only be

(Tension-overload, Torsion, Fatigue, Low-cycle-fatigue). Bending failure, which is often characterized by a plastic hinge, is excluded from the listing of probable modes.

To continue with the analysis of the pressure vessel failure, the pressure for the onset of plastic deformation on an undamaged vessel may be determined. The Pres-

Logic programming is a means for automated reasoning or using the computer to aid in problem solving that requires reasoning.

sureVessel-Ppsi rule of the knowledge base, applied to SS301 ANN gives 319 psi, far greater than the available pressure of 100 psi. Reduction of the wall thickness to 0.01 in. due to corrosion yields a failure pressure of 51 psi, well within the range of the available 100 psi.

Outlook

It is hoped that the simple example presented conveys a sense of the power of logic programming for the development of expert systems. Both PROLOG and LOGLISP are easier to learn than most programming languages. Moreover, work is in progress to provide natural language interfaces, which should further increase user friendliness. The possible scope of such expert systems is virtually unlimited, except perhaps with respect to speed of execution with increasing size of the knowledge base. However, even these limitations are expected to disappear if the rapid increase in computational speed achieved during the past few years continues. Consider a few more cases and suggestions of possible next steps.

The symptoms (SEM-striations) and (VIS-beachmarkings) observed in a hypothetical case will prompt a system response (fatigue, low cycle fatigue) as possible failure modes. For the continuation of the failure analysis, a wealth of fatigue data analysis procedures is already available in the MATISS knowledge base, or can be added readily. This includes fatigue and low cycle fatigue data and data scatter, Goodman, Gerber, or other rules for mean stress effects, and rules to estimate cumulative damage effects from service load spectra. For cases that require extensive stress analysis, it might be desirable to design hybrid systems by coupling a finite element stress analysis program to a logic programming expert system. Hybridization of a quantitative metallography or spectrography system to a logic programming expert system appears also possible with state of the art equipment.

Assemblies of several parts, for example, bearings and landing gears, can also be handled. For such an assembly, consisting of several parts, interconnections can be specified, such as:

(Is-part-of A B)
(Is-part-of A C)

stating that B and C are parts of assembly A. It might be known that, while failure of B causes assembly failure, failure of A has no effect. Such information would be contained in the knowledge base and may, if sufficiently detailed, provide a model of the assembly, which responds to inputs as the real assembly would, and thus would represent a higher level expert system than the symptom-disease system described. All logic programming expert systems have the potential to be learning systems that become more expert with use, especially in the area of expertise of the user group.*

*The continued guidance and assistance of J. Alan Robinson, the help of E. E. Sibert, and the support of Bradley J. Strait and the Center for Advanced Technology in Computer Applications and Software Engineering are gratefully acknowledged.

References

[1] L. Wos, F. Pereira, R. Hong, R. S. Boyer, J. S. Moore, W. W. Blesdoe, L. J. Henschen, B. G. Buchanan, G. Wrightson, and C. Green, "An Overview of Automated Reasoning and Related Fields," *Journal of Automated Reasoning*, Vol. 1 p. 5 (1985).

[2] A. Colmerauer, H. Kanoui, R. Pasero, and P. Roussel, "Un systeme de communication homme-machine en francais," Rapport, Groupe d'Intelligence Artificielle, Marsaille: Universite d'Aix, Luminy (1973).

[3] W. F. Clocksin and C. S. Mellish, "Programming in PROLOG," Springer Verlag, Berlin, Heidelberg, New York, (1981).

[4] J. A. Robinson and E. E. Sibert, "The LOGLISP User's Manual," Technical Report, School of Computer and Information Science, Syracuse Univ., (1981).

[5] J. A. Robinson, "A Machine-Oriented Logic Based on the Resolution Principle," *Journal of the Association for Computing Machinery*, Vol. 12, p. 23 (1965).

[6] J. A. Robinson, "Logic: Form and Function," Edinburgh University Press, (1979).

[7] V. Weiss and D. Aha, "Materials Selection with Logic Programming," in Proc. 29th National Society for the Advancement of Material and Process Engineers Symposium, Vol. 29, Technology Vectors, p. 506, (1984).

[8] V. Weiss, "Towards a Computerized Materials Characterization Expert System," Proc. 31st Sagamore Army Materials Research Conference, Pergamon Press (1986).

[9] K. L. Clark and F. G. McCabe, "PROLOG: A Language for Implementing Expert Systems," in Machine Intelligence, J. E. Hayes, D. Michie, and Y-H Pao eds. Ellis Horwood Ltd., Chichester, Eng. (1982).

DATA AND SYSTEMS REQUIREMENTS IN COMPUTER ANALYSIS OF ENGINEERING COMPONENTS

Anthony J. Barrett
ESDU International Ltd.
London, U.K.

ABSTRACT

The use of computers in the design and analysis of engineering components imposes stringent requirements on the quality of the data, the ease of use and integrity of the systems in which those data are employed. The relationships between data quality and engineering cost savings are examined. Aspects of reliability and ease of use of computer programs are discussed and an embryo set of guidelines developed. Some of the characteristics of "user friendly" systems are demonstrated in software applied to fatigue life and other structural analyses. Beyond the present generation of computer systems for the storage and manipulation of data there lie the expert systems. They may prove to be the most significant development in the holding and use of materials data since the 'invention' of the printed handbook or data sheet.

INTRODUCTION

The engineering use of computer is extensive. However, for a tool which would seem so ideally suited to the many numeric and fact handling processes in which the engineer is engaged, there is still a long way to go before it is universally accepted or available. A recent survey (Ref. 1) reveals that, in the UK, less than one third of a population of over 18,000 engineering plants have a computer available for engineering purposes. there must be many reasons for the relatively slow uptake of this tool. Prices of suitable powerful equipment have, until recently, been high. Recession has limited spending on new equipment. Hardware has been developing rapidly and in a profusion of different and often incompatible types; this has encouraged many to await the next development before making a purchase.

In addition, as we are often told, the engineer is a very conservative creature and may not quickly respond to new methods of working or readily adapt to new sources of information (Ref. 2). I would like to suggest that the conservatism of the practising engineer may not be a bad thing. In some instances reluctance to adopt the computer as a new tool is based on not illfounded, though rarely well articulated, reservations. These range from concern at the lack of visibility of the procedures and data employed in computer systems which work only in a batch processing mode, to frustration over the poor comprehensibility often encountered at the user-machine interface. It is only relatively recently that some of these issues have started to be addressed purposefully. Much remains to be done unless the engineer is to be dragged, against his or her better judgement, into the computer age there to be faced with an obligation to use systems which, except in terms of speed and the number of the design decisions they can assist, are otherwise inferior to the engineer's previous practices. In addressing these issues we shall also be supporting those who have already found ways to make the computer acceptable and to whose enthusiasm and pioneering efforts we are all indebted.

DATA AND THE DESIGN PROCESS

The long established practices of engineering design involve many phenomenological data. These are data the quality of which requires some element of subjective adjudication both in their derivation and in their application. At each stage in the conventional design processed the data to be applied come before the engineer's scrutiny when reference is made to handbooks, manuals, data sheets and other well established resources. Everything is visible and familiar. The printed form has developed over many centuries, it is randomly accessible and accommodates numerical, graphical and factual information relating to limitations on applicability and so forth, with equal facility. The engineer can apply judgment to the data so as to accept low quality data in non-critical situations and to seek higher quality in critical situations, (or at least be made aware of the degree of exposure if such data cannot be found) (Ref. 3). When data become embedded in a computer system they, and their attendant qualifications, generally become invisible to the human using that system. Accordingly, if the same integrity is to attend design decisions reached with the aid of such systems either a much higher quality of data must be made available, whatever the criticality of the application, or some means must be included in the system which will make value parameters visible to or controllable by the engineer.

COST CONSEQUENCES

The cost consequences of inadequately refined or wrongly applied data can be catastrophic. It is possible to demonstrate the cost and other consequences of inadequate data and to illustrate the variability of data used by different companies of equally high technical capability, (Ref. 3). Reference 4 has recently reproduced information originating in a 1971 NASA study of the cost consequences of the application of inadequate materials data. The five examples described entailed a total loss to the nation of at least $1.8 billion. These losses arose from dramatic failures. But data which do not give rise to a spectacular failure may still be inadequate and lead to unrevealed inefficient performance in uncompetitive products with consequent inflated operating costs and a continuous drain on company profits. The total national loss by this route is incalculable. A couple of relevant cases are described in reference 3; they illustrate losses in the region of $100,000 and $240,000 per annum respectively as the consequence of design features which could be put to rights by modest changes based on better data.

Spectacular though these figures are their impact, on the engineering community in general, has so far done little to induce widespread thirst for better and more reliable data and/or preparedness to pay for such data. One can liken the effect to that of warning a wayward child that, if it persists in some unsocial habit, it will turn into an elephant (or some equally unlikely outcome); the very scale of the consequence to a large extent undermines the credibility of its prediction!

Well argued cases, albeit in qualitative terms, of the positive value of good data may, in the end, be to better effect. The Fairfield Glade Workshop on computerised materials data systems, (Ref. 5), addressed the question of the value of data in this way. The conclusion was reached that, with the increasing application of computerised design systems, the value placed on data per unit price aid for them would be expected to rise. The optimum value would also be obtained from data at a higher degree of reliability and refinement than those accepted for manual application. Further, it was identified that the economic payoffs from systems incorporation high quality data would arise from several sources. These include:-

• Timesaving for engineers in accessing data and making materials comparisons and trade-offs.

• Improved product performance.

• Reduction in level of engineering capability required for design work.

The achievement of these benefits depends not only upon the quality of the data in the system but also on the ease and reliability of access to, and use of, the data via the computerised system. Compatibility between the system and the engineers' needs and practices will be critgical to the acceptance and widespread adoption of these systems. There are many aspects to be considered in striving for such compatibility and they may conveniently be summarised under the heading of 'user-friendliness'.

USER-FRIENDLINESS OF COMPUTERISED ENGINEERING SYSTEMS

User-friendliness is often sought by concentration on the mechanics and environment of the system hardware. The *user-machine interface* is unquestionably important but it is only part of the story and, in some ways, is more tractable than the remainder. The *user-information interface* is less manageable and although it can be examined by scientific methods most of our knowledge of its requirements is empirical or has been concluded by trial and error methods. A few observations on the user-machine interface will be in order before addressing the user-information interface since the two cannot be divorced if there is to be an efficient

flow of the traffic in which they are both involved.

In the United Kingdom, until quite recently, a phenomenon known as 'keyboard freeze' could often be observed when engineers were invited to try out a computer system for themselves. It manifested itself as an apparent aversion to handling a QWERTY keyboard! The reasons appeared to be partly social and partly educational. Such keyboards were identified only with the typewriter and, as such, were the tool of the typist and not of the professional engineer! More important, however, was the almost complete inability of most people, other than professinal typists, to handle such a keyboard. In the UK, until quite recently, the submission of school work and college exercises in other than long-hand was unacceptable. So relatively few college graduates possessed, or had been encouraged to use, a typewriter. The position in the United States has been quite different both from the social and the educational points of view and, indeed, my own observation of 'keyboard freeze' in the US has been a relatively rare event. Its incidence in the UK is already much reduced not because typewriters are more readily used but because computer literacy, particularly amongst younger people, is now at a commendably high level; I even know one young man who recently made the interesting discovery that typewriters have QWERTY keyboards very similar to that on his home computer! I relate this matter of 'keyboard freeze' simply to underscore the importance of the user-machine interface and to illustrate that divers social and psychological factors can be at play. Further consideration of this aspect is beyond the scope of this paper; a valuable recent commentary and bibliography on the user-machine interface and other human factors can be found in reference 6.

Many important aspects of the *user-information interface* are independent of the mechanism by which information is transmitted. Much of the experience of my own organisation in constructing authoritative compilations of evaluated numerical data and related factual and qualitative information has been gained with the printed form. As we have progressed in the preparation of engineering software much of that basic experience with the printed form has been found to have close analogies in the newer medium. This largely empirical basis has provided some reliable guidelines for the preparation of our computer software. A few of these will now be discussed.

GUIDELINES

Entry to the system is of first importance. If dialing up and logging on to a system require special knowledge, jargon or familiarity in a system's use they will not present a user-friendly greeting to the engineer. If these procedures are anything other than automatic, the infrequent user of the system will need to have recourse to documentation. Much of the documentation provided with hardware and with software is notoriously poor. It often contains errors, omissions and is apparently written by a computer engineer for his peers rather than for the more general engineering user. The computer itself is well capable of handling tasks such as calling up and logging on and it is surprising that so many systems are not front-ended so as to make the accomplishment of these chores all but invisible to the user.

Once the system has been provided with a friendly gateway the question of the user's experience and attitude are immediately encountered again. What level of proficiency is to be assumed in the ability of the user both in respect of the system itself and of the engineering matter which the system handles? The analogy in the printed form is clear - what is the degree of literacy required of the reader of a particular book and what is the technology attainment level to be assumed? With few exceptions (such as programmed learning books) levels have to be decided and consistently applied throughout the writing of a book. By contrast, the computer system offers much greater possibilities of being able to adapt to a variety of users of different levels of attainment; this is limited only by its power and capacity and by the imagination of the system designer.

Many different routes through the system can be offered. For example, a system may rely heavily on the offering of multiple choice menus. This may be welcomed by the first time user, who needs to be taken step by step through the process. But once the user is familiar with the process the repeated presentation of an array of information, quite extraneous to his or her purpose, can become frustrating and unnecessarily time consuming.

There are ways, of course, in which the system can be designed so that it remains informative yet is sufficiently self-contained to require even the first time user to make but minimal reference to accompanying documentation. Systems which are directed by choices from menus and by the input of information via forms or tables can be provided with alternative "express routes" to the final output for the benefit of the frequent user. These routes bypass features such as explanations of the process and its progress. They can also avoid references back to the use for en route decisions where their consequences can be anticipated by the more experienced. The incorporation of such alternative routes, however, inevitably complicates the system, adds to the overall size of the program and extends the time and effort required in its validation. Care must be taken not to allow express routes to create the impression that the process, once started, is out of the user's control or of a reversion to the batch processing mode.

A different approach is to base the program on a thesaurus of acceptable commands. These require the user to remember the available commands and, to some extent, the operations of which the system is capable. Again, the novice will have to fall back to documentation or to the repeated display of at least parts of the thesaurus in response to some form of request to the computer for help. Once more, however, this adds considerably to the multiplicity of possible routes with the consequent drawbacks outlined in the foregoing paragraph. A compromise situation, in which the system prompts the user's commands, may offer some solution. For example, a question such as:-

ARE YOU INTERESTED IN (N)ON FERROUS, (F)ERROUS, (P)LASTICS OR (C)OMPOSITE MATERIALS?

may be asked. It is less impersonal than the equivalent menu, is as quickly answered, (by keying only the bracketed letter), and leads naturally into dialogue by a confirmatory statement such as:-

WE WILL NOW LOOK FOR A SUITABLE FERROUS MATERIAL.

In addition to supporting dialogue, as this response suggests, it is an important requirement for the system to take every opportunity to underpin the confidence of the user, by confirming that the program is continuously responding to his or her directives. Even with the frequent and experienced operator the time spent, for example, in repeating the input in a table, before moving on, may be time well spent in avoiding abortive runs and in reinforcing the user's confidence.

The character of an other wise user-friendly system may be ruined by the response of the program in an unexpected way. Consistent terminology and behaviour in the system are important and particularly is this so then graphics are used extensively. For example, rigorously consistent use of line forms, symbols and colours between successive graphics frames is essential in order to boost confidence and familiarity with the system and also to avoid the possibility of extraneous errors at the user-information interface. In the past, graphical consistency has always been important to the intelligibility of such things as printed material, lecture slides and transparencies. The facility with which the modern computer graphics systems can generate wide ranges of symbols, line forms and colours needs particularly disciplined application in the engineering environment.

Flexibility in the way in which the output is presented can add much to the acceptability and usefulness of a system. The ability of the computer to make relevant calculations other than those it was specifically directed to undertake can often be used to advantage. For example, in presenting the final results of an analytical procedure it is often a simple matter to present, at the same time, what the results would have been for values of one of the critical parameters to either side of the value which was part of the input. Such a presentation, in the graphical form, can warn of perils which might closely adjoin regimes in which the engineer is working or, of course, suggest more optimal solutions which might be achieved by small variations of some of the parameters originally specified. Alternatively, a facility for the user of the system to repeat the process quickly, and thereby examine the effect of varying some of the parameters, should be offered without the necessity of starting the whole process from the beginning.

Even the most accomplished user of a computer system will make mistakes from time to time. Experienced key operators appear to achieve average success rates of about 99.5 per cent; an error rate of 1 in 200. The less experienced engineering user cannot be expected to compete with these rates but, even at these levels, there is little chance of completing any about the simplest interactive dialogue session without a few keying errors. This underscores the advisability of requiring the user to examine and confirm the input of critical numerical values as discused earlier. It also draws attention to the necessity for the system to provide simple and reliable provisions for drawing the user's attention to the erroneous input of inadmissible values (such as negative strength values) or unlikely dimensions. The system must indicate clearly the sign conventions on which it is based - whether, for example, compressive stresses are taken as positive or negative.

Simple and reliable mans for recovery from, and the

correction of, errors must be provided. No system can be made completely fool-proof. Here again, the problem facing the system designer is to decide what are the levels of technical competence, familiarity with the system and familiarity with the hardware which may be assumed and to balance these against the cost and time penalties of safety and recovery features which could be built into the software.

EXAMPLE OF A USER-FRIENDLY SYSTEM

In its COMpacs programs, ESDU has provided a number of engineering analytical processes, based upon our evaluated data resource. They also attempt to satisfy many of the user-friendly features described in the foregoing GUIDELINES. There follows an outline description of one such program (Ref. 7), which assists the prediction of the fatigue endurance of a mechanical component under variable amplitude fatigue loading.

The COMpac package comprises a tape cassette or disc, user documentation and an ESDU Data Item which fully describes the basis of the method and published references which have been drawn upon. The latter is an important component for archival purposes, to supplement the explanations which will appear on screen and to provide reference to sources of information on parameters which the infrequent user may not have readily to hand. The program is entered simply be inserting the tape-cassette and pressing an autoload key. An outline of the stages which the program will follow is then presented which the regular user may skip.

In the first stage the program offers the opportunity to recover data which was employed when the program was last used. This is of assistance if assessing the effect of modifications to a component that has recently been analysed. A choice of stress and dimensional units, (SI or customary British units), is offered and the first stafe is completed by the input of the required basic materials and component geometrical data into a table. At this stage the program will confirm the data that are to be used and/or draw attention to any which are lacking.

The next stages construct with the help of the user, or by reference to data stored from a previous run, endurance curves from a specimen containing the appropriate stress raiser and also for the plain material. These are presented graphically, of course, and the user can see how, for example, the program interpolates between discrete data points. The user is allowed to make any modifications felt to be necessary. The programme of loading on the component is input in the next stage, or recovered from a previous run; it is presented interms of nominal stress levels. Options to re-arrange the loading programme interms, for example, of descending order of severity are offered. Stress histories in terms of nominal stress in the component, theoretical elastic stress and actual stress at the stress raiser are then calculated and presented graphically.

The next stages require no further intervention by the user. The division of the sequence of stresses in the loading history into closed stress cycles and the application of the 'rainflow' diagram are shown. The final stage calculates the damage done by each stress cycle, as the stresses settle into a repeatable pattern, and the damage done in each subsequent application of the loading programme block.

After the presentation of the final results the user may assess such effects as varying the loading sequence without necessarily repeating the entire run.

In normal application the dialogue between user and program is conducted entirely in prompted language and through filling in numerical values to conventional tables. In its present form a few special commands have to be employed to restart the program if the user has to interrupt the run and in some other circumstances. These are clearly documented but might be replaced in a further development of the program depending upon user reaction to it in its present form. As noted previously, one of the decisions which has to be taken at an early stage in the design of engineering software is the degree of competence which is to be assumed in the user and the cost, validation expense and possible running time penalties of incorporating user-friendly features. These may finally only be assessed by collecting experience among users in the field. Wherever possible, therefore, it is desirable to arrange the structure of the program so that modifications can be incorporated later with minimal expense and limited need for revalidation of the overall integrity of the program.

TOWARDS THE FUTURE

The current 'state of the art' in computerised systems for engineering analysis is reasonably well exemplified by the description given in the foregoing section if, in addition, one adds the facility of accessing a numerical data bank for the performance properties of a range of materials. One could also add a facility whereby loading programme data are input directly from a suitably instrumented component in actual service or from a store of such data. Attractive though such a system may appear, to the engineer who otherwise spends some days manually accomplishing the requisite calculations, it nonetheless has a number of quite serious limitations as a tool for use in the creative work of design. Among these limitations is that the method of problem solving is predetermined. Another is that the structure of the bank of materials data, and other information upon which it draws, may constrain the use of the bank to a particular type of problem. Such data banks are extremely expensive to construct and maintain and, at present, those that exist are fairly limited scope, (Ref. 4, 5). Some of the limitations of current computerised systems may disappear, or at least be much less onerous, in the "expert system."

The expert system in the materials context of the present meeting, is to be addressed by Dr. Volker Weiss and reference 8 provides a study of the expert system in the broader context of Fifth Generation Computer Systems. In relation to issues which have been raised in this paper some of the principal features claimed for expert systems should by noted.

Two essential components to the expert system are generally specified as a "knowledge base" and an "inference engine" although a number of different terms are in use. The *knowledge base* is capable of storing a wide range of facts (or data) and rules. These are expressible in a form very similar to the way in which a human expert would express them when, for example, describing the strength of a material in a given form at a particular temperature, or when describing the requirements of an enginering standard. The *inference engine* has the capability of associating these rules and data in logical response to a design problem or in, say, a fault diagnostic procedure.

Several important features of these systems have remarkable relevance to the use of computers in engineering in particular. The knowledge base is relatively easy to extend and to update. In this connection again the inference engine, in the form of a "knowledge manager", may also operate in a knowledge acquisition mode. The method of problem solving used by the expert system is not predetermined. This contrasts with the conventional program structure which follows a single or, at best, limited number of predetermined routes. Related to this flexibility is the prospect of revealing all possible solutions to a given problem. The process is two-way. Given an input, say the response to questions asked in a fault diagnosis scenario, the system will offer all possible outputs, or causes of a fault. It may also offer some evaluation of the probability of each. With equal facility these systems are capable of offering all possible input which will produce a given output; typically this might be a search for suitable materials to suit a specified application within a given range of cost.

The user-knowledge interface of the expert system will need to be user-friendly and follow guidelines such as have been outlined in the foregoing section. Beyond this, however, the expert system has one particularly attractive feature to enhance the user-system dialogue. As it proceeds in the solution of a problem, or at the conclusion, the reasoning process which has been followed may be recalled to the user. This not only boosts the credibility of the system but also provides an exciting opportunity to advance the user's own knowledge and experience.

Before substantial expert systems for engineering application become commonly available there are numerous difficulties to be overcome. The knowledge base may be easier to construct and update, once its contents have been gathered, than the analogous numerical database as we know it today. but the gathering and refinement of that knowledge (Knowledge Engineering) requires considerable expenditure of skill, time and money. Expert knowledge is not absolute; experts differ in their views. Unless we are to proliferate competing expert systems or be content with a schizophrenic system, procedures for collecting and refining expert knowledge must be developed and applied by properly trained *knowledge engineers.* Such procedures and specialists already exist but they are generally scarce. Knowledge engineering requires a combination and concentration of skills normally found in quite disparate specialists such as people managers, psychologists and engineering analysts. Addressing these issues effectively may prove more expensive and time consuming than the development of the hardware and software for the expert system.

CONCLUSION

The acceptability and the exploited potential of existing systems, for the computer analysis of engineering problems, are far from complete. Before that stage is reached a number of issues have to be addressed.

These include the need to generate sufficient adequately evaluated data and the acceptance of the high cost alongside the appreciation of the value of such data. In addition, compatibility between the way in which these systems operate and the established engineering practices requires attention. On the one hand, user-friendliness must be given more than lip service and an attempt has been made in this paper to indicate some guidelines in this connection. On the other hand, attention needs to be given to engineering design management practices to ensure that they take proper account of the possibilities and danger areas which the accelerated introduction of this powerful, but basically unintelligent, tool offers.

Currently CAD systems primarily handle interactive styling, the production of engineering drawings and parts lists. These are impressive achievements. But it seems possible that, before such systems expand very far into the analytical and decision making aspects of design, with a concomitant requirement for extensive materials and other databases in machine readable from, they might be overtaken by the new generation of expert systems. If a reasonable proportion of the faculties claimed for them do in fact come about, expert systems will transform the now familiar practices of storing, retrieval and use of materials data and other information.

REFERENCES

1. D. Potts, "Unique Survey Takes Industry's Measure," *Engineering Computers,* pp 12-20, November 1983.
2. Ernest J. Breton, "Why Engineers Don't Use Databases," *ASIS Bulletin,* August 1981.
3. Anthony J. Barrett, "Practical Considerations and Methodology for the Evaluation of Phenomenological Data," *AGARD Lecture Series No. 130, Advisory Group for Aerospace Research and Development,* NATO, 1983.
4. W.F. Brown, et al, "Material Properties Data Management - Approaches to a Critical National Need," Report of the Committee on Materials Information Used in Computerised Design and Manufacturing Processes, National Materials Advisory Board, National Academy of Sciences, Washington D.C., 1983.
5. J.H. Westbrook, and J.R. Rumble (Eds.) "Computerised Materials Data Systems", Proceedings of the Materials Data Workshop, Fairfield Glade, TN, November 1982.
6. Redy H. Ramsey, and D. Jack Grimes, Human Factors in Interactive Computer Dialog, Annual Review of Information Sciency and Technology, Vol. 18, Martha E. Williams (Ed), American Society for Information Science, 1983.
7. Engineering Sciences Data Unit, Fatigue Damage Estimation Under Variable Amplitude Loading, COMpac 7004, (Computer Software), ESDU International Ltd, London, U.K.
8. A. d'Agapeyeff, *Expert Systems, Fifth Generation and UK Suppliers,* NCC Publications, National Computing Center Ltd, Manchester, UK. 1983.

BIOGRAPHY

Anthony James Barrett obtained his Batchelors degree at the then Northampton Polytechnic Institute (now City University, London), and his Masters in Aeronautical Enginering at Georgia Institute of Technology. Dr. Barrett obtained his PhD at University College, London. He joined the Royal Aeronautical Society's Technical Department which subsequently developed in the Engineering Sciences Data Unit under his direction. He is currently chairman of ESDU International Ltd, a company within the International Thomson Organisation. His professional contributions embrace both lightweight structural analyses and the principles and practices of what is now known as 'knowledge engineering'.

TWO DATABASE SYSTEMS FOR THE ACCESS TO MATERIALS DATA AND FOR THEIR STATISTICAL ANALYSIS

Gert Dathe
Betriebsforschungsinstitut
VDEh-Institut fur angewandte Forschung GMbH
Dusseldorf, Federal Republic of Germany

ABSTRACT

The impact of the availability of materials data on economy and innovation and recommendations for the implementation of a materials data system are pointed out. Two operational database systems for standard values and test values, respectively are described. Examples are given for the selection of materials, for the output of properties of selected materials, and for the evaluation of test values. Those examples concern but are not limited to properties of iron and steel. Some problems of standard data and possibilities for their solution are given. Those problems are the structural complexity of standards, the different scales of ranges, influencing parameters, different test methods and units. Available information products and online services are mentioned. An outlook on further developments is given.

INTRODUCTION

Materials data usually are thought of engineering data on materials as used for the industrial design process. In essence this means limiting values of mechanical properties on ferrous and non-ferrous alloys, metals, and plastics. But in reality the scope is much broader:

The origin of all data are tests performed during the production of a certain material, or in research laboratories. This raw data is evaluated for quality control and for getting design allowables and typical values for standards and specifications.

Allowables and typical values - e.g. for 0.2 percent yield strength, ultimate tensile strength, reduction in area, and percent elongation - have to be valid for all the material of a specific designation. Because of the inevitable variations of the composition and the production of technical materials each grade represents a class of materials rather than identical materials. It results that only limiting values or ranges of values with a certain probability can be given. For a specific material body, the membership of a class usually has to be tested and certified.

A good knowledge of materials properties is essential for the economic success of mass products and for the development of innovative products. For innovation, also less common properties like thermodynamic, electric and atomic ones on less common materials like composites and ceramics are needed. It is seldom the case that materials properties can be calculated for more fundamental constants of pure substances and mixtures.

Besides the properties for structural design performance, processability and economic data are of equal importance. Performance includes data on the resistance against corrosive atmospheres and liquids and against radiation. Processability information deals with machinability, weldability etc. Economic data refer to prices, supplier, etc. It follows that not one single data base can serve all purposes. So in (Ref. 1) 123 systems related to materials properties are listed, 60 of which belong to the category of "Machine Readable Files of Engineering Data on Materials." Only two data bases of this category are available on-line — one on plastics the other in the field of electronics. The here described STEELFACTS system is contained in the directory (Ref. 1) with acronym WBD but it was not offered on-line at the time when the list was compiled.

Detailed reports on the situation and recommendations for the development of a national materials data system are given in (Ref. 2) and (Ref. 3). Both reports mention international cooperation because a nearly comprehensive system might well be beyond a reasonable feasibility for one single nation. Think of foreign standards! It would be advantageous if each nation would maintain a data base on its own materials standards with international accessibility. Another important recommendation of both reports is to plan a demonstration program using existing facilities (Ref. 2, page 8) or to build up a pilot system for a few industrial application areas (Ref. 3. p. V). Those recommendations are the motivation for the following presentation of two operational data base systems. It is hoped to come to cooperatious because of the same issues of comprehensiveness and costs. Therefore an English end-user interface and a prototype data base based on ASTM and SAE standards on a steel quality group will soon be available.

A DATABASE SYSTEM FOR STANDARD VALUES

Standards contain typical values which are throughly evaluated - and negotiated - by a standardization committee. Unfortunately, there is no sufficiently rigorous national or internatinal standard of standards. So a database system has to deal with:

- properties on a specific material scattered on different standards possibly with redundant or conflicting values for the same property,
- dependency on different set of influencing parameters for the same property in different standards,
- different test methods and units for the same property especially in standards of different national or

international standardization bodies.

By using the FAMA (FActs MAnager) software system those problems are solved in the following way.

Data Model

Usually, the data in a standard is given in tables or can be converted to tables with numerical and non-numerical contents. An internal data entry type is defined together with each table. In contrast to most data base management systems, an entry type is defined at the time when the table instance is entered into the system, and there is no limitation in the number of entry types.

Each entry instance is associated with a specific material by using a unique internal identification code which is as close as possible - if not identical - to a

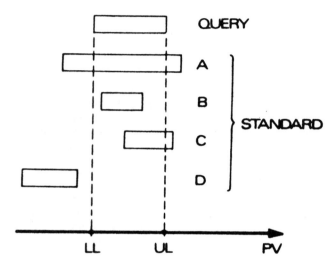

PV PROPERTIES VALUE
LL LOWER LIMIT
UL UPPER LIMIT

Fig. 1. Relative Location of Ranges of Values.

common numbering system. Thus the data on a specific material can be augmented at any time; even quite new property types can be added.

The end-user is not concerned with the relatively complex internal structure. He uses all properties and influencing parameters which are documented in an online available data dictionary without knowledge of internal entry types and internal identifiers. External identifiers like AISI or SAE numbers are treated quite the same as properties which all can be used for materials selection. So, identical materials with different numbers automatically are connected and accessible by either number.

For portions of redundant values - e.g. chemical composition - on identical materials in different standards it's up to the data base administrator if the redundant values or a reference to an already existing entry are stored. In any case the material will correctly be selected and the redundant data or the reference will be shown in the output.

All conflicting data is stored and shown in the output. So a material is selected if any or several conflicting values fulfil the selection expression.

Ranges of Values

A user with a certain range of values or a certain limiting value in mind for his application does not know hich scales have been used in the different standards. Fig. 1 shows the four possible cases which may occur: A certain range given in a standard may be contained in the range of the query (case B), may overlap with the query (cases A and C), or may be outside the range of the query (case D). Limiting values are just special cases of ranges with one boundary zero or infinite. To allow for queries of different strictness, the following comparison operators are available:

EQ - equal
GE - greater/equal
LE - less/equal
GES - greater/equal strictly
LES - less/equal strictly

If the strict operators are used only those materials will be selected the standardized range of which is completely contained within the range of the query for the property concerned, i.e. case B in Fig.1.

Influencing Parameters

All influencing parameters on a property value are connected to that value in the data base. The data dictionary shows which parameters occur anywhere together with a given property. Usually, a distinct property value is dependent on a subset of all possible parameters, only.

To cope with this uncertainty the user has two operators to connect properties and parameters in his query:

PARW parameter wide connection
PARS parameter strict connection

In any case the property value has to fulfil the specification of the query and the parameter value, too if the specified parameter is connected to that property value in the standard. If the specified parameter is not connected to a qualifying property value then the material is

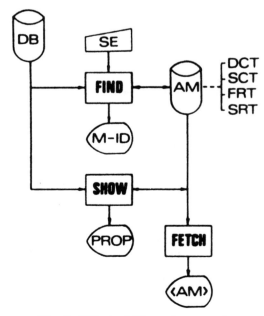

Fig. 2. Effect of Three Commands.

selected in the case of PARW and not selected in the case of PARS.

In depends on the problem which of both operators is more advantageous. Consider e.g. a thickness dependent carbon contents of the chemical composition of a steel. If the thickness of the product is given then with PARW all materials with the wanted carbon contents are selected with the exception of those materials explicitly secified for other thicknesses. PARS will be used if only steels with thickness dependent carbon contents are wanted.

Test Methods and Units

Values measured by different test methods can be defined as belonging to different properties, or the test method can be treated as a parameter. For a database on German steels it was decided to define different properties. So e.g. there are 11 different elongations after fracture according to 11 different lengths of the test pieces.

The treatment of units is a little more complicated. In the standards of most standardizing bodies there is a transition from an "old" system of any type to the International System of Units (SI). Because of standardized rounding regulations there is no 1:1 relationship of values in old and SI units. Moreover, the rounding regulations are property dependent.

Again, there are several possibilities to cope with this problem. One is to define different properties. The advantage is an exact documentation of the contents of the standards; the necessary OR connection of several properties for selection is less advantageous. Another way is storing those values given in old units redudantly also in SI units. Then SI units only are needed for selection, and the output shows SI units together with the original values in old units. Finally, there can be a decision for one system of units. Then the queries have to be formulated in that system or there can be an automatic unit conversion (not yet implemented) possibly with small inaccuracies. For efficiency the output should be -though convertible, too - in the agreed system of units.

The already mentioned database on German steels is in SI units. The planned prototype US database will contain inch-pound units together with Celsious temperatures. International cooperation could be supported if SI units also for US database users were acceptable.

End-User Interface

An end-user may define up to 9 "Applications" for each of his passwords. To each application an "Application Memory" (AM) is assigned. In Fig. 2 three commands ruling the database (DB) and the AM are shown:

- FIND interprets the "Select Expression" (SE), determines the proper material identifiers (M-ID) and shows a portion of them on the terminal. In addition, the SE together with all conditions and all M-IDs are stored in the AM in a "Select Criterion Table" (SCT) and a "FIND Result Table" (FRT), respectively. The "Dialog Control Table" (DCT) is used for an overview and the AM. Each FIND together with its results is represented by a line in the DCT. A new FIND can take on the SCT of a former FIND and modify it.
- SHOW reads data - usually values of properties (PROP) - on materials selected by a former FIND and stores the data in the "SHOW Result Table" (SRT) of the AM. A portion of data is shown on one page of the terminal.

ZNR	W.-NR.
001	1.0727
002	1.0728
003	1.4310
004	1.4406
005	1.4429
006	1.4439
007	1.4460
008	1.4582

Fig. 3. Result of the Selection According to Fig. 7.

CHEMISCHE ZUSAMMENSETZUNG IN GEW % DER STAEHLE NACH DER SCHMELZANALYSE DATUM 10.01.84 SEITE 1/1

ANWENDER : BFIT ANFRAGE : 1 NORMENWERK: BRD

ZNR	WERKSTOFF 1	2	NORM 3	JAHR 4	C 5	SI 6	MN 7	P MAX 8	S 9	SONSTIGE 10	
1	45S20	1.0727	DIN1651	1970	.42/ .50	.10/ .40	.50/ .90	.060	.150/ .250		
2	60S20	1.0728	DIN1651	1970	.57/ .65	.10/ .40	.50/ .90	.060	.150/ .250		
3	X12CRNI177	1.4310	DIN17225V	1955	.00/ .15	.00/1.00	.00/ 2.00			CR	16.00 /18.00
4										NI	7.00 / 8.00
5	X12CRNI177	1.4310	SEW400-73	1973	.00/ .12	/1.00	/ 2.00	.045	.000/ .030	CR	16.00 /18.00
6										NI	7.00 / 9.00
7	X2CRNIMON1812	1.4406	DIN17440	1972	.00/ .03	/1.00	/ 2.00	.045	-----/ .030	MO	2.00 / 2.50
8										N	.1200/ .2000
9										NI	10.50 /13.50
10										CR	16.50 /18.50
11	X2CRNIMON1813	1.4429	DIN1654T5	1980	----/ .03	/1.00	/ 2.00	.045	/ .030	MO	2.50 / 3.00
12										N	.1400/ .2200
13										NI	12.00 /14.50
14										CR	16.50 /18.50
15	X2CRNIMON1813	1.4429	DIN17440	1972	.00/ .03	/1.00	/ 2.00	.045	-----/ .030	CR	16.50 /18.50
16										NI	12.00 /14.50
17										N	.1600/ .2200
18										MO	2.50 / 3.00
19	X3CRNIMON17135	1.4439	SEW400-73	1973	.00/ .04	/1.00	/ 2.00	.045	.000/ .030	MO	4.00 / 5.00
20										CR	16.50 /18.50
21										NI	12.50 /14.50
22										N	.1000/ .2000
23	X8CRNIMO275	1.4460	SEW400-73	1973	.00/ .10	/1.00	/ 2.00	.045	.000/ .030	NI	4.00 / 5.00
24										CR	26.00 /28.00
25										MO	1.30 / 2.00
26	X4CRNIMON8257	1.4582	SEW400-73	1973	.00/ .06	/1.00	/ 2.00	.045	.000/ .030	NI	6.50 / 7.50
27										CR	24.00 /26.00
28										MO	1.30 / 2.00

Fig. 4. Standard Table on the Chemical Composition.

• FETCH is used to show portions of the AM contents (AM) on the terminal.

The language was especially designed for casual users; a language for expert users will follow later on. Currently, the user has just to enter one of ten command words, and the system will give masks for all the input needed. This is very thoroughly supported by the HELP and DISPLAY commands:
• HELP gives
a) without qualification a help adapted to the stage of the dialog,
b) with qualification -

```
S H O W F R G E R H I S - T A R F L L F              (SFT)
ZUFT  TFXT          DU ZI FKG XXXX  5FU ZFET AUSGAHF  AHZ XXXX XXXX
012   HB30+SI+MN    01 W  00R 2944 U06 AIl   NURM     0U16 2147 U0U

-----------------------------------------------------------------
ZNR  W.-NR.  N.-NAME                       NORMHEZ.        DATUM
ZIQ  FIGENSCHAFT          PARAMETER
-----------------------------------------------------------------
001  1.0727  45S20                         DIN1651         1970
002  1.0723  60S20                         DIN1651         1970
003  1.4310  X12CRNI177                    DIN17224        1962
004          X12CRNI177                    DIN17225V        1955
005          X12CRNI177                    DIN17442        1977
006          X12CRNI177                    SEW400-73        1973
007  1.4406  X2CRNIMON1812                 DIN1744U         1972
008          X2CRNIMON1812                 DIN17443V        1977
009  1.4429  X2CRNIMON1813                 DIN1654T1        1980
010          X2CRNIMON1813                 DIN1654T5        1980
011          X2CRNIMON1813                 DIN1744U         1972
012          X2CRNIMON1813                 DIN17443V        1977
013  1.4439  X3CRNIMON1715                 SEW400-73        1973
014          X3CRNIMON17155                SEW400-73        1973
015  1.4460  X4CRNIMON275                  SEW400-73        1973
016  1.4582  X4CRNIMONP257                 SEW400-73        1973
```

Fig. 5. Example of a SHOW Result Table on References to Standards.

e.g. HELP, DB - information on the qualifier, e.g. what DB is meaning. Among others each word appearing in a mask can be used as a qualifier.
• DISPLAY is used for the access to the data dictionary. It gives:
a) translations of colloquial namings of properties and parameters into formal system names and vice versa,
b) information on which parameters a certain property may depend,
c) the set of existing values on a certain non-numerical property,
d) the set of existing values on a certain non-numerical parameter with respect to a certain property.

Further output is produced by the PRINT command:
• PRINT is used for the
a) offline print of portions of the AM,
b) for the offline generation and print of "Standard Tables" (STD). STD are tables of system defined types which give a clear information on the properties of a whole set of materials by renunciation of its finest delicaces which are given in the SRTs. Fig. 4 is an example of a STD on the chemical composition.

Auxiliary commands are:
• MORE for showing more portions of FIND and SHOW results on the terminal,
• CONTINUE for transition to another stage of the dialog,
• DELETE for the deletion of portions of the AM,
• STOP for ending a session.

Examples

The following examples refer to a database on German steels. This database contains the standardized properties values on all steels - approximately 1000 grades - according to DIN standards and Stahl-Eisen-Werkstoffblatter (Steel-Iron-Data Sheets) of the Verein Deutscher Eisenhuttenleute (VDEh). From the preceding it should be clear that there are no system restrictions concerning types of materials, numbering systems, language for the naming of properties, etc.

Fig. 7 shows the conditins for selecting materials with
• Brinell hardness (HB30) 210-230 at heat treatment

(BHLZ) solution annealed (abgeschreckt) or untreated (unbehandelt), i.e. hot rolled, as finished, etc.
• Silicon (SI strictly less 1.1%)
• Managanese (MN) less 2%

Notice that further conditions could have been added before the CONTINUE command and that the condition on the tensile strength is not used in the Select Expression (Suchausdruck (SA)). The lines with a / (slash) are user input; the other lines represent the system response. The conditions have been entered in an earlier dialog step without condition numbers (BNR) and units (EIN-HEIT). These are added by the system and shown as in Fig. 7 if after a condition a CONTINUE is entered instead of a further condition.

Eight materials shown in Fig. 3 are selected.

For the complete set of these materials a SHOW Result Table of type NORM meaning "reference to standards" was generated. A portion of this SRT is given in Fig. 5.

Finally, Fig. 6 shows a cutting out of a type ALL SRT meaning the complete set of data on a specific material, here on grade 1.4310. This output was generated by a FETCH command using a filter function which filters the chemical compositions out of the complete SRT which consists of 242 lines as can be seen under ANZ in the table header. For allvalues there exists a parameter AART (type of analysis) with value SCHMELZ-ANALYSE (heat analysis).

Notice the different C contents in DIN 17225V. This is an old preliminary standard not having been withdrawn until now.

Application

The examples should have created an impression of the main fields of application for the STEELFACTS/S system:
• selection of materials in accordance with a given demand profile,
• comparison of materials by taking the most characteristic properties of a material with regard to a certain application as the selection expression,
• output of properties on materials of a known desig-

nation.

The STEEL-FACTS/S database is accessible online via Datex-P, the German packet-switched data communications network. Datex-P is accessible via a variety of other networks like Telenet, Tymnet, Euronet, etc.

A DATABASE SYSTEM FOR TEST VALUES

The evaluation of a great diversity of test values in a database sets two main tasks:
• creation of statistical populations and

```
S H O W E R G E B N I S - T A B E L L E            (SET)
ZBFT TEXT            DB ZI ERG XXXX BED ZFET AUSGABE   ANZ XXXX XXXX
012  HB30+SI+MN      01 W  008 2944 006 003 ALL        0242 3215 0000
-----------------------------------------------------------------------
ZNR W.-NR. W.-NAME                              NORMBEZ.      DATUM
ZNR EIGENSCHAFT            PARAMETER
-----------------------------------------------------------------------
001 1.4310  X12CRNI177                          SFW400-73     1973
002 C        >= .00, <= .12    AART SCHMELZANALYSE
003 MN       <= 2.00           AART SCHMELZANALYSE
004 CR       >=16.00, <=18.00  AART SCHMELZANALYSE
005 NI       >= 7.00, <= 9.00  AART SCHMELZANALYSE
006 SI       <=1.00            AART SCHMELZANALYSE
007 P        <= .045           AART SCHMELZANALYSE
008 S        >= .000, <= .030  AART SCHMELZANALYSE
078 1.4310  X12CRNI177                          DIN17225V     1955
083 C        >= .00, <= .15
084 SI       >= .00, <=1.00
085 MN       >= .00, <= 2.00
086 CR       >=16.00, <=18.00
087 NI       >= 7.00, <= 8.00
093 1.4310  X12CRNI177                          DIN17442      1977
097 1.4310  X12CRNI177                          DIN17224      1982
098 C        <= .12            AART SCHMELZENANALYSE
099 SI       <=1.50            AART SCHMELZENANALYSE
100 MN       <= 2.00           AART SCHMELZENANALYSE
101 P        <= .045           AART SCHMELZENANALYSE
102 S        <= .030           AART SCHMELZENANALYSE
103 CR       >=16.00, <=18.00  AART SCHMELZENANALYSE
104 MO       -----, <= .80     AART SCHMELZENANALYSE
105 NI       >= 6.00, <= 9.00  AART SCHMELZENANALYSE
```

Fig. 6. Example of a SHOW Result Table on All Properties of a Specific Material.

• processing of data points by statistical methods. This is reflected by the sequence of an evaluation as shown in Fig. 8.

In the first step (RETRIEVAL) data is selected out of the database (DB) and stored in a workfile (WF). The selection expression will usually contain one or several material identifiers possibly together with other properties. This is true because the values on different grades seldom will set up a population. Even the values on a certain grade often have to be decomposed into several populations. Only the values of requested properties will be stored in WF.

The WF is a usual FORTRAN file. Each WF contains a list of the stored properties. A WF can also be created by a transformation program T. This program transforms any measurement results (RES) residing on a file into the structure of the WF. There has just to be a description of the RES file structure which is interpreted by T.

The evaluation itself is performed by the second step (EVALUATION). It is based on a kind of method base including:

• further selection of data entries and properties in the WF,
• frequency distributions, also on probability paper,
• linear and polynomial regression using polynoms and their inverse until degree 4,
• multiple regression,
• correlation analysis,
• different methods for the evaluation of creep rupture data,
• other special evaluation methods, e.g. nonlinear regression for the dependence of impact strength on temperature.

Confidence and prediction intervals can be given where appropriate. Different tests of significance are performed on the coefficients.

The evaluation is performed in an interactive graphic mode to quickly obtain an optimal layout of graphs (scale, length of axes) and to eliminate obviously wrong data points.

The two step approach with the WF as an interface between both steps has some advantages:
• The WF has a much simpler structure than the database. Therefore it is easy to modify existing evaluation programs or to add new ones to the system. And a system for test values has to be prepared for changing and new applications!
• Users can easily store subsets of the database in their own WFs and join them for evaluation.

```
BNR NAME VOP1 WERT1            VOP2 WERT2      EINHEIT
-----------------------------------------------------------
001 HB30  GE   210             LE   230
002 RM    GE   700             LE   800        N/MM**2
003 BHLZ  EQ   ABGESCHRECKT
004 BHLZ  EQ   UNBEHANDELT
005 MN    LE   2                               %
006 SI    LES  1.1                             %
    ****  ***  **************   ***  ***************
/CONTINUE

SUCHAUSDRUCK (SA)  EINGEBEN
( 1 PARS 3 OR 1 PARS 4 ) AND 6 AND 5
```

Fig. 7. Example of Selection Criterion.

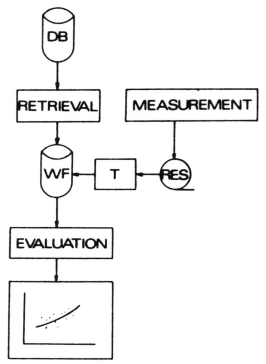

Fig. 8. Sequence of an Evaluation.

• The evaluation program package can be used independently of the database management system. so systems like ADABAS, IMS etc. can be employed for data management, and data points not intended to be stored in any database can be processed.

Examples

Two public and several proprietary databases have been built up. The public databases are STEELFACTS/T with test values on steels out of the literature and PLANTFACTS (Anlagendatenbank) with data describing production plants of the iron and steel industry. Both databases are managed by the same software system which can handle quite different schemas.

In STEELFACTS/T values on test items are stored. A test item may be e.g. a hat, a bar, or a plate. Various tests may have been carried out on such an item. All data - test values and processing history - are stored in one data entry (record) of variable length.

Each data entry consists of mandatory identifying values and the test values themselves. Mandatory values are:

• date of collection or publication,
• origin,
• steel grade,
• chemical composition,
• product,
• condition of processing and treatment,
• melting process.

Tests values are grouped according to the different tests:

• tensile test,
• notched bar impact test,
• hardening test,
• hardenability test by end quenching,
• creep rupture test,
• hot yield strength tests,
• fatique test,
• physical properties.

Mandatory values appear only once in a record, while test values may occur repeatedly or may completely be missing depending on the types and numbers of tests carried out on the item.

The database contains about 650,000 data points for 500 grades. This is the yield of almost the complete literature since 1978. Two journals with especially many results were evaluated back until 1970 and 1961, respectively. The special fields of fracture mechanics and hot yield strength were traced back to 1970 and 1957, respectively. The data is not only entered into the database but partly also published in the "Informations-dienst Werkstoffdaten Stahl-Eisen" (Information Ser-

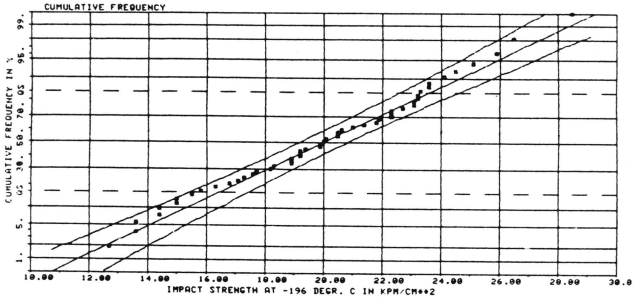

Fig. 9. Comelative Frequency of Impact Strength at −196°C.

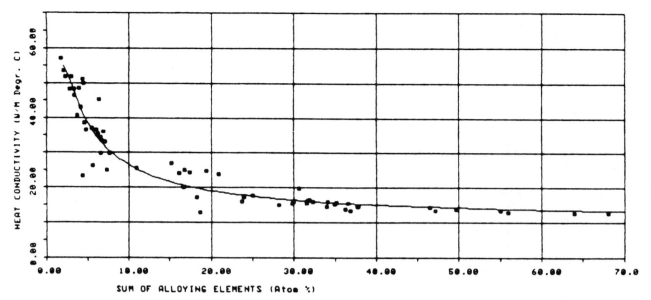

Fig. 10. Heat Conductivity for Steels at 20°C as a Function of Alloy Contents in Atom %.

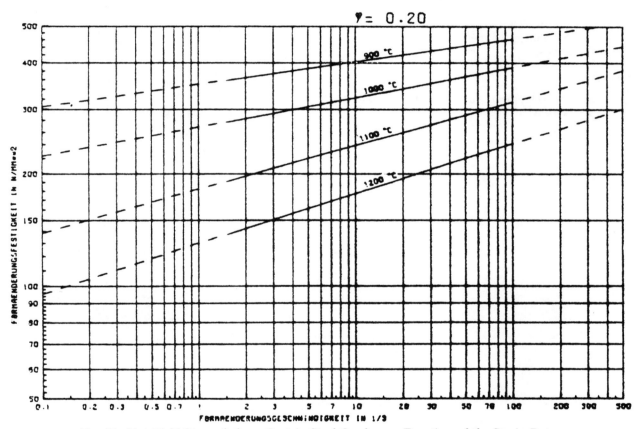

Fig. 11. Hot Yield Strength for a Certain Steel Grade as a Function of the Strain Rate.

vice Materials Data on Steel and Iron). This is a loose-leaf collection of data sheets. Since 1978 yearly 100 data sheets have been published. Fig. 9 is an example of an evaluation of a very frequent type: cumulative frequency on probability paper. The data points - in this case of the impact strength at -196 degrees C - are shown together with the cumulative frequency function and 95% confidence limits. In order to fix a limiting value it could be concluded e.g. that 90% of the values are above 14.3 kpm/cm² with a 95% confidence.

Fig. 10 represents an example of polynomial regression analysis. The dependence of the heat conductivity for steels at ambient temperature on the sum of alloying elements is shown.

Fig. 11 shows the hot yield strength in dependence on the strain rate. A lot of evaluations of this type was performed for a reference book "Formanderungsfestigkeit" (Yield Strength). For this book, data from the literature was collected and brought into a unified representation with the help of the database system. The data is required for the calculation of hot working processes such as rolling.

Application

The system is and can be used for a variety of investigations which are performed with the aid of statistics, especially if proprietary data is used. Some obvious fields of application are:
- quality control: dependence of materials or products properties and of failures on the conditions of production or on other properties,
- industrial design: generation of limiting values of different reliabilities depending on the products, also for CAD systems,
- materials science: evaluation of experiments for the generation of new materials or for the breeding of special properties,
- standardization: generation and assessment of standard values.

The online access to the system is by special arrangement, only because an expert knowledge of materials science and statistics is needed to avoid misinterpretations of the results. Those special arrangements may also include the transfer of the software system to the customer's computer.

OUTLOOK

The further development will emphasize the online services. This means:
- Development of an online product based on the STEELFACTS/T database. This product will do without statistics and graphical output. Instead, data points with references in a reasonable representation will be given.
- Augmentation of the STEELFACTS/S database by standards of other standardizing bodies and if possible by specifications concerning performance values.
- Development of a complete English user interface including all namings, HELP-texts, etc.
- Development of features of an expert system.

Especially the augmentation of the database can only be performed to a greater extent if cooperations can be installed.

REFERENCES

1. V.E. Hampel, J. Hilsenrath, J.H. Westbrook, C.A. Gaynor, and P.S. Johnson, "A Directory of Databases for Material Properties," Report Number UCAR 10099, Lawrence Livermore Laboratory, August 1983.
2. Committee on Materials Information used in Computerized Design and Manufacturing Processes, "Materials Properties Data Management - Approaches to a Critical National Need," Report Number NMAB-405, National Materials Advisory Board, National Academy of Sciences, September 1983.
3. J.H. Westbrook, J.R. Rumble, Jr., "Computerized Materials Data Systems," The Proceedings of a Workshop Devoted to Discussion of Problems Confronting Their Development held at Fairfield Glade, Tennessee, November 7-11, 1982.

SOME CONSIDERATIONS IN THE DESIGN OF PROPERTIES FILES FOR A COMPUTERIZED
MATERIALS INFORMATION SYSTEM

J.H. Westbrook

General Electric Research and Development Center

*The design of a computerized materials information system for multifunctional use
invokes a large and diverse array of properties. To gather, properly annotate, and
incorporate the requisite data into the computer memory for later search and retrieval requires careful attention to the structuring of the system and insight into external and internal variables affecting properties.*

*A picture is presented of a taxonomy of materials properties where the functional
use intended of the information defines the classes of data needed; these in turn are
divided into subclasses and data types and finally down to the specific parameters
that are to be stored. It is asserted that the data sought for inclusion in the system
should be determined by the expected application of the material. The data values
stored must be characterized in several different ways: in terms of the nature of the
retrieval expected, in terms of the defining independent variables, in terms of their
derivation from the original source, and in terms of their numerical significance.
These matters are explained in some detail and illustrated with reference to a large,
comprehensive, and multifunctional computerized materials information system now being
designed for the General Electric Company.*

Information on engineering materials is required for materials selection; for component
design; for manufacturing; and for maintenance,
repair and retrofit. Each of these functional
uses of information requires a different mix of
property considerations: property requirements
for primary application, environmental degradation information, safety and health, processibility, appearance, and material procurement.
Each of these classes in turn subsumes a variety of specific types of data, and within each
type there may be a multiplicity of individual
parameters or information units which characterize the attribute in question. The main features
of this taxonomy are shown in Fig. 1. Only selected items have been expanded for illustration;
the reader may readily detail others appropriate
to his own interests.

In designing a computerized materials information system, cognizance must be taken of what
information is to be stored, how it is to be expressed, what ties to other sets of data are required and what metadata must be recorded for
full and unambiguous interpretation. In the discussion which follows, the concentration will be
on the property requirements for primary application because of the great importance of this
data class, because the information to be stored
is more highly quantitative than in the other
classes, and in order to make this exposition
acceptably brief. Illustrative examples are
from metallic materials, although the system
under design will eventually include all classes
of engineering materials.

The size of the task confronted in the General
Electric Company is both enormous and complex.
GE's present paper system of materials information, which for over 50 years has supported
the needs of some 200 different businesses, em-
braces several thousand materials, each of which
possesses numerous variants. (1) Even this does
not include all the materials used throughout
the Company, for individual Departments will
frequently maintain separate databases on
selected materials especially critical to their
own operations. Ultimately these too should be
incorporable into a corporate-wide system of
interlinked databases.

Content Definition

Materials - The answer to the question "what
materials should be included?" might seem obvious: "those materials of interest to the system
sponsor or the hypothecated user group." However,
the answer is actually more complex than that.
In addition to materials in current use, some
obsolete materials must also be included, both
those that are no longer available and those
that have been superseded by others having more
cost-effective combinations of properties. This
result comes about because of the need to provide information for maintenance, repair and
retrofit of existing equipment or for instances,
say in lesser-developed-countries, where it may
not always be feasible to use state-of-the-art
materials technology. Wherever appropriate, inclusion of such obsolete materials in a new information system should be accompanied by the
notation "for information only, do not use for
new design."

On the other hand, some new state-of-the-art
materials must be included in the system even
though they may not yet be produced or used in
substantial quantities or where there is yet
little demand for their formal inclusion in the
system.

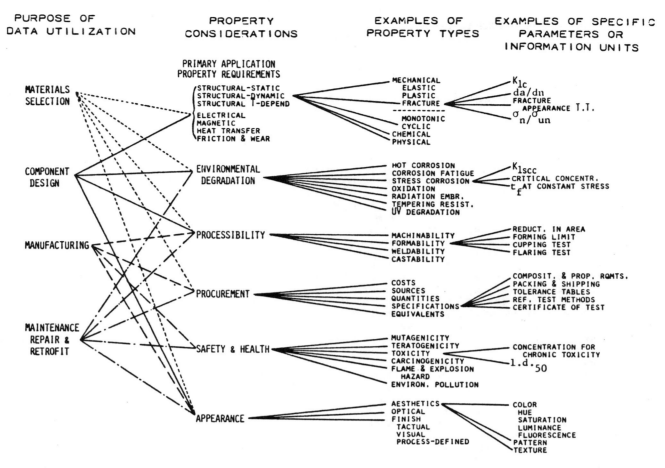

| PURPOSE OF DATA UTILIZATION | PROPERTY CONSIDERATIONS | EXAMPLES OF PROPERTY TYPES | EXAMPLES OF SPECIFIC PARAMETERS OR INFORMATION UNITS |

PRIMARY APPLICATION PROPERTY REQUIREMENTS

MATERIALS SELECTION

STRUCTURAL-STATIC
STRUCTURAL-DYNAMIC
STRUCTURAL T-DEPEND
ELECTRICAL
MAGNETIC
HEAT TRANSFER
FRICTION & WEAR

MECHANICAL
ELASTIC
PLASTIC
FRACTURE

MONOTONIC
CYCLIC
CHEMICAL
PHYSICAL

K_{1c}
da/dn
FRACTURE APPEARANCE T.T.
σ_n/σ_{un}

COMPONENT DESIGN

ENVIRONMENTAL DEGRADATION

HOT CORROSION
CORROSION FATIGUE
STRESS CORROSION
OXIDATION
RADIATION EMBR.
TEMPERING RESIST.
UV DEGRADATION

K_{1scc}
CRITICAL CONCENTR.
t_f AT CONSTANT STRESS

MANUFACTURING

PROCESSIBILITY

MACHINABILITY
FORMABILITY
WELDABILITY
CASTABILITY

REDUCT. IN AREA
FORMING LIMIT
CUPPING TEST
FLARING TEST

PROCUREMENT

COSTS
SOURCES
QUANTITIES
SPECIFICATIONS
EQUIVALENTS

COMPOSIT. & PROP. RQMTS.
PACKING & SHIPPING
TOLERANCE TABLES
REF. TEST METHODS
CERTIFICATE OF TEST

MAINTENANCE REPAIR & RETROFIT

SAFETY & HEALTH

MUTAGENICITY
TERATOGENICITY
TOXICITY
CARCINOGENICITY
FLAME & EXPLOSION HAZARD
ENVIRON. POLLUTION

CONCENTRATION FOR CHRONIC TOXICITY
$l.d._{50}$

APPEARANCE

AESTHETICS
OPTICAL
FINISH
TACTUAL
VISUAL
PROCESS-DEFINED

COLOR
HUE
SATURATION
LUMINANCE
FLUORESCENCE
PATTERN
TEXTURE

Figure 1   A taxonomy of materials information

If this is not done, system users will tend to remain in the rut of old technology and never consider new alternatives. Finally whether obsolescent or brand new, the overriding consideration for inclusion of a material should be that an essentially complete group of properties can be compiled for it. Of this, more later.

Properties - The list of possibly significant engineering properties to be compiled could be very long, easily over 100 items. Not only does the difficulty of completing the compilation task increase with the length of the list, but whatever list of desiderata is chosen, it is especially important, in a computerized system, that full data sets be entered for all materials to be included in the file. Otherwise, where application is made of the file-inversion capability of the computer (i.e., searching for materials having a specified combination of properties) the system will fail to retrieve all the "hits" it should.*

In planning a computerized materials information system for the General Electric Co. it was decided that data compilation could be facilitated if the task were focussed on individual application areas. The application areas chosen were intentionally very broad, but other system designers might well choose other areas more pertinent to their needs. The application areas considered appropriate for GE were: structural, magnetic, electrical, heat transfer, and friction and wear. These headings were expected to generate nonexclusive listings of desired properties. That is, the listing for heat transfer, for example, would include certain mechanical properties, density and electrical resistivity as well as the expected thermal properties. The structural class of primary application was so large and so important that it was deemed necessary to divide it into the following subclasses: structural-static, structural-dynamic, and structural-temperature dependent. The static class would embrace such applications as coat hangers, shelf brackets, motor housings, etc.

*In this regard it is recommended during compilation, when diligent search has failed to locate an appropriate value for some sought property datum, as good an estimate as possible be made and entered with appropriate notation that it is an estimate.

where loads are static and modest and where service conditions are near ambient. In the dynamic class we are concerned with shock and cyclic loading and will include impact, fatigue and notch effects. Temperature dependent structural applications will need to consider creep, stress-rupture, corrosion-fatigue and similar properties. When examined from this point of view, the list of sought properties for materials in each application area became much more manageable, 15 to 25 typically. Examples of the lists proposed for the structural-dynamic and heat transfer classes are shown in Table 1.

STRUCTURAL-DYNAMIC	HEAT TRANSFER
DENSITY	DENSITY
TENSILE Y.S.	TENSILE Y.S.
TENSILE U.S.	TENSILE U.S.
TENSILE ELONGATION	TENSILE MODULUS
TENSILE REDUCTION IN AREA	TENSILE ELONGATION
STRAIN RATE DEPENDENCE OF Y.S.	RESISTIVITY
DAMPING CAPACITY	SPECIFIC HEAT
CHARPY IMPACT STRENGTH	THERMAL CONDUCTIVITY
IMPACT TRANSITION TEMPERATURE	EMISSIVITY
TEMPERATURE DEPENDENCE OF IMPACT STR.	CONTINUOUS SERVICE TEMPERATURE
FRACTURE TOUGHNESS KIc	THERMAL COEFFICIENT OF EXPANSION
FRACTURE TOUGHNESS NOTCH TENSILE RATIO	POTENTIALLY CONSTRAINING PROPERTY
CRACK PROPAGATION RATE da/dn	
FATIGUE STRENGTH COEFFICIENT σ_f'	
FATIGUE DUCTILITY COEFFICIENT ϵ_f'	
FATIGUE STRENGTH EXPONENT b	
CYCLIC STRENGTH COEFFICIENT K'	
CYCLIC STRAIN HARDENING EXPONENT n'	
FATIGUE DUCTILITY EXPONENT c	
CYCLIC YIELD STRENGTH S_{ys}'	
POTENTIALLY CONSTRAINING PROPERTY	

Table 1   Typical lists of desired properties for selected application areas.

A very important parameter appearing on every list is the "potentially constraining property." This parameter is the basis of a kind of automated warning, invoked where appropriate in the file-inversion mode where candidate materials are sought having a specified combination of properties. Without such a built-in warning, a user of the computer system might well retrieve a short list of candidate materials and then simply choose for his application the cheapest, most familiar, or most accessible without further consideration. When a warning appears on the screen as in the example of Fig. 2, the user is prompted to look further to assure himself that he has not made an inappropriate selection.

```
Find:  Materials with  Y.S.        >40,000 psi
                        Resistivity <15 μΩ cm
                        Density     <10

Response:
               Y.S.     Resist.  Density   Notes
Material A  41,000      8        8.2

Material B  72,000      12       9.3       stress-corrosion
                                           susceptible in
                                           chloride environments

Material C  53,000      14       7.6       cost >$100/lb

Material D  45,000      10       8.7
```

Figure 2   Screen with an automated warning of "potentially constraining properties."

Following the philosophy outlined above, the first step in beginning the property data compilation task for each material is to determine the properties to be sought by consideration of the potential field(s) of application of the specific material in question. Often the material is obviously appropriate only for a single primary application, e.g., magnetic, and that property list alone will suffice. In other cases, say alloy steels, the primary application will be structural (static, dynamic or temperature dependent as the case might be) but with an important secondary functional use, e.g., electrical, heat transfer, etc. Experience has proven that combinations of two or three primary application areas serve to define the property requirements for almost all materials. Furthermore, the redundancy in property lists between the different areas is such that the total number of property values for any one material does not exceed 60.

Some properties, which we refer to as "derived properties," are not physically stored themselves, but are calculated on demand, by means of stored algorithms. These may be standardized parameters such as specific strength (equal to yield strength divided by density) or thermal diffusivity (equal to thermal conductivity divided by the product of density and specific heat) or they may be arbitrarily defined "factors of merit," combining fundamental properties in some algebraic fashion, possibly with weighting factors applied.

Metadata

Data descriptors have been referred to as "metadata". (2) They are, in short, data about data - systematic descriptive information about data content and organization at different logical levels of data abstraction. Metadata can themselves be retrieved, manipulated and displayed in various ways. They describe data, map logical physical storage locations and provide links between related entities, attributes and databases. Metadata are nowhere more important than in an engineering materials properties information system where few data are unique absolute values (such as density or melting point) but where, for the most part, data values must always be associated with a particular structure, processing history, specification or test method. Some of the metadata attributes particularly relevant to a materials information system are:

material metadata
    code number
    name (preferred or probable name used
        for searching)
    synonyms or aliases
    broader categorization
    composition
    form
    melting process
    heat treatment
    mechanical treatment
    grain size
    manufacturer

Source: International CODATA Conference, 1984, Jerusalem, 5 pp

The material code number may be either a unique number assigned internally by the information system builder or another generally recognized number, e.g., AISI or CDA number.

More attributes might have been needed to be added to the materials group were it not for the fact that the GE coding system already has embedded in it some of the more important of these descriptors as illustrated in Fig. 3. Thus, a unique base number, e.g., B4A51 in the figure, is assigned to each form of a particular composition. Suffix letters and numbers then distinguish different heat treatments, finishes, and special grades.

GE Material B4A51 identifies seamless low-carbon steel mechanical tubing, similar to AISI Type MT-1015, as follows:

GE designation	Condition
B4A51A	Hot-finished
B4A51A2	Hot-finished, pickled
B4A51B	Cold-finished, finish annealed
B4A51B2	Cold-finished, soft annealed
B4A51B3	Cold-finished, soft annealed, pickled
B4A51B4	Cold-finished, soft annealed, light finish pass
B4A51B5	Cold-finished
B4A51K	Special for forming

Figure 3   Example of coding for material variants from GE's EMPIS system (1).

If the system is to be useful to a diverse and dispersed group of users and, especially, if it is to be an integrated system of distributed, semiautonomous databases as is frequently advocated, (3) then it is important to add other features that may also be classified as metadata. Two of these are "helps" and "thesauri." One important "help" is a define command which should be invokable for each important term in the primary metadata list. Another is example, which would present a concise instance of the term in question. Thesauri should be available and continually expanded which provide equivalancy tables for both alternative material names and code numbers. (4) It might also be remarked that the existence of "helps" and "thesauri" can also facilitate the task of data entry when data are to be extracted and compiled from diverse sources.

property metadata
    name
    synonyms or aliases
    test method(s)
    footnotes or remarks re testing
    measurement units
    source reference

Just as with materials, it is necessary to provide synonyms or aliases for property names. The test method used must be specified as well as the units in which the property value is expressed. By adopting a single, consistent system of units for all property values to be stored in computer memory and building in a multiplicity of the appropriate conversion factors, it is possible to permit data entry, interrogation and response in any unit system with complete transparency for the user.

Another important aspect of property data is the concept of the independent variable. Many properties will not be an absolute value such as density or melting point, but will be a function of some independent variable which may be material-, geometry-, or environment-related, viz:

independent variables
    material
        heat treatment
        grain size
        etc.
    geometric
        thickness
          (section size)
        orientation
          (flat, edgewise,
          lengthwise,
          crosswise, etc.)
    environmental
        temperature
        pressure
        chemistry
        strain rate
        cycles
        time
        etc.

As will be apparent from the examples cited, some of these variables can be expressed in quantitative units while others must be described in words. Cognizance must also be taken of the possibility that some properties may be a function simultaneously of two variables, e.g., yield strength as a function of both tempering temperature and test temperature.

Data Characterization

In addition to the types of metadata already described, the data values themselves require further characterization in several respects. It will be useful to record the source format (tabular, read from a graph, or already digitized from the experimental apparatus). Another characteristic to be recorded is the data class (measured, estimated or derived from a prior compilation). Finally some indication of reliability should be made (typical value, minimum, maximum, design, etc.) and supplemented when appropriate with a quantitative expression of the reliability. Only when such annotations have been provided will the computerized information system impart to the user the confidence in the property values he requires for materials selection, substitution, or repair, or for component design. Failure to include such characterization has been widely held to be responsible for the poor utilization of many prior computerized materials information systems.

Concluding Remarks

It should be evident from the foregoing discussion that the design of an effective computerized materials information system is far more than simply the digitization of existing paper compilations of values or a clerical task of re-arranging tabular files of numbers. Much more attention must be paid to annotation and definition of values and terms, to homogenization of data, and to completeness of the files. (5) It will also be realized that the matters discussed herein, while fundamental to the specification of the computer system and helpful in guidance of the data extraction and compilation effort, still leave untouched the important tasks of software design for search, retrieval and display. Finally, it should be remarked that adherence to the suggestions and principles outlined here will not only make the information system more effective for the ultimate user, but will also provide a basis for evaluating and improving the content of the database itself.

REFERENCES:

(1) Westbrook, J.H., "EMPIS: A Materials Data Program of an Electrical Manufacturing Company," Data for Science and Technology, Proceedings of 7th International CODATA Conference, Oct. 1980, Kyoto, Japan, Pergamon Press (1981) 462.

(2) McCarthy, J.L., "Materials Management for Large Statistical Databases," Proc. 8th Int'l Conf. on Very Large Databases, Mexico City (1982) 234-243.

(3) Westbrook, J.H. and Rumble, J.R., eds. "Computerized Materials Data Systems," Proc. of the Fairfield Glade Workshop (Nov. 1982).

(4) Florczak, T., "The Computer-Aided System of Material Selection," Data for Science and Technology, Proc 8th Int'l CODATA Conf. Jachranka, Poland, North Holland Publ. Co (1983).

(5) Westbrook, J.H., "Extraction and Compilation of Numerical and Factual Data," AGARD Lecture Series #130 Development and Use of Numerical and Factual Databases, Neuilly sur Seine, France (1984).

Update on computer-aided learning

by C N Reid*

This article looks at the, large impact which computer-aided learning (CAL) has had in universities over the last few years. It represents a preliminary report of the materials CAL software known to be in current use within the UK. The separate sections report on activities by institution, and also in the article there is a summary table classifying software by subject.

Although it is some 25 years since universities were first provided with electronic computers, for most of this period only large mainframe machines have been available and most teachers have been daunted by the difficulties of using such facilities for teaching. Within the past five years there has been a profound change of attitude with the advent of cheap but powerful microcomputers. Computers have become de-centralised, providing new scope for local initiative to be taken. There is now great activity in computer-aided learning (CAL) at all levels of education, including the teaching of metals and materials at degree level. Indeed, in the USA there is now an institution which offers some degree courses organised entirely around computer-based instruction.

With these developments in mind, the General Education Committee of The Institute of Metals has set up a working party on CAL. Its remit is to find out and report on current activity in 'materials CAL', to devise ways of distributing good CAL software, and to promote the development of new software to meet perceived needs.

Surrey University

Widespread use is made of CAL with undergraduates using software developed by Dr P J Goodhew and Dr T W Clyne, and by J M Towner and M G Bader.

The majority of the software has a substantial graphical element since its main purpose is to illustrate to the student the physical significance of theoretical equations presented in lectures. The graphics abilities of the BBC micro are therefore exploited as far as is possible within the limitations of the memory of the standard Model B machine. Most of the programs can be used as demonstrations in the lecture room, and subsequently in the laboratory by students who have to answer questions based on their exploration of the available parameters.

Several of the programs were illustrated in articles in *Metallurgist and Materials Technologist* (August 1978 and 1984). The major topic areas so far covered include electron microscopy, dislocation strain fields, ordering phenomena, diffusion, heat flow and solidification. The general approach is to put several related programs on a disc, which presents a menu from which the student selects the desired program. The authors have tried to write all the programs in well-structured BBC BASIC, making extensive use of Procedures, so that the way in which the program works can be readily understood. The programs can then be modified either by staff or students, to adapt them to local requirements.

Sheffield University

Dr J A Whiteman reports that the computer programs written in the Metallurgy Department fall into two major categories. The first of these contains programs that model metallurgical processes, most of which involve the modelling of a rather complex process that is described by several relationships. Normally, the students would be able to perform calculations for only one set of conditions, but the availability of a computer program enables a greater number of situations to be investigated. Mass and heat balances of blast furnace ironmaking and basic oxygen steelmaking are good examples of this type of program. They are not sufficiently detailed to be control models of the process, but they can be regarded as policy models which enable trends to be established. By contrast, two dynamic models have been developed: One models the decarburisation of a stainless steel by using thermodynamic relationships; and the other calculates the conditions for gas evolution during the solidification of a specific steel.

The other process metallurgy programs use a very simple algorithm but perform the same calculation many times. Finite difference heat flow calculations come into this category; programs have been written for the heating and cooling of bars, billets and slabs, and to model the sump formation during continuous casting. A simple development of these heat flow models enables students to investigate reheating in different types of furnace. The availability of such programs enables students to go beyond the few iterations that they might be expected to do without a computer program, and thus obtain some feel for the temperature changes which occur as a function of time.

The other area in which a number of programs have been developed is that of crystallography and electron microscopy.

The most valuable programs in this area permit electron diffraction patterns of phases of known crystal structure to be indexed from measurements of the spacings and angles between three spots in the diffraction pattern. Other programs allow the students to predict the diffraction pattern for a particular orientation of a known crystal together with a second phase having a known orientation relationship with the first. Stereograms of known crystal structures may also be produced. Another program is useful in the measurement of convergent beam electron patterns for foil thickness determination.

Almost all of the programs mentioned above were written initially for a Hewlett Packard 9830 desk-top computer with output on either line printer or graph plotter. Some programs have now been re-written on disc for the BBC Model B, but as yet little effort has been made to use the graphics potential of the machine.

Oxford University

Dr J D Jakubovics reports that programs for interpreting x-ray powder diffraction patterns from uniaxial materials are used by the students in a second-year practical. They were originally written for mainframe computers in FORTRAN and there are versions for a VAX computer, a Research Machines 380Z, and a version for the BBC in BASIC.

After obtaining and measuring their photographs of one of three materials (Cd, Sn or Zn), the students use a program to tabulate the d-spacings.

With the data provided, the lines are indexed using an appropriate Bunn-chart, by hand. The lattice parameters are then calculated by either of two programs. These programs require identical data, which if correct (ie with accurate measurements of the lines and correct indexing), will give very similar results, although they use different methods of calculation. A minimisation technique is used in the first to look for the c/a ratio giving the smallest error, whereas the other uses an iterative technique in which a and c are calculated repeatedly with gradually improving values of c/a.

Open University

A department of the University (The Academic Computing Service) exists to assist lecturers to devise and encode learning programs, which are used by students either during a week-long residential summer school, or throughout the year via a terminal in the local study centre. Existing 'materials' programs are as follows:

Dr Reid is in the Faculty of Technology at the Open University.

Reprinted with permission from Metals and Materials, June 1985, 365-367, © 1985 Institute of Metals

AREA	PROGRAM	INSTITUTION	MACHINE	PERIPHERALS	LANGUAGE	PURPOSE
Electron microscopy/ diffraction	Ewald sphere	Surrey	BBC B	c	BASIC	DE
	Thin film	Surrey	BBC B	c	BASIC	DE
	Scattering	Surrey	BBC B	c	BASIC	DE
	Diffraction	Surrey	BBC B	cd	MICRO-TEXT	ES
	Pattern index	Sheffield	BBC B	d	–	T
	Foil thickness	Sheffield	BBC B	d	–	T
	Powder pattern	Oxford	BBC B	cd	BASIC	T
Order/ disorder	Bragg-Williams	Surrey	BBC B	cd	BASIC	DE
	Anti-phase boundary	Surrey	BBC B	cd	BASIC	DE
Dislocations	Stress fields	Surrey	BBC B	cd	BASIC	DE
	Interaction of edges	Imperial	BBC B	d	BASIC	DS
Diffusion	Interdiffusion	Surrey	BBC B	cd	BASIC	DE
	Profiles and Matano	Surrey	BBC B	cd	BASIC	DE
	Implanted layer	Surrey	BBC B	cd	BASIC	DE
	Binary couple	Birmingham	TRS-80	t	MACHINE	DE
	Random walk	Surrey	PET	t	BASIC	DE
Heat flow	Quenched bar	Surrey	BBC B	cd	BASIC	DE
	Thin sheet	Surrey	BBC B	cd	BASIC	DE
	Undercooled drop	Surrey	BBC B	cd	BASIC	DE
	Constitutional supercool	Surrey	PET	tp	BASIC	E
	Cylinders	Sheffield	BBC B	dp	BASIC	DE
	Concast slabs	Sheffield	BBC B	dp	BASIC	DE
	Reheat billets	Sheffield	BBC B	dp	BASIC	DE
Crystall- ography	Miller indices: cubic	Bell	BBC B	tp	BASIC	S
	Hexagonal	Bell	BBC B	tp	BASIC	S
	Bravais	Bell	BBC B	t	BASIC	S
	Miller indices	Manchester	BBC B	–	–	S
	Crystal vector algebra	Manchester	BBC B	–	–	S
	Stereographic projection	Manchester	BBC B	–	–	S
	Stereographic projection	Imperial	BBC B	d	BASIC	S
Corrosion	Potential-pH	Nottingham	Tek 4051	t	BASIC	S
Chemical processes	BOS process	Sheffield	BBC B	dp	BASIC	DE
	Decarburisation	Sheffield	BBC B	dp	BASIC	DE
	Gas evolution	Sheffield	BBC B	dp	BASIC	DE
Reading phase diagrams	Works metallurgist	Open	DEC20	GIGI	STAFF	S
	Binary diagrams	Bell	BBC B		BASIC	S
	Binary diagrams	Manchester	BBC B	dp	BASIC	S
	Ternary diagrams	Manchester	BBC B	dp	BASIC	S
Thermo- dynamics of phase diagrams	Regular solutions	Manchester	BBC B	dp	BASIC	E
	Ideal solutions	Manchester	BBC B	dp	BASIC	S
	Binary diagrams	Birmingham	DEC20	g	FORTRAN	D
Mechanics	Mohr's circle	Surrey	BBC B	–	BASIC	–
	Viscoelasticity	Surrey	BBC B	–	BASIC	–
	Laminated bean	Open	PET/ BBC B	d	BASIC	A
	Crack growth	Open	PET/ BBC B	d	BASIC	A
Mechanical working	Slip line fields in extrusion	Manchester	BBC B	dp	BASIC	S
	Principles of upper bound	Manchester	BBC B	dp	BASIC	S
	Applications: Extrusion and sheet drawing	Manchester	BBC B	dp	BASIC	S

A summary of software available from institutions, arranged by subject: c=colour monitor, d=disc drive, p= printer, D=demonstration, E=exploration of equations, S=self-learning/testing, T=tool for analysis of data and A=aid in design.

Works metallurgist (or 'how well can you read a binary phase diagram?'): Students use the program after they have studied a written text on phase diagrams. The program interrogates the user about the interpretation of a displayed diagram, records the user's responses and, if necessary, helps to guide the user to the correct answer. There are two versions of the program: One is concerned with eutectic diagrams; and the other considers peritectics. The duration of the program is of the order of one hour, but this depends very much on the responses of the user – incorrect answers prolong the program. These programs are run on a DEC 20 mainframe computer via GIGI terminals.

Designing a laminated beam (a design aid): This is a development of a program described in *Metallurgist and Materials Technologist* (July 1980). The program has now been adapted to run on a BBC model B.

The growth of sub-critical cracks (a design aid): Once the user has input details of the geometry, the loading, the crack growth law for the material, and the initial and final crack lengths, the program computes time taken (in cycles or seconds) for the crack to grow between these lengths. The initial crack length may be the largest crack that could escape detection by non-destructive testing, for example, or it may be the crack length which corresponds to the threshold conditions required for crack growth ($\triangle K_{TH}$ or K_{ISCC}). The program offers the option of computing the lifetime under conditions of either imposed loads or imposed displacements ('load control' or 'displacement control'). Lifetimes are computed by integrating the equation for the 'growth law', eg the Paris equation for fatigue. In a cracked body of infinite size this can be achieved analytically, but for the more practical case of a finite body this requires a time-consuming numerical integration. It is in facilitating this that the main value of the program lies. The software has been written on disc for use on a BBC model B.

Nottingham University

Dr P Brook has written a program for the construction of a Potential-pH diagram. This was written in BASIC for a Tektronic 4051 and takes advantage of the graphics available. The program takes a student through the construction of the diagram for the iron-water system and allows the student to make progress at his own rate. It is assumed that the student has some prior knowledge of the Nernst equation and of the purpose of the diagrams. The program is stored on tape.

Manchester University

Dr T A Myers reports that several programs have been written for use by undergraduates. A program on 'Regular solution binary phase diagrams' is used as a back-up to lectures. It is based upon a regular solution model, ie on an assumption that the free energy of a phase in a binary system A-B can be represented by $\triangle G = \triangle H_{reg} - T \triangle S_{ideal}$ with $\triangle H_{reg} = C.N_A.N_B$, where C is the interaction parameter. The program requires only three inputs: The two melting points and the value of C. Almost any combination will work. The program produces an iterative common tangent construction and is very slow in BASIC. The phase diagram is plotted on both the monitor and a printer.

The following influences can be studied.

The magnitude of C: C = 0 can be tried; this produces an ideal-solution phase diagram. Small positive values give a minimum in the liquidus and solidus with a miscibility gap. Increasing C causes overlap with a consequent change to a eutectic diagram. Negative values of C give a maximum in the liquidus and solidus.

The melting point difference: Increasing the temperature difference between the melting points changes the diagram from a eutectic to an approximation of a peritectic.

Computer-aided learning has also been used as a means of teaching unknown material in the following subject areas.

A Crystallography: Miller indices, crystal vector algebra, and stereographic projections.

B Phase diagrams: Binary phase diagrams and ternary phase diagrams.

C Thermodynamics: Free energy of ideal solutions.

D Mechanical working: Slip line fields for extrusion, principles of upper bound solutions, and applications to extrusion and sheet drawing.

Imperial College

Dr J Humphreys has written a tutorial program, *The Interaction of Edge Dislocations*, for use on a BBC Model B. This is a suite of short programs illustrating the forces between parallel edge dislocations and their influence on dislocation movement by climb and glide. An introductory program gives, with the aid of graphics, the background theory. There follows a series of programs in which dislocations are displayed graphically and allowed to move under their mutual interaction forces. Two dislocations may be placed anywhere on the screen and then allowed to move by glide or glide and climb; arrays of 20 dislocations of similar or mixed signs are allowed to equilibrate by glide or glide and climb.

The programs illustrate the formation of dipoles, and recovery processes such as annihiliation and low angle boundary formation. Dr Humphreys reports that other programs on slip systems, crystallography, work hardening and recrystallisation are in various states of preparation.

Birmingham University

Dr I P Jones reports details of two program packages currently in use. The first is a binary diffusion couple simulation for use with laboratory classes. This is written in machine code for a Tandy TRS-80 and loaded from tape. The user carries out an 'experiment' and checks the applicability of the error function expression.

The second is a thermodynamics package leading up to binary phase diagrams. This is implemented in FORTRAN on a DEC 20-60 and needs an ADM3A graphics terminal (monochrome). This is meant for demonstration by a lecturer and is not intended to be a self-learning aid. The program carries out:

1 Calculation of a free energy curve showing the $\triangle S$ and $\triangle H$ contributions as a function of temperature;

2 calculation of free energy curves for various phases;

3 calculation of one line of a phase diagram by the tangent construction; and

4 calculation of a full phase diagram.

Bell College of Technology (Hamilton)

Dr J E Glen wrote an article about CAL in *Metallurgist and Materials Technologist* (July 1983) and since then he has continued to build up a suite of programs, particularly in crystallography. 'Miller', 'Millbr' and 'Bravis' are teaching aids on the use of Miller indices, Miller-Bravais indices (hexagonal lattices) and on the recognition of the various Bravais lattices respectively. Number 1 in each set is a self-learning aid, number 2 is a test, and number 3 is designed for teacher use and replaces overhead projector skins and overlays.

Access to software

All the materials CAL programs known to the working party appear in the table included in this article. These programs have been written to meet the perceived needs of their authors and the authors' departments, but it is likely that some of these programs will arouse likely interest in other teaching institutions. Since writing software requires a blend of skills and can be very time-consuming, it is sensible to explore existing software before anyone writes a required program.

Discussions have taken place within The Institute of Metals on the subject of the feasibility of setting up a system to advertise and distribute selected CAL software. It has been decided to appoint an editorial board under the chairmanship of Dr P J Goodhew to review submitted software. If approved, the software will be offered for sale at prices comparable to those of software for schools. The publication and distribution of the scheme will be operational soon, when details will be announced in METALS AND MATERIALS; it will provide ready access to tested materials software. ∎

SECTION IX
Local Area Networks

SECTION IX
Local Area Networks

The concept of a local area network (LAN) dates at the earliest from the day any business or manufacturing organization finds itself with two or more computers operating within its facility. The initial advantages of tying the computers together with some kind of data communications link are obvious: first, to minimize repetitive operations and maximize computer time; second, to time-share a central database; and third, to supervise — using a host computer — units or groups of computers in a hierarchical management or control system.

The articles in this section will deal with several aspects of local area networks and how they are used. The reader will soon note that many of the articles treat the application of LAN's in factory automation use, and, indeed, this is an area of high current interest.

Some fundamentals, for the reader who is not readily acquainted with networks, are treated in "Network Primer" and "Types of Networks."

Communications between computers in factory operations is the subject of "Manufacturing Communications: Local Area Networks for the Integrated Factory."

The high level of activity in LAN theory and practice is a result of the pressing need for the technology to solve day-to-day networking problems in both business and industry. It is traditional in American business for individuals to bring out their "solution" as the best and only, and this situation is present in LAN's. The necessity of standardizing on LAN terminology and electronic parameters has long been recognized, and "Update on Local Area Network Standards" will bring the reader up to date in this area.

Network Primer

Local area networks can provide the information link needed to keep production operations under control.

By MARK HALL
Manufacturing Analyst,
Sytec Inc.
Sunnyvale, Calif.

It is no longer enough to acquire and process data. Success in manufacturing often depends on having timely production information available to the company's decision-makers.

The manufacturing equipment on today's production floor typically includes a number of devices, such as programmable controllers and numerical control systems as well as less sophisticated gages, counters, and other sensors, capable of generating a wealth of information about the manufacturing processes. Many manufacturers integrate instrumentation into production equipment, making the instrumentation essentially inseparable from the machines. Local area networks provide interactive access to this information, allowing real-time control of production processes.

A local area network is a high-speed, multiple access, private communications network. Depending on the type of cable used for the network's communication medium, it can transmit information at a rate of from 1 to more than 10 megabits per second, and it can connect—or link—from 150 to over 65,000 devices. Many local area networks are available today, each providing various capabilities and addressing different aspects of the data communications problem. Most have been installed in highly interactive, multicomputer areas, usually in offices or laboratories. Only recently have they entered the manufacturing areas, primarily because networks were originally developed for the office environment, not the elec-

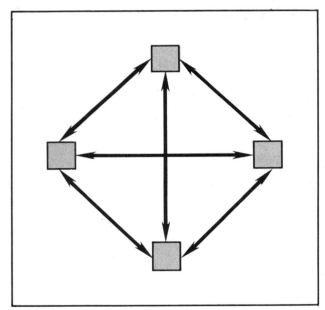

The point-to-point topology is limited in use but is generally appropriate for small networks. Each pair of devices in the network is connected via a single communications cable. This configuration can be costly and tends to be difficult to monitor because there is no central communications device. But it is more cost-effective than other topologies, especially for linking small networks with twisted-pair cable.

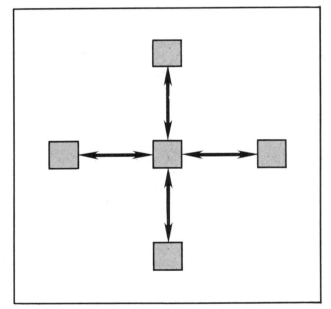

The star topology has a central node to which all the devices in the network are connected. It retransmits signals from the source to the destination. Although the star network has a complex central switching unit, it has several advantages: device connections are simple and it can often use existing wire. This topology is useful for applications that need a central data base or processing facility.

Reprinted with permission from Production Engineering, June 1983, 65-68, © 1983 Penton/IPC

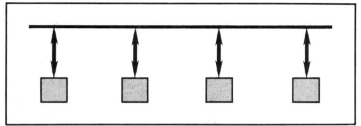

A bus topology consists of any number of devices, each connected directly to a single cable, or bus. Signals are broadcast on the bus in both directions and message collisions are controlled through software protocols. Each device has a unique address which is included with a signal and causes the device to receive the signal. At the same time, all other devices ignore the signal. Devices gain access to the bus via an interrupt priority scheme.

trically noisy, harsh environment of the shop floor.

Several different n edia can be used to carry network communications. Which medium is most appropriate for a given network depends on the maximum distance between nodes, the volume of information to be transmitted, and the transmission speed.

Twisted pairs of wires are probably the most familiar and most common medium. They are best suited for point-to-point, hard-wiring of devices to their host. At the same time, though, the mass of wiring caused by the dedicated wiring schemes of twisted pair wire configurations often creates physical problems because of conduit and plenum limitations.

Baseband is well-suited for high-speed data transfer on a dedicated cable. This dedicated cable can have a number of processors and processor-controlled devices connected to it via active taps. Although it allows only one channel for communications, it divides the access to this single channel through time division multiplexing (TDM). In TDM, each device is allotted a specific amount of time on the channel and transmits a specific amount of information during that time. Both baseband and twisted pair media handle one form of data

at a time, either digital or analog.

Broadband coaxial cable is suited for the communication requirements typically found in industry—many types of messages traveling at different rates either simultaneously or separately. It uses passive taps to connect devices to the main cable and can transmit voice, video, and digital data simultaneously by utilizing its multichannel capability. This capability, known as frequency division multiplexing (FDM) allows all of these digital or analog signals to travel at the transmission rates required by the various devices connected to the cable.

Using the standard CATV 74-ohm broadband cable, local area networks can extend for over 50 kilometers end-to-end. Broadband cannot sustain the very high rates of baseband cable, but it is a more flexible medium to work with.

Fiber optic cable is often considered the long-term solution to data communications because of its phenomenally high data rates which offer virtually immediate response. Because it's in the early stages of development, however, fiber optic cable is expensive compared with other types of cable. There are also technological problems to be overcome; for example, the difficulty of tapping into the cable.

Any of these types of cable can be combined with control devices in various configurations or, in network terminology, topologies. The most common are the point-to-point, star, ring, bus, and tree topologies.

The point-to-point topology is primarily for small network applications. The number of wiring connections that have to be made could make this topology cost prohibitive for large networks. The star topology provides centralized control of the network but does not provide a lot of flexibility. The ring topology is more flexible than the star and is best suited in decentralized organizations where coordination and communication are needed, but not on a regular basis. The bus topology is very flexible and reduces the number of cable connections considerably. This topology generally does not have a central control device that directs communications on the bus. Almost all bus configurations use broadband or baseband cable for two-way communication. The tree topology is essentially several bus networks linked together in the manner of a tree. This topology works well with communication systems that must operate over long distances, with numerous nodes, and that require flexibility.

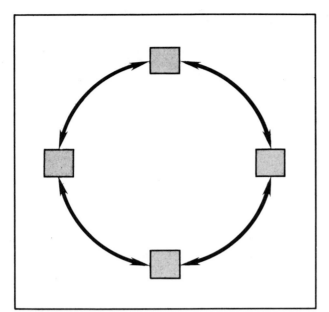

In a ring or token ring network, every processor is connected to exactly two other processors. This system is generally found in decentralized organizations where coordination and communication are needed, but not on a regular basis. A signal—a message token—is passed around the ring, offering each device in turn the opportunity to receive or request information. If a device breaks down, this network will not function until the device is identified and removed from the network.

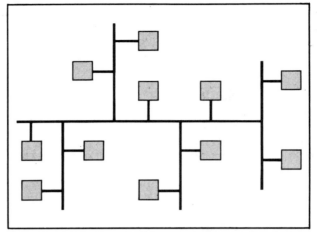

The tree topology resembles several bus networks linked together, in the same way that branches stem from a tree trunk. And it is often called a branching tree topology. The tree topology has a head-end, which is a simple frequency translator that redistributes signals to the appropriately addressed device on the network. Usually associated with broadband local area networks, tree topologies are well-suited for large systems involving longer distances and flexible systems integration.

Types of networks

Different needs require different networks.

Dr. Henrik Schutz
Senior Product Planner
Automation Controls Dept.
General Electric Co.
Charlottesville, VA

The first article in our Networking series, appearing last month, explained why networking is important in the modern factory. This month's installment describes the types of networks that are used for different purposes, and how they can be used together.

• • • • •

Industrial communications is more than just the point-to-point linkage of machine tools and controllers on the shop floor. Truly integrated industrial communications ties together all areas—shop floor supervisory workstations, factory control computers, engineering graphics systems, office systems, and telecommunications for remote sites. Each has different communications requirements, and each favors a different type of network. This article looks at the communication needs of the different areas, matches them with the appropriate network structures, and makes some suggestions for integrating them into a unified whole.

The range of requirements

The planning, designing, installation, and operation of a broad communication system requires a thorough understanding of the different types of traffic it will carry. Long-haul communications are at one end of the spectrum of communication requirements (Fig. 1). Next is the automated office, while shop floor process and manufacturing control are at the other end.

Wide-area networks

The main technology in wide-area communications is that provided by the various telephone companies, using dedicated lines, leased lines, satellites, etc. For the most part, long-haul data communications are over packet-switched networks that use the X.25 set of protocols.

To get a long message from point A to point B via packet switching, the message is broken into manageable pieces called packets. These are delivered to the network and

Fig. 1: Communications requirements range from business needs to process control. The type of data and the required network characteristics vary from one end of the spectrum to the other.

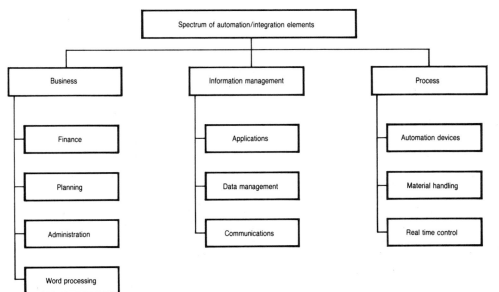

routed through it more or less independently of one another; they may take different pathways through the network, and as a consequence, may arrive out of order at the destination. They must then be put back into order and delivered to the final user. The packets are normally stored at the nodes until a way is found to send them on.

X.25 is a set of communication protocols that provide for addressing, allow nodes that are next to each other in the network to send and receive packets between them, and provide a way for moving packets between nodes that aren't next to each other on the network, and have to go by several steps. It also provides for error recovery and various classes of service.

Ordinarily X.25 would be used on the corporate communications level, e.g. between a head office and branch offices; it tends to be used for communications between host computers.

Office networks

In a single plant or cluster of factories (if they are within a few miles), local networks are used most often. Automation in the office is fairly well advanced, and highly integrated offices are now becoming increasingly common.

One of the major tasks of an office network is to make servers available. These are resources that are generally too expensive or not portable enough to distribute physically among a large number of users; examples include disk storage systems and printers. Such units serve many users over the network as if they were locally available to them.

Data traffic in the office is mainly small to medium size files—memoranda, letters, and electronic mail. Because of the large number of devices in office networks, their traffic tends to be bursty. Networks with *statistical* response are designed for this type of load. The best example is Ethernet, although there are variations on this network from a number of makers. These types of networks provide short response times under light loads, and possibly long response times under heavy loads. They offer fairly high thoughput for moderate loading, although the response time can't be predicted in advance.

The electronics in an office network are packaged for a benign environment, where moderate reliability is often acceptable. As a result, connection cost tends to be fairly low, corresponding to the relatively low cost of the equipment attached to the network.

Engineering networks

The engineering environment is quite similar to the office environment except that the network must be able to carry extremely large files of data among graphics workstations. Because network users are human design engineers, the throughput of the network, not response time, is the critical issue.

Here again, Ethernet type networks (either single or multiple channel) are a good solution. Packaging is similar to that in the office except that reliability must often be higher because of higher equipment cost.

Factory networks

As we move closer to factory floor, the requirements change. Here one must deal with the actual shop-floor computers—equipment such as programmable controllers, robots, CNCs, intelligent gauges, etc. These are usually connected with proprietary networks, like Data Highway, MODBUS, etc. Such networks are limited in physical extent, and carry much less administrative and supervisory data and a lot more process control-oriented data. A network linking factory control computers or cell control computers carries large files for downloading automation equipment as well as smaller files for shop floor status and control information.

There are both human users and electronic devices on an automated factory network. Response times are geared more to the time-sensitive control requirements of the attached devices than to the human users. Therefore, when considering a factory floor network, make sure maximum response times can be accurately calculated. Such networks are termed *deterministic*, and the most common types are the token passing bus and the token passing ring.

With these networks, industrial packaging and high reliability become important. Connection costs tend to be sev-

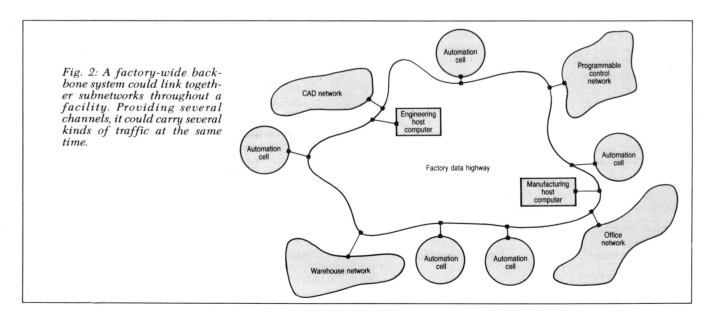

Fig. 2: A factory-wide backbone system could link together subnetworks throughout a facility. Providing several channels, it could carry several kinds of traffic at the same time.

eral times higher than office costs to cover both packaging and reliability requirements.

Process or real-time networks

Probably the most demanding network environment is in process or real-time device control. Although files and status traffic are still present, the main traffic comes from short periodic messages used to synchronize and control the system. Traffic in this segment tends to be much more stream oriented than bursty as it was in the other segments. Predictable response time versus throughput is the main consideration here.

Network electronics in this segment must operate in extremely harsh environments. Reliability is especially critical since network failure may involve expensive loss of capital intensive production capability.

Requirements analysis

It's obvious that traffic and requirements in the office are substantially different from those closer to the process environment. These differences are generally large enough that it is unwise to suppose that a single network could accommodate all conflicting requirements.

Fig. 2 is a multi-networking communications approach. It shows a common factory backbone network linking subnetworks dedicated to warehousing, process control, numerical control, office automation, and CAD functions. Each of these subnetworks has distinct requirements. Table 1 illustrates the major factors that differentiate types of subnetworks and influence their design.

Dealing with the diversity

One relatively low-cost approach for dealing with this diversity is to use a single network medium such as CATV cable that offers multi-channel or broadband capability. Different channels may be used to implement different classes of subnetworks without requiring that independent cable be strung for each subnetwork. Some channels may be allocated to office use, others for engineering, for factory control, for shop floor device control, etc. It is this flexibility (along with the potential for video and voice transmission) that favors the use of broadband networks in industrial communications over single channel schemes.

Requirements of a backbone network

The backbone network that connects the subnetworks must provide a balanced approach for both throughput and response. It must also be flexible in configuration. Some of the nodes in this network will be in an the office while others will be on the shop floor. Thus, both office and hardened packaging must be available. Similarly, all or part of the network must support redundant configurations in situations where high reliability is required.

Finally, because the primary purpose of the backbone network is to link subnetworks that may be of different kinds, the backbone network must allow for the attachment of bridges and gateways to do the linking.

The need for standards

Since the types of devices and the classes of application of these networks are very diverse, the problems for both network implementors and end users grow very quickly unless some way is found to control the proliferation of attachment schemes and protocols. Most large scale network users are pursuing networking schemes based on communication standards. In addition, the National Bureau of Standards, the European Computer Equipment Manufacturers Association, The International Electro-Technical Commission, The Institute of Electrical and Electronic Engineers, and The Electronics Industry Association have been very active recently in consolidating their standards which apply to the industrial communications market.

Another move toward industrial communication standardization is the General Motors Corporation sponsored Manufacturing Automation Protocol (MAP) Task Force—consisting of network vendors, computer equipment vendors, and automation equipment vendors. Companies in the group, along with many others, are committed to supplying networks and automation products that comply with the MAP communications specification.

It is gratifying to note that the various standards groups are more keen on converging than on identifying turf. I believe that's very useful.

Needed: Information management

What is needed to accomplish the broader level of inte-

Table 1: Industrial communications requirements

Feature	Office	CAD	Factory	Process
Traffic	Messages Memos (5Kbyte) Reports (50Kbyte)	Graphics files (1Mbyte) Screen refresh (50Kbyte)	Program files (0.5Mbyte) Status message (2Kbyte) Status monitoring (20Kbyte)	Program files (50Kbyte) Status message (2Kbyte) Process variable (50byte)
Mode	Bursty (Terminals dominate)	Bursty (Work stations dominate)	Mixed	Stream (Process variables dominate)
Throughput	High	Very high	High	Moderate
Response	2-6 sec	2-5 sec	0.5-2 sec	0.01-0.5 sec
Access	Statistical (CSMA/CD)	Statistical (CSMA/CD)	Various (CSMA/CD or token)	Deterministic (token)
Redundancy	No	No	Occasionally	Often
Broadband	Occasionally	Occasionally	Preferred	Occasionally
Packaging	Office	Office	Semi-hard	Hardened
Cost/node	$300-$600	$600-$50,000	$800-$2000	$200-$2000

Source: Instruments & Control Systems, March 1985, 35-39

gration that we've been talking about is a comprehensive approach to industrial information management. Information management is composed of three layers of disciplines: communications capability, a technique for managing and controlling data, and, finally, the applications. Without this information management link, we cannot move from basic device-level integration to the much higher level of computer integrated manufacturing (CIM) that we foresee for fully integrated factories.

Even the most sophisticated network provides only part of the total functionality required to achieve a high level of CIM. The next step in the hierarchy is the capability to control the flow of data through the network: a data management and control facility.

The requirements for data management and control have existed in paper-based factories for many years. These include the ability to identify drawings and documents; to provide these documents on a timely basis to those who need them; to provide a mechanism for modifying documents as well as a way to inform those who need to know that the changes have been made or that changes are pending; to develop dependency relationships among documents; and to provide a human interface (e.g., for drawing approval, sign-off, release).

Additional requirements stem from the fact that documents are becoming less paper-based; increasingly they are electronic, residing in multiple data bases on various vendors' computer-based products. Thus, a data management and control capability must be able to work with data distributed among several vendors' equipment. This far exceeds the usual data base management system.

Applications

Only after both a basic communications capability and a mechanism for controlling the flow of data are available, does it become possible to fully realize the benefits of integrated factory applications, such as material resource planning (MRP), computer aided process planning, tool room control, shop floor control, material movement systems, warehousing, inventory control, and the like.

An example

As an illustration of an integrated application, consider distributed numerical control (DNC).

In many respects, DNC is a microcosm of integration requirements for the whole factory. DNC systems began as little more than a wire replacement for the transfer of paper tape part programs. Now, however, the functions of DNC include data base management for parts programs, supervisory control for machine tool scheduling and performance monitoring, and real-time remote machining.

Table 2 summarizes the communications capability required to support a multifunction DNC system. There is a tremendous diversity of response times, data traffic characteristics, and throughput requirements.

At one end of the spectrum, very large part programs (of potentially unlimited length) must be transferred among computers and CNCs. On the other end, emergency conditions or alarms represented by very short packets of information must be transmitted and received with almost instantaneous response. Thus, DNC imposes some of the most demanding requirements on communications networks.

DNC networks also carry a great deal of product-related data in the form of part programs and process status information. Therefore, data management and control also are involved.

Smaller NC shops needing DNC tend to want simple, off-the-shelf systems selling for a minimal cost. On the other hand, large users, such as airframe manufacturers or the automotive industry, already have a substantial amount of DNC software implemented and require that a new DNC system conform to this installed software base. For this reason, the newest CNC products offer two classes of DNC interface. One class provides low cost, moderate data transmission capabilities comparable to those that have been offered for a number of years. The second class of DNC is oriented toward flexible manufacturing system (FMS) applications. This has considerably higher throughput capability, and is fully programmable to support either custom or standard attachment protocols. It may be used in either point-to-point or network-oriented modes. This FMS class of DNC provides not only data transfer, but application-to-application linkage through a protocol currently being standardized by the EIA (Proposal 1393).

Conclusion

CIM requires a comprehensive information management system to link devices on the shop floor with control computers, with the engineering and CAD environment, and with the automated office.

The requirements imposed on communication systems applied in various segments of industry differ widely. For this reason, it is likely that more than one network solution will be required to achieve a high level of integration.

A sound local area network strategy will likely include a CATV broadband token bus network complying with the IEEE 802.4 standard as well as networking options for various industrial segments. These network options include industrially hardened electronics, optional redundancy, a wide variety of automation device interfaces, and the bridges and gateways necessary to link these networks

Table 2: Characteristics of DNC data traffic

Message Service	Message Size (byte)	Request Interval	Response Time	Technology
Executive Software	500K	5 hr	60 s	50-100 Kbaud sync
NC Table Data	1K-5K	2 hr	10 s	4800 baud async/sync
Operator Instructions	1500	30 s	5 s	4800 baud async/sync
Shop Floor Mail	1500	Interactive	2 s	9600 baud async/sync
CNC-CNC Message	64	30 s	0.1 s	9600 baud sync/async
Supv. or Status Data	500-3000	1 hr	1 s	38 Kbaud sync
Alarm (E-stop)	32	NA	0.01 s	19-38 Kbaud sync/async
Remote Machining Data	50	0.25 s	0.1 s	9.6 Kbaud sync/async

Glossary

Bridge: A device for linking networks that have the similar topologies or protocols.

Bus: A topology in which information is made available simultaneously to all nodes.

Deterministic: A type of network in which response can be accurately predicted.

Gateway: A device for linking networks that have differing topologies or protocols.

Medium: Material over which information is passed; may be twisted wire, coaxial cable or fiber optics.

Network: An interconnected group of elements called nodes.

Nodes: A connection point on a network. Can be any of a variety of devices such as programmable controllers, numerical controls, computers or teminals.

PBX: Private Branch Exchange; a system, usually associated with telephones, that uses a central switching unit to connect different sources of voice and data within a facility.

Protocol: A procedure for formatting and interpretting data.

Response: Waiting time between request for use of the network and when the requested service is provided.

Ring: A topology in which information is passed from node to node.

Star: A topology which consists of a central node with radiating link elements.

Statistical: A type of network in which response varies probabalistically with data traffic conditions.

Throughput: Measure of the aggregate amount of data a network can carry in a given time.

Topology: Basic shape of the interconnection scheme of a network. Topologies commonly found today are the star, the ring, and the bus.

---

with one another.

Overlaying communications in the information management hierarchy is the data management and control capability. This allows the networking technology to provide a basis for achieving the paperless factory. Data management and control products continue to evolve to a fully distributed user-transparent access scheme for all types of product and process data. ■

Networking: Its importance to modern control

This critical technology is rapidly spreading from office to plant floor.

Robert J. Eaton
Vice President
General Motors Corp.

Just a few years ago, only readers of a few, select technical publications were interested in local area networks (LANs) and their role in business operations. That has now changed. People in a variety of professions need to know at least the basics of LANs, and LANs are at the top of the agenda at many technical conferences and trade shows. LANs have arrived! They will continue to grow, and to change with technology. This article looks at the impact LANs are having on manufacturing—particularly at GM.

LANs aren't a new idea, of course. They are a direct result of the development of computers and other programmable devices. And single-supplier LANs have been used extensively to permit office automation equipment to work in harmony. Computer manufacturers developed a lot of proprietary systems to allow communication between compatible devices connected to the network. Usually, these local communications systems were designed with a sole supplier in mind—the system's manufacturer.

The absence of universally-accepted standards has resulted in LANs of various types and sizes. These local communications systems can be categorized in four different ways: by the configuration of the system's wiring, commonly called topology; by the access method or the communications protocol used by the system; by the medium or type of wire used for communication; and, finally, by the market sector into which the network is sold—office versus factory, for example.

The use of computers in offices has been dominated by

(far left) Plant floor workstations will provide manufacturing personnel with process data, engineering specifications, personnel data, or real time quality feedback.

(left) Flexible machining centers combine several areas of technology. With robotics, automated guided vehicles, and computerized numerical control machines, they will implement just-in-time material control systems, rapid startup techniques, and predictable material flow.

(above) Large plant data centers have historically held plant databases for material requirements, personnel, process routings, and other data needed for plant floor operations.

(right) The manufacturing operations control room will be the nerve center and brain of the factory of the future. Here, resources will by dynamically allocated, schedules will be maintained, and production facilities will be monitored.

shared functions rather than computer-controlled processes like the plant floor. Accordingly, there has been no demand from users for standards that would permit multi-vendor local area networks without the cost of expensive software packages or add-on devices. As we'll discuss later, this is quickly changing.

The driving force behind the surge of interest in multi-vendor LAN technology is based on three things: the ubiquity of the computer, the need for quick access to additional management information, and the emphasis on tighter business and process control. Not coincidentally, this surge occurs when computer-controlled processes start to make significant inroads on the plant floor.

Office and engineering

General Motors is a good case study. Beginning with data processing, and later in the manipulation of data for computer-aided design (CAD) and computer-aided engineering (CAE), the computer has played a vital role in helping GM improve its quality and efficiency.

The computer has served us well in our offices and on the engineering drawing board. CAD has generated a 3:1 improvement in productivity, while personal computers, word processors, and other computerized office equipment in the hands of our managers and support staff are definitely making us a more efficient organization.

In one of our most recent product programs, the use of CAD accounted for up to 50% of the automobile chassis, and up to 75% of the body. Within the next decade, we expect to get close to 100%.

But office automation alone isn't enough. To fully reap the benefits of our CAD system, it must be combined with computer-aided manufacturing (CAM), to create a complete CAD/CAM system. By necessity, it must be multi-vendor; the need for computer-integrated production facilities is the driving force behind factory LANs at GM.

LANs on the factory floor

The factory floor is a challenging environment for multi-vendor LANs. The pace is fast—thousands of actions are occurring simultaneously—and signal delays, even for a fraction of a second, can be disastrous. The electromagnetic noise and the temperatures can be harsh for electronic equipment.

The climate is further complicated by economic reali-

ties. At older plants, we have many efficient computer-controlled operations that can't be scrapped just to install a modern local communications system.

Our latest generation assembly plant has over 1,200 programmable devices on the plant floor, supplied by over 10 firms. Monitoring, supporting, and communicating with all of these devices individually can be a real drag on the plant's balance sheet. However, it is a goal that must be accomplished if we are to reap the quality and efficiency benefits of factory automation. A multi-vendor LAN can provide the solution.

Levels of communication

The need for programmable device support is part of the communications requirements of an automated factory. These requirements can be divided into several levels: the first is real-time control—"on" and "off" type operations; the second is the coordination function—a robot picking a part off a conveyor or some other coordination of machines; the third is the loading of programmable controllers, the collecting of management data, and maintenance dispatching; the fourth is the data processing level of computer talking to computer. This, coupled with the need to transmit signals to inform employees and secure the plant, poses a formidable communications problem.

The level at which networking should occur is open for discussion. There is, however, no doubt that networking is definitely needed.

At GM, we plan to introduce programmable devices on the plant floor at an ever-increasing rate. By 1990 we expect to have 200,000 programmable devices within our plants. This is a significant increase over the 40,000 programmable devices already on GM plant floors, yet only 15% of these devices communicate beyond their own processes. Given this, it is imperative that we tear down the communications barriers between machines from different computer and controls companies.

MAP

Our solution to tearing down barriers on the plant floor is the Manufacturing Automation Protocol (MAP), which we adopted in October, 1982. MAP is a seven-layer, broadband, token-bus based communications standard. It incorporates standards developed by the U.S. National Bureau

Source: Instruments & Control Systems, February 1985, 27-30

(left, top) Engineering workstations provide product design information for automated machine tools and systems such as scheduling and tool management.

(left, bottom) Flexible welding and painting operations are just the beginning of flexible assembly applications. Use of flexible systems will expand to meet the need for improved quality and reduced scrap and downtime.

(right) Distributed, real-time process control and monitoring will drive and gather information from such plant floor devices as robots, vision systems, and programmable controllers.

of Standards, the International Standards Organization, the Institute of Electrical and Electronic Engineers, the American National Standards Institute, and the Electronic Industry Association. We're counting on MAP being widely accepted because it incorporates existing or emerging communications standards that are non-proprietary.

MAP specifies a coaxial cable that will be the backbone of our plant network. Eventually, this will permit us to rid ourselves of the point-to-point burden that would otherwise threaten to strangle us as programmable devices proliferate on plant floors.

To prove the feasibility of MAP, we asked seven companies to join us in a demonstration project at the GM Technical Center in Warren, MI. The demonstration includes an Allen-Bradley Vistanet Data Highway Gateway, a Digital Equipment VAX-11/750, two Gould Concepts 32/2705, a Hewlett-Packard HP 1000, an IBM Series/1, and a Motorola VME/10. This equipment communicates via a Concord Data Systems Token/Net Interface Module (TIM).

We have successfully implemented three of the seven MAP layers, and have asked several PBX manufacturers to join the project to help overcome the next hurdle.

At the 1985 Autofact Trade Show in Detroit we plan to connect a MAP LAN via PBX to a LAN of a different communications protocol, an Ethernet. In the future, we expect to have plant MAP LANs communicating with our office LANs.

We're so convinced of our approach that, as a matter of policy, GM plants in the future will not purchase programmable plant-floor devices that aren't MAP compatible.

While we demonstrate additional layers of MAP in the laboratory, we are proceeding with implementation in our plants. During 1985, at least six GM facilities will use certain layers of MAP primarily for programmable device support. We plan to implement by 1987 most functions of MAP at the Saginaw Steering Gear Factory-of-the-Future project in Saginaw, MI.

The Saginaw Factory-of-the-Future will be a multi-vendor, computer-integrated plant. It's a two-phase program: all 40 manufacturing cells will be operational by late 1985, and by 1987, we will integrate the cells into a factory control system and the plant's CAD/CAM system.

There will be more than 60 robots in the 70,000-sq-ft plant. Material will be automatically stored and retrieved,

and transported throughout the plant on automatic guided vehicles. Changeovers will be scheduled through factory control, and new programs will be automatically loaded into the machine cells. All necessary information will be electronically stored and transmitted, making it a paperless factory

One of the key ingredients is flexibility. Our goal is to be able to change any cell to another model in less than ten minutes. We also plan for multiple models of axles progressing through the plant on the same lines. We will main-

PBXs will provide the gateway to wide area networks. Integration of voice and data will allow database access and information transfer to and from world-wide data centers.

tain total part control.

Totally automated flexible assembly is proving to be one of the more difficult challenges of the project.

Each cell will verify that the work is performed to specification, before the workpiece can proceed to the next stage. We will monitor all processes.

The Saginaw Factory-of-Future is only a beginning. The program is a working laboratory to speed implementation of advanced manufacturing technology throughout GM.

It is only by taking bold technological steps that General Motors, as well as most other companies, will continue to grow and prosper in our highly-competitive, international business world. It is the pursuit of better quality and efficiency that dictates automation, and it is computer-controlled automation that dictates the need for LANs.

Multi-vendor local area networks will be the nerves leading to the heart of our business and manufacturing operations. ∎

Source: Instruments & Control Systems, February 1985, 27-30

Manufacturing Communications:
LOCAL AREA NETWORKS
for the integrated factory

Communication by means of computers claims to offer substantial opportunities for increasing manufacturing productivity. But reaping this harvest demands both an understanding of computer communication networks and the realization that their basic objective is to improve communications among people!

In any manufacturing effort and, in fact, in any human undertaking, communication of ideas, plans, or problems is essential to success. Individuals on a team cannot operate at peak efficiency without smooth communication with other members of the team and the team manager. Most

of us recognize the need for smooth communications in manufacturing, but achieving effective communications in the new world of automated manufacturing can be a significant challenge.

Although computers can solve many existing problems, they also raise questions that

demand answers. For example, how do we incorporate the computer solutions into the manufacturing environment without disrupting the team communication so necessary for success? Simply being the best-equipped team does not guarantee success.

The first step in achieving

By Terry L. Dollhoff

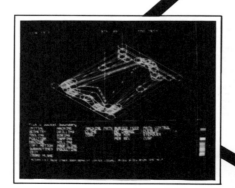

Reprinted with permission from Production Engineering, February 1985, 68-70, 74, 76, 78, © 1985 Penton/IPC

computer-based communication is to recognize the complexity of the problem. Beware of oversimplification. The problem of integrating communications is often dismissed with a glib "We solve that by supplying a local area network." Indeed, local area networks (LANs) are one of the effective tools for improving communications, but tools alone do not solve problems.

Before we discuss possible solutions, let's examine the problem itself. What is communications? Communications can be defined as: a process by which information is exchanged between individuals through a common system of symbols, signs, or behavior.

Sounds simple. But what is computer communication? Merely exchanging information between computers is not really communications. To achieve meaningful communication, information must be exchanged between individuals. The individuals might use computers to accomplish the information exchange, but it is ultimately the communication between individuals, not between computers, which improves the team performance.

Looking at communications from this viewpoint tends to put communication between computers in its proper perspective. *Individual communication can be facilitated by computer communications, but individual communication cannot be replaced by computer communications.* Thus, we might define computer communications as: a collection of hardware and software tools which are designed to improve the process by which information is exchanged between individuals.

Terry L. Dollhoff is Director of Software Development for Manufacturing Data Services Inc., Ann Arbor, Mich.

For the computer to be a useful tool for improving individual communications it must include certain attributes, among them a user interface, a database management system, and the capability to be connected to a network.

User interface. Often given short shrift as an element of computer communications, the user interface is critical to successfully using computers to transfer information between individuals. Beware of those who say they solve user interface problems with a user-friendly interface. Of course the interface is user-friendly. Who would ever consider buying a system with a user-hostile interface?

But, what really constitutes a user-friendly system? In the final analysis, only the user can judge the friendliness. But be aware that a well-polished demonstration can make any system appear to be friendly. Ask yourself, "Is it friendly to me?"

Although user-friendliness is a highly subjective issue, there are several attributes which are commonly accepted as essential to making a system user-friendly:

■ Natural language. The user should be able to communicate in his native language.
■ Industry language. The user should be able to use terminology common in his industry.
■ Application specific. The user should be presented with choices specific to the task at hand, and not with irrelevant choices which waste time.
■ Easy to learn. Every system is easy to learn to use; who would even consider buying it if the system was extremely difficult to learn to use? If the system is extraordinarily difficult to learn to use, it will be easily forgotten, which wipes out the productivity gains.

The only way to evaluate the ease of use is to try the system.
■ Handling mistakes. Today's computer systems are generally at their best when handling correct input. A good test for user-friendliness is to observe the reaction to erroneous input. What happens when you do something the computer did not expect? Many systems become downright surly. The truly friendly systems will provide detailed and helpful error responses to get you out of trouble.

Database management. The goals of communication between applications are to avoid re-entering data and to facilitate creation of similar data from existing data. Database management systems are often suggested as pat solutions. These systems let all applications access the data and automatically update stored information when new data are entered.

In addition to providing access to common data, a more complete solution converts that data when appropriate. For example, if a draftsman or engineer creates a part design using the computer, it is important to distribute that data to the manufacturing department. The manufacturing department may want to use the data to create a manufacturing process plan. But, the data may not be in a suitable format for manufacturing and may have to be changed. For example, hidden lines may be undesirable. To provide useful applications communications, the computer should change the format of the data automatically, or at least, have tools for the user to do this.

Networks. As computer technology continues to advance, we encounter increasingly more powerful desk-top computers. As these desk-top computers, or workstations, find application in manufacturing it becomes necessary to communicate information from a user of one workstation to a user of another. Therefore, tools must be available to transfer data from one workstation to another, or to a central point where both workstations can access the data.

This is the domain of a local area network (LAN). The overall definition of a LAN is simple. For data to be exchanged electronically between two computers, those computers must interconnect in some form. This interconnection is a network.

Before discussing the individual network components in more detail, we might consider the question "How local is it?"

A LAN is a network which interconnects terminals or computers (nodes) without the use of long-haul —telephone or microwave—networks. Based on the technology in use today, the distance between the farthest node is typically less than 15 mi. Thus, the LAN is useful for communications within a plant or group of nearby plants. This distance limitation allows data communications at high speeds. Today's networks operate at millions of bits a second.

A long-haul network can be used to transmit data over greater distances—worldwide, if necessary. Most of the long-haul networks utilize the existing telephone's networks. But because of the technical characteristics of the telephone network, the data transfer rate, 9,600 bits per second (bps), is slow in comparison to that of LANs.

The LAN and long-haul network are frequently used in conjunction with one another. Local data are sent through the LAN and long distance data are sent through the long-haul network. A gateway from the LAN is used to transfer data between these two types of networks.

Although the definition of a LAN is simple, selecting and installing one is not simple. There are many different ways to accomplish the electronic exchange. The only generally accepted fact is that there is no best LAN.

The reason that there are so many variations is that there are many diverse application goals. Often, these goals compete; a LAN for one application may be ill-suited to another.

Any LAN is made of several basic components, each of which has an effect on how the network operates and how well the network can handle specific problems. The basic components of any LAN or network are:

■ Media—the physical transmission media or cables.

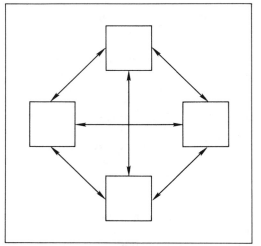

The fully interconnected topology is the simplest topology. Any two nodes which might communicate are directly connected. For a limited number of nodes this arrangement is efficient and cost effective.

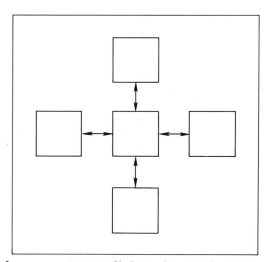

The key to a star or radial topology is that most communications occur between a remote node and a central node.

■ Transmission scheme—the arrangement of the computer data or bits which are sent on the media.
■ Topology—the way in which the individual workstations are interconnected.
■ Access scheme—the rules for gaining access to the network.
■ Protocol/interface—the software used to transmit user data from one workstation to another.

The physical hardware used to transmit data is called the media. Several varieties of media are available for net-

work service, and the media used will determine the maximum data transmission speed, immunity to noise, and basic connection cost.

The simplest media is a twisted pair of wires similar to that used to interconnect phones. In some cases, one or more of the twisted pairs will be enclosed in a shielded cable. The shield surrounding the wire will increase the noise immunity, but it has no effect on transmission characteristics. Although the twisted pair is inexpensive and simple to install, it is the slowest transmission media with a transmission limit of approximately a million bits per second (1 Mbps).

The most common media is coax cable, and it is used in LANs that are classified as either baseband or broadband. The 50-ohm coax typically used in baseband networks supports transmission to 10 Mbps and is commonly referred to as baseband coax. The 75-ohm coax typically used in broadband networks supports transmission to 150 Mbps and is commonly referred to as broadband coax. The 75-ohm coax is the same cable used in multichannel cable TV—which is itself a broadband network—and it is from this application that it first became known as broadband coax.

The latest media for network transmission is fiber optic cable. A beam of light carries the data in a fiber optic cable. This cable is capable of high data rates at low error rates; current technology allows transmission of 1,000 Mbps. The installation and connection costs are high though. However, this will probably emerge as a dominant media by the end of the decade as technological advances decrease the cost and connection complexity.

The transmission scheme indicates the electrical or physical manner in which data

are sent over the media. Basically, the transmission scheme typically refers to whether data are transmitted in a single channel (baseband) or in multiple channels (broadband).

Broadband is the most flexible scheme because several transmissions can occur simultaneously through different channels. The disadvantage of the broadband scheme is that the data are directional—all data travel in the same direction. Therefore, to enable communication between all units on a broadband network, some channels only send data and others only receive data.

Topology is the system roadmap. It refers to the manner in which nodes are interconnected. The simplest topology is a fully interconnected one, any two nodes which might communicate are directly connected. For one, two, or three nodes this topology is quite practical. But if there were 10 nodes in the network, 45 paths would be needed and each node would have 9 interconnections. Clearly, this topology quickly becomes unwieldy.

The first alternative to direct connection evolved from the timesharing systems of the early 60's and is called a star or radial topology. In this topology communications are assumed to occur between a remote node and a central node. Hence, it is very appropriate for a central processing system. Within this topology the central node is involved in any transmission, including those between two remote nodes. A disadvantage is that the central node is a single point of failure. And it is often difficult to find a good physical location for the central node which does not require long distance interconnects.

The most common topology in use today is the bus or highway. In this case, a central bus is used to transmit data. Each

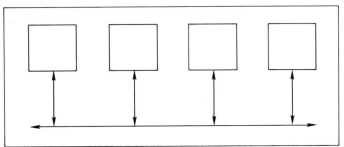

The most common local area network topology is the bus or highway. A central bus is used to transmit data and each node has a connection to that bus.

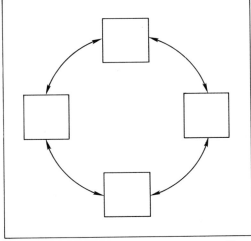

In the ring topology, all nodes are interconnected in a loop. This topology is often used for high-speed networks.

node has a connection to the bus. Advantages of a bus topology are that adding new connections is simple and there is no single point of failure which causes the entire network to fail. A disadvantage is that the rules for using the network are a bit more complicated.

Another topology is the ring topology in which all nodes are interconnected in a loop. This topology is often used for high-speed networks. A disadvantage is that the failure of any single node might cause the failure of the entire network.

The access scheme refers to the set of rules used to determine how an individual node gains access to the network. There are a number of different strategies for network access. Each one is heavily influenced by how the network is used.

In some cases communication over the network can be predicted and originates from, or is received by, one source. The access scheme most appropriate for this type of network is polling. The central source or master unit continuously asks, or polls, each node if it wishes to transmit. Transmission can occur only in response to the poll. If a node rejects a poll, it must wait until it is polled again later.

Another scheme, usually associated with the ring topology, is token-passing. A token or key to the network is circulated around the ring. When a node receives the token, it can hold it, thereby claiming the right to use the network, and begin transmission. If a node does not wish to use the network, it passes the token on to the next node. A node can only transmit a message when it has possession of the token.

The access scheme frequently associated with bus topology is called contention. It is analogous to human communication. Before using the network, a node waits until the bus is quiet. When quiet, the node starts its transmission. Usually, that's all there is. However, as can happen during a conversation between individuals, two nodes may sense the quiet and begin to transmit at the same time. This results in garbled transmissions.

The solution to a network collision follows the human conversation analogy. The two nodes detect the collision; stop transmitting; wait for a random period of time; and try again. Since the wait is random, they probably will not collide again.

The full technical term for this access mechanism is CSMA/CD which stands for Carrier Sense—wait for quiet, Multiple Access—any unit can begin, with Collision Detection—detect and handle message collisions. Ethernet is a CSMA/CD network.

One concern of the CSMA/CD access scheme is the effect collisions have on the effectiveness of the network. One can imagine a network totally consumed by collisions. In actual practice, the collisions occur to a very small fraction of the messages sent. However, collisions do occur more frequently in a heavily loaded network, but should be of no consequence in a normally loaded network.

Protocol software adds the necessary intelligence to transfer files, records, or other meaningful collections of data, as well as individual bits of data, from one node to another. The term *protocol* refers to the general conventions by which these high-level data transfers take place.

Because the goal of a LAN is to facilitate communication between individuals, the protocol software is key to how smooth or how awkward that communication will be. Protocol software is responsible for translating an application software request into lower-level commands. It is the "brain" of the network. A particular protocol is not tied to a particular network architecture. This high-level software can often operate over a variety of networks. Further, some networks like Ethernet can support many application protocols. The ability to interconnect hardware and support intelligent communication between that hardware is dependent on both the network and the protocol.

In an effort to provide some general model for development of protocol software, the International Organization for Standardization (ISO) has specified an architecture for protocol software called the ISO Architectural Model. The model breaks the protocol software into seven different layers.

The lowest level deals with the physical interconnection

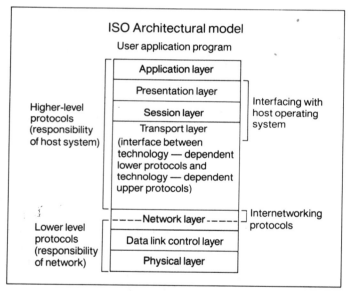

The ISO Architectureal Model provides a general format for developing protocol software. The model breaks the protocol software into seven levels. The lowest level, the physical layer, deals with the physical interconnection to a network. The highest level, the Application layer, deals with the interface between application software and the lower levels of a network software.

to a network. Each higher level becomes more abstract. The highest level, the Application Level, deals with the interface between application software and the lower levels of the software in the network. Typically, the higher levels are handled by the host operating system. It is not particularly important to understand the function of each separate layer, but to understand the goal of the ISO model: which is to simplify the interconnection of networks.

Within any LAN, network efficiency is often improved by dedicating a certain task or group of tasks to a particular node. Such a node is referred to as a server.

File management and peripheral control are common tasks frequently assigned to servers. The file server manages common files used by all nodes on the network. It usually has a larger capacity disk.

The peripheral management server handles peripheral devices used by all of the nodes. A high-speed printer or plotter is usually connected to such nodes because it would be impractical to provide a separate peripheral device for each node on the network.

Another common use of a dedicated node is to interconnect one network to another. A node that functions in this capacity is called a network gateway. It is a node on both networks. If a node on one network wishes to transmit information to a node on the second network, this transfer would be handled by the gateway node. Communications in the other direction are handled in a similar manner.

If the tasks assigned to a dedicated node do not require all of the processing capability of that node then it may be assigned more than one task.

For example, a single node may act as file server, peripheral manager, and inter-network gateway. Or the node can act as a normal workstation and handle its other tasks in the background. Determining what this node will handle depends on the size of the task and on the node's local processing capability. It is a classic price-performance trade-off. Assigning multiple tasks to a single node may degrade performance but it will lower the price of the network.

Along with the ISO, a number of groups are active in defining standards for networks. The Institute for Electrical and Electronic Engineers (IEEE) has begun to develop a set of standards for specific networks. The first such standard, called IEEE 802.3, specifies a CSMA/CD bus type network almost identical to Ethernet. In fact, the terms Ethernet and IEEE 802.3 have become synonymous. The differences between the two are relatively insignificant.

The second IEEE standard is for a token-passing bus configuration. A system based on this standard executed LAN communications between several vendors as part of the General Motors Automation Protocol (GMAP) demonstration.

Finally, IEEE is developing a standard, called IEEE 802.5, for a token-passing ring configuration. Many people feel that this will be used as a high bandwidth network.

Which network is best? The answer is simple. No single network can satisfy the requirements of all applications. There is no one perfect network. Any particular network is designed to handle some group of tasks.

It seems to be the general consensus of this industry that the future of networking will be the peaceful coexistence of many networks. Certainly, some networks will fail to survive because they do not offer any particular advantage over a more widely used one. However, there will be a number of standard networks available. And gateways between networks will become increasingly transparent.

If there is no one good network, how does the user choose the right one for his particular application? Assuming that you do not intend to develop a network from scratch, this question may be more easily answered than you might think. When purchasing off-the-shelf equipment, such as a computer workstation for interconnection to the network, the choices of network are usually limited. The key to making a choice is hardware availability. Can you purchase the necessary equipment for your application which will interconnect to the chosen network?

As a general rule, try not to select proprietary networks. With this type of network, you can be locked into buying all network hardware from a single source. If your application needs are varied and you want to include hardware from several different sources, then a standard network is mandatory.

Whatever network you choose, the key to a successful installation is planning. Be sure to install adequate network interconnections. Analyze whether network servers for file management or peripheral handling may improve overall performance. Finally, try to evaluate the network with your particular applications. Claims in networking are sometimes outlandish. The only way to judge overall performance is to try it.

Update on local area network standards

*This update describes the status of PROWAY and
IEEE 802 LAN standards, and shows how standards groups interact.*

Tony Bolton
Concord Data Systems
Waltham, MA 02154

Communications standards—including those for local area networks—set up procedures so that two or more products from different vendors can communicate with each other (Fig. 1). Unfortunately, efforts to standardize LAN's have been hampered by the popularity of existing products, advances in technology, and an ever-expanding scope of applications. But, as we'll see in this article, a few major LAN standards are being finalized, approved by sanctioning bodies, and accepted by industry.

The need for standards

If one vendor had a monopoly on computers and industrial control devices, we wouldn't need standards. Any needed communications hardware and software protocols would be supplied by the single vendor, and the issue wouldn't concern the end user. However, the industrial control world has many vendors, each with its own ideas about communications, so the need for standards is becoming critical.

Local area network standards are needed for two reasons: first, to establish a hardware-independent framework for network design; second, to establish standards for software protocols and hardware interfaces so that networks from different vendors can communicate with each other.

When a standard is accepted by vendors, semiconductor manufacturers start making low-cost, nonproprietary VLSI (very large-scale integration) chips to support it. This reduces development and production costs, simplifies the connection between devices, and helps eliminate design errors. Also, standards protect a user's investment, because they ensure that the network will be based upon a technology that is supported by many vendors. Nonstandard proprietary networks, on the other hand, tie the user to a sin-

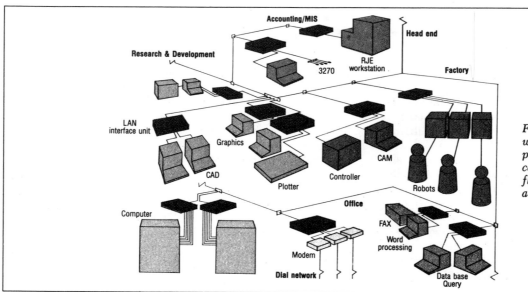

Fig. 1: Local area networks reach into every part of a plant, from the controllers on the plant floor to the mainframe accounting computers.

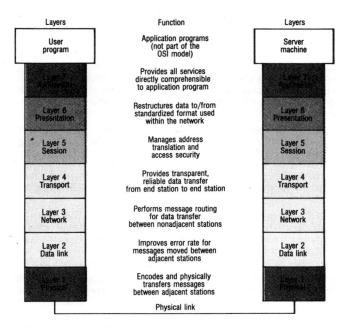

Layers	Function	Layers
User program	Application programs (not part of the OSI model)	Server machine
Layer 7 Application	Provides all services directly comprehensible to application program	Layer 7 Application
Layer 6 Presentation	Restructures data to/from standardized format used within the network	Layer 6 Presentation
Layer 5 Session	Manages address translation and access security	Layer 5 Session
Layer 4 Transport	Provides transparent, reliable data transfer from end station to end station	Layer 4 Transport
Layer 3 Network	Performs message routing for data transfer between nonadjacent stations	Layer 3 Network
Layer 2 Data link	Improves error rate for messages moved between adjacent stations	Layer 2 Data link
Layer 1 Physical	Encodes and physically transfers messages between adjacent stations	Layer 1 Physical
	Physical link	

Fig. 2: The OSI reference model has seven layers that connect the user application at Layer 7 to the actual hardware at Layer 1. Any of the layers can be eliminated if not needed in an application.

gle vendor (or to very few vendors) for equipment and services.

Much work has been devoted to developing communciations standards that apply to industrial and process control applications. (For a listing of the major standards groups in the U.S. and Europe, see box "Standards Organizations.") As we'll see in the following sections, some of the standards are ready today, while others are still in development.

The Open Systems Interconnect

In 1979, the International Standards Organization (ISO) created a foundation for the development of communcations standards. The ISO proposed a layered model called the Reference Model for Open Systems Interconnection (OSI). This establishes precise terminology, identifies the primary components of a system, defines component functions and their relationships with each other, and provides the means to functionally analyze a system and compare it to other systems adhering to the model.

Originally, the OSI model was designed mainly for long-distance telecommunications networks. But it's been widely used by local area network (LAN) designers. Because the model is *architectural*—that is, it provides a framework within which a network designer can make choices—it does not define the protocols and interfaces that must be used. These details, especially at lower levels of the model, are being handled by other standards organizations—most notably by the IEEE.

The OSI model organizes all system functions into separate "layers." Each layer provides certain services to the layer above it, and uses the services of the layer below (Fig. 2). This is consistent throughout the model, except that the top layer interfaces with the user, and the bottom layer interfaces with the transmission medium.

A layered structure allows network designers to analyze and develop a network scheme modularly. That is, the layered approach breaks a highly complex communications problem into smaller, more manageable parts. Also, if the functions of a particular layer aren't needed, the entire layer can be omitted from the network design.

The model can be used for telecommunications networks, wide area networks, and LAN's. The OSI model specifies seven layers, as described below. In general, Layers 5-7 are not as far along in standardization as Layers 1-4.

1. *Physical*—The function of this layer is to physically transmit data. It includes the connection to the transmission medium (e.g., broadband or baseband cable), interface devices, and other hardware responsible for encoding and physically transferring data between stations on a network.

2. *Data Link*—To improve the error rate for messages that move between adjacent stations, this layer detects bit transmission errors, assembles outgoing messages into blocks called "frames," disassembles incoming frames, and acknowledges that a message has been successfully transferred.

3. *Network*—This layer routes transmissions between nonadjacent stations on the network and controls "traffic" at each network station. In local area networks, however, its primary function is to connect one LAN to another.

4. *Transport*—This is the software interface between the first three layers—which are mostly concerned with physically transferring data—and the last three layers, which directly serve the user device.

5. *Session*—By synchronizing end-devices, readying data to be transferred, and managing the actual connection when it's made, this layer handles the connection between two separate devices. It can also track and manage multiple "conversations" between several terminals and a computer.

6. *Presentation*—As the name implies, this layer restructures data into the form used by the communicating device. For example, if two devices have different page formats, this layer converts the format of the transmitted data from one device to match the format of the other device. It also provides restructuring services such as translating character codes, and compressing text.

7. *Application*—The purpose of a LAN culminates in this layer. All the lower layers serve to support Layer 7, and it originates all their actions. This layer is the interface between the user device and the network.

The OSI model is very flexible, which means that many different networks can be designed, and all will conform to the overall framework. To meet the needs of a particular application, the network designer can select appropriate protocols and interfaces. However, given the large number of protocols and interface options available, it's possible to design an OSI-type network that can't interact with any other system. This leads us to the standards work being done by the IEEE and IEC

The IEEE-802 committee

It's not surprising that the most active and successful standardization work to date involves the first two layers of the OSI model, because these layers cover the more "mechanical" aspects of networks. The major U.S. group trying to resolve standards issues for these layers is the IEEE-802

Local Area Network Standards Committee. (For a detailed look at hardware developments in Layers 1-3, see "Tackling the interconnect dilemma," *I&CS*, May 1982).

When the IEEE-802 committee convened in 1979, the local area network industry already had products on the market. The most well known of these is Ethernet. To strike a balance between existing networks on the one hand, and new technology on the other, the committee developed standards covering three types of networks. It has also standardized Layer 2 (data link). Briefly, standards developed by the IEEE-802 committee are:

• *IEEE-802.2*—This standard applies to the data link layer, the second layer in the OSI model. The standard splits this layer into two sublayers: *Logical Link Control* and *Medium Access Control*. The logical link control sublayer is used in all three of the medium access control methods (802.3, 802.4 and 802.5) described below. The 802.2 standard details the services and functions of each sublayer and its relationship with the layers immediately above and below it. Essentially, 802.2 specifies a multipoint, peer communication link protocol. It operates with any compatible physical layer (Layer 1) and supports data rates from one to 20M bytes/s.

• *IEEE-802.3*—This standard covers the access method known as *Carrier Sense Multiple Access with Collision Detection* (CSMA/CD), and is nearly identical to Ethernet. The CSMA/CD method uses the principle of "listen before transmitting, listen while transmitting" to gain access to the network. All stations on the network listen to the medium and do not transmit if it is in use. If a station detects silence, it may go ahead and transmit. While this type of network is relatively inexpensive, it's limited in length to 2500 m, and its efficiency decreases as network loads increase. The 802.3 standard explains the CSMA/CD method, describes the protocol, defines the physical medium as a special low-noise baseband coaxial cable, and details the physical connections of the communicating device to the interface unit and cable. The current 802.3 standard defines a 10M byte/s baseband network; the committee also is studying a broadband version.

• *IEEE-802.4*—This standard applies to *token-passing* over a bus-type network on broadband CATV cable, or on baseband cable with either of two types of connections. Token passing access is a method in which stations on the network share access equally, but are allowed to transmit only when they have possession of the "token." The token is passed around the network in a *logical ring* (not physical), thus eliminating collisions during normal operations. This method results in a network that is highly controllable, predictable, and stable under all communication loads. The network can cover distances up to 25 miles, and network users can assign levels of priority within network stations. The 802.4 standard details the token-passing method, protocols, physical media (broadband or baseband cable), and physical connections.

• *IEEE-802.5*—This standard covers token passing using a *physical ring* structure on either baseband twisted-pair wire or cable. This network has a sequential arrangement, where data passes in only one direction, from one station to the next. While this is very efficient, failure of one station can disrupt the entire network. The 802.5 standard also defines the token-passing method, protocols, physical media, and connections.

Standards Organizations

While the IEC, ISO, and IEEE are the principal groups involved in communications standards, a number of trade groups and government agencies also play a role. Described below are the major organizations contributing to international communications standards.

• ISO—The International Standards Organization is a voluntary group composed of the principal standards groups in participating countries. The U.S. representative is ANSI. The ISO defines standards in a wide range of commercial areas, including communications, and cooperates closely with the CCITT (see below). It developed the widely used Open Systems Interconnect (OSI) model for communications networks.

• IEC—Essentially equivalent to the ISO, the International Electrotechnical Commission specializes in industrial applications. It is the parent organization for the ISA PROWAY (Process Data Highway) Committee. The PROWAY committee was chartered five years ago to develop LAN standards specifically for industrial use.

• IEEE—The Institute of Electrical and Electronics Engineers is a U.S.-based technical society whose scope of standards activities is quite wide. The IEEE functions within the ANSI framework, and its 802 Committee is actively pursuing a family of standards for local area networks. When the IEEE completes a standard, it sends it to ANSI for approval; ANSI, in turn, sends it to ISO.

• ANSI—The American National Standards Institute functions as a clearinghouse for voluntary standards proposed in the U.S. The approval process for a proposed standard is somewhat slow, because ANSI normally seeks a broad consensus before passing a standard. However, ANSI has realized the urgency for standardizing LANs, and now passes IEEE-802 standards directly to the ISO, bypassing its normal approval cycle. The ANSI X3 Committee processes most of the communications standards, including those for LANs.

• EIA—A trade organization, the Electronics Industries Association is active in developing physical interface standards, such as the commonly used "RS" standard. The EIA is not writing LAN standards itself, but has assisted the IEEE-802 committee in specifying parts of the physical layer.

• NBS—The National Bureau of Standards administers standards for the U.S. Government, and is actively developing models, architectures, and protocols for higher-level LAN functions. The NBS is involved in IEEE, ANSI, and ISO standards work.

• ECMA—The European Computer Manufacturers' Association is a trade organization of computer vendors in Europe. Strongly supported by its members, ECMA has a reputation for high quality technical work and a short approval cycle. The ECMA has been working with the IEEE-802 committee on an international level.

• CCITT—Affiliated with the United Nations, the International Telegraph and Telephone Consultative Committee is a branch of the International Telecommunications Union. The State Department represents the U.S. on the CCITT. While it deals mostly with telecommunications, the CCITT V-series and X-series (i.e., X.25) are important in any networking scheme involving the telephone system. In countries with nationalized telecommunications (most European countries), CCITT recommendations often become law.

In addition to these standards organizations, many large vendors develop their own specifications, such as those developed by Foxboro for its Foxnet LAN and by Honeywell for its TDC-2000 Data Highway. In the commercial world, Ethernet and Datapoint's Arcnet are widely used, privately-developed LANs. Popular networks always have a chance of becoming *de facto* standards, as in the case of Ethernet, which became IEEE-802.3. Most vendor architectures, however, are usually for a vendor's own equipment, making it hard to accommodate and interconnect equipment from other vendors.

The PROWAY standard

As early as 1978, the ISA PROWAY committee began developing standards aimed at the networking needs of factory and process control applications. The committee tried to develop a token-passing access mechanism with a bus arrangement, similar to IEEE-802.4.

In 1983, the PROWAY committee approached the IEEE-802 committee with a request to informally join forces to complete the initial development work begun by PROWAY. The PROWAY committee recognized that the technical sophistication and scope of the IEEE-802 committee's work was much more encompassing than their own. It felt that both groups would benefit from the joint venture.

Since then, the committees discovered several shortcomings in 802.4 for industrial applications, and enhanced the standard. The enhanced 802.4 standard provides more controlled response time and two additional protocols: the first demands an immediate acknowledgement be received before the sending station passes the token; the second asks a station to send back data immediately on receipt of request, and within the same token time frame.

However, because PROWAY is standardizing more than the first two layers, the PROWAY committee now plans to submit its own version of the standard to the American delegation to the IEC for approval in November 1983. The new PROWAY standard will conform to the appended version of IEEE-802.4.

For data communications specialists and network equipment manufacturers, the possibility of having a joint commercial and industrial standard is particularly pleasing. For industrial users, this action will reinforce the strength of the IEEE-802.4 standard for industrial and process control use.

The status of IEEE-802

While an understanding of all the standards and committees is important from a technical standpoint, the real issue for designers, manufacturers, and users of local area networks is: Which standard or group of standards will prevail? And most important for users: Which vendors currently comply or plan to comply with approved standards?

The current status of the IEEE-802 standards is:
- The 802.2 (logical link control) draft standard has received preliminary approval by IEEE and has been forwarded to the ISO. Final IEEE approval is expected in early 1984.
- The 802.3 (CSMA/CD) standard was approved by IEEE in June 1983 and is now with the ISO for review.
- The 802.4 (token bus) standard with the PROWAY enhancements was approved by IEEE in September 1983 and now is at ISO for review.
- The 802.5 (token ring) standard is about a year behind the other 802 standards, but is expected to be approved by IEEE and sent to the ISO sometime in 1984.

The choice among vendors as to which standard they

GM's Support of Network Standards

General Motors uses a wide range of products from many manufacturers for its control and computing needs. Each of these usually has its own proprietary communication protocol, making overall interconnection difficult. Thus, GM established its Manufacturing Automation Protocol (MAP) Task Force in November 1980 to address the problem. The task force's goal was to prepare guidelines and specifications that would allow communication among the products used at GM.

The task force had three objectives:
1. Define a MAP network standard to support application-to-application communications;
2. Identify applications functions to be supported by the network;
3. Recommend low-level protocols to meet GM's requirements.

GM hopes that manufacturers will design products to meet MAP protocols, but the task force has also established procedures for non-MAP devices to connect to the network. In Fig. 3, for example, a programmable controller (PC) support system connects a PC network to the MAP network.

The MAP plan incorporates emerging national and international standards, including those from IEEE, ISO, and NBS. Following the OSI model, GM has adopted the IEEE-802.4 token bus standard for Layers 1 and 2, and has strongly recommended that the PROWAY committee adopt it as well. GM is using emerging ISO/NBS standards for higher level protocols, and has recommended adopting the NBS Internet, Transport, and Session protocols for Layers 3, 4, and 5. Also, it is working with vendors to get these protocols installed in their devices. The MAP effort represents the best example to date of a powerful user forcing vendors to conform to recognized standards, while encouraging the development of standards for higher-level protocols.

A pilot project is underway at GM to implement Layers 1 and 2. Vendors involved include IBM, DEC, Hewlett-Packard, and Concord Data Systems. The pilot will use Concord's Token/Net LAN to connect equipment from the three computer manufacturers. GM hopes to demonstrate a working model of the MAP network at the 1984 National Computer Conference.

Fig. 3: General Motors' MAP network links a wide variety of assembly plant functions, from production monitoring to security.

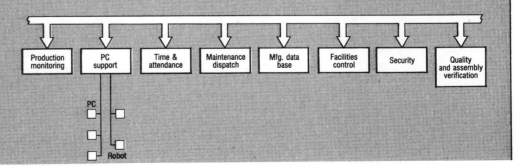

---

will comply with depends a great deal upon the anticipated application of the vendor's network. There are pros and cons to both token passing and CSMA/CD access methods, and advantages and disadvantages with baseband and broadband transmission media. A vendor also has to decide which standard will become the most commercially successful method.

To date, in the baseband CSMA/CD arena, the Ethernet system has become the leader. It is so popular, in fact, that it was used *de facto* to develop the IEEE-802.3 standard. With the full support of Xerox, Intel, and Digital Equipment Corp. behind it, Ethernet stimulated market growth and industry acceptance of CSMA/CD technology. Semiconductor manufacturers have developed standard VLSI chips for Ethernet, bringing down the cost of developing compatible products. Consequently, many vendors have jumped in to take advantage of this market.

Concord Data System's Token/Net LAN is the only broadband token-passing network on the market to date which complies with IEEE-802.4. The Token/Net system currently is being tested as part of the Manufacturing Automation Protocol (MAP) project at General Motors (see "GM's Network Standard.")

Other vendors have announced their intentions of developing products based on the 802.4 standard for token passing on broadband cable. In particular, Allen Bradley, Western Digital, and 3M have said they will jointly develop communications products based on this standard.

Also important for users is IBM's endorsement of IEEE-802.5 as its preferred method of medium access. While there are no networks available yet that comply with the 802.5 token passing ring standard, IBM is expected to develop a product along these lines.

A final note

When an engineer is called upon to put together a networking system, he or she must understand the issue of standards and the benefits they provide. However, he must also look at how each standardized network will perform under conditions existing in his application. For example, CSMA/CD networks work well when network size is small and traffic is relatively light. Token passing methods, on the other hand, are better suited to large, high-speed networks.

Baseband cable, used in both CSMA/CD and token ring networks, is relatively inexpensive to install over short distances. Broadband cable, for use in token passing bus networks, handles longer distances and combines data, voice, and video on a single cable. Thus, the engineer must consider the transmission medium as well as the networking method. Cost, speed, and capability of each must be compared to the needs of the application. ■

SECTION X
The Future

SECTION X
The Future

This section will discuss the future of the computer by presenting articles from three important and fast-developing areas in industry today: (1) the automatic factory, (2) artificial intelligence (AI), and (3) hardware.

The Automatic Factory

The concept of an automatic factory is one of the most widely and actively discussed areas of manufacturing development in the contemporary trade press. Two major publications, *I&CS* (Chilton Company, Radnor PA), and *Iron Age* (same publisher) have been publishing ongoing series on the subject since 1980. *Production Engineering* (Penton Publishing Co., Inc., Cleveland OH) publishes an annual update on computers in manufacturing. For the latest presentation from *Iron Age*, see the article "Factory 2000."

Frequent articles on the subject appear with regularity in *Assembly Engineering* (Hitchcock Publishing, Wheaton IL), *Control Engineering* (Technical Publishing, Barrington IL), *Plant Engineering* (same publisher), *Automation* (Penton Publishing Co., Inc., Cleveland OH), and many others. A recent advertising campaign by General Electric promoted its programmable controller line under the theme "Factory With a Future."

One article from the publications previously cited points out that "the common denominator for assembly factories in the future is the computer. From PC's (programmable controllers) to mainframes, these computers will make possible the multitude of tasks required for assembly automation, monitoring, control, and intelligent management."[1]

Partly automated factories are now, and have been, in existence since at least the late 1940's, when the word "automation" — dreaded by organized labor, defended by forward-looking managements — entered the American business vocabulary. Most modern discrete parts manufacturing plants now contain "islands of automation," built around CNC (computer numerically con-trolled) machine tool lines, robot welding centers, automatic paint-spray booths, and similar operations that have long since been successfully automated.

Even though the above operations are carried out not under the direct supervision of human operators, but instead as slaves to the preprogrammed instructions locked into microprocessor-based monitoring and control units, they are not regarded as computerized, as much as computer-assisted, computer-aided, or computer-integrated. By definition, then, computer-integrated-manufacturing, or CIM, is a closed-loop feedback system that uses product concepts and requirements as input, and produces finished goods as output. The term will appear often in the articles, which, it is hoped, will help the reader toward a more complete definition of the term, though not necessarily a complete one.

Artificial Intelligence

The fascinating subject of whether or not a computer can be made to think like its human user has developed into a new area of technology called "artificial intelligence," and the associated field of "expert systems." The articles included with this section treat both areas.

Hardware

How big and how fast are computers going to be in the future? Several articles in this section give a hint. "New Architectures" provides an insight into computer design considerations and points out that the basic design of electronic computers hasn't changed much since 1945! But it will!

Reference

1. "Computer Integrated Manufacturing: The Focus of Manufacturing Control," *Assembly Engineering*, May 1982.

FACTORY 2000

Between now and the year 2000 the factory will change more rapidly than it has in the past 75 years. Firms will either adapt or perish.

WRITTEN BY
ROBERT E. HARVEY
DESIGNED BY
SANDY BUTRYN
ILLUSTRATED BY
RANDY HAMBLIN

The patient is dying. In this case the patient is U.S. industry and its illness is ailing productivity. That's the popular conclusion of theorists who would have us believe that nothing can be done to revive the patient. But the fact is, this illness, this slow death, if you will, need not be fatal. Cures, although often difficult to swallow, do in fact exist.

Before today's ailing factory can be transformed into tomorrow's efficient Factory 2000, healthy doses of philosophy, strategy and technology must be ingested. Recovery can begin immedi-

362

ately because most of the elements needed for the factory of the future are available today.

According to Thomas G. Gunn, manager, Computer Integrated Manufacturing Group, Arthur D. Little, Inc., Cambridge, Mass. "We have more technology than we can implement in the next ten years."

The major technical tool, of course, is the computer, a machine with the potential to completely restructure today's manufacturing operations and launch industry into the 21st Century.

"On a global scale and by historical standard, we now find ourselves in the midst of a technological revolution on the order of the discovery of fire, the development of the water wheel, the printing press and nuclear power. This revolution will for all time change our way of life like no other change before it.... This revolution is the information revolution based on electronic miniaturization and fundamental logic," stated John R. Holland, manager, Factory Automation Laboratory, Allen-Bradley at a recent CAM-I Conference.

Factories with promising futures will be those in which management effectively harnesses this revolution. However, time is short. Computer technology is a U.S. trump card. But that card must be played soon, while it still has some value.

Rather than make effective use of this technology, however, many manufacturers are floundering amid a growing profusion of hardware, software, and computerese.

However, even more important than coming to grips with new technologies, is the need to develop philosophies and strategies that truly address the financial, human and institutional barriers that stand in the way of modernization.

The development of well-defined, long-range goals and a total commitment from upper management to pursue those goals must be the number one priority. Since manufacturing is going through so many profound changes, those firms unwilling or unprepared to move forward will not survive.

"Manufacturing will change more in the next 15 years than it has in the last 75 years," forecasted William H. Slautterback, director of manufacturing systems, Engineered Metal Products Group, Koppers Company, Inc., Baltimore, and a participant in the White House Computer Conference on Productivity. "The reasons are clear: survival and technology."

Groundwork

Dramatic change, if it is to be managed effectively, must be planned. The groundwork for Factory 2000 must be laid now. However, preparations for the factory of the future cannot be made in the dark. Some light must be thrown on what this factory is to look like, before construction plans can be made. Creating a vision of the future factory is critical. With a clear picture of where we are going, goals can be set and the steps needed to reach the goals can be formulated. Needless to say, a vision itself means little without a strong commitment by management to ensure that that vision eventually is transformed into reality.

But how to transform these visions into reality has been a concern of man from time immemorial.

A master potter bends over his wheel. The wheel spins. The clay becomes a blur beneath his hands. As the potter works, his tools, his hands, his clay, cease their separate existence. They have become but extensions of his imagination.

Because the potter's tools become extensions of himself, he is able to instantly turn his thoughts into creations. Thus, each of his pots is unique and of top quality. His extreme flexibility, his one-of-a-kind products, and his ability to instantly integrate materials, production, and management make him the antithesis of the modern mass-production-oriented, departmentally-splintered factory.

Yet Factory 2000 will in concept resemble more closely the potter than today's factory. For, say experts, in a factory at the turn of the century, the distinctions between activities will become blurred. Integration and flexibility will be pervasive, and an economic lot size of *one* will be approached.

To illustrate the relationship between present and future another way. In rural Japan, a Zen archer slowly draws back his bow, and for a timeless instant: the bowman, his bow, his ar-

Source: Iron Age, June 1984, 27-29, 31, 33-36, 39-40, 45-47, 49, 51, 53-56, 58

row, and the target are one. Then, time resumes its course: the arrow hurries toward the target's center.

The Zen archer hits the bull's-eye when he becomes one with his bow and arrow and the target. Yet in today's factory, not only can't the fragmented departments be sufficiently coordinated to draw the bow properly, close teamwork would be a valueless because the target is either ill-defined or impossible to see. Companies with ill-defined or undefined targets waste much time spraying arrows skyward, hoping that one will someday strike something valuable.

Both stories illustrate dangers that can befall the manufacturer when he enlarges his operation beyond one man. Certain economies of scale are gained through growth, but in the past there has always been a cost: inflexibility and diminishing communications and control, for example. Fragmentation sets in as departments pile upon departments.

"By breaking factories into these units, we erect boxes and walls around functions which then creates enormous political problems," explained Mr. Gunn.

Splintering

Today, in large and not-so-large manufacturing companies this splintering of departments and functions has been carried to extremes. Horizontal Towers of Babel have sprung up which make communications among the design department, the manufacturing department, the finance department, the data processing department, the marketing department, and the front office, very difficult.

A engineer, who had spent much of her life in the manufacturing arena, recently was hired by an accounting firm

Roadblocks to Manufacturing Automation Efforts

Category	Percentage
Lack of Design & Engineering Staff Knowledge (& Time)	73
Lack of Inside or Outside Programming Capabilities	56
Lack of Information on Equipment & Software Capabilities	55
Lack of Machine-to-Machine Interface Capabilities	53
Lack of Central Computer or MIS Staff Knowledge	48
Lack of Mechanical Reliability	45
Lack of Suppliers to Take Entire "System" Responsibility	44
Lack of Communications Format	44
Lack of Electrical/Electronic Reliability	43
Lack of Operator or Foreman Acceptance	38
Lack of Equipment	35
Lack of Higher Management Interest	33
Lack of Outside Consultants to Assist the Effort	27
Lack of Organized Labor Support	26

Figures were obtained by a National Electrical Manufacturers Assoc. (NEMA) survey of member companies.

The percentages relate to the number of firms that indicated a particular category was "very much" or "somewhat" of a problem.

The Booming Factory Automation Marketplace

ITEM	1967	1982	1987	1995	67/82	82/87	87/95
GNP (bil $)	799.6	3059.3	4889	9755	9.4	9.8	9.0
GNP Deflator (1972=100)	79.1	207.1	274.7	418.7	6.6	5.8	5.4
GNP (bil 72$)	1011.4	1476.9	1779.5	2330	2.6	3.8	3.4
Mfg New P&E (bil $)	32.25	119.68	205	419	9.1	11.4	9.3
Mfg FRB Index	100	137.6	186	239	2.2	6.2	3.2
factory auto/000$ GNP	5.59	2.40	3.04	3.83	—	—	—
factory auto/000$ mfg P&E	138.7	61.3	72.6	89.3	—	—	—
factory auto/FRB index pt	44.7	53.3	80.0	156.4	—	—	—
Factory Automation Sales (mil$)	4472	7337	14880	37400	3.4	15.2	12.2
Mfg Computers	2603	2280	3250	5515	−0.9	7.3	6.8
Mainframes	2541	1720	2260	3275	−2.6	5.6	4.7
Minicomputers	62	560	990	2240	15.8	12.1	10.7
CAD/CAM Systems	—	930	2800	9800	—	24.7	17.0
Machine Tools & Controls	1869	3937	7930	18085	5.1	15.0	10.9
Machine Tools	1869	3724	7100	15000	4.7	13.8	9.8
P.C.	—	213	830	3085	—	31.3	17.8
Robots	neg	190	900	4000	—	36.5	20.5
Coating	—	48	180	680	—	30.3	18.1
Welding	—	52	210	800	—	32.2	18.2
Material Handling	—	40	255	880	—	44.8	16.7
Processing	—	50	255	1640	—	38.5	26.2

Source: Predicasts

which wanted some manufacturing expertise to upgrade its management support capabilities. She emphasized that the differences between the business accountant and the manufacturing engineer went well beyond simple ignorance of the other's profession. Because each spoke a very different language and thought about production and management in very different ways, it was very difficult to communicate, even at very basic levels. Industry appears to have spawned a number of divergent cultures.

The roots of this compartment approach to manufacturing are buried in history. But the persistence of industry to continue to do things the old-fashioned way can be traced to today's outdated management techniques and mediocre management — management that lacks vision and commitment.

Divide & conquer

"Traditionally, our approach to manufacturing processes has been much like the classical industrial engineering approach to problem solving. Divide the problem into small pieces and develop the best solution for each one," explained James M. Apple, Jr., senior vice president, SysteCon, Inc., Duluth, Ga. "In doing so, we presume that optimizing each element will result in total optimization of the manufacturing plant. This approach led us to solutions that represented islands of automation."

The major rationalization for the use of this classic approach came from its ability to reduce direct labor.

"However, these systems were as inflexible as they were productive. And, they are unable to adapt readily to changes in processes and market demand," Mr. Apple said. "In order to keep efficiency measures high, we made long production runs and created large inventory within the manufacturing cycle."

To react quickly to new overseas competition and a rapidly changing marketplace, firms are finding that they have to be fast on their feet. But the weight of too much inventory, too many inefficiencies, and calcified operations has begun to take its toll. Most CEOs today have financial, legal or marketing backgrounds and know too

little about manufacturing and are of little use in formulating comprehensive rejuvenation strategies.

Executives are just beginning to realize that something drastic must be done, that new visions must be dreamed, that new goals must be set, that new commitments must be made.

However, before these new visions can become a reality, before Factory 2000 can be realized, four major obstacles must be overcome. The major barriers to Factory 2000, according to Mr. Holland, are financial, human, institutional, and technological ones.

Obviously, the financial obstruction is the first roadblock that must be dealt with, for without funding, nothing gets built.

FINANCIAL BARRIERS

Just as executives must be open to new ways of looking at manufacturing, they must also be open to ways to financially justify investment in the factory of the future. For instance, the purchase of an entire flexible manufacturing system (FMS) should not be justified in the same way that an individual machine tool is justified. An FMS system probably will have a significant strategic impact as well as an impact on productivity. Somehow that strategic impact must be evaluated.

"My experience has convinced me that this financial issue is causing some real problems for executives," said Frank T. Curtin, general manager, Industrial Automation Systems Dept., General Electric Co., at the March CIMCOM conference in Washington D.C. "These new technologies have a huge potential to increase productivity, but that potential is being roadblocked in many cases by the misuse of financial analysis tools. To be blunt, we are not getting off the dime because we are not sure how to calculate and measure what the return will be on investing that dime in flexible automation."

Shocking data

According to a recent Boston University School of Management survey, 78 pct of the respondents felt that, "Most businesses in the U.S. will remain so tied to traditional, quantitative investment criteria that they will be unable to realistically evaluate the potential value of computer-aided manufac-

On a global scale and by historical standard, we now find ourselves in the midst of a technological revolution on the order of the discovery of fire, development of the water wheel, the printing press and nuclear power. This revolution will for all time change our way of life like no other change before.

Corporation control will again be in the hands of manufacturing experts

1920-1950	Manufacturing
1950-1970	Marketing
1970-1985	Finance
1985-20xx	Manufacturing

Source: W. H. Slautterback, Koppers

turing options."

Using traditional investment formulas, management is confused about how to evaluate revolutionary computerized technologies. So, they are not buying them. But management is not buying systems for other reasons as well. Since most executives today come from financial, marketing, or legal backgrounds they often simply do not understand the impact of new technology. They know little about design and manufacturing and even less about computers. On top of that, they don't know who to turn to, to teach them what they ought to know.

"A key reason for the lack of top management action is these managers really don't know who to believe any more," said Mr. Gunn. "They seldom have complete trust in their own people because they are well aware of the vested interests within their own organization. They don't know which vendor to believe because the vendors have an inherent conflict of interest—especially since each vendor sees a different 'solution' to problems. Even consultants can be suspect because they may have a parochial or product-oriented point of view about something."

Buffeted by confusion, misinformation, and technological jargon, top management understandably has not thrown itself wholeheartedly into purchasing advanced equipment. Then, when companies have bought equipment, it has been for the wrong reason. Not knowing how they are going to use the equipment, (a robot for instance) but not wanting to see the technology pass them by, top management on a gut reaction, splurges on one piece of equipment. Sometimes equipment purchased in this way can have value as a learning tool. However, blindly purchasing advanced technology and then

expecting it to benefit the operations is extremely naive.

If one word represents the fundamental direction of computer technology it is 'integration.' The goal of integration is to get the diverse elements in manufacturing working in harmony. The primarily tool that will bring this integration is computer-integrated manufacturing (CIM), leading eventually to the computer-integrated business, the factory of the future. Buying pieces of CIM without seeing the complete picture is just asking for trouble.

However, "Rarely do CEOs think about and plan for this kind of factory automation from a sufficiently big picture or a sufficiently high level," explained Mr. Gunn. "They use their traditional capital budgeting techniques. The result is that they have not been able to justify investing in that risky long-term factory automation program.

"They really don't have the experience to deal with factoring benefits other than direct labor reduction or increased capacity into their justification calculations. So, what often happens is that even when they do look at a total system, they think, they will go in and cherry-pick the one or two most attractive parts of that system and implement those. The other parts, not as attractive, often don't exceed whatever criteria the executive is using to justify spending. As a result, the other parts of this automation package never get implemented and the firm never gains the needed synergy available from the system as a whole."

Cost/benefit formulas

What too few executives realize is that flexible automation systems radically alter the economics of automation. "What is urgently needed are new cost/benefit formulas and measurements," emphasized Mr. Curtin of G.E.

"new manufacturing economics, if you will, that go beyond the usual return-on-investment (ROI) evaluations to take into account the total impact of automation on the business. Because these technologies don't just change the factory floor. They change the entire business in new and important ways."

Those new and important ways, however, are hard to quantify. Most firms have little problem analyzing the pros and cons or automating a particular operation or machine tool. When a mixture of systems are involved, however, these firms begin to flounder. "...it is suggested that the real return from large scale automation will be incurred when it is understood that what we should be concentrating our automation efforts on is the inter-task activities which are normally ignored in terms of cost justification analysis," explained Mr. Holland. "Inter-task automation addresses the question of effectiveness as opposed to efficiency."

All the insular departments in a factory can be very effective when looked at individually but very inefficient when viewed as part of the much larger factory system. Too often integrated system solutions (where the whole is greater than the sum of its parts) are ignored because firms find they have no way of accurately determining whether the system can be economically justified. But benefits of integration are very real. Witness the separate inefficiencies but overall effectiveness of Japanese just-in-time production: In Japan, unless the work area directly ahead of a particular employee's workstation needs a part, that employee reads a book, twiddles his thumbs, or does something else with this time. His work station stays idle. The workstation, in isolation, seems inefficient. When looked at in the larger context of providing a part exactly when it is needed, the workstation becomes a very effec-

FACTORY 2000

tive force in reducing work-in-process inventory.

A recent Arthur D. Little report told of a Japanese manufacturer who decided to install a flexible manufacturing system. This system enabled the manufacturer to cut back the number of his machines from 68 to 18, the number of his employees from 215 to 12, and space requirements from 103,000 sq ft to 30,000 sq ft. Process time was dramatically cut from 35 days to a day-and-a-half. But the financial return over the first two years as only $6.9 million on an investment of $18 million. Now, despite the numerous strategic benefits such a plant would give a manufacturer, one the basis of conventional accounting principles, states the report, this low return on investment would be difficult if not impossible to justify.

"The point is," pointed out Mr. Curtin of G.E. "Many manufacturers continue to make the mistake of looking at flexible manufacturing system investments as tactical decisions with only short-term financial implications—when in fact they are strategic decisions with far-reaching competitive implications.

Long-term savings

"Automation provides permanent long-term savings and a host of intangible benefits...such as higher quality, shorter production runs, increased worker satisfaction, faster responses to

The Shrinking Percentage of the U.S. Manufacturing Workforce

| 1947 | 1970 | 1978 | 2000 | 2000 |
| 30.0 pct | 24.6 pct | 21.7 pct | 5.0 pct | 2.0 pct |

(Rand Corp.) (SRI/Cyert)

Source: M.E. Merchant, Cincinnati Milacron

Source: Iron Age, June 1984, 27-29, 31, 33-36, 39-40, 45-47, 49, 51, 53-56, 58

changes in market demand, quicker throughput, greater flexibility and on and on."

Because manufacturers continue to focus on the efficiencies of the individual work area and ignore the big picture, major cost cutting efforts are directed at replacing a man with a machine tool or a robot. These efforts are misdirected. While direct labor accounted for a majority of product costs in the past, it does not today.

"Direct labor accounts for an average of only 10 pct of manufacturing costs, while material costs are 55 pct, indirect labor is 15 pct and overhead costs total 20 pct." said Mr. Curtin. "The big hits lie in the bigger pieces of the pie: material, inventory and overhead costs."

Once a firm decides it can financially justify an integrated system, it must next turn its attention to the people who will be manning this system. The company is now confronting the two biggest obstacles on the road to Factory 2000: the human and institutional barriers to automation.

HUMAN AND INSTITUTIONAL BARRIERS

Once the financial hurdle is negotiated, two even larger barriers in the path of Factory 2000 must be surmounted. These barriers: the human and institutional barriers are so interrelated that any discussion must address both problems together. It is humans, after all, that comprise the corporate or institutional culture.

"The lack of implementation of large scale automation is often related to human factors.... It includes a resistance to change. It involves unionism. It involves people who are 'obsolete.' " said Mr. Holland.

As well as the resistance to change, a factory culture exists that is even harder to alter. Traditionally, jobs have been done a certain way, and traditional habits are hard to break. Departments usually have operated in a particular way for so long that they have substantial inertia that must be reversed if change is to occur. But if the new vision of Factory 2000 is to be embraced, if new philosophies are to be adopted, and new strategies carried out, and if integration is to be more

than a buzzword, people and departments have to change.

Communications

A major task, before operations can be integrated, is just getting departments to talk to each other.

Management information services (MIS) people have to leave their ivory tower and learn more about development and production. And as computer-aided design and computer-aided manufacturing (CAD/CAM) grows more popular, designers and manufacturing engineers will have to do more talking.

"Clearly there is a need for closer ties between our development organizations and our manufacturing organizations," stated Jim McDonald, general manager, Manufacturing Systems Prducts, IBM, "It is critical that these groups work hand-in-hand to put products out." A large gap between the two remains today. "Analyzing from a historical perspective, twenty years ago I think people came out of school knowing a lot more about how to manufacture the part that they designed than they do today. The designer needs to know how to design products that can be easily manufactured. There are tremendous training problems that need to be addressed to do that. In addition, one must continue to develop the skills of the manufacturing group so that it can provide input to the developer."

By getting the designer and manufacturer closer together, the chance of the developer designing a part that is difficult or impossible to produce is reduced. Efforts to move the two departments closer together are slow in achieving results because of the difficulty of converting the geometric information of the designer into useful information for the manufacturer as well as transforming manufacturing data into information useful to the developer. These efforts also are hindered by a corporate culture that tends to keep the to professions apart.

"There is no way that computer-integrated manufacturing will become a reality unless the people side of the scale is addressed," emphasized Dan Ciampa, executive vice president, Rath & Strong, Inc., management consultants. "People and not technology will

Traditionally our approach to manufacturing has been much like the classical industrial approach to problem solving. Divide the problem into small pieces and develop the best solution for each one. This approach led us to solutions that represented islands of automation.

My experience has convinced me that this financial issue is causing some real problems for executives. These new technologies have a huge potential to increase productivity, but that potential is being roadblocked in many cases by the misuse of financial analysis tools.

realize the marvelous potential of CIM, but unless we become smarter about managing people, CIM will go down in history as nothing more than a good idea whose time never came."

As mentioned, technological developments have advanced so rapidly that many of the pieces for Factory 2000 are already available. "But just having a lot of sophisticated, state-of-the-art options is not going to ensure the effectiveness or efficiency we need to achieve higher quality," continued Mr. Ciampa. "The only way all of these elements can produce more effectively and efficiently is if they are integrated. Machines, computers, and software cannot be integrated unless the various departments that control and use these resources are integrated. This task is more formidable than has been the task of developing computers, software, and machinery of automation. This task must be accomplished in spite of the management philosophy and climate within most firms in this country.

"The traditional manufacturing climate discourages integration, rather than encourages it. It works against a common vision by treating people at different organizational levels in different ways. It blocks unity of purpose by encouraging outdated management philosophies and approaches."

As companies move toward Factory 2000, major changes have to occur in the way people do things. Mr. Ciampa has suggested several that will occur (see table p.47).

—As manufacturing systems become more flexible, permitting them to react quickly to changing market conditions, people will have to adjust by becoming more flexible themselves. As job content changes and broadens, people will have to acquire proficiency in several skills.

—Also, "The need for better interpersonal communication will increase as we move toward CIM," said Mr. Ciampa. "This will bring the opportunity for more face-to-face contact between people and for better communication." In addition, organizational structures will have to be changed to facilitate this need for increased communication.

—Roles and relationships will

change requiring people occupying certain jobs to behave differently and to perform different tasks in those same jobs. "One opportunity of this change is to enrich and give more meaning and more challenge to jobs in the workplace. Another is to better control the decision-making process by clarifying roles and limits of authority."

—Another trend that will take place is that decision-making authority will be pushed to lower levels as equipment and systems that do more are made available on the shop floor. Being able to make more decisions and exert more control should bring greater job satisfaction to the factory floor; however, when control is moved to lower levels it must be taken from managers in higher levels. These managers could well resent this erosion of their control.

—Finally, "More involvement of the users of new equipment and systems will be necessary in determining what will be designed, and how it is to be used. The opportunity here is that new systems will better meet the needs of the people who are to use them and those people will be more committed to making them succeed," concluded Mr. Ciampa. A new system is of little value if it is not operated effectively.

Even if not planned for, the very presence of computer-integrated equipment on the shop floor will have a large effect on the way people do their jobs which will, in turn, affect the way they interact with others.

Operators
"As automation finds its way into the factory, an obvious consequence is that the operator is removed from the proximity of the machine or process. Even if the operators of machinery are not completely eliminated, they will, in general, not be as close to the machine as in the un-automated factory. As the operator backs away from the machine, the nature of his role will change from one of participating directly in the control loops at the machine level to one of supervisor and attendant," explained Mr. Holland of Allen-Bradley.

Eventually, as Factory 2000 takes shape, the operator will be entirely disengaged from the active control process. Companies that do not understand or accommodate the cultural implications of this process will eventually pay for their oversight.

With these upcoming changes in the roles and relationships of company members, it is important that top management address the potential people problems with as much concern as the technical problems. As nearly every consultant and automations expert will attest to: organizational change must begin at the top. Top executives must not only lead the way in implementing Factory 2000, they must jump into the automation effort with both feet, know what is going on, and understand the big picture. "...the CEO must describe how the company can be under CIM and in so doing paint an image of the future that is exciting and compelling. The job of creating a vision cannot be delegated," said Mr. Ciampa.

Because of the human element involved, CIM must be a step-by-step process. A manufacturer, with enough money, can plunk down $15 million for a full-blown flexible manufacturing system (FMS). But no matter what he spends, he cannot instantly enlighten the workforce. Changing a corporate culture takes time. It can take a long time: five years or more. Automation efforts which attempt to solve productivity problems in one fell swoop, which refuse to recognize the the human aspect of automation, are doomed to failure. Despite the revolutionary computer age we now live in, the implementation of computer-integrated manufacturing must be a gradual evolutionary process.

As the year 2000 approaches, the hierarchy in a manufacturing operation will be much different than it is today. With each employee assuming more control of his/her operations, fewer layers of management will be needed. In addition, the relationship between the various layers will be much less formal than today. This informality will grow as the management content of employee jobs increases and the envolvement of top management in manufacturing grows.

"In addition to various departments being integrated in this effort, there needs to be a new partnership between salaried and hourly workers," explained Mr. Ciampa. "We all recognize that CIM means enormous changes on the shop floor—and those changes cannot be successful unless the direct labor workforce is committed to success. They must feel involved and that means that they must participate in determining how CIM will take shape in their company. In order for this to happen, the traditional adversary relationship between management and labor must change. It is dysfunctional and wasteful."

Mr. Ciampa suggested that for labor to be more cooperative it needs a bigger piece of the action, "....emotionally through influencing outcomes and financially through plans such as gainsharing." Before Factory 2000 can become a reality, both management and labor are going to have to reassess their positions.

Education

If the corporate cultures are going to change and employees are going to be assigned new jobs, education will be playing an increasingly important role in tomorrow's factory.

Today there is a need to see beyond barriers, to understand the interconnecting of various disciplines. People like systems engineers will be in increasing demand. Today's systems encompass a wide variety of technologies such as electronics, mechanics, hydraulics, computers and computer programming, and laser optics to name a few.

But do the generalists who are crosstrained in many disciplines exist today?

"If I was asked that question two of three years ago if schools teaching this kind of education existed, I would have said no," responded Jim McDonald of IBM. "However a number of schools are starting to put together some excellent programs in the area of manufacturing systems engineering and many schools that do not yet have the programs are interested in starting them."

Not only will technical people with new skills be needed, people with new human relations skills will be called upon to interact with the employee of the future. Today's personnel departments are a far cry from the human resources departments soon to be needed.

"One of the biggest challenges will be to find and to develop managers and

There is no way that computer integrated manufacturing will become a reality unless the people side of the scale is addressed. Unless we become smarter about managing people, CIM will go down in history as nothing more than a good idea whose time never came.

> **Regarding standards it is time for all of us to work harder than ever to put aside short-term tactical advantages and proceed to develop the communications, electrical interfaces to allow the user to buy American. We cannot become the standard for excellence if we are non-standard.**

supervisors who can make the most of the technology that is being made available. This will require defining the areas of competence needed of managers and supervisors in the era of CIM. Further they will be responsible for developing training programs geared to raising people's knowledge, skills, and behavior so that they can perform at that level," said Mr. Ciampa. "We should look to a new breed of human resources professional to help."

The final barrier to automation that has to be addressed is a technological logical one. However, the problem is not primarily a lack of necessary technology but the inability to integrate, to connect, this technology together.

TECHNOLOGICAL BARRIERS

Technology shapes the building blocks of Factory 2000. The specific building blocks will be discussed later.

When taken as individual building blocks or parts to the factory automation puzzle, many of the pieces of that puzzle have already been developed. The main problem is not that there are not enough pieces but that the pieces don't yet fit together.

It is not the case of just a minor adjustment here and there. It is very difficult to get any two pieces of factory equipment to talk together unless they have been made by the same vendor.

What is needed are standard ways of linking equipment so they can communicate. A manufacturer ought to be able to easily attach a programmable controller with a minicomputer and a robot and have them talk back and forth. But right now this task is next to impossible.

"Regarding standards, it is time for all of us to work harder than ever to put aside short-term tactical advantages and proceed to develop the communications, electrical interface, and software interfaces to allow the user to buy American." said Mr. Holland. "We cannot become the standard for excellence if we are non-standard."

General Motors

The Company that is leading the way in standards development is General Motors. The firm is working with control manufacturers and national and international standards committees to produce a set of standards that suppliers must adhere to if they are to sell to GM. The hope is that once GM defines and adopts these standards, other firms will begin to accept the same standards for their companies.

"Productivity improvements within General Motors manufacturing generally began with the automated control of single processes such as machines, conveyors and product test stands," said Ron Floyd, who is with the advanced product and manufacturing engineering staff at the General Motors Technical Center. Over time this automated control has expanded to a functional orientation, with a control or supervisory system encompassing an entire function such as material handling or an engine head line."

Presently, GM uses approximately 20,000 programmable devices within manufacturing with projected quantities to reach 40,000 in two years. The benefits obtained from increased manufacturing automation associated with these devices is substantial.

"A major step remains to be taken however: the move from automated functions and processes to the computer-integrated 'Factory of the Future,'" said Mr. Floyd. "The factory of the future concept requires integration of separate automated processes into a sophisticated, cohesive whole. Timely information must be available for exchange between automation centers and presentation to management. This information will enable timely and accurate decisions—implemented online, for optimized manufacturing. Data communication among automated processes in some cases is not practically possible. In other cases, skilled data processing professionals are required to implement custom interconnections."

The solution to the data communications incompatibility problem at GM is establishment of standards. GM would like to set up communications networks and be able to attach equipment on that net. Today, however, vendors cannot use a common network because each vendor supplies a unique protocol which either cannot be emulated by other manufacturers or can be only at great expense.

"The lack of compatible communi-

Source: Iron Age, June 1984, 27-29, 31, 33-36, 39-40, 45-47, 49, 51, 53-56, 58

cations hardware and software results from the lack of viable networking standards," explained Mr. Floyd. "To date, computer manufacturers have generally invented their own communications protocols, many of which are proprietary. Within a facility where several computer manufacturers are used, the result is incompatibility. The need for the establishment of viable standards is here today within manufacturing. Just as important, we need wide vendor support of such standards with cost effective and compatible data communication products."

Although they still must be linked together, a lot of advanced technology is already on the market, ready to be incorporated into tomorrow's automated factory. A look at what some of this is and how it will develop over the next decade-and-a-half is is order.

PIECES OF THE PUZZLE

Moving toward Factory 2000 is not just surmounting barriers it is also assembling elements. Many of the parts for that factory, be they strategic equipment, software packages or pieces of hardware, are available today. As important as these elements are, they must be integrated into systems before their potential truly can be realized.

The goal is the creation of computer-integrated manufacturing or CIM. CIM provides the blueprint which designates how the diverse parts of the factory of the future will be linked together. "...CIM is a corporate strategy—nothing more, nothing less," explained Charles M. Stata, Manager, New Venture Technology, Wang Laboratories Inc., Lowell, Mass. at the 1984 CIMCOM Conference. "...CIM is not a new software system. It is not hardware. It is not a product which you can go out and buy off the shelf."

While hardware and software are essential to implement the technology, CIM is a way of doing business, not a pile of equipment. "CIM starts with a mindset," he continued, "which dominates the selection of equipment, the choice of software approaches, and most importantly, how we think of our business' growth."

Defined by Mr. Stata, "CIM is the methodology for automating the gathering and sharing of information among computer systems to establish a closed loop in-time feedback system for effective planning and control."

By accepting CIM as a legitimate strategy, manufacturers are beginning to recognize that the chief business of manufacturing is the gathering and manipulation of information. In today's factory much more time and effort is spent collecting and analyzing geometric and alphanumeric data (letters and numbers) than ever is spent in actual metalworking.

CIM recognizes that the success of tomorrow's factory depends on its ability to gather and share data. If data is so important, it becomes critical that a corporation "adopt a set of rules which will govern the time frame, data forms and interconnections used in the collection, storage and sharing of data among the authorized applications and users," said Mr. Stata. "Data, then, becomes a controlled corporate resource rather than an application dominated entity."

If information is such a valuable commodity, the ability to transfer the correct information to whoever needs it, becomes an important task. Building a comprehensive pipeline that carries data to all parts of the company (and even outside the company to vendors and suppliers) is a major goal for CIM users. A company's information architecture describes the shape of this data pipeline. But improving a firm's data architecture is half the problem, for once the data gets to where it is supposed to be going, it must arrive in some form that can be used and understood. The information also should arrive when it is needed. All these problems regarding where, when and how data is to be sent are being addressed by the CIM concept.

"Now, if you don't consider the data architecture as most factories don't today, you will buy a component or system that is justified on its own merits and will paint yourself into a concrete

CIM Will Bring Major Changes to Employees

Change	Opportunities	Dangers
Job Content Change	More Growth Opportunities	More Unemployment Worker Discontent
Better Interpersonal Communication	More Face-To-Face Contact Better Working Relationships	Worse Working Relationships
Roles And Relationships Change	Enrich Jobs and Offer More Challenge and Satisfaction Better Control Decisions	More Decisions Not Made More Confusion People Not Able to Adapt
Decision-Making Authority Pushed Down	Greater Motivation Faster Decisions	Threatening Vertical Checks and Balances
More User Involvement	Better Meet Needs More Commitment	Longer Development Time Complicate the Process

Source: Dan Ciampa, Rath & Strong, Inc.

corner when you want to automate up to the next level," cautioned Malcolm L. MacFarlane, vice president, Gartner Group Inc. "At this next level where you have to attach two systems together, you will find those systems mutually incompatible."

Developing a unified architecture that that takes into account how data is structured in a particular company is an important aspect of CIM. Also important is knowing what form the data is in as it travels through the pipeline.

"The data which is collected, stored,

The factory of tomorrow begins with an exciting collective vision.

Source: Iron Age, June 1984, 27-29, 31, 33-36, 39-40, 45-47, 49, 51, 53-56, 58

computed and reported resides in the computer in a form which most of us have little knowledge of," explained Mr. Stata. "Yet I submit that this is the single largest element which needs close control. Traditionally, we, the users, have not cared how the programmers coded the part number. With an integrated environment, you need to care because you want to share between previously isolated systems. Across your business a part number should be represented commonly so that all of your systems will recognize the part number the same."

According to Mr. Stata, CIM has three identifiable characteristics. The first is automated information gathering, which means information is collected by a computer not a piece of paper. The second characteristic is net working, which allows the transfer of data between systems, and the last characteristic is the shared data base.

A data base is simply the systematic organization of data files for easy access, retrieval and update. Information within a data base is not tied to a particular program but can be accessed when needed by all types specific application programs.

CIM is made up of a number of subsystems and and technologies. Some of the most important include computer-aided design and computer-aided manufacturing systems (CAD/CAM). Computer-aided process planning (CAPP), computer-aided quality inspection, robots, numerically controlled machine tools (CNC/DNC), automated materials handling, manufacturing resource planning (MRP), group technology, laser systems, flexible manufacturing systems (FMS), and manufacturing cells. In the future, artificial intelligence (AI) will be an important part of CIM.

Key ingredient

The one indispensable element in CIM is the computer itself and its accompanying software. "Clearly, computers will become smaller, more powerful, less expensive, and will therefore see increasing use. In tomorrow's factory, standards will create compatibility between computer brands as well as with other intelligent technologies. In fact, the computer will be a tool, just as

a hammer is to a carpenter; the brand name will be insignificant as long as the tool is functional and reliable." predicted Mr. Slautterback of Koppers.

Computer keyboards will probably be replaced to a large extent by touch screens and audio commands. The use of voice commands for a computer, however, present problems. It would be difficult for a manager on the shop floor to yell a command to one computer, without having another respond to that command, for voice cannot be directed to only one particular spot.

In CIM the hardware is only as good as the software that runs it and software continues to improve the computer's capabilities.

"If you look at software as having reach—that is, having the ability to reach in front of itself and in back of itself and understand what is going on, new developments are occurring that are designed to give software that extra reach," explained Ray L. Dicasali, vice president, Manufacturing Systems Div., Management Science America, Inc. (MSA). "One of the main complaints of manufacturers is that they do not get good forecasts from marketing. As a result, the factory sets up its schedules and then must change them at the last minute. We want our manufacturing resource planning (MRP) system to reach out to a customer's order management system."

By extending the reach of software beyond the loading dock and into the marketplace, manufacturing people now have an early warning system, which warns them much sooner that changes are coming.

Software is reaching into other areas as well. "Another complaint from manufacturing people is that engineering changes come through at inopportune times. Quite often engineering changes are coming through either because there is an improvement that is needed, or a part that used to be available isn't. You have to make substitutions, but these last minute production changes hurt productivity," explained Mr. Dicasali. "We think that the reach of the manufacturing materials system should be back into the engineering data base to provide early warning of an engineering change."

A third area that software is reaching into is the financial and cash management areas.

Functional integration is another direction of software development. Briefly, this means that software applications are not only integrated physically but it is also integrated logically. This means that a manufacturing software system, for example, would know its particular role and how it effected other systems down the line.

Software advances

Some of the trends in software development, according to Kenneth A. Fox, vice president, research and development, ASK Computer Systems Inc., Los Altos, Calif., include the growing use of modular software which allow manufacturers to build toward the factory of the future step by step.
Modules from ASK include MRP, shop floor control, inventory, work-in-process, purchasing, capacity and resource planning and has also added accounting packages, field service, payroll, and human resources. Linked together the modules form what is called an interactive, real time system.

"I think a major trend is the move toward standard software products (in the past, big software users developed much of their software internally). I think that this standardization must occur before these computer-integrated manufacturing technologies can be cost-effectively purchased by the average manufacturer," said Mr. Fox.

In the future most companies will use purchased software packages, but where unique or proprietary software will be needed, natural languages will exist that will allow an individual with no knowledge of programming languages to create his own software.

"Also," said Mr. Fox, "I see the need for systems like ours to be connected to many more diverse kinds of equipment for improved data collection. Today input really comes mostly from user interaction with a terminal, but there are lots of more specialized types of data collection equipment: bar code readers, for example."

Besides computer hardware and software, the next essential element of Factory 2000 is a flexible manufacturing system. FMS, which has been defined by Mr. Slautterback as the orderly, systematic arrangement of equipment (primarily machine tools, robots, measuring systems, and material handlers) to perform manufacturing operations on a group of similar parts, under computer control, without human intervention.

While there are about 30 true FMS systems in the U.S. at present, Mr. Slatterback predicted that there will be over 10,000 by the year 2000. "The logical progression from NC to CNC, to FMS, to the Unmanned Factory is in process," he stated.

The great value of FMS is that it is flexible and automated. A system can handle one type of part or a random group of related parts with equal ease. It can also run unattended for long periods of time.

The main components of an FMS are machine tools. A major problem with machine tools today and one that hinders their efficient usage in FMS systems is tool ware and breakage. In an automated system, no one is around to change a broken tool. As time goes on however, "The concern about tool wear and breakage will be reduced by using such techniques as energy beams and electrical discharge," predicted Mr. Slatterback. "Where this is not practical, exotic materials superior to ceramics and carbides will be developed. Predictive maintenance software and tool failure sensors are already being applied. Machine tool manufacturers will include coordinate measuring as a part of the normal functions of the tool; that, together with photo imaging, vision and other sensors will perform the inspection tasks."

As machine tools and robots become more sophisticated and more flexible, the material handling systems that link these systems must also become more intelligent and more versatile. The full potential of tomorrow's manufacturing systems only will be realized if they are continuously kept busy. Parts and materials must flow in an uninterrupted stream between machines, morning, noon, and night.

Islands of automation

Material handling systems will be crucial to the maintenance of this flow. Just the right amount of items must be brought to just the right place at just

Robots are a critical element of flexible manufacturing systems and cells. These robots perform operations such as material handling and changing the tooling for one or more machine tools. In the future with the assistance of accessories like the laser, robots will become increasingly involved in metalworking.

A manufacturer searching for better ways to to produce parts has to be very careful that he does not turn too quickly to technology for the quick fix he needs to pep up his sagging productivity. When re-evaluating company philosophy and startegies, the consideration of specific technologies should occur only very late in the process.

the right time. In the future, material handling will provide the critical physical linkage that integrates the islands of automation in the future factory.

"I think that the challenge for material handling is to able to interface with the islands of automation that are growing up in the factory," said Walter P. Adams, director of automated systems and controls, Portec, Oak Brook, Ill. "Today, the large changes in material handling are occurring primarily in software." Software is being developed that interfaces with material handling equipment and other machinery on the shop floor.

According to Mr. Adams, as late as the late 1970s, very few guided vehicle systems had any sort of computer control other than the controls directly on board the vehicle. Today, however, the majority of these systems have a central direct computer control.

As to where material handling is going: "We will become much more involved in system integration. In fact we will be the systems integrators, the ones who really help put the whole thing together," explained Mr. Adams."

The major challenge facing the material handling industry is finding ways to interface handling equipment with the great divergence of hardware and software that can be found in most factories. Each robot or machine tool must be catered to a little differently, since no uniformity in equipment exists. Most machine tools, for example, have different load/unload mechanisms and come in different sizes, so most have to be interfaced differently.

Robots

A more localized type of material handler is the robot, although by the year 2000 the robot and the guided material handling vehicle will be one and the same. Eventually robots will be mobile, with the capability of being programmed to go anywhere in the factory. Although today's robots are not very bright and perform simple, repetitive tasks such as spray painting, material handling, and welding, tomorrow's robots, with more intelligence and advanced sensing capabilities, will be able to perform a wide variety of tasks.

According to an executive summary of robots prepared by the Yankee

Group, Boston, there are certain dominant trends in robotics today. For one thing, while robots are becoming less expensive, they also are also developing a much greater range of capabilities. For another, robots are beginning to be looked at as advanced computer peripherals (like a printer, for example) and not as machine tools. In the future, states the report, "Robots will become more goal-directed; data will be input in plain English and converted to control procedures or subroutines. Also, human interfaces for programming will be developed for non-explicit commands."

Robots are a critical component of flexible manufacturing systems and cells. These robots perform operations such as material handling and changing the tooling for one or more machine tools. In the future, with the assistance of accessories like the laser, robots will become more involved in the metalworking and metal cutting operations.

With or without robot controls, the laser will become increasingly popular in the factory. For although it can be expensive to operate, its versatility and its accuracies make it an ideal tool to be found in an advanced manufacturing cell or system. Lasers can not only cut, weld, heat treat, and clad various materials, they can also help count, inspect, and measure.

A lot of work is being carried on today in attempting to incorporate lasers into manufacturing systems.

"One of the things we are trying to get together for demonstration is a flexible manufacturing cell. And this cell would take two large-diameter pipes, and would attempt to join those pipes at a funny angle," explained Don Bennett, president, Coherent General, Palo Alto, Calif. The company is a two-month-old joint venture between General Electric and Coherent, Inc.

To prepare the pipes for joining, the laser would cut a hole in one pipe where the other is to be joined, cut the end off that other pipe, and then put the two pipes together and weld them.

Lasers in a manufacturing cell will be increasingly common. Another trend will be the use of lasers on machining centers to do operations that the machine tool is unable to do. Coherent General also sees robots as suitable partners for it's lasers. To this end, the firm has just introduced a laser-toting robot.

"We will see robots in the future with lasers built right into their arms," said Mr. Bennett. Robots with lasers and vision systems would be able to weld, measure, and do all sorts of metalworking operations.

Future lasers will increase their flexibility by simply changing a lens.

"We have developed the concept of a laser with a tool changer. This tool changer will be loaded with what we call optical tools. These tools are set up to do different functions while the raw laser beam is the same in all cases," said Mr. Bennett.

The variety of types of equipment that will go into Factory 2000 is enormous and cannot be covered in one article. This article has not even talked about the most popular factory automation products: CAD/CAM systems, which allow the designer to produce product drawings on a computerized CRT screen. CAD/CAM significantly speeds up the work of the designer and also allows him to rotate, enlarge, dissect, and even test and analyze the physical characteristics of that part he has drawn, right on the screen.

Artificial intelligence, sure to play a major role in tomorrow's factory, has not been discussed either.

"Artificial intelligence will have a major impact on CIM within this decade," asserted Richard E. Morley, director, advanced technologies, Electronic Systems Business Section, Gould Inc., Amherst, N.H. "Simply, put, artificial intelligence is the ability to take a data base, convert it by means of appropriate rules of thumb, and put it into a system that is goal-directed—in other words, data converted by rules to accomplish goals."

New metalworking machines and new materials such as composites and ceramics have not yet been mentioned, nor have new developments in sensors and quality testing. The list is too long; yet all will play important roles in the factory of the future.

What is obvious is that there is more advanced technology around than a manufacturer knows what to do with. And even more developments are just

around the corner. Today, choosing to automate should not be contingent upon the appearance of some spectacular new technology that will solve all problems for a very low cost. But, U.S. manufacturers, fascinated by Star Wars technology, continue to dream.

What is needed is a realistic vision, not a dream, and some advice on the first steps to take in order to realize that vision.

STEPS TO INTEGRATION

A number of blueprints have been drawn up which outline the specific steps companies should take to construct the factory of the future. However, if a firm wants to build its Factory 2000, it should ignore these detailed automation plans for as long as possible.

A manufacturer searching for better ways to produce parts has to be very careful that he does not turn too quickly to technology for the quick fix he needs to pep up his sagging productivity. When re-evaluating company philosophy and strategies, the consideration of specific technologies should occur only very late in the process.

Before anything else, the first step toward Factory 2000 is an educational one. And who better to take that first step than the CEO.

Educating management

"Companies have got to start by educating senior management as to the strategic benefits of things like CIM, or total quality control or just-in-time manufacturing," insisted Mr. Gunn of Arthur D. Little. "There are powerful tools out there and that should be understood. Take a concept like CIM, it is a valuable tool that delivers proven strategic as well as financial benefits to the company. You simply have to make top management aware of that. And most aren't aware.

"You also have to get them thinking strategically, make them see that they can compete on their ability to engineer and manufacture products. That is the real basis of competition if you are a manufacturing company. And there are ways we can use CIM to change that basis of competition."

Hopefully, the result of taking step one will be a purging of the misconception that plant operations are pretty

much a given: that materials are fed into one end and a product emerges out the other side, often late and not of the highest quality; and that nothing can change that reality. Another consequence of taking step one should be a growing awareness that to do nothing is to die. "It maybe a slow death by a 1000 cuts but death just the same," commented Mr. Gunn.

Step two toward Factory 2000 involves setting goals and devising strategies to reach those goals. In developing these strategies three questions must be answered: Where am I? Where do I have to get to? And how do I get there?

"After you set the strategic objectives and realize what you want to accomplish: like reduce new product development time by 50 pct in two years, or increase inventory turns from 5 to 25 over the next two years, you can then look at how you can achieve those things and the tools to support that," said Mr. Gunn.

However, before a company can develop specifications of how modernization is going to take place, it must do some soul searching. "The company looking toward automation really has to analyze itself. Contemplate its own navel," explained Ralph Waite, vice president and general manager, ICC Business, Allen-Bradley Systems Division, Highland Heights, Ohio. "You don't do this self-analysis by merely setting up an automation task group somewhere. You do this by having management that is in place work together defining goals."

People heavily involved in manufacturing have to be part of this discussion. And their planning should be of the top-down variety. As mentioned, discussions of specific hardware still should not bar entering the conversation.

"The goal of these planners should be to create a functional specification of how they want to run their plant, not specifications of the equipment required to do it," said Mr. Waite. "They should define the task that is to be done, not how it is to be done."

But plans cannot be finalized until management really knows what is going on in its manufacturing facility. "A plan must be developed to analyze the existing plant," said Mr. Waite. "This

> Besides being exciting, the vision of your future factory should be carefully conceived, and the plans to fulfill that vision clearly spelled out. A long-range plan should start with an inventory of you current environment.
>
> Then it would be helpful to know:what technologies you have already implemented.

is really a major task and has to be documented."

Where are we going?

This documentation is necessary because before goals can be formulated, before it can be determined where a company is going, it must be known where the company is today.

Once strategies and self-evaluations have been carried out, step three is to establish a corporate-wide culture. "Factory automation cannot occur in a vacuum," explained Mr. MacFarlane of the Gartner Group. "Without an intelligent, informed, and committed management and technical staff, factory automation techniques cannot achieve their goals no matter which implementation technique is chosen."

Step four is the development of a unified communications structure or architecture, which ensures that people in the company are all pulling in the same direction.

"Consistent with the strategy and expertise of the people involved, the central architecture of the factory automation system must be established. This prevents a series of disheartening false starts and lengthy re-implementations without appreciable productivity gain," said Mr. MacFarlane.

"We see so many firms... that encourage each division to go off and do what it wants with different systems that don't talk to each other. Some day it is going to take millions of dollars and more important: years, to correct this mistake," said Mr. Gunn.

An interesting paradox arises when companies discuss moving from basic manufacturing into high-tech production. "When the implementation of modern technology is successful, it forces companies to get back to the basics, to get together and find out what are the major problems that need solving," explained Mr. MacFarlane.

If firms build their Factory 2000 on a shaky foundation, the factory will have little chance, for if inaccurate information on what is really happening in manufacturing is fed into the computers, the result will be, as they say in Silicon Valley; garbage in, garbage out.

VISIONS OF TOMORROW

Until now, this article has examined the stepping stones to be touched on and the log jams to be dislodged before Factory 2000 can be safely reached. But before any stepping and dislodging can take place, a manufacturer must have some idea where he is going. He must have the vision of Factory 2000 before him. To help flesh out that vision, we willlook at some possible shapes of this fantasy.

One problem with many modernization efforts today is that they are fragmented. Divisions and plants within companies are going their own way. Someday all of these organizations will linked together. But as the ad says: "You can pay me now or you can pay me later." If integration is ignored now, it will be heard from at a later date. Someday, when no more options are available, companies will be forced to embrace integrated manufacturing solutions. However, the longer firms wait, the more disruptive and costly the transition will be.

According to Mr. Slautterback of Koppers, Factory 2000 will be very unlike anything we have today. Because of the need for industry to survive, to remain competitive, and because of exploding computer technology, manufacturing change over the next two decades will be rapid. Because a good deal of change could occur during the next 16 years, Factory 2000 probably will look quite different from anything we have today. Although the Japanese have already built two or three plants that are operated by only a few people.

By the time the year 2000 rolls around, "The distinctions between the process industries and discrete manufacturing; between the manufacturing of electronics and machinery; between assembly and fabrications; between engineering and manufacturing will all tend to blend (or blur) as an economic lot size of one is approached," speculated Mr. Slautterback.

More specifically, tomorrow's factory "...will have defined generic work centers depending upon the process or automation requirements," suggested Mr. Morley of Gould. "These process centers will have generic manufacturing, or certainly FMS approaches to manufacturing. Some of the processing will, of course, be robots inside each of these FMS centers. These centers will

FACTORY 2000

in turn be interconnected by materials handling components such as material handlers, and intelligent vehicles and conveying systems.

Mr. Morley continued,"Robots will be used as local material handlers to put parts in and manipulate parts within these flexible machining centers. A command and control center will be used to modify, control, monitor, and in general, run the automation system of the future."

Most futurists say that tomorrow's factory will not be devoid of people. These people, however, would be performing much different jobs than factory workers today. The people primarily would be the decision makers who would use the timely decision support information prepared by computers using artificial intelligence.

"In the year 2000, computer applications will make possible the unmanned factory," said Mr. Slautterback. "The individuals remaining in manufacturing will be primarily top managers, engineers and skilled maintenance personnel. Hierarchical organizations will become obsolete, because dynamic responsiveness will be required to maintain competitive leadership. This means that organizations currently with 11 to 15 levels of employment will streamline to 4 to 7 levels. The pyramid will flatten or perhaps become diamond shaped.... It is interesting to note that over 70 pct of the jobs in the year 2000 cannot be identified or described today."

Generalist

The future employee will be a generalist, he will be cross-trained in several jobs. The staff engineer, for example, will design and instruct the manufacture of products from his graphic workstation.

"The engineer will analyze, schedule, instruct the robot, create the process plan, create the NC machining instructions, define inspection criteria, simulate the complete process, as well as many other tasks," concluded Mr. Slautterback. "He therefore will become line management. The line manager, on the other hand, will monitor the manufacturing environment for un-expected problems and become functionally staff."

Besides being exciting, the vision of your future factory should be carefully conceived, and the plans to fulfill that vision clearly spelled out. A long-range plan should start with an inventory of your current environment.

Then it would be helpful to know: what technologies you have already implemented, "The availability and status of new technologies. The logical sequence of implementation. The rate of change that your company can tolerate. The real cost and potential benefits. Who is the champion to lead each area of charge. And what is the implementation status of your global competitors," outlined Mr. Slautterback.

Despite the complicated technical tools that will be used in tomorrow's factory, operations actually will be much simpler. Already procedures have started to get simpler in a few factories.

"Parts now are being designed to be made easily; thousands of parts have been replaced by hundreds of families of parts; hundreds of machines have been replaced with several flexible manufacturing systems.

"Large inventories and long lead times have been replaced with small inventories and short lead times that are manageable; confusion created by scrap and rework has been drastically reduced; automated material handling has improved the material flow and part location; fewer employees have created less confusion; less paper has reduced transposition errors, lost copies, and filing mistakes.

"Redundancies have been eliminated; and the walls between manufacturing, engineering, marketing and finance have been replaced with one cohesive system that works: computer-integrated manufacturing," said Mr. Slautterback.

Even now far-seeing companies are taking their first tenative steps toward the 21st century. A century where the factory manager, much like the early potter, will again have integrated mind, tools and materials into an integrated operation. □

Source: Iron Age, June 1984, 27-29, 31, 33-36, 39-40, 45-47, 49, 51, 53-56, 58

Artificial intelligence and expert system technologies provide means of simulating human behavior in applications outside the domain of conventional mathematics and logic. Systems combining these techniques with deep knowledge of process chemistry and physics are attractive candidates for many monitoring and control tasks now done by human operators. The concept promises to bring expertise to every shift, *while providing constant surveillance and eliminating much of the uncertainty that has traditionally caused fluctuations in plant performance.*

KEYWORDS: artificial intelligence, expert systems, heuristics, production rules, knowledge bases, process monitoring and control.

Expert Systems: Are They the Next Generation of Process Controls?

ROBERT L MOORE, LOWELL B HAWKINSON, CARL G KNICKERBOCKER
LISP Machine Inc

Artificial intelligence (AI) provides a means to simulate human problem-solving capabilities in applications outside the scope of formal mathematics and logic. Expert system technology, a branch of AI, uses searching or *heuristic* techniques based on *production rules* or strategies provided by skilled practitioners — much as troubleshooting guidelines might be presented in a technical article or manual. The heuristics typically take the form of chaining procedures. Backward chaining traces sequences of events leading to faults or other phenomena responsible for observed conditions; forward-chaining predicts consequences of measured phenomena or alternative proposed actions.

Systems combining traditional deep knowledge of process chemistry and physics with expert production rules are attractive candidates for many monitoring and control tasks now done by human operators. The concept promises to bring expertise to every shift, while providing constant surveillance and eliminating much of the uncertainty that has traditionally caused fluctuations in plant performance. Even at the present state of the art, such systems can be effective to aid process engineers and operators by continuously monitoring the plant and identifying situations needing manual attention.

INDUSTRIAL APPLICATIONS

Expert systems have a 20-year history of development. The Heuristic Programming Project at Stanford University in 1965 provided the impetus for much of the work now being done. Recent success in developing expert systems for medical diagnosis (Ref 1), oil well data interpretation (Ref 2), and computer hardware configuration (Ref 3) has stimulated interest in the potentials of the technique for other applications.

Process monitoring and control includes several classes of applications for which expert systems promise significant benefits. As experience with this technology grows, the list can be expected to broaden.

Alarm diagnosis is an important potential expert system application. A modern control console might have thousands of measurement and alarm points. During upsets, how are operators to

Reprinted with permission from InTech, May 1985, 55-57, © 1985 Instrument Society of America

know the status of the plants and to decide which alarms should be considered first, which are the most critical, and what actions should be taken? Expert systems, acting as alarm advisors, could answer these questions by analyzing events, distinguishing causes from effects, determining underlying process problems, and suggesting reasonable alternative actions under current circumstances (Ref 4).

Determination of critical measurement failure is another key potential application of expert systems in process plants. A faulty sensor might cause a control system to act incorrectly, yet not result in an alarm. A system using deep knowledge and heuristic rules could detect inconsistencies, alerting operators to the lack of measurement integrity and suggesting where to look for the bad elements.

Requirements for real-time monitoring and control impose stringent conditions on expert systems

---

Expert systems could also provide significant economic benefits through plant optimization. A system could scan operating conditions, calculate performance indices, and estimate how various criteria would change if different adjustments were made.

Researchers in Japan have been particularly active in working on expert systems for real-time control. One such system applies heuristic production rules to diagnose faults in electric power systems and determine recovery procedures (Ref 5). Another system employs heuristic pattern matching to compare physical component performance descriptions with measured symptoms, and uses results to identify faulty elements in a nuclear power plant (Ref 6). Work has also begun on a system for automated performance diagnosis of large scale systems (Ref 7); progress in these applications was reported to be limited by insufficient computer speed and memory, lack of an effective knowledge acquisition method and a proper knowledge representation scheme, and difficulty in detecting inconsistencies in the knowledge base — problems that are being resolved by current efforts.

HARDWARE AND SOFTWARE STRUCTURES

In conventional software-oriented systems, information about the problem domain is usually buried in the program code. In expert systems, the production rules comprising the knowledge base are separate from the inference structure.

Heuristic rules are normally implemented in chunks. This facilitates starting with a small knowledge base and augmenting rules through experience. It also provides modular organization that simplifies identifying the knowledge associated with a particular application area and implementing explanation facilities that permit users to trace the paths followed in reaching conclusions.

Processor efficiency

Except in highly structured problem domains, expert systems and other AI programs are being written in languages developed especially for type checking and data structure manipulations. LISP is most common, although work is also being done in PROLOG. Specialized processors, different from those employed for arithmetic data manipulation, are needed for these programs to run efficiently. The emergence of appropriate hardware with large complements of low-cost memory and accompanying software tools enhances expert system feasibility (Ref 8).

Execution efficiency

Requirements for real-time monitoring and control impose stringent conditions on expert systems for efficient operation. A major part of the efficiency will depend on the heuristic procedures incorporated in the inference programs. A combination of forward- and backward-chaining appears to be desirable for many applications. For instance, a scanned forward-inference technique provides a facility to monitor key process variables, alarms, and performance indices to detect problems which may not cause explicit alarms. This involves scanning heuristic rules which determine events of possible significance, and triggering alert messages if conditions are matched. Conversely, backward-chaining would be employed when the expert system detects a combination of alarm points, and diagnoses relationships to find the cause of the problem.

High efficiency can also be achieved using multi-processor systems with hardware and system-level software modules optimized separately for arithmetic and heuristic tanks. Front-end processors designed for conventional data manipulation would then perform computationally-intensive tasks such as calculating unmeasurable variables, checking the reasonability of raw data, or performing material and energy balances. The high-level results would be made available to AI-oriented machines for pattern matching and chained-inference logic. Trade-off opportunities could be investigated as a means of maximizing overall throughput in particular environments. For instance, the front end processor could perform enough logic to identify a limited set of potentially significant events; the expert system could then operate under normal conditions by routinely scanning a table of alerting conditions and initiating analysis only when triggered by appropriate high-level input signals.

Efficiency in real-time environments can be further enhanced if the gateways between the expert systems and the instrument data buses transmit information only when requested to do so. This would avoid overloading the expert systems with measurements and alarms not required for current processing, while also minimizing burdens placed on networks for fetching and transmitting data.

A further requirement imposed by real-time operation is that the system accommodate dynamic changes in the knowledge base and input structure. This can be achieved efficiently using goal-directed computational routines, brought into the environment at execution time under the supervision of a high-level module (Ref 9). Such a facility allows the expert system to set scan rates and specify algorithms for particular situations.

As an example, a backward-chaining diagnostic expert system might reach a point where an inference test is required. The expert program could then direct the front end to gather and analyze the associated measurements. Dynamic resource allocation can also be employed to scan the plant in a background mode, directing attention to any elements that trigger alerts based on heuristic inferences.

KNOWLEDGE BASE CAPTURE

An expert system for process monitoring and control applications would most likely have at least a minimal pre-engineered knowledge base. For instance, graphical representations, parameterized mathematical models, and physical constants or attributes relevant to the process domain but not specific to the structure of any particular plant might be included in a component knowledge base. This could provide capabilities such as detecting measurement noise or inconsistencies, identifying incipient upsets, or characterizing thermocouples, valves, chromatographs, or other common pieces of equipment. In some cases, production rules might also be developed automatically from deep knowledge of the plant (Ref 10); such an approach has been employed in nuclear generating stations to aid operators in directing coolant to critical units under various equipment failure conditions.

A convenient interface would be needed for capturing knowledge pertaining to details of each installation. An extension of the menu-driven configuration capabilities being employed with current distributed control systems appears to be appropriate. This implies an interactive facility with graphic and structured natural language features to capture plant structure, attribute, rule, and model information.

The overall process knowledge base might be organized as a multi-level plant schematic. The complete plant would be at the top and individual components at the bottom; various instrument loops, unit operations, and processes would be at intermediate levels. The schematic showing the elements and relationships among them would be built interactively. As parts of the schematic are specified, the expert configuration system would generate questions, check for inconsistencies, and perform other functions to aid in acquiring knowledge from expert personnel.

REFERENCES

1. **Barr, A; Feigenbaum, E;** *Handbook of Artificial Intelligence, Vol 2,* Wm Kaufmann (Los Altos CA), 1982.
2. **Davis, A; et al;** "The Dipmeter Advisor: Interpretation of Geologic Signals," *Proceedings of the 7th International Joint Conference on Artificial Intelligence,* Wm Kaufmann (Los Altos CA), 1981, pp 846-849.
3. **McDermott, J;** "R1: A Rule-Based Configurer of Computer Systems," *Artificial Intelligence,* No 19, 1982, pp 39-88.
4. **Fortin, D A; Rooney, T B; Bristol, E H;** "Of Christmas Trees and Sweaty Palms," *Learning Systems and Pattern Recognition in Industrial Control,* Technical Publishing Co — *Control Engineering Magazine* (Barrington IL), 1983, pp 49-54.
5. **Sakuguchi, T; Matsumoto, K;** "Development of a Knowledge Based System for Power System Restoration," *IEEE Transactions on Power Apparatus and Systems,* Vol PAS-102 No 2, Feb 1983, pp 320-329.
6. **Yamada, N; Motoda, H;** "A Diagnosis Method of Dynamic System Using the Knowledge on System Description," *Proceedings of the 8th International Joint Conference on Artificial Intelligence,* Wm Kaufmann (Los Altos CA), 1983, pp 225-229.
7. **Yawakami, J;** "Overview of Artificial Intelligence Applied to Control Technology in Japan and in Hitachi," Hitachi Research Laboratory (Japan), 1984.
8. **Hawkinson, L;** "High-Level Languages Based on LISP Aid Expert System Design," *Computer Technology Review,* Winter 1983, pp 13-21.
9. **Sauers, R; Walsh, R;** "On the Requirements of Future Expert Systems," *Proceedings of the 8th International Joint Conference on Artificial Intelligence,* Wm Kaufmann (Los Altos CA), 1983, pp 110-115.
10. **Lusk, E; Stratton, R;** "Automated Reasoning in Man-Machine Control Systems," *Learning Systems and Pattern Recognition in Industrial Control,* Technical Publishing Co — *Control Engineering Magazine,* (Barrington IL), 1983, pp 41-48.

THE TECHNOLOGY OF EXPERT SYSTEMS

BY ROBERT H. MICHAELSEN, DONALD MICHIE, AND ALBERT BOULANGER

Transplanting expert knowledge to machines

THE PURPOSE OF this article is to introduce expert systems. Initially, we'll define these systems. Next, we'll discuss methods for building them, including the advantages and disadvantages of each method. Finally, we'll review the computer resources needed to build and run expert systems.

DEFINITION

Expert systems are a class of computer programs that can advise, analyze, categorize, communicate, consult, design, diagnose, explain, explore, forecast, form concepts, identify, interpret, justify, learn, manage, monitor, plan, present, retrieve, schedule, test, and tutor. They address problems normally thought to require human specialists for their solution. Some of these programs have achieved expert levels of performance on the problems for which they were designed (see reference 6).

Expert systems are usually developed with the help of human experts who solve specific problems and reveal their thought processes as they proceed. If this process of protocol analysis is successful, the computer program based on this analysis will be able to solve the narrowly defined problems as well as an expert. (For a discussion of successful expert systems, see reference 2.)

Experts typically solve problems that are unstructured and ill-defined, usually in a setting that involves diagnosis or planning. They cope with this lack of structure by employing heuristics, which are the rules of thumb that people use to solve problems when a lack of time or understanding prevents an analysis of all the parameters involved. Likewise, expert systems employ programmed heuristics to solve problems. Figure 1 is an example of a complex heuristic used by TAXADVISOR, an expert system that gives estate-planning advice (see reference 17).

Experts engage in several different problem-solving activities. For instance, the following problem-solving activities have been identified in MYCIN (see figure 2): identify the problem, process data, generate questions, collect information, establish hypothesis space, group and differentiate, pursue and test hypothesis, explore and refine, ask general questions, and make a decision (see reference 11).

Experts are capable of

● Applying their expertise to the solution of problems in an efficient manner. They are able to employ plausible inference and reasoning from incomplete or uncertain data.
● Explaining and justifying what they do.
● Communicating well with other ex-

(continued)

Robert H. Michaelsen is an assistant professor of accounting at the University of Nebraska (Lincoln, NE 68588-0488). He received his Ph.D. in accountancy from the University of Illinois. Donald Michie is Director of Research at the Turing Institute (36 North Hanover St., Glasgow G1 2AD, Scotland). Formerly a professor at the University of Edinburgh, he is the author of numerous books and articles on artificial intelligence. Albert Boulanger is a scientist for Bolt Beranek and Newman Inc. (10 Moulton St., Cambridge, MA 02238). He has a master's degree in computer science from the University of Illinois at Urbana-Champaign.

perts and acquiring new knowledge.
- Restructuring and reorganizing knowledge.
- Breaking rules. They have almost as many exceptions as they have rules. They understand both the spirit and the letter of a rule.
- Determining relevance. They know when a problem is outside their expertise and when to make referrals.
- Degrading gracefully. At the boundaries of their expertise, they become gradually less proficient at solving problems, rather than suddenly incapable (see reference 4).

Expert systems have modeled only the first three expert capabilities to any extent, and even explanation and knowledge acquisition have just begun.

Expert systems, like human experts, can have both deep and surface representations of knowledge. Deep representations are causal models, categories, abstractions, and analogies. In such cases, we try to represent an understanding of structure and function. Surface representations are often empirical associations but are sometimes "compiled" from an understanding of structure and function. In the former case, the association between premises and conclusions of rules is based on empirical observation of past association. Causality is implicit in the rule, rather than explicit.

Deep representations enhance the explanatory powers of expert systems. With surface representations, all the system knows is that an empirical association exists; it is unable to explain why, beyond repeating the association. Where more fundamental insight is available, deep representation will enable the system to respond more substantively. If computer induction is used for knowledge acquisition, a model for understanding events in the domain (a deep representation) often guides the induction of rules from examples by distinguishing meaningful hypotheses from coincidences in the data. It is also likely that deep representation will enhance the incorporation of the last four previously listed expert capabilities into expert systems. Surface representations have offered little in this regard.

However, surface representations have their advantages if the only concern is problem-solving performance, empirical associations, or compiled understanding. They should be less costly to formulate than causal models. This lower cost can provide a reasonable level of explanation along with a primitive form of knowledge acquisition. If a domain's expertise is based on empirical association, as in many areas of medicine, surface representations are the only kind available (see reference 4).

The best approach to expert-system building is probably to use deep representations when they are cost-effective and surface representations for the rest of the system. This approach has already been explicated in a paper by Hart (reference 12) and implemented in Digitalis Advisor, a system that provided advice on digitalis dosages for cardiac patients (see reference 29).

BUILDING EXPERT SYSTEMS

An expert system is able to make decisions on a par with an expert primarily because its structure reflects the manner in which human specialists arrange and make inferences from their knowledge of the subject. The system is driven by a database of inexact and judgmental knowledge that is typically made up of if-then rules when surface representation is used, or frames and semantic nets when deep representation is used (see "A Glossary of Artificial Intelligence Terms" on page 138). Domain knowledge is processed in a strict order of deductive inference and is invoked by a pattern match with specified features of the task environment. Figure 3 is an example of pattern matching by TAXADVISOR. Because uncertainty is usually involved in expert judgments, expert systems must allow

---

RULE 216

(This rule applies to clients and is tried to find out whether a short-term trust should be recommended.)

If:
1) The client and/or spouse do wish to shift property income to another (not for legal support), etc., for at least 10 years or until the death of the beneficiary,
2) The client and/or spouse do desire to eventually reclaim control of this property (for retirement, estate liquidity, etc.),
3) The client and/or spouse are in a higher income bracket than the beneficiary,
4) The client and/or spouse are willing to relinquish control of the beneficial enjoyment of the property,
5) The client and/or spouse are able to provide for their living needs without this income, even in the event of disability or unemployment,
6) The client and/or spouse do not plan to have the trust income used to pay life-insurance premiums on his/her life without the consent of an adverse party,
7) The client and/or spouse do not plan to use the trust for a leaseback of assets, and
8) A: The client and/or spouse have a person (e.g., a parent) they are supporting without legal obligation with this property income (will lose a dependent if trust is formed),
   B: The client and/or spouse have a child, not a minor, that they will be putting through college with this property income (can set up early and accumulate income without tax problems), or
   C: The client and/or spouse are using some of their after-tax income for the benefit of some other taxpayer (child's marriage and/or home purchase, etc.),

Then: It is definite (1.0) that client should TRANSFER ASSETS TO A SHORT-TERM TRUST.

Figure 1: *An example of a TAXADVISOR rule.*

conclusions to be reached with less than certainty. Figure 4 illustrates how TAXADVISOR copes with uncertainty during a consultation. (For more information on uncertainty mechanisms in expert systems, see reference 32.)

The type of computer program that is used to develop an expert system cannot have its flow of control and data utilization rigidly fixed because such a structure is ill-adapted for simulating a human's responses to a complex, rapidly changing, and unfamiliar environment. Instead, such a program must examine the state of the world at each step of the decision process and react appropriately because new stimuli continually arise. The type of program that has been developed to cope with this constant change is a loosely organized collection of pattern-directed modules (PDMs) that detect situations and respond to them (see reference 31). The rule in figure 1 is a PDM from TAX-ADVISOR.

Each PDM examines and modifies data structures that model critical aspects of the external environment. In TAXADVISOR, the client's financial-planning situation and objectives constitute the environment. A PDM should be written as a single and separate unit that is independently meaningful within the task domain of the program. This aids incremental program growth and debugging, since revision of one PDM does not affect the others. It also provides explanation power; a single PDM can be used to explain a recommendation by the system.

Any system composed of several PDMs, one or more data structures that may be examined and modified by the PDMs, and an executive program to schedule and run the PDMs is called a pattern-directed inference system (PDIS). In effect, a PDIS factors complex problems into manageable, largely independent subproblems.

SURFACE REPRESENTATIONS

Rule-based systems (RBSs) were originally used in cognitive modeling of short-term memory. Since expert

(continued)

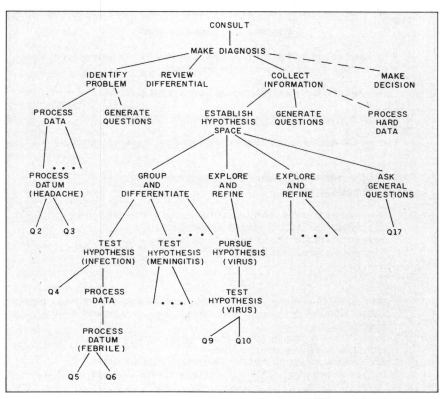

Figure 2: *The MYCIN problem-solving hierarchy. Question numbers (Q2, etc.) correspond to questions asked in the consultation. Solid lines show tasks actually done, dashed lines those that might be done. (Figure used with permission; see reference 11.)*

Necessary Attributes for Short-Term Trust	Prestored Value	Client's Value
1) Does client wish to shift property income to another (not for legal support) for at least 10 years or until the death of the beneficiary?	yes	yes
2) Does client desire to eventually reclaim control of the property?	yes	yes
3) Is client in a higher income-tax bracket than the beneficiary?	yes	yes
4) Is client willing to relinquish control of the beneficial enjoyment of the property?	yes	yes
5) Is client able to provide for his living needs without this income even if disabled or unemployed?	yes	yes
6) Does client plan to have trust income pay life-insurance premiums on his life without consent of an adverse party?	no	no
7) Does client plan to use the trust for a leaseback of assets?	no	no
8) Does client have a person he is supporting without legal obligation?	yes	yes

Figure 3: *An example of pattern matching done by TAXADVISOR, performed to determine if the client should be forming a short-term trust. The Prestored Value column shows the pattern of attribute values that a client must have before TAXADVISOR will recommend a transfer of assets to a short-term trust. Since the client's pattern matches the prestored one, the trust will be recommended.*

Source: BYTE Magazine, April 1985, 303-307, 309-312

```
(begin)
                    —Formation of a Temporary Trust—

1)  Does client wish to shift property income to another (not for legal support) for at
    least 10 years or until the death of the beneficiary?
**YES
2)  Does client desire to eventually reclaim control of the property?
**YES
3)  Is client in a higher income-tax bracket than the beneficiary?
**YES
4)  Is client willing to relinquish control of the beneficial enjoyment of the property?
**WHY

This will aid in determining whether client should form a temporary trust. It has
already been established that

    1)  Client wishes to shift property income to another (not for legal support) for at
        least 10 years or until the death of the beneficiary,
    2)  desires to eventually reclaim control of the property, and
    3)  is in a higher income-tax bracket than the beneficiary.

Therefore, if

    1)  client is willing to relinquish control of the beneficial enjoyment of the property,
    2)  is able to provide for his living needs without this income, even if disabled or
        unemployed,
    3)  does not plan to have trust income pay life-insurance premiums on his life
        without the consent of an adverse party,
    4)  does not plan to use the trust for a leaseback of assets, and
    5)  A:  has a person (e.g., a parent) he is supporting without legal obligation with
            this property income,
        B:  has a child, not a minor, that he will be putting through college with this
            property income, or
        C:  is using some of his after-tax income for the benefit of some other
            taxpayer,

then

            client should form a temporary trust
            (back to question 4 . . .)

**YES

5)  Is client able to provide for his living needs without this income, even if disabled
    or unemployed?
**YES (8) [Whenever a response is made with less than certainty, the system user
enters a number betwen 1 and 9 indicating his degree of certainty in that response.]
6)  Does client plan to have trust income pay life-insurance premiums on his life
    without consent of an adverse party?
**NO
7)  Does client plan to use the trust for a leaseback of assets?
**NO
8)  Does client have a person he is supporting without legal obligation?
**YES
I recommend that the client form a short-term trust.
[The degree of certainty that the system has in this recommendation is .8. This
certainty factor (CF) was calculated as follows. The temporary trust rule's action CF
was 1.0 and it had an "AND" premise. In such a case, the rule's CF is the minimum
CF used in the responses, or .8. Since the system's threshold CF is .2, the
recommendation was made.]

(end)
```

Figure 4: A *partial interactive consultation with* TAXADVISOR. *The user's input is in uppercase.*

systems attempt to imitate people, it was natural that RBSs would also be used in their development. To date, RBSs are by far the most common structure for expert systems. Among the successful rule-based expert systems that have been developed are the following:

- MYCIN—diagnoses infections (reference 26)
- HEURISTIC DENDRAL—identifies organic compounds (reference 8)
- PROSPECTOR—aids geologists in evaluating mineral sites (reference 5)
- PUFF—analyzes pulmonary function tests (reference 15)
- INTERNIST—performs diagnosis in internal medicine (reference 22)
- XCON (formerly R1)—configures the VAX-11/780 computer system (reference 16)
- SACON—provides engineers with advice on structural analysis (reference 1)

Because of the popularity of RBSs, several domain-independent systems have been developed to make it much easier to build rule-based expert systems in many fields. The following is a partial list of domain-independent systems:

- EMYCIN (reference 30)
- AGE (reference 21)
- OPS5 (reference 16)
- ADVISE (reference 18)
- Hearsay-3 (reference 7)
- AL/X (reference 23)
- Expert-Ease (Human Edge Software Corp., Palo Alto, California)
- KS 300 (Revamped EMYCIN; from Teknowledge Inc., Palo Alto, California)
- KES (Intelligenetics Inc., Palo Alto, California)
- Personal Consultant (Texas Instruments Inc., Dallas, Texas)

An RBS is composed of PDMs called rules, each with a left-hand side (the antecedent, a logical combination of propositions about the database) and a separate right-hand side (the consequent, a collection of actions). An RBS separates data examination (done by the left-hand side) from data modification (done by the right-hand side of the rule).

Most RBSs are production systems (PSs), in which matching and scheduling are explicitly defined by the operation of the executive (control) program. The control schema can be characterized as having four basic parts:

1. Selection: select relevant rules and data elements. Selection may be trivial (e.g., on each cycle all rules and all data elements can be considered) or quite complex (e.g., special filters can be designed to eliminate from consideration many rules that could not possibly match the current data). In TAXADVISOR, rules are organized in a hierarchy to narrow the rules considered.
2. Matching: compare active rules against active data elements, looking for patterns that match, i.e., rules whose conditions are satisfied. Figure 3 is an example of pattern matching.
3. Scheduling: decide which "satisfied" rule should be "fired." "Firing" consists of accessing and executing the procedures associated with the pattern elements that matched the current data. If more than one rule is satisfied, conflict-resolution heuristics are used to decide which rule to fire.
4. Execution: fire the rule chosen during the scheduling process. The result of execution is a modification of data elements or structure. With TAXADVISOR, execution results in an estate-planning recommendation for a client. This is illustrated in the test consultation in Figure 4 (see reference 31).

PSs are either consequent-driven systems or antecedent-driven systems. A consequent-driven (backward-chaining) system, which is the type used in TAXADVISOR, uses rule consequents (which represent goals) to guide the search for rules to fire (with TAXADVISOR, estate-planning actions to recommend). The system collects those rules that can satisfy the goal in question and tries to satisfy the consequents of those rules, which usually represent the values of variables. In order to find these values, the values of the rule antecedent must

be found. To satisfy each antecedent, which represents a subgoal, the system collects those rules whose consequents satisfy its value. The process of working backward through the rules from consequents to antecedents to consequents in search of a causal chain that will satisfy the goal is called backward chaining. (For a simple backward-chaining program written in BASIC, see "Knowledge-Based Expert Systems Come of Age" by Richard O. Duda and John G. Gaschnig, September 1981 BYTE, page 238.)

With antecedent-driven (forward-chaining) systems, program execution consists solely of a continuous sequence of cycles terminating when a rule's action dictates a halt. At each cycle, the system scans the antecedents and determines all rules with antecedents that are satisfied by the contents of the database. If there is more than one such rule, select one by means of a conflict-resolution strategy. Perform all actions associated with the selected rule and change the database accordingly. For example, with R1 (XCON), you enter all the information on the problem into the database, and the system then applies the rules to reason forward from the data to the conclusions. In summary, forward chaining consists of putting the rules in a queue and then using a recognize-act cycle on them.

Some forward-chaining systems try to control the search for rules in the recognize cycle by grouping rules into packets. These rule groupings are appealing conceptual structures, since they group rules according to the subtopic that they deal with. Object-oriented programming can also be used to organize collections of rules. In object-oriented programming, we give objects behavior, and thus we can distribute the control of rules into rule, rule-packet, and domain objects. This approach, which has been taken in LOOPS, a domain-independent system (see reference 27), also allows multiple instantiations of the same set of rules to solve subproblems of the same type within one session.

The primary difference between backward and forward chaining is a top-down versus bottom-up style of linking rules together. Though the most common, these are not the only control structures for rule-based systems. For example, rules are represented as an "inference" network in PROSPECTOR (see reference 5).

DEEP REPRESENTATIONS

Frame- and network-based approaches allow the implementation of "deeper-level" reasoning such as abstraction and analogy. Reasoning by abstraction and analogy is an important expert activity. You can also represent the *objects* (e.g., "pump" in figure 5) and *processes* (e.g., the "start" instructions in figure 5) of the domain of expertise at this level. What is important are the *relations between objects*. Deep-representation expert systems perform inference using relations represented by networks or frames. A semantic network is a graph of the relations. A frame or script system (see references 20 and 24) organize the objects and their relations into *entities* (recognizable collections of objects). Frame systems also provide a system to inherit attributes from a taxonomy of entities. Thus, a frame system implements the semantics of some of the relations between objects. With a semantic-net or frame system you can represent objects of the domain of expertise as well as the process, strategies, etc., that are also part of the domain. The control of frame or semantic-net systems is usually much more involved than with surface systems and is implemented in a way that an explanation facility can't get at. But surface systems are "shallow"; a surface system may be viewed as a projection of deep-level knowledge of a domain for a specific expert activity.

One type of expertise that has been represented with a deep-level approach is tutoring (see "The LISP Tutor" by John R. Anderson and Brian J. Reiser on page 159). Here we want to convey to the pupil domain knowledge that is best represented at the deep level: concepts, abstractions,

Some systems have a built-in capability for taking a file of expert decisions and generalizing from this knowledge an executable rule.

analogies, and problem-solving strategies.

Steamer is a training aid developed jointly by Bolt Beranek and Newman Inc. and the Navy Personnel Research and Development Center. Its goal is to teach operating procedures of shipboard steam plants. These procedures consist of a series of steps on subcomponents of the plant. The components and procedures are represented as frames in Steamer, as are the abstractions of components and procedures that experts use in teaching steam-plant operations. The steps of a procedure come from the abstractions and subcomponents of the device the procedure applies to. The ordering of the steps comes from a third represented entity: operating principles. These principles are culled from experienced operators and represent "compiled" knowledge of steam-plant operation (although they are not represented as rules but frames).

KNOWLEDGE ACQUISITION

The following are ways of acquiring knowledge in a form that can be used by an expert system (reference 19):

- being told
- analogy
- example
- observation, discovery, and experimentation
- reasoning from deep structure

The manual acquisition of knowledge

from human experts is a very labor-intensive process. There is an acknowledged need to have aids for knowledge acquisition as part of the system.

Methods to speed knowledge acquisition are now becoming available in the form of machine learning of rules from examples. Systems such as Expert-Ease have a built-in capability for taking a file of expert decisions from you and generalizing from these an *executable rule*. In a sense, you are able to transplant chunks of decision-making skill from your own brain to the personal computer, a possibility foreseen as early as 1966 by Earl Hunt and his colleagues.

The machine procedure that allows this skill transplant was developed from a Pascal-coded program called ID3 (Iterative Dichotomiser 3) due to Professor Ross Quinlan of the New South Wales Institute of Science and Technology.

A number of conclusions follow from Quinlan's work:

1. It is possible, using such a program, to generate machine-executable solutions for complex decision problems in a fraction of the time a programmer would need for developing a solution by conventional hand coding.
2. The resulting solutions are super-efficient as compared with those obtainable by the old hand methods.
3. It is important to make up your mind in advance whether super-efficiency is all you demand of a machine-executable solution, or whether you *also* want the resulting rule base to be understandable on inspection.

If the answer to the third statement above is that user transparency of induced rules is desired, then (unless it is a very small one) do *not* treat your problem as one big superproblem with a single associated file of ex-

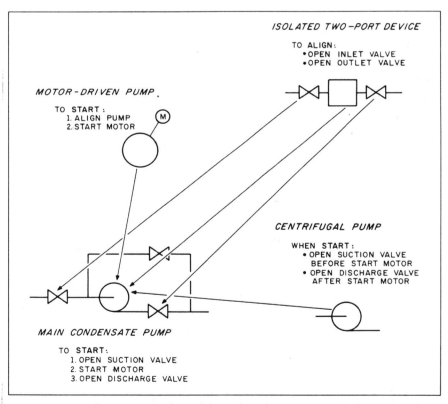

Figure 5: *Procedure steps are obtained from the subcomponents and abstractions of an object, here a main condensate pump. This example comes from Steamer, a tutorial system designed to teach operating procedures of shipboard steam plants.*

amples. Instead, first break it down into a main problem and a set of subproblems, even going further (to the level of sub-subproblems) if the complexity of the problem domain seems to call for it. The originators of this style, which is known as "structured induction," are Drs. Shapiro and Niblett (reference 25). Corporations enjoying the use of powerful inductive generators such as ITL's FORTRAN-based EX-TRAN system or Radian Corporation's C-coded RuleMaster have applied the approach to the building of complex systems for troubleshooting large transformers, severe-storm warning, circuit-board fault diagnosis, and user-friendly guidance to set up numerical batch jobs in seismic analysis in the oil industry. Rates of production of compact installed code in excess of 100 lines per worker day are now commonly reported.

Any robust expert system takes a tremendous amount of resources to develop. Once developed, the knowledge along with the control structure can be "compiled out"; that is, the system of rules is rewritten into a piece of code that performs the same function on a personal computer. For example, some expert systems (AD-VISE, EMYCIN, OPS5—see reference 10) can generate code or other primitive forms of the knowledge for use on a personal computer. (Systems run on a personal computer are usually referred to as "delivery systems.")

KNOWLEDGE REPRESENTATION
As AI researchers point out, a robust expert system that can explain, justify, acquire new knowledge, adapt, break rules, determine relevance, and degrade gracefully will have to use a multitude of knowledge representations that lie in a space whose dimensions include deep/surface, qualitative/quantitative, approximate (uncertain)/exact (certain), specific/general, and descriptive/prescriptive. Systems that use knowledge represented in different forms have been termed *multilevel systems*. Steamer is an example of one such expert system. Steamer uses the following represen-

tations:

1. A graphical (icon) representation of the objects of the Steamer domain, such as valves, pumps, tanks, and systems composed of these.
2. A frame representation of Steamer objects, procedures, and operating principles. This is used for describing, explaining, categorizing, abstracting, and referring.
3. An assertional database where assertions about Steamer entities can be made and retracted.
4. A quantitative numerical simulation of the steam plant that is used in illustrating cause and effect and ramifications of the application (or misapplication) of procedures.

Work is just beginning in building such multilevel systems, and they will be a major research topic for this decade. Work needs to be done in studying and representing in a general way the different problem-solving activities an expert does (see reference 3). When you build expert systems, you realize that the power behind them is that they provide a regimen for experts to crystallize and codify their knowledge, and in the knowledge lies the power.

RESOURCES NEEDED
Before resource needs are discussed, you must precisely define the type of expert system you want to build. If you wish to build a large, "custom" model expert system (i.e., it is not feasible to use many of the smaller domain-independent systems that are available), you will need substantial resources: large memory, good language support, and fast execution of the code. You may need to develop such a system in LISP on hardware specialized to processing the language, or on time-sharing machines with a large address space. Such "custom" systems are usually referred to as "prototype" or "development" systems. They can either be developed for a specific domain (e.g., MYCIN) or be domain-independent (e.g., ADVISE).

If you are able to build a less com-

plex expert system using an existing domain-independent system or if the system has a rule-compilation facility that allows applications to be run on personal computers, then a personal computer (preferably with 512K bytes) is sufficient. If all you need are resources to run an existing expert system, a large personal computer should nearly always be sufficient.

There is no obvious line of demarcation for a given project. However, certain barriers make personal computer use less desirable as system size and complexity increase.

SYSTEM BARRIERS
Many high-level languages do not offer the right primitives (i.e., programming-language statements) for developing expert systems. Among the desirable primitives are

• A parser or interpreter that parses statements during program run time. Without this, you have to write a parser for the rules.
• List and nonnumeric processing primitives.
• A language design that allows incremental compilation and other fast prototyping facilities. Incremental compilation enables you to recompile a function or other portion of a file without recompiling the entire file.

The view that many people in the field are adopting is that high-level languages like Pascal, Ada, and C are acceptable for the delivery system, but for prototyping, a language like LISP or Prolog is preferred. Program-generation tools are then used to write the system in the delivery language.

The knowledge-intensive approach to expert systems implies that the memory will be highly utilized in all but the most nontrivial applications. AL/X is one example that ran on a 64K-byte machine, but it was a small expert-system shell. Since memory prices have gone down and many small machines have broken the 64K-byte barrier, we can expect that more expert systems can be developed, at least for the delivery system, on per-

Some researchers predict that memory needs of advanced expert systems will drive development of encyclopedic memories.

sonal computers. Some researchers predict that the memory needs of advanced expert systems will drive the development of encyclopedic memories for personal computers.

Conclusion

Expert systems can be built in many ways, involving rules, networks, frames, and combinations thereof, with all sorts of variations within these categories with respect to knowledge representation and control. We could not begin to cover all possible approaches to building expert systems, since new ones are being developed almost daily.

Even if the most efficient approach has been ascertained for the domain in question, the most cost-effective computer resource must still be determined. In most cases, approach selection at least narrows the choice for resources; in some cases, approach and resources can be selected together. However, this hardly reduces the complexity of the choice. To make matters worse, computer resources are changing as rapidly as the new system-building approaches are being developed. The best we can hope to convey is an awareness of the opportunities and complexities involved in the development of expert systems. ■

REFERENCES
1. Bennett, J. S., and R. S. Englemore. "SACON: A Knowledge-Based Consultant for Structural Analysis." *IJCA*, 179, 1979, page 47.
2. Bramer, M. A. "A Survey and Critical Review of Expert Systems Research." *Introductory Readings in Expert Systems*, D. Michie, ed. London and New York: Gordon and Breach, 1982.
3. Chandrasekaran, B., and Sanjay Mittal. "Deep Versus Compiled Knowledge Approaches to Diagnostic Problem-Solving." *International Journal of Man-Machine Studies*, #19, 1983, page 425.
4. Davis, R. "Expert Systems: Where Are We? and Where Do We Go From Here?" *AI Magazine*, Spring 1982, page 3.
5. Duda, R., J. Gaschnig, and P. Hart. "Model Design in PROSPECTOR Consultant System for Mineral Exploration." *ESMA*, 1979, page 153.
6. Duda, R. O., and E. H. Shortliffe. "Expert Systems Research." *Science*, April 1983, page 261.
7. Erman, L. D., P. E. London, and S. F. Fickas. "The Design and Example Use of Hearsay 3." *Proceedings of IJCA* no. 7, 1981, page 409.
8. Feigenbaum, E. A., B. G. Buchanan, and J. Lederberg. "On Generality and Problem Solving: A Case Study Using the DENDRAL Program." *Machine Intelligence 6*, B. Meltzer and D. Michie, eds. New York: Edinburgh University Press and Halsted Press (Wiley), 1971, page 165.
9. Forbus, Kenneth D. "Qualitative Process Theory." *MIT Technical Report 789*, MIT AI Laboratory, May 1984.
10. Forgey, C. L. "Rete: A Fast Algorithm for the Many Pattern/Many Object Match Problem." *Artificial Intelligence*, September 1982.
11. Hasling, Diane Warner, William J. Clancey, and Glenn Rennels. "Strategic Explanations for a Diagnostic Consultation System." *International Journal of Man-Machine Studies*, January 1984, page 3.
12. Hart, P. "Direction for AI in the 80's." *SIGART Newsletter*, November 1981, *page* 11.
13. Hollan, James, Edwin Hutchins, and L. Weitzman. "Steamer: An Interactive Inspectable Simulation-Based Training System." *AI Magazine*, Summer 1984, page 15.
14. Hutchins, Edwin, Terry Roe, and James Hollan. "Project STEAMER: VII. A Computer-Based System for Monitoring the Boiler Light-Off Procedure for a 1078-Class Frigate." *NPRDC Technical Note 82-85*, August 1982.
15. Kunz, J. C., et al. "A Physiological Rule-Based System for Interpreting Pulmonary Function Tests." Heuristic Programming Project, Memo HPP-78-19, Stanford University, 1978.
16. McDermott, J. "R1: A Rule-Based Configuror of Computer Systems." Computer Science Department, Carnegie-Mellon University, 1980.
17. Michaelsen, R. H. "An Expert System for Federal Tax Planning." *Expert Systems: The International Journal of Knowledge Engineering*, October, 1984, page 149.
18. Michalski, R. S., A. B. Baskin, A. Boulanger, R. Reinke, L. Rodewald, M. Seyler, K. Spachman, and C. Uhrik. "A Technical Description of the ADVISE Meta Expert System." Department of Computer Science, University of Illinois at Urbana-Champaign, 1983.
19. Michalski, R. S., J. Carbonell, and T. Mitchell, eds. *Machine Learning: An Artificial Intelligence Approach*. Los Altos, CA: Tioga Publishing Company, 1983.
20. Minsky, M. "A Framework for Representing Knowledge." *The Psychology of Computer Vision*, P. Winston, ed. New York: McGraw-Hill, 1975.
21. Nii, H. P., and N. Aiello. "AGE (Attempt to Generalize): A Knowledge-Based Program for Building Knowledge-Based Programs." *IJCA*, 179, 1979, page 645.
22. Pople, H. E., J. D. Myers, and R. A. Miller. "Dialog: A Model of Diagnostic Logic for Internal Medicine." *IJCA*, 175, 1975, page 848.
23. Reiter, J. "AL/X: An Expert System Using Plausible Inference." Intelligent Terminals Ltd., University of Edinburgh, 1980.
24. Schank, R. C., and R. P. Abelson. *Scripts, Plans, Goals, and Understanding*. Hillsdale, NJ: Larrence Erlbaum Associates, 1977.
25. Shapiro, A., and T. Niblett. "Automatic Induction of Classification Rules for a Chess Endgame." *Advances in Computer Chess 3*, M. R. B. Clarke, ed. Oxford: Pergamon, 1982.
26. Shortliffe, E. H. *Computer-Based Medical Consultations: MYCIN*. New York: American Elsevier/North-Holland, 1976.
27. Stefik, Mark, Daniel G. Bobrow, Sanjay Mittal, and Lynn Conway. "Knowledge Programming in LOOPS: Report on an Experimental Course." *AI Magazine*, Fall 1983.
28. Stevens, Albert, and Bruce Roberts. "Quantitative and Qualitative Simulation in Computer Base Training." *Journal of Computer Based Instruction*, volume 10, numbers 1 and 2, Summer 1983, page 16.
29. Swartout, W. R. "A Digitalis Therapy Advisor with Explanations." *Technical Report 176*, MIT Lab for Computer Science, February 1977.
30. Van Melle, W. "A Domain-Independent Production Rule System for Consultation Programs." *IJCA*, 179, 1979, page 923.
31. Waterman, D. A., and F. Hayes-Roth, eds. *Pattern-Directed Inference Systems*. New York: Academic Press, 1978.
32. Whalen, Thomas, and Brian Schott. "Issues in Fuzzy Production Systems." *International Journal of Man-Machine Studies*, #19, 1983, page 57.

New Architectures

Novel computer designs promise a dramatically faster generation of computers

BY ED TEJA

DESIGNING computers that run faster than existing models is a never-ending challenge facing computer scientists. The task is largely a matter of looking at the places where a little more speed will make a lot more difference in the overall system performance.

A block diagram of conventional computer architecture is shown in Fig. 1. What can be done to speed it up? Without changing the architecture, about all a designer can do is choose newer and faster microprocessor chips. Since the processor does the bulk of the work in this design, faster processors usually mean faster computers. As a result, a great deal of research effort has been dedicated to finding new manufacturing processes and materials that will provide us with faster processors.

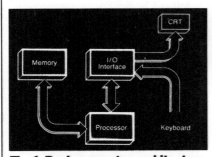

Fig. 1. Basic computer architecture.

But the newest processors are always the most expensive. This is one strike against relying on state-of-the-art components, but not the most important one. Using a fast new processor also demands comparable improvements from other system components. This might mean faster memory chips as well as disk subsystems that transfer data at higher rates (and this is a major bottleneck in many small systems) just to keep up with the processor's new-found speed. Im-

Ed Teja writes frequently for C&E on computer electronics.

proving the performance of these components will raise the system's cost, too. And you might find that even the system bus itself needs improvement. In this case, it's back to the drawing board to create a new system bus. Even after all this effort, the computer might still prove too slow. What then?

Adding Processor Power

Fortunately, faster processors and memory chips aren't the only solution to increasing a computer's speed. Designers can choose another tack—creating systems that use multiple processors. Each processor you add to the system has the potential of adding to its overall performance.

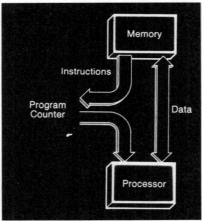

Fig. 2. The Von Neumann design.

The basic idea behind using extra processors to gain speed isn't a new one at all. In fact, it comes from one of the oldest adages applied to any problem solving task—divide and conquer. Whether you are talking about armies or mathematics problems, breaking a big problem up into a bunch of easily solved smaller ones makes a great deal of sense. And your computer already does that much. When you program it to compute:

$$x = (34*4) + (2*6)$$

the computer doesn't try to handle it as one large problem, rather it breaks the problem into three smaller ones. Properly divided, the problem is now to:

(1) solve: 34*4
(2) solve: 2*6
(3) add result of (1) to result of (2).

The point here is that steps one and two can be executed in any order, or, and this is important, they can be executed at the *same time* without adversely affecting the result. So instead of the system waiting the time it takes to perform two multiplications sequentially, the system could assign both at the same time. In an 8086 processor, the simplest possible addition, a register-to-register immediate, takes three clock cycles. Doing all three steps in sequence would require three complete clock cycles for each addition.

Performing the two multiplications simultaneously would mean that both could take place during the same clock cycles. If you are in a hurry for the result (in a real-time industrial control situation, perhaps), saving clock cycles can be significant.

Adding Specialized Processors

But this describes the solution—what does the hardware look like? To put several processors to work in a single system requires a new architecture. The simplest approach is to extend the basic architecture somewhat, using one or more specialized processors that are optimized to repetitively perform a specialized task. Take, for example, the math coprocessor. No one, even computers it seems, likes to do arithmetic. Math is the processor's most difficult task. Therefore, nearly every one of the popular 16-bit microprocessors has a math chip available to work with it. This chip, the coprocessor, sits on the bus along with the microprocessor and handles all math processing—at much higher speeds than the processor could handle them.

Using the math coprocessor to save system time is fantastic if the application is sure to encounter certain types of math problems over and over again. But while the math processor is working, the main processor takes a vacation. It will pass the math problem to the math processor and wait for the answer; it won't do part of the math itself. So you still wind up doing all three steps of our problem in sequence. It is more of a divide-and-send-the-parts-to-a-specialist approach than it is divide-and-conquer.

We are stuck with what is termed a control-flow or control-driven computer architecture; a design developed in 1945 by John Von Neumann. The diagram shown in Fig. 2 describes the flow of data

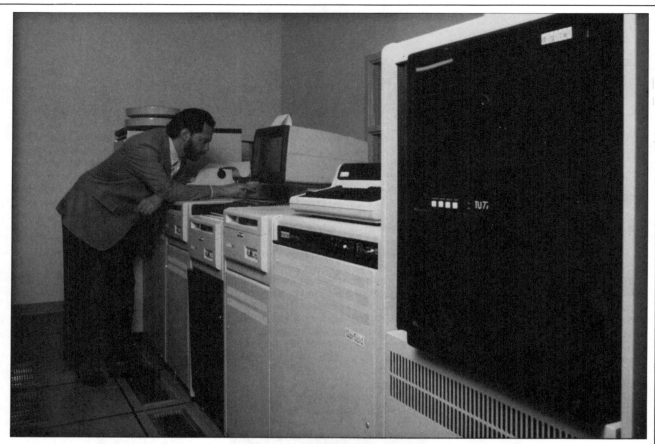

Columbia University's research facility, where the Non-Von 1 is being developed.

and instructions that defines the basic Von Neumann computer architecture.

The program counter (PC) plays an important role in Von Neumann's design—it controls the passage of instructions from memory to the processor. The PC tells the processor the next activity that should take place. The processor itself decodes and executes each instruction. The instruction will contain the coded address of the data. Before the processor can execute the instruction, therefore, it must first fetch the data from memory. Another part of the instruction tells the processor where to store the results when it is through executing the instruction.

Adding a coprocessor doesn't really change the computer's design—it extends it. We have modified the flow of data and instructions a bit so that the path appears as shown in Fig. 3. When the PC indicates that an instruction must be executed, the processor must still decode the instruction, but the math processor can access the data directly.

Extending Control-Driven Computers

You can gain a great deal more speed by extending the control-driven architecture until it resembles a tree, with a central processor passing tasks off to independent processing elements running concurrently (Fig. 4.). Each processor has access to the data directly, allowing it to run independently of the other processors. In the case of one such machine, the Non-Von 1 currently under develop-

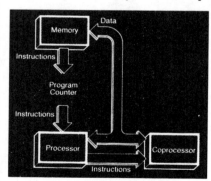

Fig. 3. Adding a coprocessor.

ment by a group headed by Dr. David Shaw at Columbia University in New York, each processor has its own path to memory via an intelligent head unit—a disk head that will read just the data needed by the processing element that it serves. (For more information on this project, see the accompanying sidebar.)

All control-driven machines suffer from one weakness—the programmer must decide, and then specify, how each problem will be divided among the processors. The central processor has no inherent way of knowing which pieces of the job can be performed concurrently. The programmer must subdivide the work; the central processor will assign it. Programming, therefore, becomes more of a chore with the enhanced architecture than it was on simpler machines.

Data-Driven Machines

But extending the existing architecture isn't the only way to improve performance. A new generation of designs might manage to gain speed and avoid the problems and limitations of control-driven machines by abandoning the Von Neumann approach altogether. Unlike the coprocessor approach, they will configure the processors as equal general-processing units that operate concurrently. And unlike the Non-Von-1, these machines will automatically uncover the concurrency within a problem.

Instead of using a program counter to control the flow of data and instructions, these machines will be either data- or demand-driven architectures. Let's see what that means.

Data-driven machines quite simply call for a processor to execute an instruc-

394

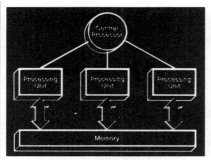

Fig. 4. Extended design.

tion whenever the data which that instruction needs becomes available. To implement this, data is packaged with its destination and information about how it is to be processed, into units termed tokens.

A network of input/output switches (Fig. 5) distributes the tokens to independent processors. Within each processing unit, the token flows through a circle where it is processed and then returned, via the network, to the system. If

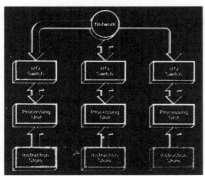

Fig. 5. Data-driven architecture.

the information in the token indicates that other data is needed, the processor will monitor the network for data addressed to the same location as the current data. If no other data is needed, the processor will fetch the appropriate instruction and execute it. The processing unit then transmits the results via the network, to the destination required by the tokens.

The University of Manchester has developed a version of the data-driven machine that uses 12 processors. It executes 1.7 million machine instructions per second (MIPS).

Demand-Driven Machines

The last technique being used to make a computer's operations run in parallel is the demand-driven machine. Here, a task isn't executed until the results of that process are needed. Effectively, this prioritizes the parts of the problem.

Known as symbolic reduction machines to artificial intelligence researchers, the demand-driven engines automatically reduce a complex symbol to a simpler one, repeating this task until it can reduce it no further. In semantic terms, the machine evaluates symbols until it reduces them to actual values. The expression $3+4$, for example, would become 7. The expression $2(3+4)$ would first reduce to $2(7)$, then 14.

The tree diagram shown in Fig. 6 illustrates how a demand-driven machine might execute the simple math problem we used earlier. The more complex a problem is, within limits, the more time the demand-driven engine saves when compared to Von Neumann machines.

Apparently then, systems using parallel architectures resemble networks of computers more than conventional computer architectures. Each processor does its job independently of the others the same way that each user on a network does his or her job, while sharing valuable system resources, such as a hard disk. Each user has access to the same data, the same programs, the same system capabilities. The success of the overall operation depends on how well management has divided up the problem parts and assigned them to the users.

State of the Art

Perhaps nowhere more than in artificial intelligence applications, where the computer is called upon to do large quantities of calculations quickly, is parallel processing awaited so excitedly. And the future is already beginning to appear. Architectures become more like networks and less like conventional configurations all the time. In other words, they become more parallel. The Lambda machine, for example, from LISP Machines, Inc. (Culver City, CA) uses a 10-MHz 68000 microprocessor and a 4-board 32-bit LISP processor that runs at 20 MHz. Each processor has its own memory and its own bus. The LISP processor uses the NuBus with its 40M-byte/s peak transfer rate while the 68000

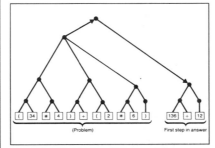

Fig. 6. Reduction machine.

plugs into a conventional Multibus.

The two processors are loosely coupled; the design is more parallel than many, but serves more as a practical example of the trend toward parallelism than as the first article of any particular architecture. The processors are nearly independent, but networked together.

The dual-processor environment uses the 68000 (running Unix) to handle normal system tasks; the faster LISP machine runs the AI algorithms—directly executing instructions written in the LISP programming language.

Speaking in Tongues

Part of the process of dividing problems and assigning tasks will reside in the system software. At least some of the success of parallel computers will depend on the ability of system software designers to develop high-level languages that take adequate advantage of the hardware. And this is no mean feat by any estimation.

At the University of Illinois in Champaign-Urbana, for example, researchers have spent 10 years working on making versions of the FORTRAN programming language that will support parallel, pipelined and multiprocessor systems. It's important that FORTRAN run on the next generation of machines, because it is the most popular language for large-scale scientific programming. Programmers want to be able to use the vast numbers of programs already written in FORTRAN. There's no pain like that of starting completely over. And most institutions won't be willing to do it.

But new languages are also being developed; languages that uniquely allow a programmer to get the most from a highly parallel structure. Many of the algorithms used in artificial intelligence have been contorted to fit the demands of conventional programming languages. Perhaps languages based on those algorithms would help break new ground in artificial intelligence.

Getting There from Here

Currently research and development projects to create what is called the "Fifth-Generation Computer" are underway in the United States, Japan and Europe. The Japanese goal, in fact, is to have a commercial model available by the 1990s. This drive to build a non-Von Neumann, parallel processing computer should not only give new meaning to the word "fast," but should also create a computer the likes of which we've never seen before. ◇

WORKING FASTER TOGETHER

*C&E talks with
Dr. David Shaw about
his efforts to build
a new kind
of computer*

By Joe Desposito

THERE are various ways to squeeze more speed out of a computer, as discussed in the accompanying article. One of them is being tried by Dr. David Shaw, whom we spoke with at Columbia University in New York City. Shaw has designed a machine, called the Non-Von computer which uses massive parallelism as its underlying architecture. He contrasts his approach with other parallel processing designs, saying, "There are certainly various interesting relationships between it (Non-Von) and other architectures, but I think that most people in the field will agree—especially my critics—that it has very little to do with other approaches to parallelism.

"For one thing, many people today are trying to design parallel machines with one hundred, one thousand or at most a few thousand processing elements. Our machine, on the other hand, is really designed to ultimately use a million processors or something on that order. And you have to use those processors in completely different ways. Some of the influences on this architecture were, first of all, thinking about humans and how they work, and the associational nature of human memory. We seem to come up fairly rapidly with many associations to a single concept. Also, there's the notion of massive parallelism. We

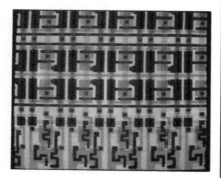

CAD terminal display.

Dr. Shaw is a pioneer in parallel processing design.

know that the brain is made up of a tremendous number of neurons, and that there are a tremendous number of interconnections between them. Even though the neurons work quite slowly—in comparison to a computer a brain works extremely slowly, or at least an individual neuron does—when you put them all together, the net processing power of the brain seems to be sufficient to do all the things that we do. It's not that Non-Von is an attempt to be a brain by any means. And it's not that I believe that the brain is organized that way. But those are some of the subtle influences that pushed me in that direction."

Dr. Shaw's knowledge of the brain's structure doesn't come from casual reading about the subject. He studied experimental cognitive psychology at the University of California at San Diego (UCSD). His advisor there was very interested in using the computer to model how people think and remember things, what attention is all about, and so forth. That's what first got Shaw interested in

making computers do things that seem to require intelligence. He went on to do graduate work at Stanford University, one of the two important research centers for that type of research at the time.

Shaw studied at the artificial intelligence laboratory there, trying to make computers do things that we consider intelligent. Near the end of his graduate career he realized that, even if he and his colleagues could figure out what thought is about and how you might trick a computer into imitating it, available computer power would be a limiting factor. That is, even if they were able to figure out how to get a computer to understand people speaking English and figure out what they were talking about—and there were still a lot of problems as to how they would go about achieving this—it would take existing computers hours to do things that ought to happen in seconds. That led Shaw to look at radically different ways of organizing computers to enable them to do work much faster. Most of his research over the last five years has focused on this problem. He has also done work on artificial intelligence, database management, and other applications of computers. But his main concern is how these tasks can be done dramatically faster using completely different types of computers.

Dr. Shaw's perspective is not only theoretical; it's practical, too. Before he got his PhD at Stanford, he took out some time to start his own corporation and he ran it for three years. This gave him a slightly more practical perspective on the question of computation. It was really the first time that he stopped thinking about how "thought" could be accomplished on a computer and started to consider questions like efficiency and what computers really are and how they are organized.

What's the current status of the project? Dr. Shaw says, "There have been two significant milestones to date. One was having a chip that was 100% functional. The chip had a single processor on it, not the eight it will ultimately have, but it executed all of the instructions perfectly. That was just recently accomplished. The next big milestone will be producing our first prototype, a very small one. What we'll be doing is hooking up 64 of these single-processor chips in the same way that they will be hooked up in the final machine. So it will be a real Non-Von machine, but it will only have 64 processing elements. It won't be enough to demonstrate anything useful in terms of speed, but it will demonstrate that the approach works and give us the signal that our further work will be likely to produce a machine that could do something—funding permitted. The time that we're shooting for is the end of this year. At that time we should also have software to run on the machine. It won't be very extensive— just some toy programs—but we would be able to demonstrate a working machine.

As director of the project, Shaw oversees both the hardware and software development of the machine. However, the key person involved with the hardware effort is research engineer Ted Sabety, who has three other full-time engineers on his staff. In addition, he has a fairly large number of students working either for him or one of the other engineers. Although they aren't full-time staff members, per se, many of them spend huge amounts of their spare time on the project. "Because it's an exciting project for them," says Dr. Shaw. On the software side, most of the work has been done by PhD students doing theses, who report directly to Shaw. "We have had at least 60 people thus far writing software for Non-Von, or rather simulating it," says Shaw, "since we don't have a computer actually working."

Creating a new computer means using existing software and hardware as your primary tools. Dr. Shaw's group first defined what they wanted their computer to be and then went to great lengths to simulate it at various levels of detail. In fact, they have five levels of simulation for the machine. The first is to see whether or not programs can be run on it. Then there is a more detailed level to find out if the data is transferred between the registers the right way. The levels go all the way down to testing the primitive physics properties (device physics) that described the behavior of the transistors. Much of the work was done using ordinary computers, mostly from Digital

The big prize in the field will be a successor to the Von Neumann

Equipment Corporation (DEC). Software simulation is one part of the project and it is proceeding for designing later versions of the computer and for writing new programs for the one being implemented now.

The other part is the actual hardware design, which involves design of the chip. For that they use color graphics displays connected to DEC VAX computers. They also use a whole set of design tools imported from places such as Berkley and MIT and software that they wrote themselves. The final part of the process is submitting integrated circuit chips to a manufacturer, first with various components of the processor, then with larger and larger pieces until a complete processing element is built. With this final step of processor design already completed, they have begun ordering all the components for the first prototype, Non-Von 1. They now have to put the whole system together. "And find out why it doesn't work the first time, because it never does," commented a smiling Shaw. "At that point we'll be able to perform some really interesting experiments with it."

Where does the broad range of abilities needed to build Non-Von 1 from scratch come from? "No one person on my staff has all those abilities," says Dr. Shaw. "For example, Ted Sabety came to us from Hewlett-Packard's integrated circuit labs where he was involved with several things. He was himself an IC designer, but also a consultant within the company, telling other people various things about how IC design works. This is, in fact, one of our advantages over many universities—most of them don't have a Ted Sabety. They have mostly university people who have learned about the process of IC design through a very effective system of teaching it, which has been spread through the country during the past three years. But the system doesn't help you get out of serious unexpected binds that come up when you're building a production chip. And then we have people with different sorts of experience in other technologies.

"We also go outside for help—often to our competitors. For example, Chuck Seitz of Cal Tech, who I think is one of the most gifted researchers in this field, is building some very interesting machines that are organized according to completely different principles. He doesn't really agree with our basic architectural principles, but we enjoy talking to each other, and he has helped us tremendously in thinking out how we were going to do our timing and solve other system problems."

Does Dr. Shaw see commercial viability down the line? "I'm not sure about that," he says. "There seems to be some evidence in selected areas to indicate that it ought to be commercially viable. But we won't know until we actually build one. But I guess I should say that we wouldn't be doing any of this unless we thought it had the that kind of potential. Not that *we* want to do it here, but we're really interested in producing a machine that will have a substantial economic impact. It's just a long process of research and trying it out on various applications until one knows that it's possible."

When asked whether he thinks his approach to parallelism is the best one, Shaw responded, "Again, it's hard to say. I think it's premature to say whether Non-Von or any of the other machines is using the strongest approach. Most of my serious colleagues have some ingenious ideas, and we aren't yet able to compare them. We really won't know until we see a completed machine, find out how much it costs, and see if it's maintainable, produceable, and so forth. Then we can actually run some applications programs generated by real users, instead of generated just as test programs.

"One other thing worth mentioning," said Dr. Shaw, "is that there is at least one other project that we on the Non-Von project view as being of a kindred spirit, even though it's very different in some ways. That's the Connection machine up at MIT. One of the main people there is Dan Hillis. Like ours, the Connection machine is based on massive parallelism.

"They, too, are trying to put a number of processors on a chip. Like ours, it's based on something called SIMD (Single Instruction Multiple Data) execution, which says that all—in our case, it's actually large subsets of—the processing elements are at any given time all doing the same thing instead of working as completely independent processors. And many of our colleagues don't believe that's the right way to go. In fact, probably the majority of our colleagues think that it's a mistake. They think it's better to use a different approach where the processors are independent. But this approach to the problem doesn't allow you as many processors.

"So things are very controversial, but I think it's too early to say who's right. What would be exciting to all of us, I think, would be if one type of parallel architecture emerged ten years from now that turned out to be better than the Von Neuman machine. And it also turned out to be better for most purposes, even if not all, than most of the other kinds of machines.

"It's clear to everybody that, if you want to do a particular problem as fast as you can, and you have a large amount of money to spend (there are such applications), it will be best done by a special-purpose machine. None of us is going to come up with a general-purpose machine that will do as well. On the other hand, the big prize in the field, will be to come up with a successor to the Von Neuman machine.

"That's what we're all trying to achieve. It will be some new way of organizing computers that, although it isn't perfect, is good enough for people to adopt it for 90% of what they do. And it can be capable of being mass produced and sold to a very wide set of markets.

We don't yet have any evidence of whether that can be done. I rather suspect that it can. It would surprise me greatly if, by the year 2000, there weren't some other kind of machine that was general purpose enough for people to use it in a wide range of applications. But we still don't know what that is at this point."   ◇

With the need for a standard industrial local area network growing steadily as the amount of automation increases, the Manufacturing Automation Protocol developed by GM leads to a low cost, off the shelf solution. The migration strategy provides interim steps which meet short-term critical manufacturing needs and reduces vendors risk.

GM's Manufacturing Automation Protocol

William J. Riker, Senior Project Engineer, Advanced Product and Manufacturing Engineering Staff, Technical Center, General Motors Corp., Warren, Mich.

IN the next five years, GM will spend hundreds of millions of dollars on applied computer technology. Computer technology in the hands of GM designers is already tripling their productivity—thereby shortening the design time, increasing the number of designs considered and also yielding efficiencies in design integration. In addition, over 40% of the product engineering design and drafting workload is presently being accomplished using computers. This percentage doubled in the past three years and is projected to more than double again by the end of the decade.

Productivity improvements within GM began with automated control of single processes such as machines, conveyors and product test stands. Today, there are approximately 20,000 programmable controllers and 2000 robots installed at GM. The number of intelligent devices installed in the manufacturing area totals more than 40,000 with a projected 400 to 500% increase in the next five years. The benefits associated with this increased automation are substantial, but full benefit is not achieved until the automated processes are integrated with manufacturing data bases. Only about 15% of these processes communicate outside their own process, creating thousands of automation islands. Simply increasing the quantity of automation is insufficient. Integrating these islands of automation is essential to GM's Factory of the Future and success in a world market demanding high quality, cost competitive products.

Current integration difficulties

While integration is necessary, current communication methods are expensive. A 1981 study of incoming appropriation requests for new automation revealed that up to 50% of the costs were directly related to communication: wiring, hardware interfaces, custom software and training expense were major contributors to the excessive communication costs.

Wiring is the most visible communications expense. Recent plant installations provide an appreciation for the cabling problem. Communication links evolved into a maze of point to point connections. Costs are incurred when new systems are installed and again each time the process changes. In the automotive industry where retooling for new models occurs annually, rearrangement costs become significant.

Custom software and hardware interfaces are the second major cost factor brought about by vendor unique communication methods. Custom software is usually required to interface two process applications. To make matters worse, incompatible software performing similar functions usually exists to connect two different process applications located elsewhere in the plant.

Single-sourcing has advantages and solves some incompatibility problems. One vendor's development and installation projects are easier to coordinate than multiple vendors.

However, each system type presents a particular set of device requirements that often are best met by a particular supplier. GM manufacturing plants tend to avoid single-sourcing any given capital equipment. Utilizing multiple manufacturers and promotion of a competitive environment applies to computing and control equipment. It also allows the use of the best, most cost-effective hardware available.

Manufacturing Automation Protocol

Given a multi-vendor environment, common communication methods are required for integration. Realizing the need for standards within GM, the Manufacturing Automation Protocol (MAP) task force was created in 1980. The charter of MAP has been to identify communications standards which provide for multi-vendor data communications in manufacturing. The task force is comprised of representatives from approximately 15 GM divisions. Divisional participation assures that plant needs are met and that appropriate computer and controls vendors are involved.

Soon after its formation, the task force identified the International Standards Organization's (ISO) 7-layer model for Open Systems Interconnection (OSI) as a basis for standardized networks. Even in 1981, the model was gaining wide support in the communications industry. However, since the model specifies function rather than protocols, compliance to the model does not assure multi-vendor communication. The strength and current support of the MAP specification is the selection of existing or emerging standard protocols. All seven layers of the MAP specification will eventually be international standard protocols.

The MAP task force has concentrated primarily on evaluation of IEE Project 802 and emerging ISO/NBS (National Bureau of Standards) specifications. The MAP protocol choices to date are illustrated in Table I. GM specific upper layer protocols are necessary for a functioning interim MAP network. These interim specifications are jointly developed by GM and participating vendors with standards organizations' working papers as a base. Both GM and participating vendors are active in standards groups to influence future

TABLE I   Current MAP specification

ISO Layer	MAP Protocol
7. Application	(ISO)/GM
6. Presentation	(ISO)/GM
5. Session	ISO/NBS
4. Transport	ISO/NBS Class IV
3. Network	ISO-NBS Internet
2. Data link	IEEE 802.2 Class I
	802.4 Token Medium Access
1. Physical	IEEE 802.4 Broadband

direction and encourage adoption. Standard upper layer protocols will be added to MAP as they mature.

Implementation strategy

Developing an efficient communications network, even under ideal circumstances, requires many years. For instance, both IBM and DEC invested several years developing SNA and DECNET, respectively. Implementation of a multivendor MAP communications network requires even greater time and effort. Since GM operating divisions cannot halt automation efforts to wait for a complete MAP, a graceful migration plan is essential to MAP's success. Existing divisional systems must be accommodated, without scrapping all previous installations. Vendor risk must be minimized to promote investment with some reasonable expectation of return.

The Manufacturing Systems Development Center at GM's Technical Center in Warren plays a major role in the migration plan. In 1983, the Integrated Communication Networks (ICN) staff of the Development Center expanded to meet the following objectives:

- Install a testbed network to alpha test prototype MAP implementations with participating vendors.
- Develop specifications for protocol layers not addressed by standards.
- Promote standards development on IEEE, ANSI, IEC and EIA Committees.
- Identify production pilot sites and provide assistance for installation.

The migration strategy is divided into five steps:

- 1. Centralized communication network: multivendor connections via a centralized computer node; terminal emulation via a centralized computer node; completion 2nd quarter 1984.
- 2. Local area network: multivendor connection via a distributed LAN; gateway to selected programmable controllers; completion 3rd quarter 1984.
- 3. Application services: enhance Step 2 LAN with additional application services; gateway to wide area network; completion 2nd quarter 1985.
- 4. Low cost hardware: reduce ISO layers 1-4 to hardware; proliferate Step 3 on multiple processors; add ISO layers 5-6; completion 1986.
- 5. Complete network utility: plug compatibility by majority suppliers; completion 1988.

Centralized communications network (Step 1) — Application programs exist within GM plants on processors such as the IBM 4300, DEC PDP/11 or the HP 1000. Most computers offer serial communications capability but the three computers mentioned cannot communicate with each other as purchased. In Step 1, an IBM Series/1 mini-computer is used to connect existing systems with minimal impact on application software or communications software. The Series/1 is used as a centralized communications node which provides terminal and protocol emulation (Fig. 1).

Without a local area network, Step 1 solves the immediate problem of incompatible computers which cannot communicate. A significant item is the Step 1 connection protocol, ISO Transport, which is required for Step 2.

Local area network (Step 2) — Dissimilar computers from different vendors are attached to a distributed local area network (LAN). Each participating computer vendor implements the same ISO Transport that IBM installed in Step 1. Additionally, all Step 2 computers interface to an external unit that provides ISO layers 1 and 2. The combination of ISO layers 1, 2 and 4 allows each MAP computer to connect and pass data to any other MAP attached computer (Fig. 2).

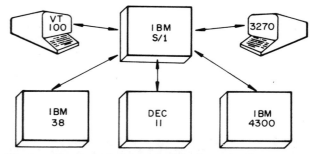

Fig. 1 — Step 1. Centralized communications network.

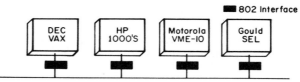

Fig. 2 — Step 2. Local area network.

Since the centralized Step 1 processor was a stand-alone system, there is an added benefit of the Step 1 to Step 2 migration. The Series/1 now exists as a node on the MAP network system but still provides connection services to incompatible computers. For example, the DEC PDP/11 does not support the ISO transport in Step 2 but will have access to the MAP network through the Step 1 centralized node.

Selected programmable controller access to MAP is the second objective of Step 2. Gateways support the existing vendor specific programmable controller nets on one side of the gateway and ISO layers 1, 2 and 4 on the other side (Fig. 3). The gateway concept allows for graceful migration since a number of installed programmable controllers are already connected by vendor specific network products. The gateway supports the existing systems and further provides the additional connection capability to each of the MAP attached computers.

Application services (Step 3) — Hardware configuration is similar or identical to Step 2 but significant software enhancements are added. The final goal of a network is application to application communication such as remote file access, file transfer and message exchange. Protocols specified by ISO layers 1 through 4 provide connection and reliable data transfer between computers but application compatibility is left to layers 5 through 7. Step 3 software is concentrated at layers 6 and 7.

Fig. 3 — Local area network and gateways.

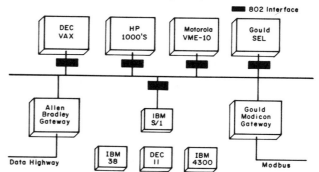

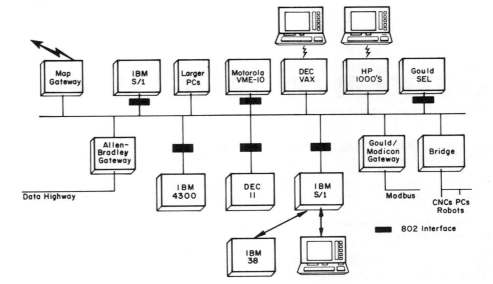

Fig. 4 — Step 4. Low-cost hardware.

Current MAP work on ISO layers 6 and 7 is based on working drafts from various standards groups. The interim MAP specifications for file transfer, messaging, virtual terminals, directory service and network management are GM unique because no international standards exist. The Step 3 software will be a mix of GM specific and standard protocol based on pilot needs and the approval process of standards groups.

Low cost hardware (Step 4) — The reason for selecting standard protocols is most apparent in Step 4 where VLSI hardware is expected. Acceptance of standards generates a high volume market necessary to encourage VLSI products. Silicon implementation will lower networking costs and, in turn, lead to more installed networks. In the 1986 Step 4 timeframe, MAP networks will proliferate in GM manufacturing plants.

Some computer nodes without Step 2 external hardware which perform ISO layers 1 and 2 functions are illustrated in Fig. 4. Board level hardware for various computer backplanes supporting ISO layers 1 through 4 will exist. A low-cost communications board will allow more processors and programmable devices to participate directly in a MAP network.

Network utility (Step 5) — With an installed utility, computers supporting MAP will plug-in and communicate with existing MAP systems. The concept of a totally transparent network utility is idealistic but, nevertheless, it is the goal of Step 5. In the 1988 to 1990 timeframe, most computer and control equipment for GM manufacturing should participate on MAP networks.

Summary

As the amount of automation increases, the need for a standard industrial local area network also increases. The MAP specification leads to low-cost off the shelf solution. The migration strategy outlined provides interim steps and meets short-term critical manufacturing needs. The migration plan reduces vendor risk which makes the entire project realistic.

Canadian GE Builds an Automated Plant

HENRY M. MORRIS, CONTROL ENGINEERING

Today's high efficiency jet engines demand high precision parts—formed to within 0.0001-in. tolerances. To manufacture the blades (rotating) and vanes (fixed) used in the compressor sections of GE's CFM 56, F101, and F110 families of turbofan jet engines, Canadian GE and General Electric Co.'s Aircraft Engine Business Group jointly designed, constructed, and are in the process of bringing on-line what will be a highly automated, paperless facility in Bromont, Québec, Canada. We recently were allowed to tour this facility. Here's what we saw and learned.

Nestled in a valley in ski country, approximately 70 km east by south (ESE) of Montreal, PQ, Canada, lies the 128,000 ft² CGE (Canadian General Electric) airfoil plant in Bromont. While it's quite ordinary on the outside, the doors of this $100 million factory open onto what promises to be one of the most modern, highly automated manufacturing facilities in existence. At peak capacity, over 300 people will be employed in the complex manufacturing process that turns slugs of titanium into finished jet engine compressor airfoils (blades and vanes) used in the Canadian Air Force's CF-18 Hornet fighter, Boeing's 737-300 and 747, McDonnell-Douglas' DC-10, and Airbus' A310 and A320 airliners, and to re-engine the DC-8 (renamed the DC-8 Series 70). Designed and built from the ground up to be an automated facility, Bromont serves as an example of what can be done when you've got the time, talent, resources, and money.

The organization

Bromont is not organized in the typical pyramidical hierarchy. The human, technical, and management systems are integrated circumferentially. Hence, the approach to automation is somewhat untypical too. One must understand the organization to appreciate the implementation.

The plant is a joint-venture of Canadian General Electric Co. Ltd. (Toronto, Ontario, Canada) and the General Electric Co. Aircraft Engine Business Group (Cincinnati, OH). It was created as part of a technology transfer agreement brought about by the Canadian Air Force's (Ottawa, PQ) decision to purchase the F-18 Hornet from McDonnell-Douglas (St. Louis, MO). Initially, the plant was to be a clone of GE's Rutland (VT) airfoil facility. However, during the planning stages it was decided that the most cost-effective approach would be to make Bromont as automated and integrated as practical. Lessons learned could also be transferred to Rutland.

There are three main processes involved in making airfoils—forge, machining, and pinch and roll. Each is considered a manufacturing team and operates as a fairly autonomous business, vending its services to the others as needed.

Surrounding the three production businesses are the four support teams. The support teams include maintenance, administration, tooling, and M.A.Q.E.T. (manufacturing and quality control engineering team). The first three support teams' names are self evident. M.A.Q.E.T. is responsible for a variety of functions including the robotization of various production stages and the design of jigs and fixtures.

The senior team surrounds the support teams, interfaces with the outside world, and provides overall leadership.

The result of this organization is a team approach where every individual worker contributes equally. In fact, every employee has not only the right, but the responsibility, to be involved in decisions affecting the plant's operation.

The plan

At the onset, it was determined that a five-year plan (1984-1988) held the most chance of success. The task force planned that Bromont should be a paperless plant, characterized by on-line, real-time systems which are user-friendly and menu-driven. Software would be acquired and integrated where possible to get in business quickly. In keeping with the paperless theme, bar code readers, voice-entry systems, optical scanners, and factory data collection terminals would be used wherever practical.

The automation strategy, as set down, called for the progressive auto-

Voice data entry is one of the techniques Bromont is using to achieve a "paperless" factory. Naturally the system understands French.

mation of sections of the manufacturing processes, then to integrate them as the business evolved.

One of the most important initial automation efforts was to tie a Calma CAD/CAM system internally between automatic inspection equipment and Burgmaster NC machining centers and externally to the engineering department at Rutland. Both Rutland and the automatic inspection equipment provide the raw data needed by the tooling designer who then uses the Calma system to generate the new tooling required and downloads the design to the Burgmasters where the tools are fabricated.

There are several different sizes (1¼-5 in.) of airfoils (blades and vanes) used in the compressor section of a modern jet engine. General Electric's designs call for 894 airfoils (⅔ blades, ⅓ vanes) in a set and one set of airfoils per turbofan compressor. Bromont's plans call for the plant to put out 700 sets of airfoils in 1986. Because working with titanium wears out machine tools quickly, and due to the extremely tight tolerances required on every piece part, there is a constant need to retool. Hence, the savings brought about by the integration of tooling inspection, design, engineering, and fabrication provides one of the greatest paybacks.

At the end of the five-year plan, Bromont will be on-line with 46 robots, 18 programmable controllers, 16 CNC machines, several automatic guided vehicles (AGV), and 14 work cells containing 51 production machines. Three computers, Prime 250s and 9950s, will coordinate the factory management system (FMS) and shop activity management systems (SAM) and will provide general purpose computational power as needed. Quality systems, automated material handling, and controlling the water treatment plant will be done by four HP 1000 Series Es. The Calma CAD/CAM system, NC programming, and the DNC link are to be handled by a DEC VAX 11/780.

Plant-wide and out-of-plant computer communications have been placed under the control of a Gandalf PACX 4 communications controller. A Honeywell DPS 8 Plus is located at GEC Toronto and GE Lynn (MA). The Toronto Honeywell is used for financial systems while the Lynn Honeywell has the manufacturing database needed for SAM routing sheet generation.

Many IBM PC-compatible personal computers are used throughout the plant for a variety of purposes, and each can access appropriate external data via a local area network, as needed.

Measurements taken by the fixture in front of the inspector (bottom, right) are electrically transmitted to the Calma CAD/CAM system (top). The final design data is then electrically fed to the Burgmaster machining center (bottom, background, left).

Hooking it all up

The overall organization of the plant's communications is divided into three interconnected hubs of activity. One is the acronym CIMAP, which stands for Computer Integrated Manufacturing & Administrative Plan. The second, FMS, stands for Factory Management System. The third hub of activity is called CAQ, which means Computer-Aided Quality. All facets of the plant's operation and communication between these facets take place via one or more of these hubs of activity.

Robotization efforts

Robotization was selected over hard automation for several reasons:
- to retain flexibility needed to adapt to future products;
- faster return on investment;
- minimal changes in the plant layout for future changes;
- faster engineering and implementa-

Source: Control Engineering, February 1986, 87-89

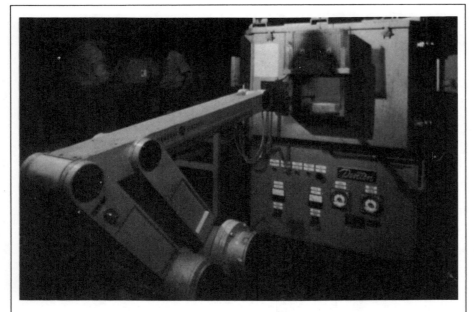

CGE uses GE robots to handle parts in hazardous locations such as this example where the robot is putting an airfoil into a furnace for heat treating.

Robots were installed in this work cell only after the operation had first been implemented as a manual process and all of the bottlenecks and bugs were worked out.

tion than hard automation;

- and, less initial investment compared to hard automation (such as transfer machines, flexible manufacturing centers, and the like).

The initial task force used the following criteria to identify which of the possible manufacturing operations were to be robotized:

- labor intensive operations;
- hazardous operations;
- operations which can be easily combined;
- easy to robotize;
- and bottle necks.

According to the original automation plan, no process or workstation will be robotized until it is first running smoothly on a manual basis. To regain some of the initial costs, as the technology is developed it will be replicated at GE's Rutland plant.

The task force identified 24 robotic opportunities using the above criteria and assessments of economics, risk, quality, productivity, and the ability to integrate the "islands of automation" into the overall automation plan.

Automation implemented

During the past two years, Bromont has engineered and implemented a substantial amount of the automation called for in the five-year plan. There are six robotized systems with 13 robots, 2 vision systems, 12 cameras, and 5 programmable controllers on line performing such tasks as hot forming, cold forming, fatigue enhancement, and inspection/identification.

Four chemical processes have been automated, virtually eliminating the operator's exposure to the chemicals.

Seven of the 13 electronic gaging units scheduled for installation in 1985 were in place during our tour. These units obtain information on blade dimensions electronically, replacing the earlier handwritten hard copies. This data can be quickly transmitted and subsequently manipulated by statistical analysis software. The data can also be used by the Calma system for tool design.

There are three voice entry systems installed at visual inspection stations. Results of the visual inspection are verbally entered (in French) into the system, thereby eliminating hand-written hard copies of the data.

Direct labor has been reduced by 23 hours per set due to the installation of a semiautomatic deburring system.

Establishing an automation laboratory enables the development and evaluation of systems and equipment as applied to the manufacturing and quality control processes.

Additional major systems now partially or completely installed include: office automation; project planning and control; tool & gage control; factory management (FMS-SAM); Chubb 3-level security; purchasing and receiving; integrated cost and accounting; scheduling; operator control program; shipping and invoicing; and the quality decision management system.

Automation still coming

While much has already been accomplished, Bromont is intended to be fully automated. Hence, there's still more coming. The deburring and benching operations are scheduled for robotization. A flexible manufacturing system will be installed in the machining area for the production of variable vanes. The flexible manufacturing system will contain 7 machine tools, 5 robots, and a material transport system.

Noncontact flexible gaging systems employing laser micrometers and laser probes are being examined presently with the first prototype already in the automation laboratory. These systems are scheduled for introduction on a production basis by the end of 1985.

Plans for further automation of the manufacturing processes include block and final forge, trimming operations, broaching, and benching. Finally, the process monitoring and control system and the quality decision management system will be connected to the data bases to furnish information in a paperless mode. □

Bibliography:
Books on Computer Subjects

Personal Computers

1. **Exploring the World of the Personal Computer**, Jack M. Milles, Prentice-Hall, Englewood Cliffs NJ, 1982.
2. **A Practical Guide to Small Computers**, R. M. Rinder, Monarch Press, New York, 1983.
3. **The Personal Computer Book**, Third Edition, Peter A. McWilliams, Prelude Press, Los Angeles CA, 1983.
4. **Big Things for Little Computers**, Dale Peterson, Prentice-Hall, Englewood Cliffs NJ, 1982.
5. **Bowker's 1985 Complete Sourcebook of Personal Computing**, R. R. Bowker, New York, 1985.

Guides and Directories

1. **The McGraw-Hill Computer Handbook**, Harry Helms, McGraw-Hill, New York, 1983.
2. **The Illustrated Computer Dictionary**, Revised Edition, Donald D. Spencer, Charles E. Merrill Publishing Co., Columbus OH, 1983.
3. **Encyclopedia of Computer Science and Technology** (16 volumes), edited by Jack Belzer, Albert G. Holzman and Allen Kent, Marcel Dekker, Inc., New York, 1979.
4. **User's Guide to Microcomputer Buzzwords**, David H. Dasenbrock, Howard W. Sams & Co., Indianapolis IN, 1983.
5. **The Reader's Guide to Microcomputer Books**, Second Edition, edited by Michael Nicita and Ronald Petrushka, Garden-Lee Books, New York, 1984.
6. **Standard Dictionary of Computer and Information Processing**, Second Edition, Martin H. Weik, Hayden Book Co., Rochelle Park NJ, 1977.
7. **The Encyclopedia of Microcomputer Terms**, edited by Gail Christie and John Christie, Prentice-Hall, Englewood Cliffs NJ, 1984.
8. **Omni On-Line Database Directory**, edited by Mike Edelhart and Owen Davies, Macmillan, New York, 1983.
9. **Encyclopedia of Computer Science**, edited by Anthony Ralston and Chester L. Meek, Van Nostrand Reinhold, New York, 1976.

Databases and Systems

1. **Computerized Materials Data Systems**, edited by J. H. Westbrook and J. R. Rumble, Jr., published by the Steering Committee of the Computerized Materials Data Workshop, Fairfield Glade TN, 1983.
2. **Numeric Databases**, edited by Ching-Chih Chen and Peter Hernon, Ablex Publishing Co., Norwood NJ, 1984.

Software and Programming

1. **Directory of Microcomputer Software for Mechanical Engineering Design**, 1985 Edition, edited by Colette O'Connell, Jean K. Sheviak, Alice R. Browne and Samuel V. Johnson, Marcel Dekker, Inc., New York, 1985.
2. **Microcomputer Software Buyers Guide**, edited by Tony Webster and Richard Champion, McGraw-Hill, Sydney, Australia, 1984.
3. **The Software Handbook**, Dimitris N. Chorafas, Petrocelli Books, Princeton NJ, 1984.
4. **The Language of Computer Programming in English** (English for Careers Series), John C. Keegel, Regents Publishing Co., New York, 1976.
5. **The Complete Guide to Writing Software Users Manuals**, Brad M. McGehee, Writers Digest Books, Cincinnati OH, 1984.
6. **Computer Programming for the Complete Idiot**, Donald McCann, Design Enterprises of San Francisco CA, 1979.
7. **The Software Catalog — Micros, Minis**, Elsevier Science Publishers, New York, published four times a year.
8. **Engineering Softwhere**, published by Moore Data Management Services, Minneapolis MN, 1984.

Historical

1. **Memoirs of a Computer Pioneer**, Maurice V. Wilkes, MIT Press, Cambridge MA, 1985.
2. **The Micro Millenium**, Christopher Evans, Viking Press, New York, 1979.
3. **Thinking Machines**, Irving Adler. New American Library, New York, 1964.

Computer Control

1. **Elements of Computer Process Control with Advanced Control Applications**, Pradeep B. Deshpande and Raymond H. Ash, Instrument Society of America, Research Triangle Park NC, 1981.
2. **Minicomputers in Industrial Control — An Intro-**

duction, edited by Thomas J. Harrison, Instrument Society of America, Research Triangle Park NC, 1980.

Manufacturing Automation

1. **Computers in Manufacturing**, editors of *American Machinist*, McGraw-Hill, New York, 1983.
2. **Laboratory Minicomputing**, John R. Bourse, Academic Press–Harcourt, Brace, Jovanovich, New York, 1980.
3. **CAD/CAM Productivity Equipment Series**, Second Edition, published by the Society of Manufacturing Engineers, Dearborn MI, 1985.

The Computer's Future

1. **Next Generation Computers**, edited by Edward A. Torrero, IEEE, Inc., Spectrum Series, New York, 1985.
2. **Advances in Computers**, edited by Marshall C. Yovik, Academic Press–Harcourt, Brace, Jovanovich, Orlando FL, 1985.

3. **Proceedings of the Annual Conference, Association for Computing Machinery**, Denver CO, October 14-16, 1985, Association for Computer Machinery, Inc., New York, 1985.
4. **Comoputers, Today and Tomorrow**, Tom Logsdon, Computer Science Press, Inc., Rockville MD, 1985.

Miscellaneous

1. **The Digital Villain — Notes on the Numerology, Parapsychology, and Metaphysics of the Computer**, Robert M. Baer, Addison-Wesley, Reading MA, 1972.
2. **Computer Consciousness — Surviving the Automated '80's**, H. D. Covvey and N. H. McAlister, Addison-Wesley, Reading MA, 1980.
3. **Computers and the Cybernetic Society**, Second Edition, Michael A. Arbib. Academic Press–Harcourt, Brace, Jovanovich, Orlando FL, 1985.
4. **Scientific Analysis on the Pocket Calculator**, Jon M. Smith, John Wiley & Sons, New York, 1977.

INDEX

407

414